Regression Modeling for Linguistic Data

Regression Modeling for Linguistic Data

Morgan Sonderegger

The MIT Press
Cambridge, Massachusetts
London, England

This book was set in Times and Syntax by the author. Printed and bound in the United States of America.

Library of Congress Cataloging-in-Publication Data

Names: Sonderegger, Morgan, author.
Title: Regression modeling for linguistic data / Morgan Sonderegger.
Description: Cambridge, Massachusetts : The MIT Press, [2023] | Includes bibliographical references and index.
Identifiers: LCCN 2022026919 (print) | LCCN 2022026920 (ebook) | ISBN 9780262045483 (paperback) | ISBN 9780262362467 (epub) | ISBN 9780262375030 (pdf)
Subjects: LCSH: Linguistics–Statistical methods. | Regression analysis. | R (Computer program language) | LCGFT: Textbooks.
Classification: LCC P138.5 .S66 2023 (print) | LCC P138.5 (ebook) | DDC 410.72/7–dc23/eng/20220822
LC record available at https://lccn.loc.gov/2022026919
LC ebook record available at https://lccn.loc.gov/2022026920

10 9 8 7 6 5 4 3 2 1

To Katie, Ariel, and Cleo

Contents

Preface

This book introduces applied regression analysis for analyzing linguistic data, using R. It aims to provide both conceptual understanding and practical skills through extensive examples, using three different kinds of linguistic data:

1. *Preliminaries to regression modeling* (chapters 1–3): assumptions, inferential statistics, hypothesis testing, power, and other errors.
2. *Regression models for nonclustered data* (chapters 4–7): linear regression, model selection and validation, logistic regression, and practical topics (e.g., contrast coding, post hoc tests, nonlinear effects).
3. *Regression models for clustered data* (chapters 8–10): linear and logistic mixed-effects models, and practical topics (e.g., model predictions, convergence, model selection).

The book started as a minor revision of *Quantitative Methods for Linguistic Data* (Sonderegger, Wagner, and Torreira 2018), co-authored with Francisco Torreira and Michael Wagner, which itself grew out of lectures for a one-semester graduate quantitative methods course at McGill Linguistics that is taught most years. I ended up rewriting the entire manuscript, including adding several new chapters. So this book is best understood as a new text, which incorporates aspects of *QMLD*. I thank Michael and Francisco for their understanding, and letting me incorporate their work here.

I see this book as a "frozen" version of an evolving document. Any feedback is welcome (morgan.sonderegger@mcgill.ca) and will hopefully be incorporated in a future version. I hope to update the book's website, which also contains all code and datasets, as the text is updated in the future.[1]

Audience This book is for graduate students and researchers in linguistics and other language sciences, who work with quantitative data. This includes "data-heavy" subfields of linguistics (e.g., experimental syntax/semantics/phonology, phonetics, psycholinguistics, corpus linguistics, language acquisition, sociolinguistics), as well as communication sciences and disorders, psychologists or cognitive scientists of language, and so on. I have taught this material since 2013 to students and faculty (about 50% from each group). I use

1. The website is currently at https://osf.io/pnumg/. Please check my personal web page if the OSF link no longer works when you read this.

linguist and *language scientist* interchangeably in this book for brevity, similarly for *linguistics* and *language sciences*.[2] The text assumes knowledge of elementary linguistics, using terms such as *phone* or *lexical item* without comment, but should be understandable regardless (you'll just have to think of some variables as y or x rather than as voice onset time or place of articulation).

Background This book joins many existing texts on quantitative analysis/statistics for linguists, which are also mostly practical introductions using R, including Brezina (2018), Eddington (2016), Garcia (2021), Gómez (2014), and Gries (2021, 2013), Johnson (2008), Levshina (2015), Rietveld and Van Hout (1993, 2005), Vasishth and Broe (2011), Vasishth et al. (2021), and Winter (2019). Baayen (2008) has been particularly influential. These texts differ in many respects, such as the type of linguistic data assumed and the theory/practice balance, but mostly share two aspects: starting from scratch and broad scope (statistics or quantitative analysis generally).

The approach of this book, described in more detail in chapter 1, is different. Its goals are narrower—conceptual understanding and practical experience with regression modeling of linguistic data—and I assume you have some experience with statistics and R. It does not cover other (important) quantitative tools, such as classification methods or exploratory data analysis. While I don't assume you have read any of the preceding books, the current book can be seen as complementary to them.

I focus on regression models because these are the main form of statistical analysis in papers (using quantitative data) published in major journals, but they are complex tools to use in practice. I give practical and detailed treatments of a smaller number of topics, describing decision points and the pros and cons of different methods, conventions in the current language sciences literature, and how to report your analyses. The goal is to equip you to use these methods in practice and understand how they are used in current work. This book will not cover the full range of possible regression models (e.g., Poisson or multinomial models), but extending to them should be straightforward after you've used this book.

Other differences from existing texts are a greater focus on data from (laboratory) phonology, phonetics, and language variation and change (though other areas are represented) and equal emphasis on observational data (especially from speech corpora) and data from controlled experiments. These reflect my own background as a linguist working primary in these areas, often using corpus data—but I have tried to keep the presentation useful for language scientists generally.

Regression modeling is used across many scientific fields, and we can learn from best practices in other fields to better analyze linguistic data. At the same time, it's easier to learn data analysis with data that looks like your own, and analysis tools become specialized for particular kinds of data as they are used in a field over time. I try to give context and places to read more for particular topics of interest—from (statistics for) language sciences, as well as behavioral sciences, social sciences, and ecology—both in the text and in Further Reading sections at the end of each chapter. No statistical method used in this book is new, so it is neither possible nor useful to give comprehensive citations; I give references I am

2. The different connotations of the two terms are not relevant for statistical analysis.

familiar with and have found helpful, including particularly detailed treatments of linguistic data by other authors (in books, cited above, plus articles). My goal is to provide useful entry points to the vast literature on statistical methods for when you want to learn more.

What you need to know This is an intermediate text. It assumes previous exposure to quantitative methods, such as from one of the books above, but aims to be useful for readers with different backgrounds.

You should be familiar with secondary school math concepts, such as algebra, logarithms, exponentials, summation notation, and basic linear algebra (vectors, matrices, matrix multiplication); as well as some probability theory, specifically what the following terms mean: *probability distribution*, *normal distribution*, *random variable*, *binomial distribution*, and *conditional probability*. Most important is familiarity with statistics, at the level of a first course:

- Descriptive statistics: data summarization and visualization: the meaning of concepts such as *mean*, *mode*, *standard deviation*, *quantile*, and *correlation*; how to make and read common statistical plots (e.g., boxplots, histograms, density plots, scatterplots).
- Inferential statistics: the idea of sample and population, basic hypothesis testing concepts (p-values, test statistics), and tests (t-tests, χ^2-tests), and maybe basic analysis of variance.

The book focuses on conceptual understanding and practical skills by working in R, but without actually providing instruction in R. Thus, I assume a working knowledge of R and R programming, including both base R and some familiarity with tidyverse functionality (packages such as ggplot2, dplyr: https://tidyverse.org).[3]

In practice, dozens of graduate students with a variety of backgrounds in these areas (including "not much") have done well in the course using this book's material, with some doing extra work to catch up. In particular, many students have learned R from scratch at the same time as using this material, using online tutorials for base R and *R for Data Science* (Grolemund and Wickham 2016) for tidyverse, and students with less math background than described have done well.

Some resources for math/probability/statistics are Khan Academy videos, Sharon Goldwater's math tutorials, general statistics books listed in chapter 2, Further Reading, and those for linguistics listed earlier; for descriptive statistics and visualization, Grolemund and Wickham (2016) (general), Gries (2021, chapter 3), and Garcia (2021, part 2) (linguistics) are particularly thorough.

Caveat I am not a statistician, but a self-taught practitioner with some masters-level math/statistics training. Books written by practitioners can be useful because of the perspective of working with this kind of data, but they can also contain errors. If you frequently use a particular tool for data analysis, I recommend (eventually) consulting a more authoritative source; one goal of this book is to equip you to do so. This is particularly important because statistical practice is not static: this book emphasizes practical skills (e.g., which package to use, best practices for fitting and interpreting models) for which best practices

3. Some people have strong opinions about base R versus tidyverse. If you don't like one idiom or the other, it will be perfectly possible to follow along, I just won't be explaining particular functions in detail.

are constantly evolving. My aims are for this book to be useful for analyzing linguistic data and as (technically) correct as possible.

How to use this book This book is ideally *read while executing the code* shown in code blocks on your own computer, for example, by pasting them in to RStudio or the R console. The code in each chapter is independent, so you should always be able to start reading/coding at the beginning of a chapter. I often refer to objects that have been created in previous code, and sometimes the output of running code is not actually shown (I'm assuming you can see it in your console).

Most code is shown in the actual PDF; this is the code I assume you are able to run. An important exception is code for creating plots, which is usually omitted because code to make decent plots is verbose. You can find *all code* (including plotting code) in the code file for each chapter, on the book's website.

Each chapter consists of main text and boxes, inspired by McElreath (2020)'s *Statistical Rethinking*. Boxed text is not essential for understanding the main text and gives extra information. There are two types of boxes: Broader Context boxes provide more in-depth explanations of technical concepts or math, connect to other approaches, discuss common misunderstandings, and so on. Practical Note boxes discuss aspects of statistical analysis you're unlikely to delve into until you actually use these methods in your own work, such as statistical reporting or R details.

Offsetting materials in boxes is intended to make the book easier to use. On a first read you can focus just on the essentials by skipping boxes, and when analyzing your own data later or learning more about a particular method, the boxes call out relevant material.

Datasets This book uses publicly available datasets, all of which are available on the book's website or in the languageR package. I am grateful to dataset authors for their willingness to post data publicly: Michael Wagner (`givenness`); Timo Roettger and Bodo Winter (`neutralization`); Francisco Torreira, Seán Roberts, and Stephen Levinson (`transitions`); Michael McAuliffe and Hye-Young Bang (`turkish_if0`); Max Bane and Peter Graff (`vot`); and R. Harald Baayen and co-authors (`english`, `regularity`).

Acknowledgments Many people deserve thanks for the long journey to a book. Above all, students in McGill classes from 2013 to 2021 provided feedback, as did students who continued to use the materials for their research—especially Hye-Young Bang, Amélie Bernard, Guilherme Garcia, Claire Honda, Oriana Kilbourn-Ceron, Donghyun Kim, Bing'er Jiang, Jeff Lamontagne, James Tanner, and Connie Ting. Substantial editing and typsetting work have been done by David Flesicher, Claire Honda, Jacob Hoover, Vanna Willerton, and especially Michaela Socolof. Colleagues who have offered encouragement and comments over the years include Meghan Clayards, Jessamyn Schertz, Tyler Kendall, Michael Wagner, Paul Boersma, Christian Di Canio, Volya Kapatsinki, Roger Levy, Jane Stuart-Smith, Tim O'Donnell, Alan Yu, Bodo Winter, Tania Zamuner, and five anonymous reviewers. Special thanks go to James Kirby for comments on the entire manuscript and reminders to keep going. I am grateful to you all, including those I have forgotten. Finally, I am thankful to Katie for her companionship, support, and especially patience—it's finally done.

Morgan Sonderegger
January 2022

1 Preliminaries

1.1 Our R Toolset

This book exclusively uses R, which has become the de facto standard across language sciences. R is free, relatively easy to learn, and incorporates very broad functionality through packages. It is an excellent default for visualization and statistical analysis.

A downside of using R is that you must decide what "dialect" to use: the core set of base packages, which are stable but clunky, or a better alternative set of packages that have been developed, which may become obsolete.[1] The currently dominant dialect is tidyverse, a family of packages that offer wide functionality and an elegant implementation based on the "tidy data" philosophy (Wickham et al. 2019; Grolemund and Wickham 2016), but that evolve over time. The world being what it is, base R and tidyverse zealots can be easily found online.

I agree with Winter (2019) that the most realistic option is to learn both dialects: you need to know base R to function, while tidyverse functionality is often superior in practice, and both are widely used in online resources (e.g., StackOverflow pages). This book uses both, leaning toward tidyverse functionality when available, but often showing both base and tidy ways to do the same thing, in line with my general philosophy of showing you alternative methods so you can choose for yourself. This book also often uses data and functions from the languageR package associated with Baayen (2008)'s *Analyzing Linguistic Data* (Baayen and Shafaei-Bajestan 2019).

The book's code will be kept updated on its website, currently osf.io/pnumg (or see my personal website). Check there if any code doesn't work using your (future) R/tidyverse version. Appendix B shows the exact R and package versions used to compile this book.

1.2 Our Approach

The primary goals of this book are conceptual understanding and practical experience with regression modeling, in R.

Conceptual understanding My general philosophy is that *understanding* the statistical/data analysis methods you use ("why do *X*," "what is *X*") is of primary importance, and the best way to do this is through practical demonstration ("how to do *X*"). Developing conceptual

1. For example, Baayen (2008) uses Lattice packages that are now obsolete.

understanding also often requires math, whether via equations or simulation, because this is the underlying language. Conceptual understanding is the most practical thing you can learn for data analysis, because statistical methods (i.e., best practices for doing X) change over time, but new methods always build on old ones. For this reason, the book focuses on fewer topics—such as covering linear and logistic regression, but not other types of linear models—in greater detail, and in a cumulative fashion: linear mixed-effects models build on linear regression, which builds on t-tests.

Practical experience Conceptual understanding alone will not let you analyze data. There are often little tricks, or best practices that you learn with experience, which are essential in practice, but don't come up unless you actually spend a lot of time analyzing data. So this book contains a lot of working code, integrated into the main text, showing how to do everything discussed, using a good set of R packages for 2022.[2] The only exception is code for making plots, which is not shown in the text because it is verbose.

I strongly recommend actually running the code as you read, because "just as one cannot learn martial arts by watching Bruce Lee movies, you can't learn to program statistical models by only reading a book" (McElreath 2020, xiii). To facilitate this, R files of just the code for each chapter (including plotting code) are posted on the book's website.

Statistical modeling Regression analysis is a *statistical modeling* approach to data analysis, where we seek to interpret some data with respect to research questions or hypotheses we have, by building and interpreting a model. A different approach to data analysis, which language scientists often learn, is a *hypothesis testing* approach: the researcher applies one of a fixed set of tests (e.g., t-test, one-way analysis of variance), depending on the type of data and the question being asked. Hypothesis testing is one foundation of statistical modeling (see chapter 2), but the underlying philosophy is quite different (see, e.g., Rodgers 2010; Gelman and Hill 2007). The statistical modeling approach is more flexible, but harder to learn—it involves making choices, and thinking about the data (box 1.1).

The statistical modeling approach recognizes three central facts about data analysis. First, we are not building a model of the actual generating mechanism of the data (e.g., neurons, vocal tract muscles), which for linguistic data is usually unknown; at best we are building a "process model" to gain insight into the research questions motivating the analysis (McElreath 2020, section 1.2).[3] It follows that while there are incorrect ways to analyze one's data, there is never a single "right" way—data analysis requires an educated choice of method, and different choices carry different risks and rewards. Finally, data analysis always takes place in a scientific context: the hypotheses or research questions motivating your analysis are fundamental to choosing the analysis method, because the goal of the analysis is to address these questions.[4]

2. Note that there is no "best" set of tools: R packages change rapidly, and different tools work for different people. For example, I prefer customizing my plots carefully; you may be fine with just using existing prewritten functions. You may strongly prefer base R or tidyverse functionality.

3. This is simplified somewhat: McElreath distinguishes between process models, which are well-specified quantitative causal models of the process (e.g., how speakers parse sentences, how vowels are realized acoustically), and statistical models, which are the actual models we fit to data.

4. On the centrality of scientific questions for statistical analysis, see Speed (1986) and McElreath (2020, chapter 17) (who cites it).

These points inform this book's presentation. Rather than showing you the "right way" to do an analysis (e.g., fit and report a linear mixed-effects model), I show decision points, the pros and cons of different paths, and what is done in current practice. I try to introduce methods in the context of concrete research questions; a corollary is that this book uses fewer datasets than many other books do.

Box 1.1
Broader Context: Trade-Offs versus Flowcharts

Many researchers, including language scientists, just want to know how to analyze their data—they see statistical analysis as "an onerous task that one would rather leave to others" (Baayen 2008, viii)—and don't want to have to choose among different methods depending on the pros and cons. This is understandable, and to meet this demand, statistics textbooks often use a flowchart/recipe approach to guide researchers in choosing a method given their data (e.g., compare two normally distributed groups ⇒ two-sample *t*-test). This approach is simple, and requires less conceptual understanding, but it has serious disadvantages in practice: one's data often does not fit neatly into a flowchart box (e.g., you can't assess normality, or a reviewer asks for a different method), in which case you don't know what to do. Also, no intuition is developed for the consequences of using different analysis methods, as in different papers in the literature. If you understand the pros and cons of different methods, you are better able to address the scientific questions you want to ask about your data, and you will be a more informed consumer of the literature.

1.3 Context

This section defines terminology and notation used throughout the book, including classic oppositions among experimental/observational studies, correlation/causation, and exploratory/confirmatory analysis.

1.3.1 Types of Data and Study

In this book *linguistic data* is often used as a shorthand for any quantitative dataset produced in a linguistic study.[5] These come most often from laboratory experiments or linguistic corpora (of speech or text), but they could also be from typology (typological frequencies), computational studies (e.g., output of a speech recognizer as parameters are varied), observation of language acquisition (what words children know at age X) or language change, lexicons, and many other sources.

These sources can be divided into *experimental studies*, where the researcher constructs a world and manipulates some variables (x) to observe the effect on other(s) (y), and *correlational* or *observational studies*, where the researcher observes the real world (e.g., Field, Miles, and Field 2012, section 1.6.1). Linguistic data can be *experimental* or *observational*, for example, respectively, from controlled laboratory experiments versus corpus studies.[6]

This distinction is closely related to *causation* versus *correlation* (i.e., data description). In the classic formulation, causality ($x \rightarrow y$) can only be inferred from an experimental study, but it is unclear whether the results generalize to the real world, while observational

5. This shorthand is just for convenience, because only quantitative data are relevant for us.

6. In linguistics proper, "experimental [data/linguistics]" is commonly used in two senses: a laboratory experiment, or any study that primarily uses quantitative data, which would include most sources of what I am calling *quantitative data*. I avoid usage like "experimental linguistics" because of this ambiguity and only use *experimental* as defined in the text.

studies are always correlational ($x \sim y$), but the results have greater "ecological validity". In reality, inferring causality is hard even for experimental data (e.g., there may be unobserved confounders, a nonrandom sample), and we must just always assume that we are fitting correlational models and bear the "correlation is not causation" adage in mind when interpreting our results. Section 4.2 discusses further for regression models.

1.3.2 Exploratory and Confirmatory Analysis

Traditionally, data analysis can be *exploratory* or *confirmatory* (EDA, CDA; Tukey 1977): exploring the data, often by visualization, to generate hypotheses versus testing known hypotheses in novel data, for example, fitting a regression model and calculating p-values for each coefficient. An alternative characterization is that CDA/EDA is any data analysis that does/doesn't involve statistical modeling. The ideal is that EDA precedes CDA. In realistic data analysis, the boundary between the two is often unclear (Tukey 1980), especially for modern regression models (e.g., Gelman 2004). It is common to explore and confirm using the same dataset, including going back and forth, and in many linguistic studies (especially observational) the exact statistical model to be run cannot be *fully* specified in advance. The exact balance is tricky and depends on the context.

Rather than discussing exploration versus confirmation in depth, this book assumes that you are familiar with the basic issues relevant for regression modeling:

- EDA (read: "making many plots") is critical before any statistical analysis, even if the statistical methods have been completely prespecified: to find problems with the data, get intuitions about what the data "say" that the fitted model can be checked against, and so on.
- Testing hypotheses (CDA) suggested by the data (EDA) is dangerous: this makes it less likely your results generalize to new data, and you can easily hypothesize after the research question is known (sometimes referred to as HARK) after you see what terms are significant.
- Both exploratory and confirmatory phases are valuable for regression analysis: fitting a prespecified model to data will often miss important aspects of the data, while a model whose structure comes (only) from examining empirical plots may not generalize to new data.
- A (published) statistical analysis can be exploratory or confirmatory or both. Confirmatory analyses have higher status, but this is just convention—exploratory studies are very important and should be reported as exploratory (not written as if they are confirmatory). Most realistic data analysis is actually both exploratory and confirmatory, and it is important in your writeup to specify which part of your analysis was performed in exploratory mode versus confirmatory mode. (For example, some terms in a regression may be based on scientific hypotheses and some from examining empirical plots.)

Some places to read more are Winter (2019, chapter 16), Gries (2021, chapter 4, section 5.5), Nicenboim et al. (2018) and Roettger (2019) (for linguistic data); Baguley (2012, section 1.3), and references already discussed.

Most regression models that this book covers are confirmatory, but they are used in analysis pipelines including exploratory steps. These include making empirical plots (before modeling) and checking fitted models (model validation) to detect problems that can lead to refitting.

Box 1.2
Broader Context: Assumptions about Our World

This book assumes a classic "frequentist" framework , in an idealized world:

1. The goal is to estimate parameter values (e.g., the mean voice onset time for stops produced by American English speakers).
2. These parameters have true *(population)* values, which are approximated by taking a sample.
3. The population from which the sample is taken is infinitely large.
4. Samples drawn from the population are *representative* and *random* (e.g., samples are from all American English speakers, randomly).

Assumptions 1 and 2 are assumptions of frequentist data analysis, as opposed to Bayesian (see box 2.1). Assumptions 3 and 4 are idealizations about the sample that are assumed by most statistical methods; in reality a researcher is usually working with a *convenience sample*, which she hopes is representative/random enough. While assumptions 3 and 4 are important, they form part of the general issue of how the data were obtained, which I abstract away from in this book. Note that one prerequisite for assumption 4 is that observations are assumed to be *independent*. This is sometimes true for linguistic data, in which case methods from chapters 4 to 7 can be used, but often not; this is the primary motivation for mixed-effects regression models (chapters 8–10).

For more on these points, see Navarro (2016, chapter 10), Kline (2013, chapter 2), or other statistics textbooks (some are listed in section 2.8)

1.3.3 Mathematical Notation

This book's notation largely follows Gelman and Hill (2007). Values of parameters (the population values) are written with Greek letters: μ, σ. Estimates of parameters are written with a "hat": $\hat{\mu}$, $\hat{\sigma}$. Random variables corresponding to observed data are typically written with lowercase Roman letters, with subscripts denoting individual observations. For example, y could be "observed reaction time" in data from a laboratory experiment, and $y_1, \ldots, y_n$ are the values of the n observations. However, parameters that are proportions are written with p (whose estimate would be $\hat{p}$), and some Greek letters (ε, δ, γ) are used for error terms, which are discussed when they are first used (chapters 4, 8). Sample means are written with a "bar": $\bar{y}$ is the mean of $y_1, \ldots, y_n$.

The notation $\sim$ is used to describe how a random variable is distributed. For example, "individual observations y_i follow a normal distribution with mean 1 and standard deviation 5" is written "$y \sim N(1, 5)$." $N(\mu, \sigma)$ means "a normal probability density with mean μ and standard deviation σ."

2 Samples, Estimates, and Hypothesis Tests

This chapter and the next cover basics of *inferential statistics*: going from a finite sample of data from a population to inferences about the population, with the goal of "[drawing] conclusions about which parameter values are supported by the data and which are not" (Hoenig and Heisey 2001, 4).

Regression modeling is a type of inferential statistics that builds on concepts covered in these two chapters. They first cover estimation of population values and differences using sample statistics (section 2.2), uncertainty in these estimates (section 2.3), and assessment of the reliability of conclusions that we reach about population values/differences (sections 2.4–2.7) using *hypothesis testing*. Chapter 3 covers the size of estimates and different kinds of errors that we can make in assessing the size and reliability of an effect.

I assume you already have some exposure to the topics in the current chapter, which are covered in depth in many sources; some are listed in section 2.8. However, these topics are covered in very different ways in different settings (e.g., in a statistics class vs. an R tutorial). For language scientists learning regression modeling, it is useful to establish a common set of concepts, terminology, and practical guidelines, using linguistic data examples. This is the goal of this chapter.

2.1 Preliminaries

2.1.1 Packages

This discussion assumes that you have loaded the tidyverse and languageR libraries (section 1.1).

```
library(languageR)
library(tidyverse)
```

`library(tidyverse)` is a shortcut to load a set of tidyverse packages (section 1.1).[1]

You can alternatively just install and load single packages as needed. For this book, the dplyr, ggplot2, and tidyr packages are the most important (Wickham et al. 2021; Wickham 2016, 2021).

1. As of late 2021: ggplot2, dplyr, tidyr, readr, purrr, tibble, stringr, forcats.

2.1.2 Data

The `transitions` dataset

This discussion assumes that you have loaded the `transitions` dataset:

```
## change path for your computer
transitions <- read.csv("data/transitions_rmld.csv")
```

This dataset (described in more detail in appendix A.1), comes from a study by Roberts, Torreira, and Levinson (2015) that examines approximately 20,000 transitions between conversational turns in a corpus of telephone calls. Each conversation (column `file`) is between two different speakers. Of interest is what factors affect transition durations (column `dur`): how long after one speaker finishes speaking before the other speaker begins. The "before" and "after" speakers for each turn are called speaker A and speaker B (columns `spkA`, `spkB`). For example, in conversation SW3154.EAF (the first rows of the dataframe), the two speakers are SPKR1290 and SPKR1288, and which one is speaker A or B alternates:

```
# Show the first few rows of transitions for some columns:
select(transitions, file, spkA, spkB, sexA, sexB, dur)
##              file     spkA     spkB sexA sexB  dur
## 1     sw3154.eaf spkr1290 spkr1288    F    M  323
## 2     sw3154.eaf spkr1288 spkr1290    M    F  250
## 3     sw3154.eaf spkr1290 spkr1288    F    M   99
## 4     sw3154.eaf spkr1288 spkr1290    M    F  333
...
```

(Here, ... indicates omitted lines of R output. You can always run the code yourself to see full output.)

Observations from the same conversation are not independent—because individual speakers probably have characteristic durations—but independent observations are assumed by methods introduced in this chapter. Thus, we take a small subset of the data where observations are plausibly independent, by choosing a random observation from each conversation:

```
# Makes random sample replicable
set.seed(201)

# Randomize rows of dataframe, take first row for each conversation
random_rows <- sample(1:nrow(transitions))
transitions_sub <- transitions[random_rows, ] %>%
  filter(!duplicated(file))
```

I assume that you have run these commands, so the dataframe `transitions_sub` exists ($n = 349$), and you are using the same "random" dataframe.

2.1.3 Notational Conventions

This chapter refers to individual datasets and R objects. R libraries (e.g., tidyverse, ggplot) are kept in plain text, while datasets are referred to in `teletype`, such as the `transitions` dataset. Teletype is used for objects in R code, such as the `transitions_sub` dataframe, or individual columns of the dataframe. A fundamental data type in R is the *factor*, a categorical variable that takes on discrete values. Factors (typically columns of a dataframe) are written using teletype and individual levels with SMALL CAPS. For example, the factor `sexB` in the `transitions` dataset has levels F and M.

2.2 Point Estimation

In a quantitative study we are often interested in estimating single numbers (called *point estimates* in statistics) that characterize an aspect of the world. For example, in the `transitions` data, we may be interested in the effect of speaker B's gender (column `sexB`: values F, M).[2] This could be quantified by three numbers:

- How long are transitions if speaker B is male?
- How long are transitions if speaker B is female?
- What is the effect of gender on transition duration (the difference between male and female durations)?

2.2.1 Population and Sample

In quantitative studies we are typically interested in *population* values of a parameter—their "true" values in the world, under the model of the world we are assuming (box 2.1).

Typically population values refer to a context beyond just the setting for the study—we are probably interested in gender effects on transition time among all speakers of American English, not just all American English speakers who volunteered to be recorded for this corpus. However, in the real world we never observe population values; we can only take a *sample* of size n and make an inference about the population values.

For example, to estimate the preceding three quantities using the `transitions_sub` data ($n = 349$), we could use:

- The average value of `dur` when `sexB` is F or M (91 msec, 259 msec)
- The difference between these averages (–167 msec)

These values can be calculated for the `transitions_sub` data using functions from dplyr:

```
transitions_sub %>%
  group_by(sexB) %>%
  summarise(mean(dur))
```

These estimates are not the same as the population values, for several reasons: the sample may (a) not be representative of the desired population (e.g., all American English speakers), or (b) truly random, and (c) the sample is finite. While (a) and (b) are important, they form part of the general issue of how the data were obtained, which we are abstracting away from

2. I assume that this variable reflects the speaker's gender presentation rather than biological sex, but this discussion uses the column name provided in the dataset.

in this book (section 1.3, box 1.2). We thus assume we do have a random sample from the population of interest. This leaves (c), which is a fundamental issue addressed by statistical inference: estimation of (population) quantities of interest, whose true values we will never actually know, based on a finite sample.

Box 2.1
Broader Context: Frequentist and Bayesian Statistics

There are two major approaches to statistical inference, corresponding to different philosophies of what *probability* means. The assumption that true values of parameters exist implies we are doing *frequentist* statistics rather than *Bayesian* statistics, where inference results in a probability distribution describing degrees of belief over possible values of the parameter. This is simply a pragmatic choice—frequentist methods are vastly more common in behavioral and social sciences, though Bayesian methods offer some serious advantages and are making inroads. Many sources describe the general differences between Bayesian and frequentist approaches (e.g., Dienes 2008, chapter 4; McElreath 2020, chapter 1), and Nicenboim and Vasishth (2016) and Vasishth, Nicenboim, et al. (2018) are good starting points for Bayesian methods for analyzing linguistic data in particular.

2.2.2 Sampling Distribution of the Sample Mean

In inferential statistics, the general setup is that we have a data sample from a quantitative study, which we assume is representative and random. We use this sample to calculate *sample statistics*, which are estimates of the *population values* of quantities we care about—typically parameters of a statistical model.

Ideally a sample statistic should be an *unbiased estimator* of the population value: the statistic's average value should be the same as the population value, meaning that if we kept repeating the study and computing the statistic, averaging these values would get us closer and closer to the true value.

The most basic sample statistic is the *sample mean*, which is the average of n observations (written $x_i, \dots, x_n$):

$$\hat{\mu} = \frac{\sum_{i=1}^{n} x_i}{n}. \tag{2.1}$$

The sample mean approximates the population mean, which we write μ. To understand how the sample mean is related to the population mean, we can explore using simulations where we know the population distribution.

Suppose that durations of transitions (`dur`) to female speakers in the `transitions` data were in fact drawn from a normal distribution, with mean $\mu = 200$ and standard deviation (SD) $\sigma = 450$, which we write $N(200, 450)$ (see section 1.3.3 on notation). These are the (made-up) population values.

We are interested in the *sampling distribution* of the sample mean: How likely are we to calculate different values for μ if we kept drawing random samples? We can plot a good approximation of this distribution as follows:

1. Draw a sample of n observations from the distribution $N(200, 450)$.
2. Calculate $\hat{\mu}$ for the sample.
3. Repeat steps 1 and 2 many times (n_{sim}), and plot a histogram showing the distribution of $\hat{\mu}$ values.

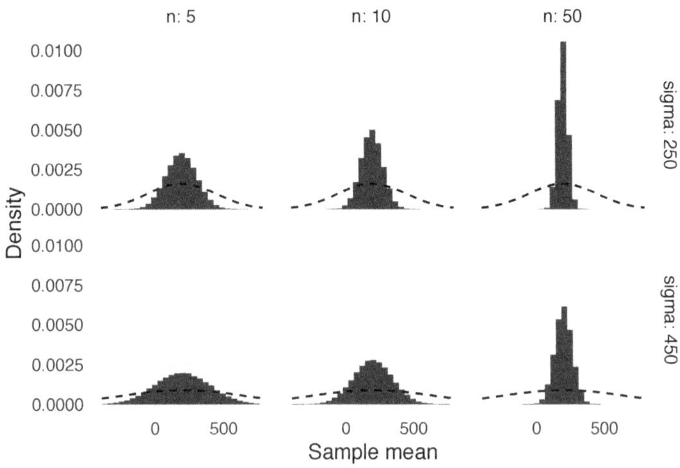

Figure 2.1
Sampling distribution of the sample mean (histograms), calculated over n observations drawn from a normal distribution with $\mu = 200$ and SD σ, for varying n and σ. Dotted lines show the probability distribution of observations [$N(200, \sigma)$].

Figure 2.1 (top row) shows these histograms when the sample mean is calculated over 5, 10, and 50 observations (with $n_{sim} = 100{,}000$). The distribution of the sample mean gets narrower for larger n. Thus, how certain we should be about our observed sample mean (91 msec) depends a lot on sample size: if $n = 5$, we would be likely to calculate a sample mean that is at least this far (109 msec) from the true value, just by chance.

The distribution of the sample mean is also narrower if the quantity that we are estimating is less variable (that is, smaller σ), as illustrated in the bottom row of figure 2.1. Thus, the more observations in the sample or the less variable the quantity we are estimating, the more precise (= less variable) is the mean value that we calculate based on the sample.

Box 2.2
Practical Note: Normal Distributions Refresher

It is useful to know some properties and notation for normal distributions that come up frequently in regression modeling (and in R output). The probability density for a normal distribution is

$$P(x|\mu, \sigma) = \frac{1}{\sqrt{2\pi\sigma^2}} \exp -\frac{(x-\mu)^2}{2\sigma^2}.$$

This is often abbreviated as $N(\mu, \sigma)$ distribution. (Or as $N(\mu, \sigma^2)$ depending on the author.) σ^2 is the *variance*, and the inverse variance ($1/\sigma^2$) is the *precision*.

A normal distribution with mean 0 and SD 1 is called a *standard normal distribution*, written $N(0, 1)$. It is common (in statistics texts or in R output) to use z (or Z) to refer to any random variable that is expected to follow a standard normal distribution. If you draw an observation z from such a random variable, the probability that $|z| < 1$, $|z| < 2$, or $|z| < 3$ is 0.68, 0.954, and 0.997 (respectively). That is, about 66% of probability lies within one σ from the mean, about 95% lies within two σ, and almost all probability lies within three σ. Given the ubiquity of the 95% significance criterion in language sciences, it is also useful to remember that exactly 95% of probability lies within 1.96 σ from the mean. But in general 2 is "close enough" to 1.96 to represent "95% probability."

A very useful property of normal distributions is closure under linear combination:

- If $Z \sim N(\mu, \sigma)$ and a and b are constants, then

$$aZ + b \sim N(a\mu + b, \sqrt{a}\sigma).$$

  That is, adding a constant increases the mean and multiplying by a constant multiplies the variance (which is now $a\sigma^2$).
- If $Z_1 \sim N(\mu_1, \sigma_1)$ and $Z_2 \sim N(\mu_2, \sigma_2)$ and Z_1 and Z_2 are independent, then

$$Z_1 + Z_2 \sim N\left(\mu_1 + \mu_2, \sqrt{\sigma_1^2 + \sigma_2^2}\right).$$

  That is, the mean and variance of the sum are just the sums of the individual means and variances.

One application is normality of the sample mean of n normally distributed (and independent) observations. This follows from the last two equations, because the sample mean is just a sum of normally distributed random variables divided by a constant.

As suggested by the shape of the distributions in figure 2.1, the sample mean is itself normally distributed (box 2.2). The mean of this distribution is μ—because the sample mean is an unbiased estimator of the population mean—and its SD is $\sigma/\sqrt{n}$. This can written more succinctly as

$$\hat{\mu} \sim N(\mu, \frac{\sigma}{\sqrt{n}}) \tag{2.2}$$

using notation for describing the distribution of a random variable (section 1.3.3). The $\sigma/\sqrt{n}$ term quantifies the observation from figure 2.1: either higher sample size or lower variability (in the data we're analyzing) leads to a more precise estimate.

2.2.3 Nonnormal Distributions and the Central Limit Theorem

Much of regression modeling boils down to estimating mean values, as we did for estimating the mean of a normal distribution (as well as quantifying uncertainty in the estimates, as we'll do in section 2.3). But in general, we'll want to analyze data beyond just continuous variables drawn from a normal distribution—much linguistic data is discrete (e.g., yes/no responses in an experiment, syntactic construction A vs. B observed in a corpus), and much continuous-valued linguistic data isn't normally distributed (e.g., word frequencies, phonetic parameters such as voice onset time, reaction times). What happens if we take the sample mean for observations from a nonnormal distribution?

For example, consider the Dutch verb regularity data (dataframe `regularity`) from the languageR package, described in more detail on its help page (type `?regularity`). This dataset, originally from Baayen and Prado Martin (2005), lists 700 Dutch irregular and regular verbs (column `Regularity`) and includes lexical and distributional variables that may help predict whether a verb is regular, including the verb's frequency (column `WrittenFrequency`) and which auxiliary verb is used to form certain past tenses (column `Auxiliary`: levels HEBBEN, ZIJN, ZIJNHEB). In this sample, 159 verbs (23%) are irregular.

Suppose we are trying to estimate a single (population) probability, p: how often Dutch verbs are irregular. (If we picked a random verb in a large Dutch dictionary, how likely would we be to select an irregular one?) We observe n Dutch verbs, $x_1, \ldots, x_n$, each of which is 0 (regular) or 1 (irregular),

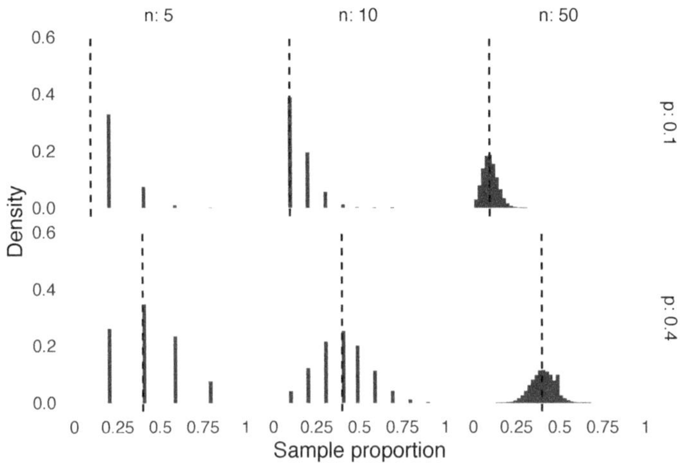

Figure 2.2
Sampling distribution of the sample proportion ($\hat{p}$) for n observations of a Bernoulli random variable with probability p (histogram), varying n and p. Dotted lines show the value of p.

Our estimate for p is the *sample proportion*:

$$\hat{p} = \frac{\sum_{i=1}^{n} x_i}{n}, \tag{2.3}$$

which is the proportion of verbs that were irregular. Equation (2.3) looks the same as equation (2.1), because they are both sample means, but each x_i in equation (2.3) follows a Bernoulli distribution rather than a normal distribution. The numerator of equation (2.3) is thus a count, which follows a binomial distribution, not a normal distribution.

To examine the probability distribution of the sample proportion, we can use the same simulation procedure as in section 2.2.2. Figure 2.2 shows the distribution for different sample sizes, assuming that $p = 0.1$ or 0.4 (made-up values), with a dotted line showing p.

The distribution of the sample proportion looks somewhat normal for $n = 10$, and by $n = 50$ looks perfectly normal—even though the distribution of the actual random variable whose mean is being estimated is not normal (it only takes on values 0 or 1). This illustrates one of the most important results of probability theory, the *central limit theorem*: for a large enough sample from **any** random variable with mean μ and SD σ, the sampling distribution of the sample mean is approximately normally distributed with mean μ and SD of $\sigma/\sqrt{n}$ (equation (2.2)).[3]

The central limit theorem essentially says that the larger the sample you collect, the closer to normally distributed the sample mean is. This remarkable result is frequently used in inferential statistics, because it allows us to apply the same tools (for dealing with normally distributed data) to many different kinds of data. Nonetheless, as the preceding example shows, it is important to bear in mind that any "normal approximation"

3. We are assuming a fairly weak version of the central limit theorem, where observations are independent and identically distributed. See Rice (2006, section 5.3).

is still an approximation, which depends on sample size and the exact distribution being approximated.

Box 2.3
Broader Context: Variability in Bernoulli and Binomial Distributions

When estimating the sample mean of a normally distributed quantity we can vary n and σ, but for the sample proportion (figure 2.2) there is no σ. This is because for a Bernoulli random variable (e.g., a coin flip) the only free parameter is the probability of a success (p). The distribution still has an SD σ, it is just a function of the free parameter p:

$$\sigma_{\text{Bernoulli}} = \sqrt{p(1-p)}.$$

This quantity is maximized when $p=0.5$ and approaches 0 as p gets closer to 0 or 1. The (population) standard error of the sample proportion is just $\sigma_{\text{Bernoulli}}/\sqrt{n}$, which decreases either for higher n or p further from 0.5—the pattern seen in figure 2.2. Intuitively, the further p is from 0.5, the more certain you can be about the outcome of an individual observation. (If $p=1$ or 0, there is no uncertainty.)

The fact that Bernoulli distributions don't have an independent variance parameter will be important for understanding logistic regressions (chapter 6).

2.3 Uncertainty and Interval Estimation

Almost as important as estimating the value of a quantity is estimating the uncertainty in our estimate, measured either by a single number or by a range of values (called an *interval estimate*).

2.3.1 Standard Error

For our estimate of the sample mean, we saw that the width of the distribution in figure 2.1, $\sigma/\sqrt{n}$, quantifies how much uncertainty there is in our estimate of the sample mean. But in general we do not know σ (the population value) and must estimate it. An unbiased estimator for σ is

$$\hat{\sigma} = \sqrt{\frac{\sum_{i=1}^{n}(x_i-\hat{\mu})^2}{n-1}}, \tag{2.4}$$

which looks almost identical to the formula for calculating an SD of a sample, except with $n-1$ in the denominator instead of n (which corrects for finite sample size [Rice 2006, section 7.3]).

We can then define the *standard error* (SE) of the sample mean, which is an unbiased estimator of $\sigma/\sqrt{n}$:

$$SE = \frac{\hat{\sigma}}{\sqrt{n}}. \tag{2.5}$$

The SE estimates how much error there is, on average (across many samples), in our estimate of the population mean μ using $\hat{\mu}$.

One consequence of the central limit theorem is that (for large enough n) we can use $\hat{\sigma}/\sqrt{n}$ as an approximate SE when estimating any sample mean (equation (2.5)), just by replacing $\hat{\sigma}$ by an estimate of the SD. Intuitively, whatever we're trying to estimate, our estimate will be more precise for larger sample size or lower variability.

For example, when estimating a proportion, an unbiased estimator for σ is

$$\hat{\sigma}_{\hat{p}} = \sqrt{\frac{\hat{p}(1-\hat{p})}{n-1}},$$

and the standard error is

$$SE_{\hat{p}} = \frac{\hat{\sigma}_{\hat{p}}}{\sqrt{n}}.$$

Box 2.4
Practical Note: Standard Error and Sample Size

Technically, we should call $\frac{\sigma}{\sqrt{n}}$ the "standard error" and $\frac{\hat{\sigma}}{\sqrt{n}}$ the "estimated standard error" (section 7.3). This book almost always is referring to the latter, so it will use SE to mean "estimated SE" except where there is ambiguity.

Something useful to remember is that *error scales as the square root of sample size* (because the standard error has a $\sqrt{n}$ in the denominator). This is the reason why collecting more data has diminishing returns: doubling sample size only decreases error by a factor of 1.41, and to halve error you need four times as much data! Note that the (estimated) SE won't be exactly halved if you collect four times the data, because this number ($\hat{\sigma}/\sqrt{n}$), is itself an estimate, and s will change for a new sample. Nonetheless, the true SE will be halved. (See exercise 2.1.)

This basic relationship (error $\sim \sqrt{n}$) holds for many kinds of errors and is useful in practical settings, such as planning a new study (how much more data should you collect to observe a hypothesized effect?) or critically assessing empirical patterns in published work (how much should you trust the means of cells A and B if cell B contains half as much data?).

2.3.2 Confidence Intervals

In isolation a standard error is not an intuitive measure of uncertainty, because it does not give a sense of which values are likely. One commonly used notion is a *confidence interval* (CI): a range of values that is $X\%$ likely to contain the population value. Most often, $X = 95\%$.

2.3.2.1 Z-based confidence intervals

Consider the simplest case, where the sample mean is normally distributed. Because the difference between the sample mean and the population value lies within $[-1.96 \cdot \text{SE}, 1.96 \cdot \text{SE}]$ with 95% probability (box 2.2), this interval contains μ with 95% probability:

$$\left[\hat{\mu} - 1.96 \cdot \frac{\sigma}{\sqrt{n}}, \hat{\mu} + 1.96 \cdot \frac{\sigma}{\sqrt{n}}\right],$$

We can define it as the 95% CI.

This formula assumes that σ is known, but this isn't usually the case, so the interval is approximated by substituting in our (estimated) SE (equation (2.5)):

$$\left[\hat{\mu} - 1.96 \cdot \frac{\hat{\sigma}}{\sqrt{n}}, \hat{\mu} + 1.96 \cdot \frac{\hat{\sigma}}{\sqrt{n}}\right]. \tag{2.6}$$

This second interval should be called an "approximate 95% CI," but it is often reported in publications as just "the 95% CI" (sometimes rounding up to 2 instead of 1.96). This interval approximates a true CI for large enough n (by the central limit theorem), even when the data are not normally distributed.

2.3.2.2 Bootstrapped confidence intervals

Another way to approximate a CI for an estimate of a quantity T uses *bootstrapping* (Efron and Tibshirani 1994): use the sample we have to simulate many (fake) new random samples (say $n_{sim} > 10{,}000$), each one drawn by sampling with replacement from the original sample (meaning we can draw the same original observation multiple times). For each new sample, we recalculate T. The resulting set of values $T_1, \ldots, T_{n_{sim}}$ can be used to construct a *bootstrapped CI* in different ways, the most intuitive of which is just to take the 2.5% and 97.5% quantiles of the distribution of the T_i as the lower and upper bounds.[4]

> **Box 2.5**
> **Practical Note: Bootstrapping**
>
> Bootstrap methods are powerful tools for regression modeling—they can be used to calculate not just CIs, but almost any measure of accuracy, uncertainty, or reliability (e.g., p-values), all with minimal assumptions about the data (beyond independence of observations). The major tradeoff is computation time, but if you can wait long enough, a bootstrap estimate is often the gold standard. On contemporary computers, bootstrapping is fast enough to be the default option for many simple cases, such as CIs for estimated means (discussed in the text), but too slow for everyday use for more complex cases, such as p-values for mixed-effects models (section 9.4.1.3).
>
> Although the basic technique of resampling with replacement is intuitive, exactly how the bootstrap sample is used to compute such measures can be subtle and generally is not relevant for interpreting the results. Thus, for most researchers it is best to just use existing R implementations of bootstrap methods (in packages such as boot [Canty and Ripley 2017]), and read their documentation for more details if needed. To learn more about bootstrapping methods, Good (2006) is a readable introduction, general statistics books (e.g., Baguley 2012, section 3.5; Kline 2013, 54–57) contain short introductions, and many online resources cover bootstrapping methods in R.

2.3.2.3 Interpretation of confidence intervals

CIs are less intuitive than they seem (Hoekstra et al. 2014; Cumming 2014). A common (and natural) misconception is that the CI means we can be 95% sure that the population value lies in this interval. What it really means is that if we collect many samples, about 95% of the time the CI we calculate would contain the true (population) value—which is unintuitive, because we typically only have one sample. The unintuitive meaning arises because the CI is itself a sample estimate (of an interval), which changes with every sample. For large enough n, the two interpretations give similar intervals, but in general we must bear in mind that we could simply have drawn an (un)lucky sample—the population value may not lie in our 95% CI at all. This point holds whether CIs are computed using a parametric method (e.g., z-based) or a bootstrap method.

To discuss CIs in an intuitive yet correct way, Kline (2013, 41) suggests the phrasing: "The interval [96.28, 103.72] estimates μ, with 95% confidence."

---

4. This "percentile bootstrap" method has problems, and typically more complex methods are used to construct an estimate (Good 2006, chapter 2).

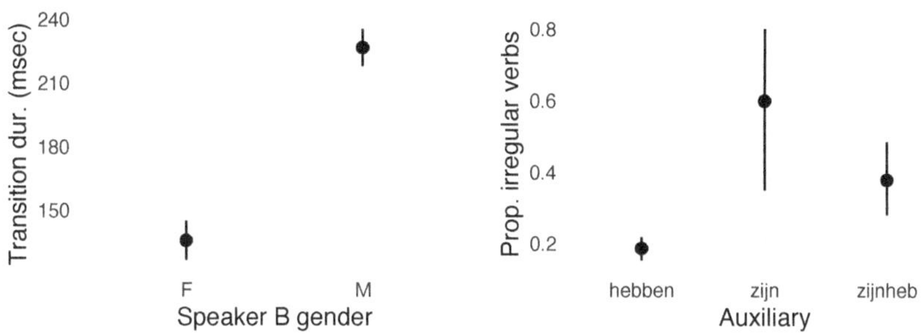

Figure 2.3
Left: Mean transition duration as a function of speaker B's gender for the `transitions` dataset, with approximate 95% CIs (mean $\pm$ 1.96 $\cdot$ SE). *Right*: Proportion of irregular verbs as a function of auxiliary class for the `regularity` dataset, with bootstrapped 95% CIs.

2.3.2.4 Example

Let's look at the (sample) mean and (approximate) 95% CIs for the two examples we've considered so far: transition duration as a function of speaker gender in the `transitions` dataset and the likelihood of a verb being irregular in the `regularity` dataset.

Figure 2.3 shows these data summaries, using the two methods for CIs discussed. (The ggplot2 code is not shown, as always, but is in the code file for this chapter.)

The left plot suggests that transitions to male speakers are longer than the transitions to female speakers. The right plot suggests that verbs with auxiliary HEBBEN are less likely to be irregular than other verbs, but it is not clear whether ZIJN and ZIJNHEB verbs differ in how likely irregularity is. The rule of thumb behind these interpretations is that when errorbars represent 95% CIs, a difference in estimates is "real" if the errorbars do not overlap.[5]

2.3.3 The *t*-Distribution

When analyzing quantitative data, we are usually interested not just in estimating values, but also in assessing whether they differ from some reference level (usually 0) or from each other. This means we should use uncertainty in the estimate of the values—the SE—as the scale on which distances are measured. This implies dividing means or differences in means by $\sigma/\sqrt{n}$ (the population SE); the most basic distance we could measure on this scale is $\mu/(\sigma/\sqrt{n})$ (distance of a single mean from 0). For example, in figure 2.3 (left), this distance could measure how many SEs from 0 the FEMALE point lies, letting us assess whether transitions to female speakers have mean duration different from 0.

To estimate $\mu/(\sigma/\sqrt{n})$, because we cannot directly observe μ or σ, we substitute in our estimates of them (equations (2.1) and (2.5)):

$$t = \frac{\hat{\mu}}{\hat{\sigma}/\sqrt{n}}.$$

5. This intuition, though widespread, is actually too pessimistic. For two independent means, a more useful heuristic is Cumming and Finch's "overlap rule of eye" (Cumming and Finch 2005; Cumming 2012): nonoverlapping errorbars mean $p < 0.01$, just-touching errorbars mean $p \approx 0.01$, and "moderate overlap" of errorbars (about 1/2 of errorbar length) corresponds to $p = 0.05$. The heuristic works best for larger ($n > 10$) and more similar samples (in terms of n, σ).

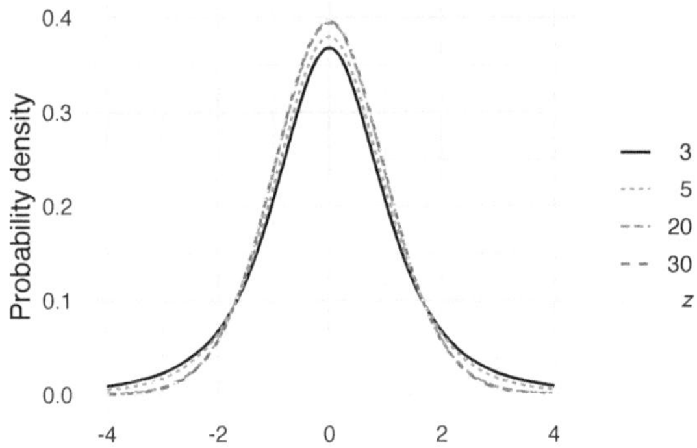

Figure 2.4
Probability distributions for a t_{n-1} distribution with varying *df* ($n-1$ = 3–30) and a normal distribution with $\mu = 0$ and $\sigma = 1$ (the z line).

Although $\mu/(\sigma/\sqrt{n})$ is normally distributed (with mean 0 and SD 1), t is not. As the sample size becomes smaller, s becomes more uncertain (note the $n-1$ in equation (2.5)), making the distribution of t wider than a normal distribution for small n. This distribution is called the *t-distribution*, and it is sensitive to sample size. For a sample of n observations, our quantity of interest (how many SEs the mean is from 0) follows a t-distribution with $n-1$ *degrees of freedom* (Rice 2006, section 6.3), which is written

$$\frac{\hat{\mu}}{\hat{\sigma}/\sqrt{n}} \sim t_{n-1}. \tag{2.7}$$

The t-distribution looks similar to a normal distribution, but has fatter tails for smaller sample sizes ($n < 30$), as shown in figure 2.4. Once n is larger than about 30, the t-distribution is very similar to a $N(0, 1)$ normal distribution.

This is the first time we have encountered the notion of *degrees of freedom*, often abbreviated *df*, which measures the number of pieces of *independent* information that go into estimating some parameter(s). It is hard to give a definition of *df* that is both precise and intuitive (the Wikipedia page is a good place to start), but *df* is an important concept that will often come up in our discussion of different hypothesis tests and regression models.

2.3.4 *t*-Based Confidence Intervals

t-Distributions are common in regression analysis, and it is worth knowing that they are like normal distributions, but a bit wider. To get a sense of how much wider, we can consider the common case of working with a 95% CI. Recall that for a normal distribution, 95% of the probability mass lies within 1.96 SEs from the mean. To see how many SEs from the mean 95% of probability lies for a t-distribution for different sample sizes, we use one of R's built-in functions for working with t-distributions, `qt()`, which gives a *quantile* value: the point on the distribution below which X% of the probability lies. (Here, X = 97.5%.)

```
# Corresponds to 95% significance level (1-alpha)
alpha <- 0.05
```

```
# mutate: a function from dplyr which adds columns
data.frame(n  = c(5, 15, 25, 50, 100, 500)) %>%
  mutate(ciUpper = qt(1 - alpha / 2, n))
##     n ciUpper
## 1   5     2.6
## 2  15     2.1
## 3  25     2.1
## 4  50     2.0
## 5 100     2.0
## 6 500     2.0
```

Because a t-distribution is a little wider than a normal distribution, t-based 95% CIs based on the t-distribution are slightly larger than CIs based on a normal distribution, especially when $n < 30$. For large enough n, the CIs are the same as for a normal distribution ($1.96 \cdot$ SE). Even for $n = 5$, a very low sample size, the t-distribution is only about 25% wider than a normal distribution.

2.4 Hypothesis Testing

Null hypothesis significance testing (NHST) is the most commonly used framework for assessing the evidence for an effect across scientific fields and is what is usually meant by "hypothesis testing." This book assumes that the reader has basic exposure to NHST hypothesis testing (see section 2.8 otherwise). However, the way hypothesis testing is often presented in introductory courses—as a large set of tests to be applied for different kinds of data, often aided by a flowchart—is not good background for how hypothesis tests play into regression modeling. This requires a good *conceptual* understanding of NHST methods and how they are applied in practice—because the underlying concepts are fundamental to most regression models used in current practice in language sciences. If you are comfortable with these things, you can skip to the next chapter.

The remainder of this chapter reviews NHST terminology and concepts that are important as background for regression models introduced in later chapters. The discussion is mostly about t-tests and z-tests, the most common NHST tests, which are central to regression models. These tests are a good vehicle to review NHST terminology and concepts, which are largely shared with other hypothesis tests used for regression modeling (e.g., likelihood ratio tests), which are introduced in later chapters as needed.

Box 2.6
Broader Context: NHST—Ubiquity and Controversy

NHST has been the dominant paradigm for assessing evidence across a variety of fields for decades, and this is unlikely to change anytime soon. For just as long, NHST methods have been dogged by controversy, which has increased in recent years with the "replication crisis" and associated criticism of p-values. An extensive literature discusses problems with NHST, ranging from its philosophical foundations (e.g., logically inconsistent assumptions) to the nonintuitive nature of its underlying concepts (e.g., repeated sampling is unintuitive), to rampant misinterpretation of NHST results (section 2.6). Criticisms of NHST often oppose it to Bayesian methods, which avoid some of the problems, and have become increasingly widely used as computational power increases. It is beyond the scope of this chapter to even begin to discuss these issues;

the American Statistical Association's (ASA) official position paper on the topic (Wasserstein and Lazar 2016), which is a good place to start, gives a "brief" bibliography of 40 key references.

What do these issues mean for regression modeling for linguistic data, which make extensive use of NHST methods? At least in language sciences, problems with NHST in practice seem to result more from a lack of understanding among researchers and mindless application of NHST methods than from foundational issues with the paradigm itself. (Navarro 2016, section 11.9 gives a similar view, for psychologists.) NHST methods can be unintuitive for thinking about data and the world, and reporting NHST tests uses a lot of numbers and technical terms—yet because NHST has been the dominant paradigm for decades, it is widely used and rarely explained in scientific papers. It is crucial to have a good conceptual understanding of NHST for regression modeling—both to spot common fallacies (e.g., "$p > 0.05$ means the null hypothesis is true") and avoid them in your own work. It is less important to understand the math, though if you find it helpful many good treatments are available (see section 2.8). This book tries to build informed and critical understanding of (some) NHST methods, but does not generally go into broader foundational issues or alternative methodologies (e.g., Bayesian methods).

2.4.1 Evaluating One Sample

To introduce hypothesis testing, it is useful to have a running example in mind.

Example: One speaker in the `transitions` data

Let's assume the following setting, using a subset of the `transitions` data:

We observe $n = 89$ floor transitions to a single male speaker (SPKR1540). The sample mean for this speaker is $\hat{\mu} = 83$ msec. The sample mean is approximately normally distributed (by the central limit theorem), with (population) mean μ and SD $\frac{\sigma}{\sqrt{n}}$, so the interval $\mu \pm 1.96 \cdot \frac{\sigma}{\sqrt{n}}$ contains approximately 95% of the probability mass. We don't know μ, but suppose we have a hypothesis about it that is meaningful, with respect to a value μ_0—say $\mu = 0$ ($\mu_0 = 0$), which in the current example would mean "on average, this speaker starts talking exactly when his interlocutor stops." (This would mean that the speaker often interrupts his interlocutor.)

If we knew the population SD σ, we could answer the question "is the sample mean far enough away from 0 to be 95% sure that $\mu \neq 0$?"—in the sense that if the true mean were 0, we'd be unlikely to observe a sample mean this large. For the moment, suppose we magically knew that $\sigma = 300$—so the (true) SE is 32 ($300/\sqrt{n}$), and the sample mean (83) is 2.6 SEs from 0. We could then operationalize our question as "how likely is the sample mean to be at least this far from 0?", using a $N(0, 32)$ distribution (figure 2.5).

2.4.2 *z*-Scores and *z*-Tests

To answer this question it is useful to introduce the notion of a *z-score*, which measures how many SDs an observation x_i is from a mean:

$$z_i = \frac{x_i - \mu}{\sigma}.$$

If the observations are normally distributed, the z_i values follow a standard normal distribution ($\mu = 0, \sigma = 1$).

We can test our hypothesis by calculating how many SEs the observed sample mean lies from μ_0:

$$z = \frac{(\hat{\mu} - \mu_0)}{\sigma/\sqrt{n}}, \tag{2.8}$$

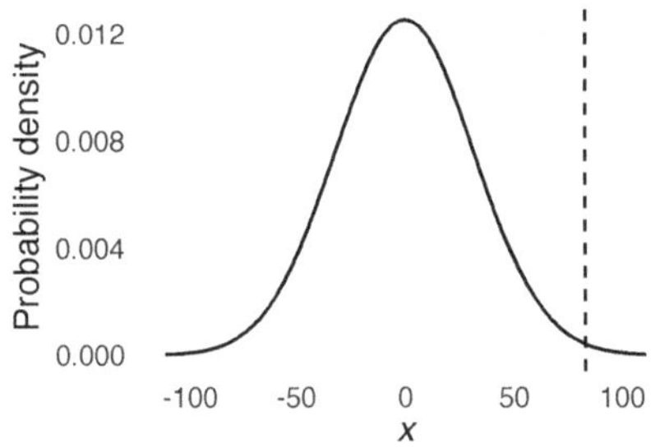

Figure 2.5
The distribution of the sample mean (solid line) for 89 observations from $N(0, 300)$. The dotted line shows the observed sample mean (83) for one sample.

which in our example ($\hat{\mu} = 83$, $\mu_0 = 0$, $\sigma = 300$, $n = 89$) gives $z = 2.6$. We then calculate how much probability lies at least this far from 0 for an $N(0, 1)$ distribution:

```
transitions_spkr1540 <- filter(transitions, spkB == "spkr1540")

mu_hat <- mean(transitions_spkr1540$dur)
s_hat <- sd(transitions_spkr1540$dur)
n_sub <- nrow(transitions_spkr1540)
se_hat <- s_hat / sqrt(n_sub)
# Assumed population value of SE
se_true <- 300 / sqrt(n_sub)
2 * (1 - pnorm(mean = 0, sd = 1, mu_hat / se_true))
## [1] 0.0089
```

This is the probability of observing a sample mean at least this far from 0.

This procedure is called a *z-test*, after the test statistic.

2.4.3 One-Sample *t*-Tests

In the previous example, we assumed that we knew the population SE, but this isn't usually true. Instead, we estimate the SE from the sample as $\hat{\sigma}/\sqrt{n}$, using equation (2.5), and use it to make an estimate of how far the sample mean is from μ_0 in units of SE (equation (2.8)):

$$t = \frac{\hat{\mu} - \mu_0}{\hat{\sigma}/\sqrt{n}}.$$

This is the *t-statistic*, an estimate of z that can be calculated just using the data in the sample.

The t-statistic follows a t-distribution (section 2.3.3) with $n - 1$ degrees of freedom, which is slightly wider than the normal ($N(0, 1)$) distribution of z, reflecting greater uncertainty in estimating the SE (figure 2.4).

In this case, t is the *test statistic*—a value we compute based on the sample, which we will then use to decide between two hypotheses: the *null hypothesis* (H_0) and the alternative hypothesis (H_a). Here, these are

- Null hypothesis: $\mu = \mu_0$
- Alternative hypothesis: $\mu \neq \mu_0$

That is, the population mean is either the null value (usually 0) or it isn't. For our example, the logic of hypothesis testing is as follows:

- If H_0 were true (under the null hypothesis), t would follow a t_{n-1} distribution.
- Calculate how likely we are to get a value of t at least as extreme as the value we observed, using the t-distribution. This is p (the *significance*, or *p-value*).
- If p is less than the *significance level* α (a number between 0 and 1), we reject the null hypothesis. Otherwise, we fail to reject the null hypothesis.[6]

This procedure is called a (one-sample) *t-test*.

For our example, we would calculate p as follows (using $\alpha = 0.05$):

```
# Estimated standard error
se_hat <- sd(transitions_spkr1540$dur) / sqrt(n_sub)
t <- mu_hat / se_hat

# Probability that t lies at least this far from 0
2 * (1 - pt(mu_hat / se_hat, df = n_sub - 1))
## [1] 0.016
```

Or we could just use R's built-in `t.test()` function:

```
t.test(transitions_spkr1540$dur)
...
## t = 2, df = 88, p-value = 0.02
## alternative hypothesis: true mean is not equal to 0
...
```

p is slightly higher than for the z-test ($p = 0.016$ vs. 0.0089), as expected due to increased uncertainty from estimating the SE.

2.4.4 Hypothesis Testing in General

More generally, (NHST) hypothesis testing follows these steps:

- Choose a significance level, α.
- Formulate a null hypothesis, H_0.
- Formulate an alternative hypothesis, H_a.
- Gather data, calculate a test statistic, T.
- Determine the probability of obtaining T or a more extreme value under H_0, the p-value.
- If $p \leq \alpha$, reject H_0. Otherwise, fail to reject H_0.

6. The latter is sometimes informally stated as "accepting the null hypothesis," although this is not technically correct.

This procedure underlies most inferential statistics commonly applied to linguistic data—*t*-tests, analyses of variance, linear regressions, mixed-effects regressions—though the steps are not usually explicitly stated. In particular, it is often assumed that

1. $\alpha = 0.05$: we have to be "95% certain" to reject the null hypothesis.
2. The null hypothesis is a *nil hypothesis* (Kline 2013, 69): H_0 corresponds to "no difference" or "no effect." (For a *t*-test, this is $H_0 : \mu_0 = 0$.)
3. A two-sided test is used (for tests where one- and two-sided options are possible).

Why these assumptions? Assumption 1 is purely convention and is widely used in behavioral and social sciences. There is no inherent justification for this particular value of α, which means that researchers accept making a type I error 1/20 of the time. Type I errors and the consequences of different α values will be discussed in chapter 3.

Assumption 2 is also convention of a sort: common applications of hypothesis tests assume a null hypothesis of this form, such as assessing whether a predictor has an effect in a linear regression model (H_0: no effect) or whether two categorical variables are associated (H_0: no association).

In assumption 3, two-sided tests are the default principally because they are "more conservative" (less likely to falsely reject the null hypothesis; see section 3.3.5.1) than one-sided tests are, and they do not require choosing an effect direction. Instead they remain agnostic on the direction of the effect.

However, when the research hypothesis calls for it, you would use a one-sided test. For example, to assess whether transitions to speaker 1540 in the `transitions` dataset are at least 0 msec long, you'd conduct a one-sided version of the *t*-test in section 2.4.3, with null hypothesis H_0: $\mu \leq 0$:

```
t.test(transitions_spkr1540$dur, alternative = "greater")
...
## t = 2, df = 88, p-value = 0.008
## alternative hypothesis: true mean is greater than 0
...
```

Box 2.7
Broader Context: NHST Is a Hybrid of Two Philosophies

The logic of NHST, presented at the beginning of section 2.4.4 and ubiquitous in current practice, is a hybrid of two very different philosophies—Neyman-Pearson and Fisherian—for assessing evidence for an effect relative to a null hypothesis.

In the Neyman-Pearson approach, hypothesis testing is a binary decision process (reject/do not reject), leading to H_a or H_0. All that matters for the decision is whether you find $p < \alpha$, and there is no conceptual difference between the two outcomes—rejecting or failing to reject the null hypothesis. Thus, there is no notion of "significance" (except as a descriptive term for $p < \alpha$), and describing results as "highly significant" versus "significant" (e.g., $p = 0.002$ vs. $p = 0.04$, when $\alpha = 0.05$) makes no sense.

In the Fisherian approach, there is only a null hypothesis, and testing assesses *significance*, which is a gradient measure of confidence against the null hypothesis (assessed by p). In this interpretation lower values of p offer stronger evidence against H_0. The significance level α is a cutoff point defined for convenience, but it is not essential. An alternative hypothesis is not specified; implicitly it is just "not H_0." All interpretation relates to whether the null hypothesis can be rejected. Thus, there is no notion of a low p-value supporting an alternative hypothesis (even though this is usually the hypothesis of interest to the researcher)—it just gives high confidence against the null.

In practice, NHST as it is taught to and applied by researchers is a hybrid of these approaches, which results in sometimes thinking in Neyman-Pearson (binary decision) and sometimes in Fisherian (gradient significance) terms. Roughly, many researchers "follow Neyman-Pearson procedurally but Fisher philosophically" (Perezgonzalez 2015). For example, it is common to write in the same paragraph that an effect is "highly significant" (which implies that p is a gradient measure) and that we can "reject the null hypothesis" (a binary decision).

It has long been pointed out that this hybrid nature is a problem—the two approaches rest on fundamentally different premises and lead to different interpretations of results, and the logical inconsistency is at the root of many common NHST misconceptions (e.g., Perezgonzalez 2015; Gigerenzer 2004; Kline 2013, chapters 1 and 3; and references therein).

Whether these problems mean researchers should overhaul a hybrid approach completely or be better-informed users is a matter of debate, but in practice a hybrid approach is deeply ingrained in current practice in language sciences, as in most fields. To orient you in current practice, this book will follow the hybrid approach, but will highlight common pitfalls and how to avoid them.

2.4.5 Paired *t*-Tests

A useful application of a one-sample *t*-test is for *paired* data, where the data consist of pairs of observations from two samples, and the difference between the two samples is of interest.

For example, for the `transitions` data, we could assess whether turns get shorter over the course of a conversation, on average, by computing for each conversation the difference between mean `dur` during the first minute and mean `dur` after the first minute, and then submitting these differences to a one-sample t-test (exercise 2.4).

2.4.6 Two-sample *t*-Tests

The one-sample *t*-test is mostly used as part of more complex statistical procedures (e.g., linear regression; section 4.3.3). More commonly the *two-sample t-test* is used on its own, to examine the difference in means between two groups.

For example, we can return to the question posed in section 2.2 for the `transitions_sub` data: What effect does the gender of speaker B have on transition duration? The empirical data (figure 2.6) suggest that transitions to male speakers may be slightly longer on average. A two-sample *t*-test lets us test whether this difference is significant.

2.4.6.1 Equal variances

A two-sample *t*-test assumes two *independent* samples: the values drawn for one sample cannot be predicted at all from the values drawn from the other sample. The two samples are assumed to be drawn from normal distributions with (in the simplest setting) the same SD:

- Sample 1: n_1 observations $(x_1^1, \ldots, x_1^{n_1})$ from $N(\mu_1, \sigma)$
- Sample 2: n_2 observations $(x_2^1, \ldots, x_2^{n_2})$ from $N(\mu_2, \sigma)$

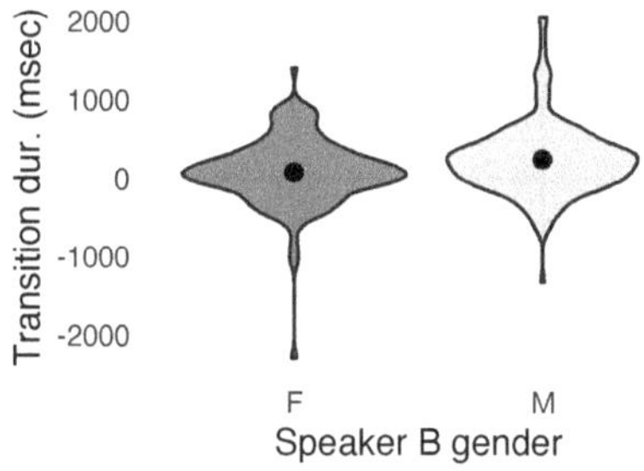

Figure 2.6
Distribution of transition duration as a function of speaker B's gender for the `transitions` data (violin plots), with mean value (dots).

Let $\hat{\mu}_1$ and $\hat{\mu}_2$ be the sample means—the averages of the observations in each sample. We are interested in the difference in sample means: $\hat{\mu}_1 - \hat{\mu}_2$.

Each sample mean is normally distributed, so their difference is as well (because the sum of normally distributed random variables is normally distributed; see box 2.2). We want to estimate a measure analogous to z for the one-sample t-test, but now defined as

$$Z = \frac{\text{difference in sample means}}{\text{SE of difference in sample means}}.$$

As with the one-sample t-test, we must estimate the SE. The sample SDs $\hat{\sigma}_1$ and $\hat{\sigma}_2$ are calculated as in equation (2.4). Because we are assuming the two samples have equal variance, we can estimate a single sample SD $\hat{\sigma}$:

$$\hat{\sigma} = \sqrt{\frac{(n_1-1)\hat{\sigma}_1^2 + (n_2-1)\hat{\sigma}_2^2}{n_1+n_2-2}}.$$

That is, we are estimating the variance σ^2 as the average of the two sample variances $\hat{\sigma}_1^2$ and $\hat{\sigma}_2^2$, weighted by their summed degrees of freedom (n_1-1, n_2-2), which is an unbiased estimator of σ (e.g., 42–43).

After further math, we get the following test statistic for the difference in sample means:

$$t = \frac{\bar{x}_1 - \bar{x}_2}{\hat{\sigma}\sqrt{(1/n_1 + 1/n_2)}}, \tag{2.9}$$

which follows a t distribution with $n_1 + n_2 - 2$ degrees of freedom.

We then can carry out a two-sample t-test with

- Null hypothesis: $H_0: \mu_1 - \mu_2 = 0$ (equal means)
- Alternative hypothesis: $H_a: \mu_1 - \mu_2 \neq 0$ (means are not equal)

by computing how far out t is on a t-distribution with $n_1 + n_2 - 2$ degrees of freedom.

2.4.6.2 Unequal variances

When testing whether two samples differ, there is not usually any reason to think that the samples have equal variances. In practice we should assume by default that the samples have unequal variances—as R does in the `t.test()` function. This default is the *Welch t-test*, which corrects the *df* and SE calculations for unequal variances. (The formulas get complicated, which is why it is useful for exposition to first assume equal variances.)

This variant of the *t*-test—two-sample, unequal variances—is by far the most commonly used in practice, and it is often just called "a *t*-test" in papers.

Box 2.8
Broader Context: Uncertainties Add

The SE and *df* estimates for a Welch *t*-test illustrate useful broader points (Kline 2013, 47; Baguley 2012, section 3.4.2).

The estimated SE is

$$SE_{Welch} = \sqrt{\frac{\hat{\sigma}_1^2}{n_1} + \frac{\hat{\sigma}_2^2}{n_2}},$$

where $\hat{\sigma}_1/\sqrt{n_1}$ and $\hat{\sigma}_2\sqrt{n}$ are the estimated SEs for each sample. So we can think of this formula as $SE^2_{\mu_1-\mu_2} = SE^2_{\mu_1} + SE^2_{\mu_2}$: the error in estimating a difference in two quantities is the sum of the error in estimating either quantity. More generally, *uncertainties in means add*: when estimating any sum/difference of interest in a quantitative study, your estimate is at least as noisy as each individual condition.

The estimated *df* are

$$df_{Welch} = \frac{SE^4_{Welch}}{\frac{SE^4_1}{df_1} + \frac{SE^4_2}{df_2}},$$

where df_1 and df_2 are the degrees of freedom for each sample ($n_1 - 1$, $n_2 - 1$). Although this formula is not intuitive, you can convince yourself that the lower df_1 or df_2 are, the lower df_{Welch} is. Because lower *df* means "this *t*-test involves more uncertainty in estimating a variance," the formula can be understood as *uncertainties in variance add*: the uncertainty in estimating a variance in a difference in two quantities is a function of the error in estimating the variance in either quantity.

2.4.6.3 Example

To test the effect of speaker B gender on transition duration in the `transitions_sub` data, we can carry out a Welch two-sample *t*-test:

```
t.test(dur ~ sexB, data = transitions_sub)
...
##  Welch Two Sample t-test
##
## data:  dur by sexB
## t = -4, df = 329, p-value = 5e-04
## alternative hypothesis: true difference in means between
## group F and group M is not equal to 0
## 95 percent confidence interval:
##  -261  -74
...
```

Durations to male speakers are longer than durations to female speakers; the difference is significant at the $\alpha = 0.05$ level ($p < 0.001$), with 95% CI [–261, –74] for the male/female difference. The first argument to `t.test` is a "formula," using R's formula notation; roughly, `Y ~ X` means "predict `Y` as a function of `X`."

This notation is explained further when we get to regression models (box 4.3).

2.5 Parametric and Nonparametric Tests

In *parametric* tests, such as z-tests and t-tests (sections 2.4.2 and 2.4.3), we assume sample(s) drawn from a distribution with a particular *functional form*—an equation characterized by a few parameters, such as a normal distribution (μ, σ, n) or a binomial distribution (p, n). For example, a two-sample t-test assumes that the samples are normally distributed with means μ_1 and μ_2 and variances σ_1 and σ_2. The null hypothesis ($\mu_1 = \mu_2$) and alternative hypothesis ($\mu_1 \neq \mu_2$) are defined with respect to these parameters.

In contrast, *nonparametric* hypothesis tests do not assume that the sample(s) being tested are drawn from a particular type of distribution (e.g., normal) and tend to have fewer other assumptions. There are nonparametric analogues to all commonly used parametric hypothesis tests, of which we discuss one example here.

2.5.1 Wilcoxon Tests

The most commonly used nonparametric analogue to t-tests are the *Wilcoxon signed-rank test* (for one sample or paired data) and the *Wilcoxon rank-sum test* (for two samples).[7] It is easiest to think of all these as variants of "Wilcoxon tests," analogous to "t-tests," which have the same variants: one-sample, two-sample, and paired. (Because nonparametric tests don't assume anything about population variances, there are no equal or unequal variance versions.) Wilcoxon tests have the same independence assumptions as t-tests: for a two-sample test, this means independent samples and independence of observations within each sample. (Section 3.5 discusses this further.)

2.5.1.1 Two-sample Wilcoxon test

Perhaps the most commonly used nonparametric test is the (two-sample) Wilcoxon rank-sum test, which checks whether two independent samples were drawn from populations with the same distributions, up to a "location shift" μ_0 (shifting one distribution over horizontally by μ_0). This test can be informally thought of as testing whether the samples have different medians (as opposed to means, in the two-sample t-test).

The null hypothesis for this test is that the two samples are drawn from identical population distributions (up to the location shift). The test statistic is related to the sum of ranks of the data in sample 1, when all observations from both samples are combined and put in order.[8]

2.5.1.2 Example

For the `regularity` dataset, a variable measured was `NcountStem`, the number of words which are very close in spelling to the verb (i.e., orthographic "neighborhood density"), which may be higher for regular verbs. We are interested in assessing whether neighborhood density does significantly differ by verb regularity. The empirical pattern is shown in figure 2.7.

---

7. The one-sample test is sometimes called "Wilcoxon t-test" (because it is analogous to a one-sample t-test), and the two sample test is also called a "Mann-Whitney U-test" or a "Mann-Whitney-Wilcoxon test" or a variation.

8. For more detail and intuition, see e.g., Rice (2006, section 11.2.3) and Field, Miles, and Field (2012, section 15.4).

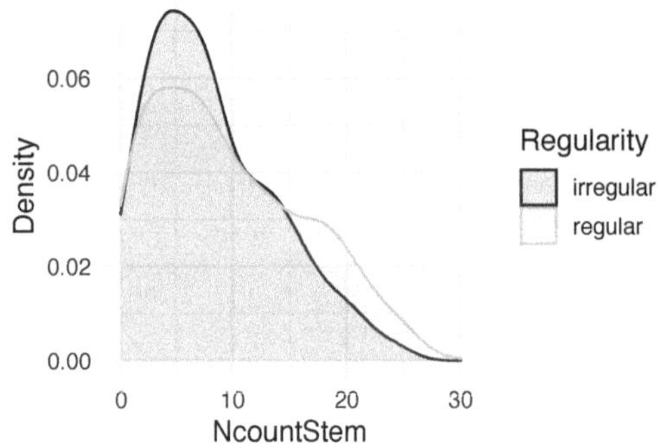

Figure 2.7
Distribution of orthographic neighborhood density as a function of verb regularity for the `regularity` data, shown as a smoothed density estimate.

The distribution of `NcountStem` is right-skewed and clearly not normal. We therefore assess whether `NcountStem` differs by verb regularity (`Regularity`) using a Wilcoxon rank-sum test (with location shift of 0). All variants of Wilcoxon tests are executed in R using `wilcox.test`, which automatically selects the correct Wilcoxon test depending on the arguments it is given. In this case, the two-sample test is used:

```
wilcox.test(NcountStem ~ Regularity, data = regularity)
...
##  Wilcoxon rank sum test with continuity correction
##
## data:  NcountStem by Regularity
## W = 38418, p-value = 0.04
## alternative hypothesis: true location shift is not equal to 0
```

The result suggests that regular and irregular verbs significantly differ (at the $\alpha = 0.05$ level) in orthographic neighborhood density ($p = 0.04$).

2.5.2 *t*-Tests versus Wilcoxon Tests

Testing differences in means is very common in practice; when should a parametric versus nonparametric method be used, and does it matter? The main difference between these methods is the assumption by a *t*-test that data are normally distributed. Sometimes it is clear a priori that data cannot be normally distributed, for example, for a rating task with upper and lower bounds. Otherwise, you can check whether data look normally distributed using visual or quantitative methods, provided the sample is large enough (as described in section 2.5.3).

In practice, *t*-tests are relatively robust to deviations from normality (box 2.9). But certain deviations from normality can strongly affect the results of *t*-tests, such as when one or both samples contain clear outliers or have highly-skewed distributions—as in the example that follows.

As such, standard advice for one's own data is just to use a nonparametric test (such as Wilcoxon rank-sum) if the data are clearly nonnormal, or you cannot assess normality

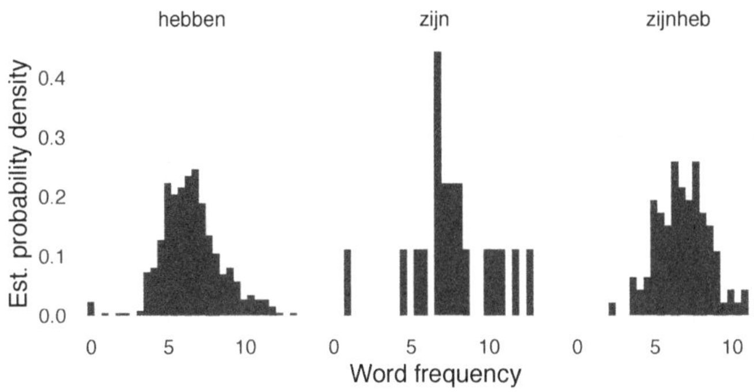

Figure 2.8
Histograms of `WrittenFrequency` for verbs in each `Auxiliary` class (e.g., HEBBEN), for the `regularity` data.

(small sample size). Because Wilcoxon tests are computed using ranks, they are robust to outliers and skewed distributions—like medians are when compared to means.[9]

Box 2.9
Broader Context: Does Normality Matter for *t*-Tests?

If you have taken a statistics class, you have probably heard that *t*-tests can only be applied to normally distributed data, so you need to check normality before applying a *t*-test. This is sensible as default advice. However, in practice, *t*-tests are relatively robust to moderate deviations from normality—especially for larger sample size, where the central limit theorem quickly kicks in, and especially for deciding whether to reject the null hypothesis (e.g., whether $p < 0.05$, as opposed to determining *p*'s exact value). This is useful to know both for confidence in linear regression models presented in later chapters (where *t*-tests are often used) and in reading the literature. Many papers report *t*-tests on (probably) nonnormal observation distributions, in part because carrying out nonparametric tests was computationally difficult before the 1990s. One shouldn't immediately discount the results of *t*-tests for the many types of linguistic data where normality is unlikely (such as Likert scores, reaction times, durations of sounds/words) or where sample size is small, but do view these results critically—especially where significances are near the α cutoff. Because *t*-tests involve comparisons of means, a good rule of thumb for assessing whether a *t*-test might be affected by a particular case of nonnormality is just to ask, "how would this nonnormality affect the estimated mean?"

Note that there is no magic cutoff for what "large" or "small" sample sizes are or what a "moderate deviation" is. By a common rule of thumb, when $n = 30$ the central limit theorem kicks in (the *t*-distribution looks normal), but the required *n* actually depends on the shape of the data's distribution.

See Baguley (2012, 49–51, sections 3.4.2, 10.2) for further discussion of these issues.

Example: Nonnormality effects on *t*-tests versus Wilcoxon tests For the `regularity` data, figure 2.8 shows the distribution of frequency for verbs in each auxiliary class. Let's consider whether HEBBEN and ZIJN verbs differ in frequency (left and center panels). The histograms suggest that ZIJN verbs have slightly higher frequency.

From a *t*-test (Welch's, two-sample), we would conclude that the difference in frequency is significant at the $\alpha = 0.05$ level, but just barely ($p = 0.048$): the 95% CI almost contains 0.

9. Wilcoxon tests are also invariant under monotonic transformations of the data (e.g., log-transforming), but *t*-tests are not.

```
reg_hz <- filter(regularity, Auxiliary != "zijnheb")
t.test(WrittenFrequency ~ Auxiliary, data = reg_hz)
...
## t = -2, df = 20, p-value = 0.05
## alternative hypothesis: true difference in means between
## group hebben and group zijn is not equal to 0
## 95 percent confidence interval:
##  -2.473 -0.014
...
```

In contrast, from a Wilcoxon rank-sum test, we would again conclude that the difference in frequency is significant at the $\alpha = 0.05$ level, but with higher confidence ($p = 0.006$), reflected in a 95% CI that does not contain values near 0.

```
wilcox.test(WrittenFrequency ~ Auxiliary, conf.int = TRUE,
            data = reg_hz)
...
## W = 3686, p-value = 0.006
## alternative hypothesis: true location shift is not equal to 0
## 95 percent confidence interval:
##  -2.18 -0.41
...
```

The Wilcoxon test is more reliable here because the distribution of `WrittenFrequency` is not normal—for HEBBEN verbs, there is clear right skew and some outlying points to the left of the main distribution, while for ZIJN verbs normality cannot be assessed, but assuming normality forces a high estimated variance for ZIJN verbs (and thus a lower *t*-value for the ZIJN/HEBBEN differences) due to a single point with frequency near 0.

If this single point is excluded as an outlier, a *t*-test gives a very different *p*-value and CI:

```
reg_hz_1 <- filter(
  regularity,
  Auxiliary == "hebben" | (Auxiliary == "zijn" & WrittenFrequency > 4)
)
t.test(WrittenFrequency ~ Auxiliary, data = reg_hz_1)
...
## t = -3, df = 19, p-value = 0.005
## alternative hypothesis: true difference in means between
## group hebben and group zijn is not equal to 0
## 95 percent confidence interval:
##  -2.64 -0.55
...
```

Whereas the Wilcoxon test gives a similar result to the test where the outlier was not excluded (compare the *p*-values and CIs):

```
wilcox.test(WrittenFrequency ~ Auxiliary, conf.int = TRUE,
            data = reg_hz_1)
...
```

```
## W = 3116, p-value = 0.001
## alternative hypothesis: true location shift is not equal to 0
## 95 percent confidence interval:
##  -2.4 -0.6
...
```

Intuitively, the *t*-test's result is strongly affected while the Wilcoxon test's result is not because removing this point changes the difference in sample means much more than the difference in medians.

2.5.3 Checking Normality

It is often useful to check whether data are normally distributed, for example, when checking the normality assumption for *t*-tests or for linear regression residuals (section 5.3.2). The first point to note is that "checking normality" only makes sense for a sufficiently large sample. For small samples (e.g., $n < 20$), no method can assess whether data look normal (Rice 2006, section 9.9).

2.5.3.1 Visual methods

Histograms The simplest method is to examine a histogram or smoothed density estimate plot (e.g., figure 2.7) and eyeball whether it looks like a bell curve (= normally distributed). For example, the distributions of `NcountStem` in figure 2.7 are clearly not normal: the distributions are right-skewed compared to a bell curve.

Other cases are less clear. For example, in figure 2.8, the HEBBEN and ZIJNHEB histograms seem slightly nonsymmetric and thus not normal. But in exactly what way are they not normal? Perhaps they are very close to normal? For the center histogram, there are too few data to assess normality—the data could plausibly come from a normal or nonnormal distribution, and if we used different histogram bin sizes, the histogram's shape will change a lot. Histograms are fine for detecting gross deviations from normality but not for understanding more subtle deviations or for small samples.

Q-Q plots A better visual method for assessing normality is a *quantile-quantile plot* (or *Q-Q plot*; see Baguley [2012, section 9.4.1]). For each observation, a Q-Q plot shows the percentile of each observation in the sample (the *sample quantile*) against the quantile that would be expected from a normal distribution with the same mean and SD as the sample. If the sample were normally distributed, these two things would be the same, and the plot would just show a straight $y = x$ line. The degree of deviation from this line thus captures the nonnormality of the sample and allows us to see which data points are responsible for deviations from normality.

For example, figure 2.9 (left) shows a Q-Q plot for the frequency data for words with auxiliary HEBBEN, with the $y = x$ line; the right panel shows a histogram of this data, with a superimposed normal distribution. The parts of the Q-Q plot that are furthest from the line correspond to parts of the histogram that deviate from a normal distribution: there are fewer data in the left tail (due to the cluster of points at `WrittenFrequency` = 0) and more data in the right tail than are expected in a normal distribution.

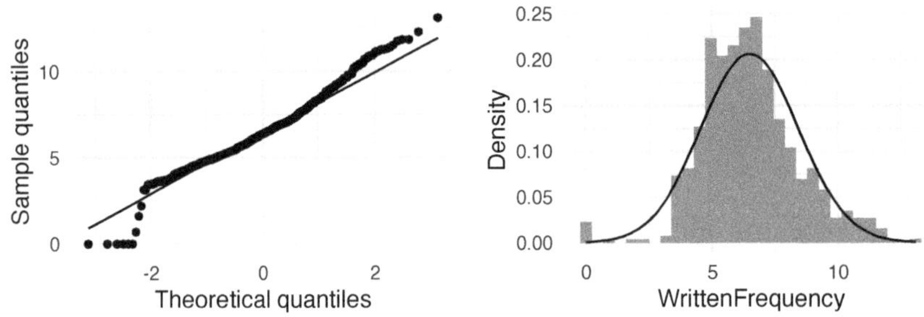

Figure 2.9
Left: Q-Q plot of word frequency for words with auxiliary HEBBEN in the `regularity` data. *Right*: empirical distribution of this data, with superimposed normal distribution using sample mean and SD.

The deviations from the line correspond to parts of the histogram that make the distribution nonnormal: there are fewer data in the left tail and more data in the right tail than are expected in a normal distribution, and the cluster of points at `WrittenFrequency` = 0 especially stand out.

2.5.3.2 Hypothesis test

Normality of a sample can also be assessed using hypothesis tests, such as the Shapiro-Wilk test, whose null hypothesis is that the sample is drawn from a normal distribution. For example, we can use this test (`shapiro.test()`) to confirm our intuition from visual methods that HEBBEN verb frequencies are not normally distributed (meaning, low p-value):

```
shapiro.test(regularity_hebben$WrittenFrequency)
...
## data:  regularity_hebben$WrittenFrequency
## W = 1, p-value = 1e-08
```

However, such tests are less useful than they appear (box 2.10), and visual methods are in general preferable for assessing normality.

Box 2.10
Broader Context: Do Normality Tests Tell Us Anything?

Shapiro-Wilk and other hypothesis tests for assessing normality are widely used, especially to assess whether samples meet assumptions for t-tests or linear regression. Such tests can be useful as a companion to visual methods or when visual methods are infeasible (for example, automatically assessing normality of hundreds of distributions). But in general these tests are less useful than they may seem, for several reasons—which point to important larger issues. (See, e.g., Baguley 2012, section 9.3.5, as well as an excellent ongoing discussion of these issues in this post: Cross Validated [2016].)

First, any sufficiently large sample will show some deviation from normality—even when a Q-Q plot suggests the deviation is very small. (For this reason, `shapiro.test()` in R only works when $n < 5,000$.) This is because real-world data aren't perfectly normally distributed: nature doesn't provide such data.

Second, a nonsignificant result doesn't let us conclude that the data *are* normally distributed. Nonsignificant hypothesis tests do not give evidence for the null hypothesis (section 2.6).

Finally, a significant result doesn't give any information about what the (nonnormal) distribution looks like. When a statistical procedure assumes the data follow a normal distribution, the important question is not whether they are normally distributed (they are not), but whether the observed nonnormality is of concern for applying the procedure (e.g., in terms of type I or type II error). For example, for *t*-tests, normality essentially does not matter for sufficiently large samples (thanks to the central limit theorem).

2.6 Common Misconceptions about *p*-Values

Perhaps the most important issue with NHST is *p*-values, which are widely reported but notoriously difficult to interpret correctly, resulting in widespread misconceptions among researchers (Greenland et al. 2016, list 25). For example, Nieuwenhuis, Forstmann, and Wagenmakers (2011) found that 50% of a sample of neuroscience articles made error 4 in the following list, and Haller and Krauss (2002) found that 33% of a sample of psychology academic staff *who taught statistics* believed a statement similar to error 1. Language scientists are probably no better. Every researcher who reports *p*-values should be aware of common misinterpretations, and reading more is highly recommended (see section 2.8).

These are four true statements that address misconceptions, listed by Vasishth and Nicenboim (2016) and Winter (2019, section 10.1), which seem most common in language sciences and, therefore, most consequential for interpreting results:

1. *p*-Values are not the probability that the null hypothesis is true—in particular, a "nonsignificant" finding (e.g., $p > 0.05$) says nothing about whether the null hypothesis holds.
2. *p*-Values do not measure strength of an effect. (This is "effect size," which will be introduced in chapter 3.)
3. A smaller *p*-value does not mean greater confidence in our specific alternative hypothesis. (It only means greater confidence that the null hypothesis is false.) It also does not mean the alternative hypothesis is true in reality. It just means that if we followed the NHST procedure, across many studies, we'd incorrectly reject the null with 0.05 probability.
4. Significant and nonsignificant effects of X in two conditions of a study does not mean the conditions significantly differ in the effect of X. (For this we need a hypothesis test for the difference in the effect between the two conditions—an "interaction" [section 4.5].)

The following list shows some of these misconceptions in practice:

- 1: "A two-sample *t*-test showed that male and female speakers do not differ in speech rate ($t(10) = 1.5$, $df = 9$, $p > 0.05$)." (Correct: "found no evidence that..." or "do not significantly differ....")
- 3: "The noun/verb difference in reaction times was highly significant ($p < 0.001$; nouns mean RT = 400 msec, verbs mean RT = 300 msec), suggesting that nouns have longer reaction times." (Correct: "Nouns (mean = 400 msec) have longer reaction time than verbs (mean RT = 300 msec), and this difference was significant....")
- 2, 4: "The effect of speaker age differed between conditions: age had a strong effect in condition A ($p = 0.01$) but only a weak effect in condition B ($p = 0.06$)." (Correct: a test

of the AGE:CONDITION interaction would be needed, and the p-value says nothing about effect size.)

These kinds of errors should be avoided when writing up results, even if you know what you mean but are being brief. For example, "A and B are the same" is often used in papers as shorthand for "A and B are not significantly different [by some hypothesis test]." But even if the author knows what was meant, many readers will not. Most misconceptions involve "wishful thinking about things that researchers really want to know" (Kline 2013, 103), making it easy to read the conclusion one wants (e.g., this nonsignificant finding supports the null hypothesis) into imprecise language.

2.7 Reporting Hypothesis Tests

The most commonly used format for reporting statistical findings is APA style, as described in the American Psychological Association's publication manual (7th edition) (American Psychological Association 2020), especially the journal article reporting standards for quantitative research in chapter 3 (JARS-Quant), which are freely available online.[10] APA style prescribes particular notation for reporting numbers and statistical analyses, which this book mostly follows, but you can look up as needed. The more important part of APA style is not exactly what numbers you report, but *why*. One important principle is the following:

> When reporting the results of inferential statistical tests or when providing estimates of parameters or effect sizes, include sufficient information to help the reader fully understand the analyses conducted and possible alternative explanations for the outcomes of those analyses. (section 3.7)

Such principles help bridge the gap between what the APA manual actually recommends (extensive reporting) and common practice (which is less verbose). This book tries to distill the recommendations and give multiple possibilities for statistical reporting.

2.7.1 Parts of a Report

When reporting a hypothesis test, you must report minimally (APA manual section 3.7) the following information:

1. The observed value of the *test statistic*.
2. Its *degrees of freedom*.
3. The exact *p-value*: probability of getting a value of the test statistic at least as extreme as the value observed, under the null hypothesis.
4. The *effect size*, including magnitude and direction.

The significance level α must also be given, but it should already have been fixed and reported in your writeup; let's assume $\alpha = 0.05$ here. It is conventional to abbreviate any p-value below a very small value (often 0.001) as, for example, $p < 0.001$ because there is no practical difference between very small p-values (Greenland et al. 2016, 341).

10. At https://apastyle.apa.org/jars/quantitative, including a useful three-page summary PDF, summarizing the full article by Appelbaum et al. (2018).

All of (1)–(4) are needed to interpret a reported effect. Without any of (1)–(3) we don't know how confident we should be in the result versus the null hypothesis. Without (4) we don't know how large the effect is, which is arguably as important as significance. The following example is a "minimal" report for the t-test in section 2.4.6.3 that covers (1)–(4):

> Transitions to male speakers were longer than transitions to female speakers by 167 msec, and the difference was significant ($t(329) = -3.5$, $p < 0.001$).

It is also recommended to give the following:

5. Basic *descriptive statistics* corresponding to the hypothesis test (e.g., mean and SD of each group, for two-sample t-tests).
6. A *measure of variability* for any reported point estimate (e.g., a SE for each reported group mean).
7. *CIs* for any parameter estimate; these are especially useful because they provide information on both precision and location.

Descriptive statistics help convey effect size and allow the interested reader to construct a mental image of the distribution of the empirical data (and thus "possible alternative explanations..."). From this perspective it is also ideal to include the following:

8. A *plot* showing the distribution of the data (e.g., like the plots in figure 2.6).

However, (8) is not always possible when writing results up due to space limitations.

With (1)–(8) in mind, a "medium" report of the preceding t-test could be the following:

> On average, transitions to male speakers ($M = 259$, $SE = 32$) were longer than transitions to female speakers ($M = 91$, $SE = 35$). The difference (167 msec) was significant ($t(329) = -3.5$, $p < 0.001$).

A "maximal" report of the same result could be the following:

> On average, transitions to male speakers ($M = 259$, $s = 454$, $SE = 32$) were longer than transitions to female speakers ($M = 259$, $s = 427$, $SE = 35$); the empirical distribution of transition durations is shown in figure 2.6. The difference (167 msec) was significant ($t(329) = -3.5$, $p < 0.001$, 95% CI [−261, −74]) at the prespecified $\alpha = 0.05$ level, by a two-sided Welch t-test.

2.7.2 Context

Reporting all this information in text leads to lengthy and unreadable prose, especially if many tests are reported. It is often better to report hypothesis tests (and descriptive statistics and similar information) in tables and just highlight relevant aspects in the text.

Whether in tables or the text, reporting (1)–(8) for every hypothesis test in a paper (or every row of a regression table) would be overkill. What is most important is to give the reader sufficient information to understand effects of primary interest in the research being reported. For example, in a paper on how properties of speakers (e.g., gender, age) affect turn-taking dynamics, you'd report more information for the previous t-test, but less for a Shapiro-Wilk test assessing whether the data are normally distributed enough to apply this t-test.

Box 2.11
Practical Note: Reporting Guidelines, Reality, and Reproducibility

The temptation when writing up results is to just search "APA style" online or look in published papers in your research area, which will give you many examples of "how to write up a *t*-test" and so on. Unfortunately, many of these examples violate at least one core recommendation of the APA manual, and often recommendations from the ASA (Wasserstein and Lazar 2016; Wasserstein, Schirm, and Lazar 2019) as well. (The same holds for reporting regression models.) The APA and ASA guidelines are excellent reads, but they contain sometimes contradictory recommendations, an overwhelming degree of detail, and many qualified statements (e.g., "confidence intervals are strongly recommended"). Even the distillation into recommendations (1)–(8) in the text is unrealistic to follow in many cases. What is a researcher who wants to report their findings in a correct yet readable format to do?

As in so many aspects of statistics, there is no fixed correct practice, just general principles, and what makes sense to do depends on the particular data and analysis being reported. But this book can highlight some common incorrect and correct practices that illustrate these principles, which you can adopt immediately to improve your statistical reporting.

Don't:

- Report a hypothesis test ((1)–(3)) without an associated effect size (4) for an effect of primary interest: this omits important information for the reader to judge the effect's importance.
- Report a *p*-value alone (just (3)): this collapses independent dimensions of uncertainty (conveyed by the test statistic and its *df*) into a single number.
- Report just whether the *p*-value is above or below a threshold (e.g., "$p < 0.05$"): this loses even more information about the result and encourages dichotomous thinking about results as important/not, which is highly problematic, especially when based just on a *p*-value without an effect size.

Do:

- Always report effect size and variability in estimates (points (4) and (6)) for effects of primary interest—or at least give the information from which an interested reader can reconstruct these things.
- Always report exact *p*-values (abbreviating small enough *p*-values as desired: e.g., "$p < 0.001$"), for all effects.
- Make your analysis as replicable as possible, so another researcher can simply look up any detail not reported in the text. Always post your analysis code publicly, and whenever (ethically/socially) possible post the raw data. Do this even if your code is messy or if you can only post a small and/or anonymous subset of the data (so at least your code runs).

The last point relates to the more general ideals of open and reproducible research, which most language scientists (including myself) don't currently meet in our work. We should aspire to improve to avoid a "replication crisis" of our own (Winter 2019, section 2.10).

2.8 Further Reading

Sampling, uncertainty, point estimation, interval estimation, and hypothesis testing are core topics covered in many statistics textbooks, often using R examples, which have been cited in passing in this chapter (e.g., Kline 2013, chapter 3; Baguley 2012; Field, Miles, and Field 2012; Crawley 2015; Navarro 2016), as well as every book for language scientists in particular (listed in preface); Gries (2021, chapter 1) is particularly extensive. Vasishth and Nicenboim (2016) is an introductory article for linguists.

On issues with *p*-values and NHST generally, there is a vast literature, distilled and summarized by Cumming (2012), Greenland et al. (2016), and Wasserstein and Lazar (2016);

I found Kline (2013) especially helpful. Vasishth and Nicenboim (2016), Nicenboim and Vasishth (2016), and Winter (2019, chapters 9, 10) discuss using linguistic data.

Exercises

2.1. Consider an imaginary dataset of measurements where participants say "hi" every time they hear a beep, and we measure the latency between the beep and the beginning of "hi." There are 100 participants, each of whom is measured once. The sample mean and SD are $\hat{\mu} = 200$ msec and $s = 75$ msec.

(a) What is the (estimated) SE of the sample mean?

(b) Suppose we wanted to repeat this experiment, aiming to get a more precise estimate, with SE of 2.5 msec. Based on your answer to (a), how many participants do you think we should use in this replication?

(c) What assumption(s) underlie the calculation in (b)?

2.2. For the Dutch verb `regularity` data, use the code below to make a plot showing the distribution of word frequency for each of the three `Auxiliary` values. In this exercise you carry out hypothesis tests to assess the difference between each pair of auxiliaries.

```
ggplot(regularity, aes(x = Auxiliary, y = WrittenFrequency)) +
  geom_boxplot() +
  geom_point(position = "jitter", size = 0.75, alpha = 0.25)
```

(a) Carry out two-sample Wilcoxon tests to assess whether HEBBEN and ZIJN verbs differ in frequency and similarly for HEBBEN/ZIJNHEB and ZIJNHEB/ZIJN. Does it matter whether a Wilcoxon test or *t*-test is used?

(b) Write up the results of these tests in a few sentences, as you would in a paper.

(c) In the plot you should see that the difference in sample means is *larger* for the ZIJN/HEBBEN comparison than for the ZIJN/ZIJNHEB comparison. However, you should have found that the *p*-value is higher for the ZIJN/HEBBEN comparison—suggesting we can be less confident in rejecting the null hypothesis. Explain what is going on—how can one difference in means be larger than another, yet also be less clearly different from 0?

2.3. We skipped a couple steps in the explanation of this code used in the text to compute 95% CIs:

```
mutate(ciUpper = qt(1 - alpha / 2, n))
```

(a) The `qt()` command computes a 97.5% quantile. Explain what the value returned by `qt()` means, intuitively.

(b) Why do we compute the 97.5% quantile, rather than the 95% quantile—given that this code is part of constructing a 95% CI?

2.4. Carry out a paired *t*-test for the example mentioned in section 2.4.5 using the `transitions` data.

(a) First, set up a dataframe showing the mean transition duration (`dur`) before and after the first minute, for each conversation:

```
d <- transitions %>%
  group_by(file) %>%
  summarise(
    # 60000 msec = 1 minute
    meanDurFirst = mean(dur[time < 60000]),
    meanDurAfter = mean(dur[time > 60000])
  )
```

(b) Carry out a paired t-test, using the `t.test()` function. (You'll have to figure out exactly which options and arguments to use.)

(c) You should find that there is a significant difference at the $\alpha = 0.05$ level. Is it positive or negative? Explain why the direction makes intuitive sense.

(d) There are two different answers to (b) that give the same results (in terms of the t-value, p-value, CIs) using different arguments to the `t.test()` function. What are these two ways, and why are they equivalent? (Hint: only one of the two ways results in R output beginning with `Paired t-test`.)

3 Effect Size, Power, and Error

Our goal in analyzing data is to assess evidence for effects of interest. Chapter 2 introduced the most common metric used for this purpose: the *p*-value/significance. In practice, in many fields (including linguistics) researchers interpret results primarily based on significance. However, a *p*-value only gives limited information about a result. Two other numbers are arguably just as important and do not follow from a *p*-value: *power* of the statistical analysis used to find an effect of a given size and the *effect size* (its magnitude and direction) (sections 3.2–3.3).

These three things (*p*, power, effect size) correspond to different but related kinds of errors: (a) finding that an effect exists when one doesn't, (b) failing to find an effect when one exists (type I and II error: section 3.3), and (c) estimating the wrong magnitude or sign of the effect size (type M and type S errors; see section 3.4). Understanding these different kinds of errors is broadly useful for data analysis.

3.1 Preliminaries

3.1.1 Packages

We assume that you have loaded the same packages as we used in chapter 2, as well as the pwr and effsize packages (Champely 2018; Torchiano 2018):

```
library(languageR)
library(tidyverse)
library(pwr)
library(effsize)
```

3.1.2 Data

The `neutralization` dataset

We also assume that you have loaded the `neutralization` dataset:

```
neutralization <- read.csv("data/neutralization_rmld.csv")
```

This dataset, described in more detail by Kirby and Sonderegger (2020), originally comes from a phonetic study by Roettger et al. (2014) examining neutralization of German stops (their experiment 1).

In German, words ending in underlying voiced and voiceless stops (reflected in the orthography) are traditionally described as *neutralized* at the end of words—they are all pronounced with a voiceless stop. One result is that many pairs of words sound the same in singular but not plural forms. For example:

- *Rat* [ʁaːt] 'council', *Räte* [ʁɛːtɐ] 'councils'
- *Rad* [ʁaːt] 'wheel', *Räder* [ʁɛːdɐ] 'wheels'

In the experiment, German speakers were asked to produce singular forms of nonsense words ending in stops (like *Frup*, *Frub*), where they had previously heard the plural forms (*Frupe*, *Frube*). There are 24 such voiced/voiceless pairs (indexed by `item_pair`). Of interest is whether (underlying) voiced and voiceless stops (column `voicing`: levels VOICED, VOICELESS) are actually pronounced identically, which is assessed by measuring the preceding vowel's duration (column `vowel_dur`). Traditional phonological theories predict no difference (*complete neutralization*), whereas some previous phonetic studies suggest there might be a small but significant difference (*incomplete neutralization* [IN]), which may be explained as an artifact of the structure of the mental lexicon. (Nicenboim, Roettger, and Vasishth [2018] review the literature on IN.)

IN is a good phenomenon to exemplify power and effect size (including errors) because:[1]

- Whether IN exists at all (i.e., whether the null is true) is controversial, with both existence and nonexistence corresponding to reasonable theoretical positions.
- There is a substantial literature on IN (especially for German), with some supporting the existence of IN and some null results. Interpreting these studies requires thinking about both type I and type II errors (and thus power).
- It is scientifically important what the exact effect size is, beyond just "0 or not." The size tells us whether the effect is meaningful and differentiates between different theories. Type M and S errors are thus of interest.

3.2 Effect Size

3.2.1 Introduction

Effect size is a measure of the direction and magnitude of an effect. There are multiple ways to quantify effect size, ranging from a raw difference in means to standardized scores such as Cohen's *d*. First, this section introduces some examples to give a conceptual overview, then it discusses metrics commonly used to quantify effect size.

Examples Let's consider the top-left panel of figure 3.1 as just showing the difference in how x is distributed in two samples. Think of a measure of "how different are the two groups" that is just based on visual inspection of the graph—like how far apart the modes

1. Several existing articles use German IN to discuss the concepts covered in this chapter, using the `neutralization` dataset among other datasets from psycholinguistics and phonetics (Kirby and Sonderegger 2018a, 2018b; Nicenboim, Roettger, and Vasishth 2018; Vasishth, Mertzen, et al. 2018; Vasishth and Nicenboim 2016). This presentation is indebted to these sources (especially Kirby and Sonderegger 2018a, 2018b) but tries to be more accessible.

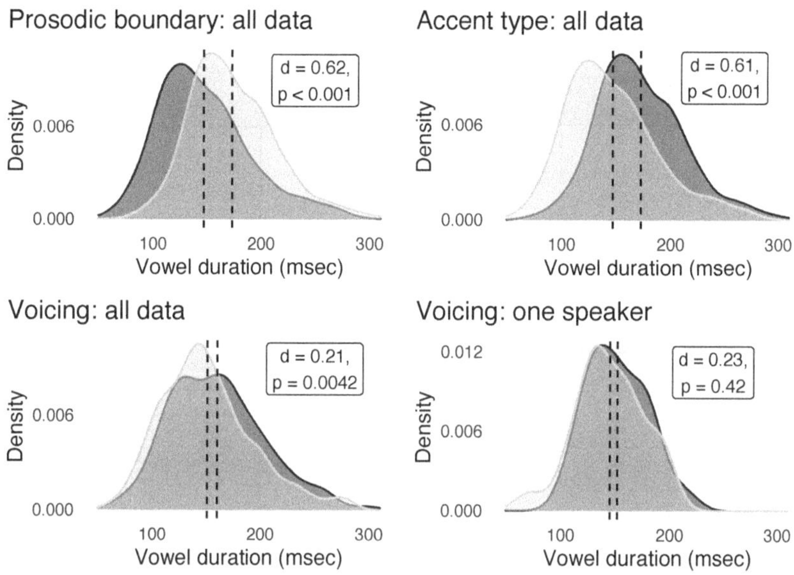

Figure 3.1
Empirical distribution of vowel duration in the `neutralization` data as a function of two prosodic variables (darker/lighter = no/yes `prosodic_boundary`, nuclear/prenuclear `accent_type`), and following consonant voicing (*bottom row*: darker/lighter = VOICELESS/VOICED). *Bottom-right plot* is for a single subject ($n = 48$), other plots are across all subjects ($n \approx 750$). Vertical lines indicate sample means, and boxes show effect size (Cohen's d) and significance (p for a two-sample t-test) of the darker/lighter difference.

are, or how far apart they are relative to the width of the distributions. Suppose I now tell you the figure is based on $n = 5$ points rather than $n = 750$ points (the reality). Your measure hasn't changed (it's just based on the figure), but the p-value will:

Now suppose we were interested in whether people in Montreal speak faster than those from Toronto, so we measured speaking rate for n speakers from the two cities and found that they differ by 0.01 syllables per second on average. This is clearly an extremely small effect. This judgment shouldn't change whether $n = 5$ or 1,000,000, but the p-value will: with a million observations, anything is significant ($p < 0.001$).

Finally, consider the bottom-left and bottom-right plots in figure 3.1, which plot the same thing (distribution of `vowel_dur` as a function of `voicing`) for $n = 750$ and $n = 48$ observations from the `neutralization` dataset.[2] Intuitively, the plots show a similar pattern in terms of effect size—the groups mostly overlap, with vowel duration maybe a bit larger for VOICED stops; in each case the group means are 7–8 msec apart. But a two-sample t-test gives a much lower p-value/narrower CI for the bottom-left plot ($p = 0.0042$) than for the bottom-right plot ($p = 0.42$), due to sample size:

```
t.test(vowel_dur ~ voicing, data = neutralization)
...
```

2. Note that what's important here is that the samples differ in the number of observations, not that they differ in the number of subjects the observations are from.

```
## t = 3, df = 747, p-value = 0.004
## alternative hypothesis: true difference in means between
## group voiced and group voiceless is not equal to 0
## 95 percent confidence interval:
##   2.9 15.3
...
```

```
neut_7 <- filter(neutralization, subject==7)
t.test(vowel_dur ~ voicing, data = filter(neutralization, subject == 7))
...
## t = 0.8, df = 45, p-value = 0.4
## alternative hypothesis: true difference in means between
## group voiced and group voiceless is not equal to 0
## 95 percent confidence interval:
##  -10  24
...
```

These examples demonstrate intuitively that an effect's size, which is independent of the sample size, is not the same as an effect's significance. Informally, an *effect size* is any quantitative measure of the size of something we are interested in (an *effect*) (Cumming 2012, 34).[3] The measure must be independent of sample size (so it purely describes the observed effect) and appropriate for the research question(s) being addressed, but otherwise the definition is intentionally broad.

More formally, in almost any NHST hypothesis test, the test statistic T can be expressed as

$$T = ES \cdot f(N),$$

where ES is an effect size measure and $f(N)$ is a function of sample size (Kline 2013, 76). For example, for a two-sample t-test in the simplest case (equal variances and sample sizes; see equation (2.9)), the test statistic is

$$t = \underbrace{\frac{\mu_1 - \mu_2}{\hat{\sigma}}}_{ES} \underbrace{\sqrt{\frac{n}{2}}}_{f(N)}.$$

Because p follows directly from the test statistic, it follows that for any hypothesis test the p-value confounds effect size and sample size. Large effects may not be significant in small samples, and tiny effects will be significant in large samples. For this reason, it is ideal to report a corresponding effect size whenever a p-value is reported—as recommended in the APA manual (and the guidelines in section 2.7).

In this book "effect size" is usually used to mean a quantity calculated from the sample, and "true effect size" is used to refer to the population value of the quantity, if needed. This

3. Kelley and Preacher (2012, 140) give a more nuanced definition, which makes clear that an effect size measure is always tied to the question(s) it is being used to address: "Effect size is defined as a quantitative reflection of the magnitude of some phenomenon that is used for the purpose of addressing a question of interest."

book first introduces effect size as a descriptive quantity and then later discusses it as an estimated quantity (section 3.4).

3.2.2 Standardized versus Unstandardized

There are two broad types of effect sizes. Measures such as differences in means, or (unscaled) regression coefficients, are called *unstandardized*—they are on the scale used in the study and may allow for direct comparison with other concrete, interpretable metrics.

Unstandardized effect sizes have the primary advantage of interpretability and so are better used for outcomes measured on a meaningful metric (such as vowel duration, in msec) or comparing studies measured on the same metric. For example, in the IN study, the vowel duration difference could be interpreted relative to the smallest duration difference humans can perceive or relative to other duration differences observed in the study.

Measures where effect size is normalized in a way that abstracts away from the scale of measurement are called *standardized*. This includes all measures of association/variance explained (such as correlation coefficients, R^2, or η^2 for analyses of variance), as well as standardized differences of means, such as standardized regression coefficients or *Cohen's d*: the difference in group means divided by an SD that (intuitively) reflects the "amount of variability" in the data, given the study's design (Cohen 1988). Standardized effect size measures are useful when measurement units are not intrinsically meaningful (e.g., rating scales, acceptability judgments) or easily interpretable (e.g., log-transformed reaction times, word frequencies). They have the advantage of comparability across different metrics, which makes them better suited for comparing results that are conceptually similar but may be on different scales: for example, studies evaluating IN in different languages, or different types of sounds (where observations would be of a different phonetic parameter from vowel duration).

Example Consider the effects of following consonant voicing on vowel duration for the `neutralization` data (figure 3.1, bottom-left), which corresponds to the research question for the experiment. Two measures of effect size that we could compute are the difference in means $\Delta\mu$ and Cohen's d—this difference is scaled by within-group variance. $\Delta\mu$ is an unstandardized measure and d is standardized. The following code computes these measures, giving $\Delta\mu = 9.1$ msec and $d = 0.21$.

```
mean(filter(neutralization, voicing == "voiced")$vowel_dur) -
  mean(filter(neutralization, voicing == "voiceless")$vowel_dur)
## [1] 9.1
cohen.d(vowel_dur ~ voicing, data = neutralization)$estimate
## [1] 0.21
```

Here we used the `cohen.d()` function from the effsize package, which automatically calculates the correct amount of variability for the denominator of d.

These effect size measures are for the plot using the whole dataset (figure 3.1, bottom-left). We can compute the same measures computed just for subject 7 (bottom-right plot) using similar code:

```
mean(filter(neut_7, voicing == "voiced")$vowel_dur) -
  mean(filter(neut_7, voicing == "voiceless")$vowel_dur)
## [1] 6.9
cohen.d(vowel_dur ~ voicing, data = neut_7)$estimate
## [1] 0.23
```

Thus, the standardized effect size (*d*) is similar for the two plots, confirming our intuition from the preceding visual inspection.

3.2.3 Interpreting Effect Size

Calculating an effect's size is usually straightforward, whether by hand or using functions from R packages (such as effsize). The number by itself says nothing about the research question, for which the researcher must determine the effect's *substantive significance* (i.e., its meaningfulness, importance, theoretical/practical value), which is more qualitative. For example, clearly $d = 5$ is a large effect and $d = 0.01$ is a very small effect, but how should one assess a finding of $d = 0.5$?

3.2.3.1 Rules of thumb

For standardized measures, the easy option is to use a preexisting rule of thumb for a particular effect size measure. Rules of thumb originating with Cohen (1969) (summarized in Cohen 1992, table 1) are commonly used to interpret effect size measures based on their magnitudes—most often, Cohen's *d* and the correlation coefficient *r*:

	Effect size			
	Negligible	Small	Medium	Large
$\lvert d \rvert$	<0.2	0.2–0.5	0.5–0.8	>0.8
$\lvert r \rvert$	<0.1	0.1–0.3	0.30–0.5	>0.5

For example, for the preceding `neutralization` example, the effect would be small, bordering on negligible ($d = 0.21$).

Despite being widely used, these cutoffs are arbitrary; Cohen intended them as educated guesses, to be used when previous work in an area wasn't available to estimate what small/medium/large meant. Although tutorials on effect size strongly recommend against such cutoffs (e.g., Cohen 1992; Kline 2013, chapter 5; Cumming 2014), they can be useful when effect sizes from previous work aren't available, which (anecdotally) seems to be common for linguistic studies.

3.2.3.2 In context

Whenever possible the harder option is preferred: determining substantive significance based on "the particular area, research design, population of interest, and research goal" (Kelley and Preacher 2012). In other words, what should count as a large/important effect for the particular study questions/goals based on previous work or theoretical considerations?

For the case of incomplete neutralization, one option is to assess the effect as negligible if humans can just perceive it (the *just-noticeable difference* for vowel duration), and large if it has a value similar to languages where voicing is not neutralized word-finally; these values are about ~5 msec and >30 msec (see discussion in Kirby and Sonderegger 2018a). From this perspective, the effect (9.1 msec) is very small—probably almost imperceptible.

However, we might also assess the effect relative to 0, the value predicted by standard theories. From this perspective, even a tiny effect is scientifically important, because it shows that the two classes of words somehow do differ for German speakers, and the effect size may help differentiate between theoretical accounts for the effect.

Another option would be comparing closely related effect sizes in the same study. For this example, the effect of `voicing` can be compared to the effect of control predictors, which are expected from previous work to affect vowel duration: the presence of a prosodic boundary or word prominence and speaker gender. Their effect sizes are:

- `prosodic_boundary`: $d = -0.62$
- `accent_type`: $d = 0.61$
- `gender`: $d = 0.34$

(see Exercise 3.1). Voicing again seems to have a small effect on duration—smaller than anything else measured in the experiment.

In language sciences, *reporting* effect sizes is becoming more common—especially for regression models (where the coefficients are effect sizes)—but *discussing* effect sizes (what they mean for the study's research questions) is still not standard. We will return to different effect size measures, and their interpretation in particular cases, throughout this book.

3.2.4 Correlation Coefficients

Let's digress to review a very common type of effect size:[4] the *correlation coefficient* describing the degree of association between two continuous variables, x and y, for a sample of n observations.

The *Pearson's correlation coefficient* measures the degree to which there is a *linear association* between x and y in the population ("looks like a line"):

$$\rho_{xy} = \frac{\sigma_{xy}}{\sigma_x \sigma_y}, \tag{3.1}$$

where σ_{xy} is the *covariance* of x and y, and σ_x and σ_y are the variances of x and y. So Pearson's correlation is a "standardized covariance."

The *sample Pearson's correlation coefficient* approximates each of σ_{xy}, σ_x, and σ_y using the sample

$$r_{xy} = \frac{\hat{\sigma}_{xy}}{\hat{\sigma}_x \hat{\sigma}_y}. \tag{3.2}$$

This is often just written r, and called "Pearson's correlation" (or even "the correlation").

4. This book assumes you have seen correlation coefficients before. If not, parametric and nonparametric correlation coefficients are covered in, e.g., Field, Miles, and Field (2012, chapter 5), Navarro (2016, section 5.7), or many online sources.

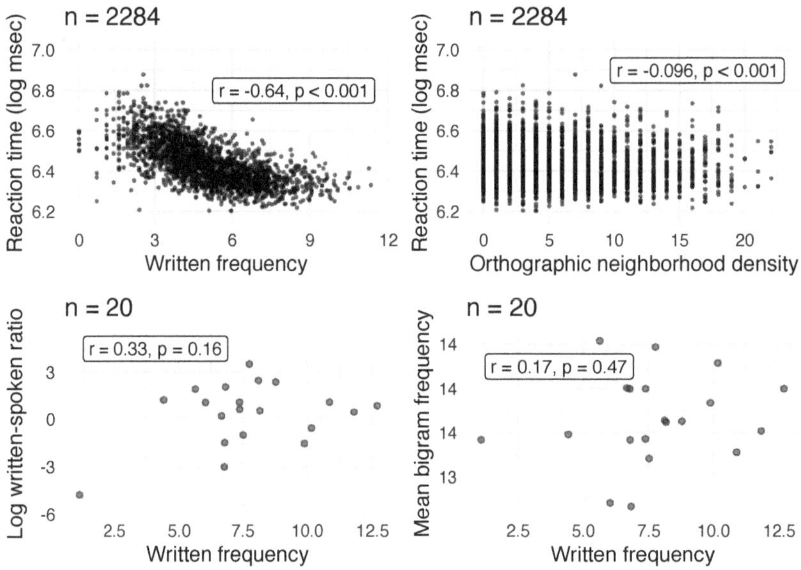

Figure 3.2
Examples of scatterplots with differing Pearson correlation strengths ($|r|$) and p-values. *Top row*: `english` dataset, `AgeSubject=YOUNG` observations only. *Bottom row*: `regularity` dataset, `Auxiliary=ZIJN` observations only.

There are different ways to obtain a p-value (null hypothesis: $\rho_{xy} = 0$) and 95% CIs for r; they usually give very similar results, so you can just use R's default (via `cor.test`).[5] The important point to remember is that (whatever method is used), r is subject to everything we have learned about hypothesis tests, including

- p and the 95% CI depend on sample size—the smaller the sample, the less reliably different from 0 the same r value is.
- When many r values are computed, p and the CI probably should be corrected for multiple comparisons (section 3.3.6).

These points are important to bear in mind because correlations are often reported as *descriptive* statistics when summarizing data—without p-values/CIs—and sometimes without even sample size.

To get a sense of what different r and p "look like," figure 3.2 shows some scatterplots between pairs of variables in a large subset ($n = 2284$) of the `english` dataset (see section 4.1.2) and a small subset of the `regularity` dataset ($n = 20$). Here again we see the independence of effect size (r) and significance: a correlation can be medium (using Cohen's rule of thumb) but nonsignificant, due to low sample size (bottom-left), or negligible but with $p < 0.001$ (top-right), because with high sample size anything becomes significant.

5. This default obtains p-values from a t-test with $df = n - 2$ with a test statistic computed from r_{xy}. Section 4.3.6 gives some intuition on why a t-test applies here.

Different *nonparametric correlation coefficients* measure the degree of *monotonic association* between x and y: does increasing x consistently result in an increase (positive correlation) or decrease (negative correlation) in y? The most common nonparametric correlation coefficients are Spearman's ρ and Kendall's τ. Both are based on ranks/medians rather than values/means.

Nonparametric correlation coefficients address important shortcomings of r, which only measures the degree to which x and y show a linear relationship. First, r isn't sensitive to whether the relationship is actually linear. The relationship in figure 3.2, top-left, is monotonic, but not linear, so r is misleading: it averages across part of the relationship that is roughly linear (`WrittenFrequency` < 6) and part that is not (`WrittenFrequency` > 6). Second, r is sensitive to outliers. In figure 3.2, bottom-left, removing the bottom-left-most point changes r drastically ($r: 0.33 \rightarrow -0.06$), whereas Spearman's ρ changes less ($\rho: 0.09 \rightarrow -0.06$).

3.3 Type I/II Errors and Power

In the NHST paradigm introduced in chapter 2, to decide whether an effect exists we apply a hypothesis test with significance level α, resulting in rejecting or failing to reject the null. The null may be true or not in reality. Thus, there are two kinds of errors we can make:

- *Type I error* (or "false positive"): falsely concluding there is an effect when none exists.
- *Type II error* (or "false negative"): falsely concluding there is no effect when one in fact exists.

And there are four possible outcomes:

		Statistical analysis result (**sample**)	
		Reject H_0	Don't reject H_0
Reality	H_0 is true	Type I error (α)	Correct decision (null result)
(**population**)	H_0 is false	Correct decision (significant)	Type II error (β)

The rate of a type I error is denoted α. For an NHST test, this is just the significance level, which we have set in advance, defined as how often we will make the wrong decision over the long term if the null is true. Type I errors are arguably more familiar, and everyday statistical practice places considerable emphasis on avoiding them, as reflected in the convention to set α to 0.05 or 0.01. The rate of a type II error is denoted β.

Although statistical practice focuses largely on type I errors (via p-values), this is odd because we get type I errors by assuming the null is true—yet the null hypothesis is typically neither true nor interesting: "Researchers typically test a null hypothesis with the hope of being able to reject it so that they may infer that their data better fit another, more interesting hypothesis" (Hallahan and Rosenthal 1996, 490). Thus, it makes sense to consider the types of errors that we make assuming the null is *not* true: type II errors and errors in the sign and magnitude of the (nonnull) effect. These errors are the focus of the rest of this chapter.

3.3.1 Power

Rather than considering type II error directly, it is conventional to consider *power*, defined as $1-\beta$: the probability of rejecting the null hypothesis when it is false. Power is one way to capture the "sensitivity" of a statistical procedure to find an effect when one actually exists (Dienes 2008, 68).

Before proceeding, it's worth asking: why should we care about power/type II error? Statistical practice in language sciences tends to focus mostly on type I error (especially via the *p*-value), but power is not usually considered—mirroring the situation in most behavioral and social sciences, despite calls for reform dating to the 1960s (e.g., Cohen 1969).

A lot has been written about why this is a problem, and considering both type I and type II errors are important—especially when planning new studies (e.g., any reference discussing power in section 3.6, Further Reading).

Here are some key reasons to consider power after data are collected—in data analysis and reading published studies—which are unintuitive if you are used to thinking in terms of just type I error/*p*-values:

Usual thinking	Reality
"A null result doesn't tell me anything."	Whether null results are informative depends on power.
"A statistical method is better if it is more conservative (fewer spurious results)."	Type I and type II errors usually trade off for different methods—more conservative methods (good) have lower power (bad).
"A low *p*-value means an effect is reliable."	Significant results can be misleading, depending on power.

We expand on these points using examples in the following sections (3.3.3, 3.3.5, 3.3.6, 3.4).

Power calculations Power is a function of several factors:

1. The sample size
2. The true effect size
3. The amount of variability in the data
4. The significance level (α)

An effect is more likely to be detected in a study with a larger sample, if the (true) effect size is larger, if the data are less noisy, or with a less stringent cutoff (higher α). (If a standardized effect size is used, (1) and (3) are combined into one measure such as Cohen's *d*.) Power also depends on

5. The statistical procedure used

This point is discussed in section 3.3.4.

Assuming the statistical procedure is fixed, power can be calculated given values of (1)–(4), or the necessary value of one of (1)–(4) to achieve a given power level can be determined. Power calculations can be done in study planning or interpreting results. A priori power calculation, for example to assess the sample size necessary to detect an effect of a given size is an important topic and the focus of most introductions to power (see works listed in section 3.6). This book assumes that all data have been collected and focuses on power for interpreting results.

Although power and type I error are related, one does not determine the other, because of the roles of (1)–(3). A common critical value for β, analogous to $\alpha = 0.05$ for rejecting the null, is $\beta = 0.2$, giving power of 80%.

Box 3.1
Broader Context: The Problem with Observed Power

Computing power for a study that has already been conducted seems simple: just use the effect size and variability observed in the study to calculate *observed power* for the statistical test used. Although widely used in behavioral/social sciences, observed (or "retrospective") power analysis is pointless (e.g., Cox 1958; Hoenig and Heisey 2001; O'Keefe 2007). This is because power can be computed directly from the p-value, (given the study design, observed effect size, and variance observed effect size, and variance); thus, power gives no additional information once the p-value is known. Intuitively: if a significant effect ($p < 0.05$, say) was/was not found, you will compute that observed power must have been high/low to give an effect of the observed size. A common (mis)use of observed power is to explain away a null result, by showing that power to detect the observed effect was low (e.g., <0.8). Compared to true power, observed power will tend to be too high for significant results and too low for null results.

However, power calculations made after the study has been conducted can still be informative—as long as we don't just plug in our observed effect size/variability. (Design analysis will be discussed in section 3.4.3 and box 3.3).

First, let's do a basic power calculation to show how it can be computed and then go on to several examples that demonstrate the importance of considering power in data analysis.

3.3.2 Example

To give a sense of how power varies as a function of the true effect size and sample size, let's consider the case addressed in the `neutralization` data: detecting differences in vowel duration before voiced and voiceless consonants in a given language.

We will compute power for a two-sample t-test, assuming that the two samples (VOICED and VOICELESS) have equal size. To compute power we need to specify (1)–(4) from the power calculations list. Let's assume that the degree of variability σ is fixed, at 45 msec—roughly the value observed in the `neutralization` data—and hold α fixed at 0.05.[6]

We will consider three possible true effect sizes, using the difference in means between VOICED and VOICELESS as the effect size, based on the discussion in section 3.2.3.2:

1. 5 msec: the just-noticeable-difference for vowel duration differences
2. 10 msec: approximate size of the incomplete neutralization effect in German (from Nicenboim, Roettger, and Vasishth [2018] meta-analysis; see box 3.2)

6. If we're going to use these power calculations to reason about the `neutralization` data (we are), it is not ideal to set σ based on our sample (box 3.1). We are just assuming this number to keep the example simpler. In a real application we would want some independent motivation for the value(s) of σ used.

3. 30 msec: the rough size of the minimum difference in languages where the voicing contrast is not neutralized (e.g., English)

Conceptually, we can think of these three effect sizes as “complete,” “incomplete,” and “no” neutralization.

We vary sample size between 25 and 500. Note that 25 is roughly the sample size per subject in the `neutralization` data, and 375 is the sample size for the whole dataset. (There are 375 VOICED/VOICELESS pairs, corresponding to $n = 750$ rows.) This range is arbitrary, but it could be thought of as sample sizes ranging from a very small to a large study of German incomplete neutralization.

Power calculations for all basic types of hypothesis tests can be done using the pwr package. The `power.t.test()` function computes power for a t-test.

For example, for our sample when effect size = 10 and $n = 50$:

```
## Default: two-sample, two-sided, alpha=0.05
power.t.test(n = 50, delta = 10, sd = 45)
##
##      Two-sample t test power calculation
##
##               n = 50
##           delta = 10
##              sd = 45
##       sig.level = 0.05
##           power = 0.19
##     alternative = two.sided
##
## NOTE: n is number in *each* group
```

To compute power across different effect sizes and sample sizes, we use a method from the `map()` family in purrr (which applies a function to each element of an object):

```
## Vary effect size and sample size as discussed
crossing(d = c(5, 10, 30), n = seq(5, 500, by = 5)) %>%
  ## Compute power
  mutate(power = map2_dbl(n, d, ~ power.t.test(.x, .y, sd = 45)$power))
##       d    n power
## 1     5    5 0.036
## 2     5   10 0.042
...
```

The results are plotted in figure 3.3. The general pattern is as expected: power increases as the sample size or the true size of the effect increases. The three curves shown in the plot are useful for interpreting the results of studies examining each kind of neutralization: complete, incomplete, and no neutralization. For example, consider points on these curves corresponding to two samples sizes of interest: one subject ($n = 25$) and all subjects ($n = 375$) in the `neutralization` experiment:

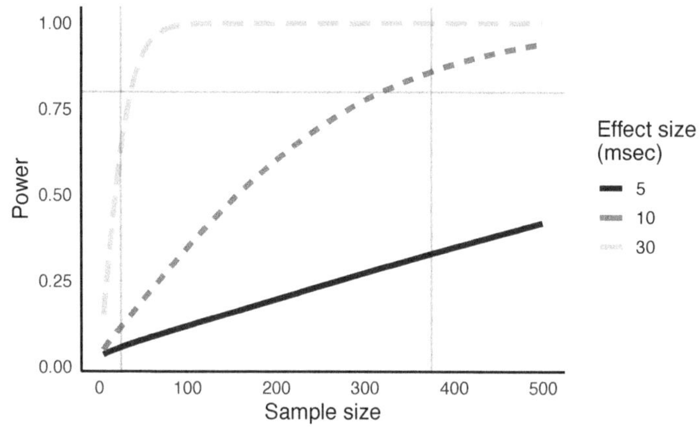

Figure 3.3
Power for a two-sample t-test as true effect size (difference in means) and sample size are varied, with variability ($\sigma = 45$) and significance level ($\alpha = 0.05$) held fixed. The horizontal line indicates 80% power, and the vertical lines indicate two sample sizes (25, 375) referenced in the text.

- Power to detect complete neutralization (5 msec) is low— <50% across sample sizes. For example, a study with $n = 375$ would be 67% likely to give a null result even though an effect exists.
- Power to detect incomplete neutralization (10 msec) ranges from very low (when $n < 100$) to high (above the 80% threshold for $n > 300$). Thus, if this is the true effect size, a null result is unsurprising if $n = 25$, but (relatively) surprising if $n = 375$.
- Power to detect no neutralization (30 msec) is very high, as long as sample size is above about 50. A null result is not surprising if $n = 25$, but very surprising if $n = 375$ (power ≈ 1).

3.3.3 Interpreting Null Results

3.3.3.1 Using power

The preceding example illustrates an important application of power: interpreting null results ($p > \alpha$). Concluding from nonsignificance that there is "no effect" is a common fallacy (section 2.6)—the p-value is not the probability that the null hypothesis is true. To avoid this pitfall, it is sometimes taught or propagated in practice, that null results are uninterpretable. However, this depends on power. When power is low, a null result is not informative: it would be likely to occur whether the null were true or not. It makes sense to have this interpretation as one's default, because most studies are underpowered (box 3.2).

However, *when power is high, a null result is meaningful*: if there were an effect of this size, we would be surprised to not detect it. For the `neutralization` case, we could conclude that any effect that exists is <10 msec.

There is an important caveat to this reasoning: computing power requires knowing the true effect size (and degree of variability), but in reality these are unknown. So reasoning about null results requires at least specifying what the smallest interesting ("substantively significant") parameter value is (see section 3.2.3)—functionally, this is the same as 0

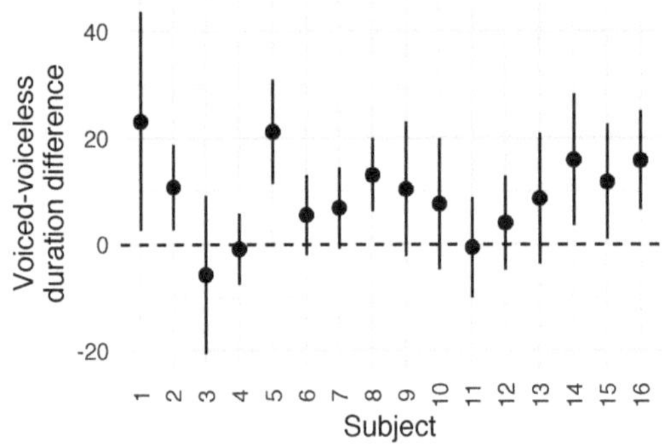

Figure 3.4
Summary of voiced-voiceless duration differences for each of the 16 speakers for the `neutralization` data. Points/errorbars are means ± 2 SEs.

(box 3.3). For the `neutralization` example, $\Delta = 5$, the value we have decided indicates complete neutralization.

3.3.3.2 Using confidence intervals

However, calculating power can be difficult (e.g., for mixed-effects regression models, where simulation is required; see section 10.8), and interpreting power can be unintuitive. In practice a more approximate measure of sensitivity than power calculation is often easier for interpreting a null result. Let's examine the 95% CI (which includes 0 for a null result) and apply this rule of thumb (Hoenig and Heisey 2001; Dienes 2008, 72–73):

- If the CI is narrow (it includes 0 but not δ or $-\delta$), the null is (tentatively) "effectively true."
- If the CI is wide (it includes 0, and either δ or $-\delta$), the result is inconclusive.

As an example, consider figure 3.4, which shows approximate 95% CIs for the effect of voicing on each subject. There are 9 subjects with null results; for each subject the CI overlaps ±5 msec ($\delta = 5$). So whether or not the effect is 0 is in fact inconclusive for every one of these subjects.

Box 3.2
Broader Context: Interpreting "Mixed" Results—Power and Meta-Analysis

The text shows that, in addition to the p-value, power is important for interpreting the results of single studies, including null results. Power is also crucial for answering: what can we conclude from a literature of studies of (more or less) the same phenomenon? Strictly speaking this is the domain of *meta-analysis* (Cumming 2012), a methodology for pooling the results of previous studies to determine likely parameter values. But even in the absence of formal meta-analysis, the same considerations of power that apply for individual studies can be used to interpret a literature.

Across language sciences there are theoretically important phenomena where the literature gives "mixed" results—both significant and null findings. (Or significant results in conflicting directions.) The usual approach in linguistics and psychology is to simply count the percentage of significant findings; if not near 100% or 0%, the literature is said to give ambiguous evidence on whether the effect exists. This reasoning is incorrect, especially given that most studies in linguistics are likely underpowered (Winter 2019, chapter 10), similarly to psychology where the median power is about 0.50 (e.g., Maxwell 2004). With power of 0.50, the number of null and significant results will be roughly equal even when there is a true effect! Thus, even thinking about power at a rough level can help in summarizing a mixed literature. For example, if studies with null results have low sample size (suggesting lower power), this is consistent with a true effect; if some such studies have high sample size, the literature is not consistent with a true effect.

In the German incomplete neutralization literature, it turns out that the null results all come from low-powered studies, so the literature is consistent with there being a small but real effect. Kirby and Sonderegger (2018b, 2018a) summarize the logic in this case using power calculations, whereas Nicenboim, Roettger, and Vasishth (2018) conduct a formal meta-analysis of all (14) studies to date and find strong evidence for a small but real effect (voiced/voiceless vowel duration difference of $\sim$10–12 msec)—despite null results in 50% of individual studies. Meta-analyses are still fairly new in linguistics, but examples so far in psycholinguistics have shown that mixed evidence for a long-studied phenomenon sometimes strongly supports a true effect and sometimes strongly suggests there is no effect (Vasishth et al. 2013; Mahowald et al. 2016; Jäger, Engelmann, and Vasishth 2017).

3.3.4 Maximizing Power

It should be clear by this point that we want to maximize power when carrying out quantitative studies and analyzing their results, because a high-powered study gives useful information regardless of the p-value.

Consider factors (1)–(5) affecting power, from section 3.3.1. The most obvious way to increase power is to carry out a higher-power study ((1)–(3): larger sample size, less variability, or larger effect). This route is outside the scope of this book, which focuses on data that have already been collected (see readings in section 3.6). The convention in data analysis is to control type I error by fixing α (factor (4)). This leaves different analysis methods (5).

Sometimes it is possible to increase power without affecting type I error by using a different analysis method. For example:

- Parametric tests are usually more powerful than their nonparametric equivalents (e.g., t-test vs. Wilcoxon test), if the assumptions of the parametric test are met (e.g., normally distributed data). In general, the more assumptions a test makes, the higher its power (when these assumptions are met).[7]
- The most popular method for correcting for multiple comparisons (Bonferroni) can be replaced by another method with higher power, without increasing type I error (see section 3.3.6).[8]

3.3.5 Type I/II Error Trade-Off

When analyzing any dataset to address some research questions, there are always multiple analysis methods to choose between which are plausibly "right." What are the consequences

7. This is because as a rule, the more estimates/fewer assumptions go into a test statistic, the broader its distribution under the null hypothesis. This implies a larger p-value and (assuming the null hypothesis is in fact false) means the test has less power to detect differences.

8. This book assumes you have previous exposure to multiple comparisons and at least one procedure to correct for them (probably Bonferroni), which are covered in most introductory applied statistics texts (e.g., Navarro 2016, section 14.5; Field, Miles, and Field 2012, section 10.5).

of using one method versus another? Sometimes one method is unambiguously "better" in the sense of higher power without affecting type I error (as in the preceding examples). But in practice, there is typically a trade-off between type I and type II errors.

3.3.5.1 Conservative and anticonservative

This trade-off is easier to think about if we first define terms to discuss errors in the p-value (which we are used to thinking about), rather than type I error per se (which we never observe directly).

Applying different NHST methods to the same data will give different p-values (e.g., p_1, p_2). For example, suppose the null hypothesis is true, and we have fixed the significance level at 0.05. If the experiment is repeated many times, different methods will give different rates of false positives:

$$\alpha_1 \approx \text{how often } p_1 < 0.05, \quad \alpha_2 \approx \text{how often } p_2 < 0.05.$$

These are the real type I error rates of each method, which we will denote with α_i. These rates are not the same as the significance level (typically written α)—the target (or *nominal*) type I error rate—because each hypothesis test relies on many assumptions, which may be violated to some degree, and the test may also give higher type I errors in some situations (e.g., small sample size, high variance) than in others.

Formally, a hypothesis test is

- *Conservative* if $P(p \leq \alpha | H_0) < \alpha$,
- *Anticonservative* (or "liberal") if $P(p \leq \alpha | H_0) > \alpha$,

whenever its assumptions hold (Good 2006, section 4.4.5).

Informally, a "conservative" test tends to give p-values that are too high, and an anticonservative test tends to give p-values that are too low (and CIs that are too narrow/wide, respectively). This is the sense in which the terms are often used in data analysis—a "more conservative" test makes false positives less likely. Because most researchers are used to thinking in terms of p-values, it is common to talk about different statistical procedures in terms of conservativity rather than type I error. It is also common in practice (although this isn't strictly allowed by the preceding definition) to use conservativity to discuss the effects of violating an assumption of a procedure, for example, "the normal approximation to the test statistic gives a conservative p-value."

For example, suppose we analyze the same data using a one-sided versus two-sided t-test, with $\alpha = 0.05$. p for the one-sided test (H_0 : effect is not positive) is always 50% of its value for the two-sided test (H_0: effect is 0), so the false positive rate is 50% lower for the one-sided test. In reality, in almost all cases the parameter being estimated is either positive or negative, not 0, so the two-sided test is conservative. (For example, in our running `neutralization` example, we know that any effect will show VOICED>VOICELESS.)

3.3.5.2 The trade-off

Different statistical methods applied to the same data tend to trade off in conservativity and power: more conservative (good) tends to mean lower power (bad), and vice versa. This is a more intuitive way of saying "type I and type II errors tend to trade off."

Once you know it exists, this trade-off comes up frequently in choosing between methods. For example:

- The two-sided t-test in the preceding discussion will fail to detect some small differences between two samples that would have been detected by a one-sided test. The two-sided test is more conservative but has lower power.
- Chapter 10 discusses a debate over how to analyze grouped data (whether "maximal" random effect structure should be used) that generated significant discussion throughout the 2010s and that essentially comes down to a choice between power and conservativity.
- Different ways of correcting for multiple comparisons, as discussed in the next section.

In such cases, choosing among different analysis methods often means choosing how much you care about type I versus type II errors. The convention in linguistics and psychology, as in many fields, is to prioritize minimizing type I error over type II error (e.g., Nickerson 2000). This motivates the use of NHST testing (where p-values are the focus) and the lower "reference" level for α than for β (commonly 0.05 or 0.01 vs. 0.2) used by convention, which implies that we can live with missing 20% of true effects, but we only want a 1%–5% chance of a spurious effect. However, the choice to minimize type I error at the expense of power is essentially tradition, and depending on your research questions, it may make more sense to use a method that prioritizes power.

For example, in any study where you want to argue *for* the null, you want to prioritize power, given that null results are only interpretable if power is high. Vasishth and Nicenboim (2016) discuss this point for psycholinguistics examples, and a nice example from phonetics is Warner et al. (2006), who examine IN in Dutch and argue that neutralization is in fact complete. Their study has high power (>0.8) to detect a very small voiced/voiceless difference (<3.5 msec); they also do not find any significant voiced/voiceless difference for other acoustic cues.

3.3.6 Multiple Comparisons

One important place the conservativity/power trade-off comes into play is when correcting for multiple comparisons: the more hypothesis tests you conduct, the less meaningful the p-value for each test is.

To see why, recall that the significance level α is set by deciding our level of tolerance for type I errors. This interpretation ($\alpha = 0.05$ means "5% chance of falsely rejecting the null") assumes that we are carrying out just one hypothesis test. If we carry out multiple hypothesis tests, this interpretation no longer makes sense: we are more and more likely to find at least one significant result—even though the null is (always) true. Intuitively, this is why "data dredging" works: the more plots you make of two things you have measured, the more likely one of them will look like there is a relationship.

To quantify this intuition, consider the *family-wise error rate* (FWER), the probability of making at least one type I error for a family of k tests. If the results of the k tests were independent, then

$$FWER = 1 - (1 - \alpha)^k.$$

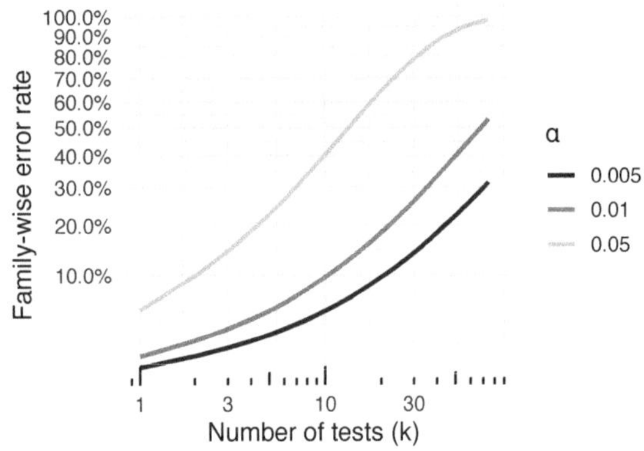

Figure 3.5
FWER for k independent hypothesis tests with significance level α. The x/y-axes are square root/log-transformed.

Figure 3.5 shows FWER for k tests for common significance levels ($\alpha = 0.05, 0.01, 0.005$). For example, when $\alpha = 0.05$, FWER is 9.7% for 2 tests, 23% for 5 tests, and 97% for 50 tests.

Different multiple comparison procedures attempt to hold FWER (or another metric, such as *false discovery rate* (FDR)—the proportion of significant tests that are in fact type I errors) at a specified level (α, e.g., 0.05) when conducting many tests, by "correcting" the p-value for each individual test.

The simplest option is just *no correction*—use the original p-value.[9] The most common actual "multiple comparison procedure" is the *Bonferroni method*, which just multiples the p-value for each test by k. The Bonferroni method is the most conservative multiple comparison procedure—FWER is always less than α, with no assumptions about the data—but it has very low power: in most situations, at least one test will fail to detect a difference that actually exists. "No correction" is obviously the least conservative procedure possible, but it also has the highest power.

There are many other multiple comparison methods, which trade off between these options (conservativity vs. power), making different assumptions about the data, and controlling either FWER or another metric (such as FDR). Some common methods are

- The *Holm-Bonferroni method*: adjusts lower p-values more severely than higher ones (also controls FWER without assumptions).
- The *Tukey honestly significant difference* method (controls FWER): used in the (common) special case of *pairwise comparisons*—testing for a difference between each pair of m groups (so, $k = m(m-1)/2$), using modified t-tests.
- The *Benjamini-Hochberg method* (or FDR method): controls FDR, but it makes assumptions about independence of tests.

9. For example, we rarely correct all the p-values reported for unrelated analyses in a paper.

Methods minimizing FDR (such as the FDR method) tend to have higher power than FWER methods but are less conservative. This may be why the conventions in cognitive/behavioral sciences (which focus on type I error) are to use Bonferroni corrections and Tukey tests for pairwise comparisons. But it should be kept in mind that both methods are very conservative, and the Bonferroni correction should arguably never be used (section 3.3.6.2).

3.3.6.1 Example

One application where correcting for multiple comparisons is common is examining *individual differences*—what effect different people in the same study show. In this example we look for individual differences in the effect of voicing in the `neutralization` data by carrying out a *t*-test for each subject, correcting for multiple comparisons in different ways.

Figure 3.6 (left), which repeats figure 3.4, summarizes the empirical voiced/voiceless duration difference for each subject. If we went from this plot alone, we would conclude that 7 subjects have nonzero effects (those where the errorbar does not overlap 0). But how would this conclusion change if we correct for multiple comparisons?

Within each speaker, the data are organized into voiced/voiceless pairs by `item_pair` (see section 3.1.2). We first organize the data so each subject/item pair is in one row, with one column each for `vowel_dur` for `voicing`=VOICED and VOICELESS:

```
neut_paired <- neutralization %>%
  select(subject, item_pair, vowel_dur, voicing) %>%
  spread(voicing, vowel_dur) %>%
  # Remove rows with NA for voiced or voiceless
  remove_missing()
```

To get the uncorrected *p*-value, we then apply a paired *t*-test ($\alpha = 0.05$) to the voiced/voiceless pairs from each subject:

```
p_df <- neut_paired %>%
  nest(data = c(item_pair, voiceless, voiced)) %>%
  mutate(
    test = map(data,
      ~ t.test(.x$voiced, .x$voiceless, data = .x, paired = TRUE)
    ),
    ## Extract p-value
    p = map_dbl(test, "p.value")
  ) %>%
  select(-test, -data) %>%
  unnest(p)
```

This code uses functions from purrr: `nest()` to split up the dataframe by subject, `map()` to apply a function to each subset, then `unnest()` to put the dataframe back together.[10] The resulting *p*-values are

10. This kind of "nest-map-unnest" workflow is useful and made easier by tidyverse functionality (Grolemund and Wickham 2016, chapter 25).

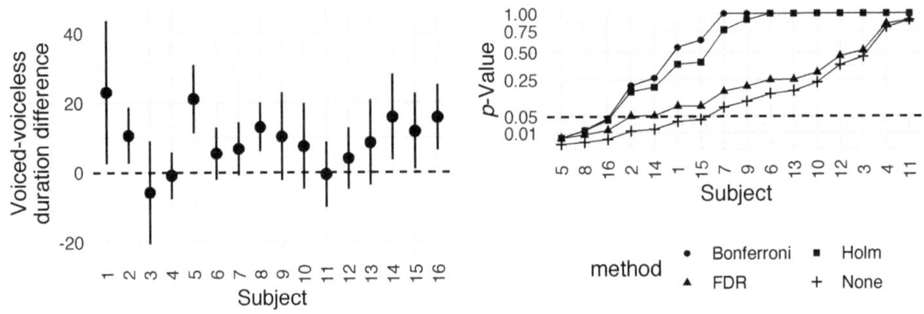

Figure 3.6
Left: Summary of voiced-voiceless duration differences for each of the 16 speakers for the `neutralization` data. Points/errorbars are means ± 2 SEs. *Right*: p-Values for one-sample t-tests (on square root-scale) examining the effect of stop voicing on vowel duration within each subject. Subjects are ordered by p-value. p-Values shown are uncorrected (none) or corrected using the Bonferroni, Holm-Bonferroni, or FDR method. Note that some points overlap for subjects where different calculation methods give similar p-values. Dotted line is significance level ($\alpha = 0.05$).

```
round(p_df$p, 4)
##  [1] 0.0346 0.0131 0.4515 0.7983 0.0002 0.1551 0.0848 0.0010 0.1133
## [10] 0.2291 0.9032 0.3709 0.1731 0.0165 0.0399 0.0028
```

We now correct these p-values for multiple comparisons using the Bonferroni, Holm, and FDR methods, using the `p.adjust` function:

```
p_df <- p_df %>% mutate(
  ## No correction
  None = p,
  Bonferroni = p.adjust(p, method = "bonferroni"),
  Holm = p.adjust(p, method = "holm"),
  FDR = p.adjust(p, method = "fdr")
)
```

Figure 3.6 (right) plots these p-values. First, note how for each subject, $p_{FDR} < p_{holm} \leq p_{bonferroni}$. This is consistent with the Bonferroni correction being the most conservative and the FDR comparison the least conservative. Second, we see that the conclusion we draw about "how many speakers show incomplete neutralization?" changes depending on whether we correct for multiple comparisons and to a lesser extent how we do so:

- $\alpha = 0.05$ significance level: 7 subjects using uncorrected p-values ($p < 0.05$ for the 7 subjects whose errorbars don't overlap 0); 3 subjects using corrected p-values (any method).
- $\alpha = 0.01$ significance level: 3 subjects using uncorrected p-values, versus 2 subjects using FDR or 1 subject using Holm or Bonferroni.

3.3.6.2 Discussion

As the example showed, either lowering α or adjusting for multiple comparisons using a more conservative method will lead to fewer significant results (more possible type II errors) but also to fewer spurious results (fewer possible type I errors). The point here is not which method is right, but that they trade off in conservativity versus power.

That said, some practical advice can be given:

- The minimally conservative method of *not* correcting for multiple comparisons, especially when performing many (say >5) conceptually related hypothesis tests, is almost never a good idea.[11] You will be very prone to spurious findings.
- The maximally conservative method, Bonferroni, is never a good idea, because the Holm-Bonferroni method always has higher power, without increasing type I error (Aickin and Gensler 1996; see also `?p.adjust.methods` in R).
- If the conclusions you reach about a research question depend greatly on which multiple comparison method is used, they are shaky.

3.4 Error in Effect Size: Type M and Type S Errors

Type I and II errors are defined with respect to a binary decision: is the parameter null or not? We will now discuss errors in (unstandardized) effect size—the actual value of the parameter we're estimating.

3.4.1 Bias

First, there is some terminology we'll need going forward. Recall that a statistical estimator is *unbiased* if on average (across many samples) it gives an estimate that is the same as the population value (section 2.2.2). The estimator is *biased* otherwise: too high or too low, or toward or away from 0 (*shrunk/inflated*). The *variance* of an estimator measures how "spread out" the estimates are across many samples. (For example, the width of the distributions in figure 2.1 is the variance of the sample mean.) *Precision* is the inverse of variance (low variance$\Rightarrow$high precision).

We try to always use estimators that are unbiased, but this property typically only holds in ideal settings (large sample size, model assumptions met) and breaks down for small enough sample size or if certain model assumptions are violated. A simple example we have already seen (section 2.3.1) is using $\sqrt{n-1}$ in the denominator to estimate a population's SD from a sample. If we instead use $\sqrt{n}$, the estimate is biased downward.

In analyzing data we usually implicitly assume that our estimators are close enough to unbiased, so that on average, the estimate we make of a parameter of interest is correct. Because bias is defined as an average over many samples, this assumption only makes sense if we pay equal attention to all results and report whatever the statistical model says.

---

11. An exception is when "proving the null" is of interest. Warner et al. (2006) do not correct for multiple comparisons in their Dutch incomplete neutralization study, which maximizes power to find any voiced/voiceless difference that does exist and strengthens their argument for complete neutralization.

3.4.2 Type M and S Errors

However, in practice, research in cognitive and behavioral science doesn't give equal weight to all results: in publications it is common to only discuss significant effects ($p<\alpha$) and interpret them based on their sign and (sometimes) magnitude. Nonsignificant results are often not discussed or (worse) not published at all. Whatever the other consequences of this practice,[12] the implicit assumption is that the significant results are reliable. But how accurate are the parameter estimates in reality?

Let β_x denote the true size of an effect of interest and $\hat{\beta}_x$ the estimated effect size when an experiment is done to estimate it. Because $\hat{\beta}_x$ is a random variable, which will be different each time the experiment is run, the estimated effect size can be incorrect relative to the true effect size in either magnitude or sign. Gelman and Carlin (2014) define two corresponding measures of error of the estimated effect size (in *magnitude* or *sign*), across different replications of the same experiment:

- The *type M error* is the value of $|\hat{\beta}_x/\beta_x|$ in one replication.
- The *exaggeration ratio* is the average type M error across many replications.
- The *type S error* is the probability that the estimated effect has the wrong sign ($\text{sign}(\hat{\beta}_x) \neq \text{sign}(\beta_x)$).

These quantities are all measures of bias, typically defined as conditional on significance. How biased would the estimate's magnitude be, and how often would its sign be wrong, *if only significant results are considered*? The idea is to quantify two kinds of error made in realistic data analysis, where only results that are made "with confidence" are taken seriously (Gelman and Tuerlinckx 2000).[13]

Like power, calculating type M and S errors requires assuming a true effect size (and degree of variability), and the concepts are closely related. In general (for unbiased estimators) type M and type S errors decrease as power increases, and type M error increases faster than type S error as power decreases.

When power is low, type M and S errors can be surprisingly high—even for statistically significant results (Gelman and Carlin 2014; Lu, Qiu, and Deng 2019). The existence of type M and type S errors means that it is not always the case that significant findings are correct, in the sense of providing accurate estimates of the effect of interest.

Let's illustrate these points in an example using the `neutralization` data, using simulation to calculate type M and S errors for t-tests, assuming different true effect sizes.[14]

Example Let's assume the same setup as in the power example in section 3.3.2—two-sample t-test to detect a VOICED/VOICELESS difference, σ fixed at 45 msec, and so on—but now we will examine how the exaggeration ratio and type S error vary as a function of sample size n (ranging 25–500 msec) and (true) effect size Δ (of 5, 10, or 30 msec).

---

12. Reporting only significant results is highly problematic. See Vasishth, Mertzen, et al. (2018) for discussion for (psycho)linguistics in particular.

13. Note that statistical significance is only one way to define which claims are made with confidence; the general points hold for other criteria (e.g., Bayes factors or likelihood ratios).

14. Because the points are primarily conceptual, this book does not go into how to actually calculate type M and S errors. They can be calculated for simple cases using calculators (e.g., t-tests; see Gelman and Carlin 2014) or via simulation for more complex designs (e.g., Kirby and Sonderegger 2018a).

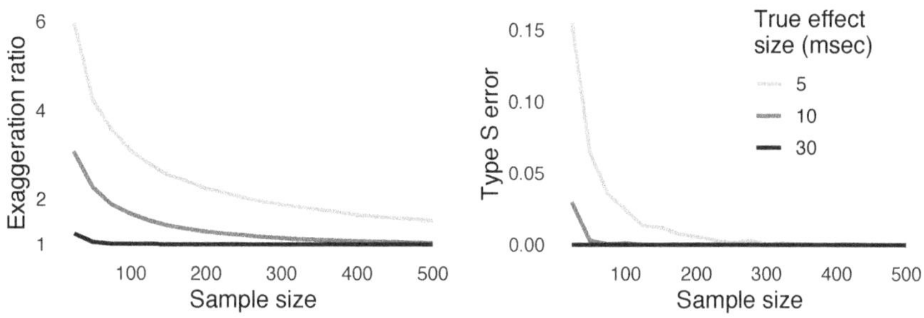

Figure 3.7
Exaggeration ratio (*left*) and type S error (*right*) for a two-sample *t*-test as true effect size (difference in means) and sample size are varied, with variability ($\sigma = 45$) and significance level ($\alpha = 0.05$) held fixed.

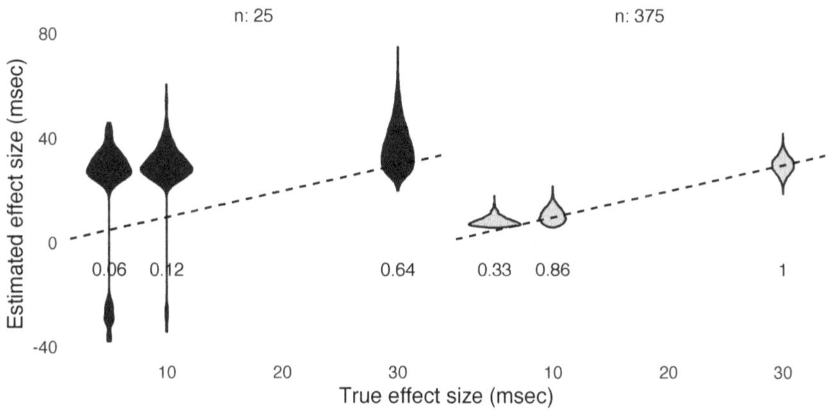

Figure 3.8
Distribution of estimated effect sizes (difference between sample means: violin plots) for 5000 simulations of n observations from two normal distributions, with means separated by a true effect size (x-axis: 5, 10, 30 msec). Inset numbers show power for a two-sample t-test in each case.

Let's calculate type M and type S errors by simulation (code shown in source file only): for a given Δ and n, we will draw normally distributed data from two groups many times, perform a t-test for each draw, and calculate the exaggeration ratio and type S error using the estimated effect sizes from draws where the t-test was significant.

Figure 3.7 shows the exaggeration ratio and type S error as a function of true effect size and sample size. The general pattern is as expected: as power increases (via larger n or Δ), the effect's magnitude is less exaggerated and less likely to have the wrong sign, and the type S error decreases faster than the exaggeration ratio does.

To make this more concrete, figure 3.8 shows the distribution of estimated effect sizes (for individual draws) in the same illustrative cases as in the power example: complete/incomplete/no neutralization effect sizes ($d = 5$, 10, 30) and small and large sample sizes ($n = 25$, 375).[15] The distributions let us think more easily about what could happen in an individual

15. This figure is inspired by figure 8 of Vasishth and Nicenboim (2016).

study (as opposed to on average). It is useful to think in terms of the power of different cases, which ranges from 0.06 to 1:

High power For high power (>0.8), the estimate for a significant result is guaranteed to have the right sign and will likely have a magnitude near its true value—bias is low and precision is high. For example, in a study of incomplete neutralization with $n = 375$ (assuming $\Delta = 10$ msec), where power is 0.86, a significant result will give an estimate in the right direction (VOICED>VOICELESS) and an estimate precise to about 5 msec.[16]

Medium power As power decreases (here, 0.33–0.64), the magnitude of the effect is increasingly likely to be positively biased, and precision decreases drastically, though the effect's sign is very likely to be correct. For example, in a study of a nonneutralized contrast (assuming $\Delta = 30$ msec) with a small sample size ($n = 25$), a significant result will give wildly differing estimates (~20–70 msec), which on average will be inflated from the true value by 24%.

Low power For lower power (here, <0.2), significant effects will have greatly inflated magnitude and sometimes the wrong sign—thus, high bias and low precision. For example, consider the case of using a small sample per subject ($n = 25$) to detect a contrast of $\Delta = 5$ or 10 msec. Many subjects will give nonsignificant results (because power is low), the estimates for subjects with significant results are guaranteed to be greatly inflated (by a factor of ~5–6 on average!), and there is a reasonable chance these significant effects will have the wrong sign (5%–10%).

Box 3.3
Broader Context: "True Effect Size"?

The calculations behind design analysis (power and type M and S errors) depend on specifying a range of true effect sizes to consider. This is arguably the most difficult part of design analysis—or power analyses for study planning, for that matter—because the true effect size is unknown. Following the same logic for why observed power is useless (box 3.1), you can't just use your observed effect size.

However, delimiting a range of plausible effect sizes of the phenomenon is usually possible. Ways to do this include a meta-analysis or literature review of previous studies, if literature exists, or, if not, computational modeling, theoretical considerations, soliciting expert opinions, or reasoning based on studies of related phenomena. Gelman and Carlin (2014), Nickerson (2000), and Vasishth and Nicenboim (2016) discuss these options with some practical examples from different fields (social sciences, psychology, linguistics). Section 3.2.1 and Kirby and Sonderegger (2018a) provide an example for German incomplete neutralization in particular.

3.4.3 Summary: Power, Type M/S Errors, and Design Analysis

This example, and the one in section 3.3.2, show the consequences of power and effect size errors in a concrete case: the effect of consonant voicing on vowel duration in the `neutralization` dataset. The exact numbers (for power, type M/S errors) are specific to this particular case, but the general points raised in both examples hold for interpreting any study (or for planning a new one):

16. Even this precision may not be enough to differentiate between theories. Recall that $\Delta = 5$ would mean "imperceptible," whereas $\Delta = 10$ would mean "in line with other incomplete neutralization effects."

- When power is low, null results are uninformative, effect size estimates are imprecise, and significant results are virtually guaranteed to give an effect size inflated in magnitude and possibly with the wrong sign (the "winner's curse;" see Button et al. 2013; Ioannidis 2008).
- When power is high, null results are informative, effect size estimates are precise, and significant results give effects that are minimally biased and have the correct sign.

In sum, in a high-powered study results give useful information regardless of the p-value. In contrast, in a low-powered study neither the p-value nor effect size gives much useful information.

These considerations can be used for *design analysis* (Gelman and Carlin 2014; Kirby and Sonderegger 2018a): using statistical tools to interpret the results of a study by reasoning about likely outcomes if the study were replicated—which is generally what we are interested in (not the results of the study itself)—by considering a range of plausible effect sizes (box 3.3). Design analysis can tell us what we can learn from a study with a given design and sample size about likely values of a parameter of interest regardless of whether the parameter's estimate was significant in our study.

For example, consider the example in section 3.3.6.1, where we tested for the effect of `voicing` within individual subjects ($n \approx 25$) and found that about half of the subjects showed null effects, while others showed effects estimated at 11–22 msec. This kind of result is sometimes discussed in the phonetics literature as evidence for individual differences: some subjects have an effect, some don't (in reality). However, if in reality $\Delta = 5$–12 msec for all subjects (no individual differences), the pattern of results observed would be unsurprising. Power is low to medium, resulting in the type II errors (subjects with null results), and significant effects are biased upward from the true value. But this result would also be unsurprising if individual differences did exist—say some subjects have $\Delta = 0$ and others have $\Delta = 5$–12—so our results are just inconclusive as to whether individual differences exist.

3.5 Assumptions of Hypothesis Tests and Consequences

Any statistical method makes assumptions about the data and how they were generated. In practice, one's data never meets all assumptions, or some assumptions cannot be verified. Thus it is important to understand how serious (or not) violations of each assumption are and what effect they could have on the results. The error concepts introduced previously (bias, type I/II and M/S errors) provide a precise way to do so.

Later chapters (especially chapter 5) discuss the assumptions of regression models in detail. This section gives a couple examples using two-sample t-tests and Wilcoxon tests, because they share much in common with regression model assumptions but in a simpler setting.

3.5.1 Data Distribution

Two-sample t-tests assume that data (within each sample) are normally distributed, whereas Wilcoxon rank-sum tests make almost no assumptions about the distribution.

Because *t*-tests calculate two means, they can be strongly affected by deviations from normality that affect means—such as clear outliers or highly skewed data (in one or both samples). The example in section 2.5.2 showed how a type II error could occur in this way. Outliers or skew can also cause type M error (because they bias the estimated means).

Type I or type S errors can also result from applying a *t*-test to nonnormal data, especially in a small dataset. For example, consider the plot in figure 3.1 (right), based on ~25 observations per group. The *t*-test carried out on this data gave $p = 0.42$, 95% CI $[-10.1, 23.9]$ msec. Now suppose we add 5 `voicing`=VOICELESS observations with duration of 1,000 msec; this could have happened by experimenter or annotator error. The result of the *t*-test then changes dramatically (*p*-value and 95% CI):

```
neut_7 %>%
  select(voicing, vowel_dur) %>%
  add_row(voicing = rep("voiceless", 5), vowel_dur = rep(1000, 5)) %>%
  t.test(vowel_dur ~ voicing, data = .)
...
## t = -2, df = 28, p-value = 0.03
## alternative hypothesis: true difference in means between
## group voiced and group voiceless is not equal to 0
## 95 percent confidence interval:
##  -266  -15
...
```

If the null hypothesis is in fact true (no difference between groups), the hypothesis test carried out in section 2.5.2 gave the right decision, whereas the test after adding the 5 points gives a type I error.

If the null hypothesis isn't true, we know on scientific grounds (section 3.1) that it is very unlikely that voiceless stops have longer duration—as estimated by this *t*-test—so the estimated difference in means gives a type S error.

3.5.2 Independence Assumptions

Both *t*-tests and Wilcoxon tests make the same independence assumptions:

- Within-sample: observations within each sample are independent.
- Between-sample: the two samples are independent (i.e., every pair of observations from samples 1 and 2 is independent).

The independence assumptions do not hold for many basic questions addressable with linguistic datasets, because most linguistic data involves multiple observations per person or per linguistic object (e.g., sentence), which we don't expect to be independent. Violations of independence assumptions are very important, and can easily lead to type I or type II errors; they are probably the most important assumptions underlying statistical analysis of linguistic data more generally (box 3.4). We will discuss this further when we turn to clustered data in chapter 8. For now, here are a couple examples for *t*-tests, using the `neutralization` data.

Box 3.4
Broader Context: Pseudoreplication and *t*-Tests

Applying a *t*-test to data where the independence assumptions are not met is an example of *pseudoreplication*: analyzing nonindependent observations as if they were independent. Pseudoreplication is a much larger issue than *t*-tests are, because similar independence assumptions hold for most basic statistical methods: correlations, (non-repeated-measure) analyses of variance, linear and logistic regressions, and so on. Violations of independence assumptions are very important because they can easily lead to type I/II (and M/S) errors; they are very common when analyzing data not just in linguistics but in many fields (e.g., ecology, neuroscience; see Hurlbert [1984], Lazic [2010]). In linguistics, Winter (2011) introduced the term *pseudoreplication* in the context of analyzing phonetic data.

Because pseudoreplication is so common, it is useful to know how to detect it in published work. For *t*-tests, this is simple: as long as degrees of freedom are reported, one can check *df* against the number of independent units in the study. For example, the two-sample *t*-test for the effect of voicing on vowel duration in the `neutralization` data in section 3.2.1 has $df = 747$. This is far closer to the number of observations in the dataset (750) than the number of subjects (16) or items (24) in this dataset, showing that pseudoreplication occurred.

This points to a couple broader implications for reading the literature and analyzing data. Pseudoreplication is very common in linguistic research, especially in older papers in subfields with long histories of quantitative work (e.g., phonetics, acquisition, sociolinguistics); psycholinguistics is a notable exception in having adopted methods for analyzing nonindependent data from the early 1970s (see box 8.3). It is important to be aware of this issue when reading the literature (especially older papers), and if pseudoreplication is present, think critically about what effects it may have. This point applies just as much for regression models, and we will return to this issue in chapters 5 and 8–10.

Yet it is also important to not immediately discount a study's results because of pseudoreplication—similarly to other assumptions made by statistical models (e.g., normality: box 2.9). As important as independence violations are, in many cases the qualitative results of an analysis do not change when a more correct analysis method is used. For example, most *t*-tests conducted in the current chapter violate independence assumptions—yet the results are very similar (in terms of effect size, whether *p* is greater than or less than 0.05) to when a "correct" method is used. Often it makes sense in data analysis to use simpler methods—whether as a first step toward a more complex method or for pedagogical purposes.

3.5.2.1 Example: Type I error from nonindependence

Let's group subjects in a way we are sure does not in reality affect vowel duration—whether the subject was in the first or second 50% of people recruited for the experiment (`early_subject`):

```
neutralization$early_subject <-
  ifelse(as.numeric(neutralization$subject) < 9, "early", "late")
```

Table 3.1
Results of two-sample *t*-tests corresponding to the left and right panels of figure 3.9

Data	Estimate	*p*	*t*	*df*	95% CI
Observations	−10.32	0.00	−3.27	745.39	[−16.5, −4.1]
Speaker means	−9.77	0.50	−0.69	13.64	[−40.1, 20.5]

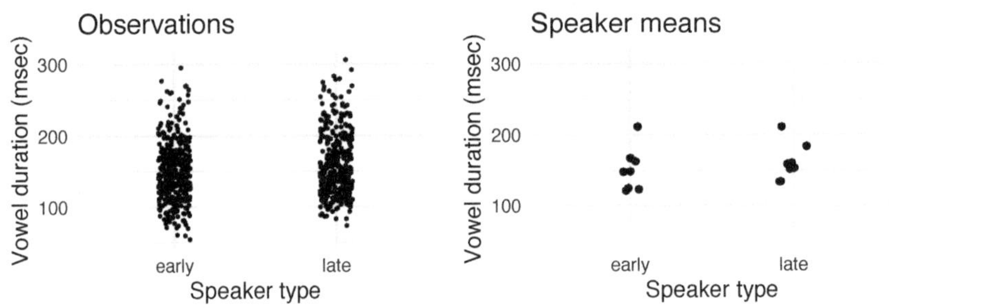

Figure 3.9
Distribution of vowel duration as a function of speaker "earliness" (from the first or second 50% of the `neutralization` data) for all observations (*left*) and for speakers' mean durations (*right*).

Plotting the data (figure 3.9, left) suggests that LATE subjects have longer vowel durations, and we can quantify this intuition with a two-sample *t*-test (results summarized in table 3.1, "Observations"):

```
neutralization %>% t.test(vowel_dur ~ early_subject, data = .)
```

But this must be a spurious finding (type I error, anticonservative *p*-value)—the null hypothesis of no difference between groups is presumably true.

A two-sample *t*-test is not appropriate here because observations are not in fact independent—different observations from the same subject (e.g., the red dots in figure 3.9, left) tend to cluster together. Intuitively, each subject should only give one data point toward answering the question, "do early subjects have shorter vowels?" We can do this by examining the mean duration for each subject, in figure 3.9 (right), which suggests weak support for a LATE/EARLY difference. This intuition is confirmed by a two-sample *t*-test on subjects' mean durations (table 3.1: "Speaker means"):

```
neutralization %>%
  group_by(subject, early_subject) %>%
  summarise(m = mean(vowel_dur)) %>%
  t.test(m ~ early_subject, data = .)
```

This test gives a similar estimated early/late difference but (correctly) concludes this difference is not significant.

Here, not accounting for grouping structure (one kind of nonindependence) causes a type I error because of the model's assumption about how much independent information the data contain (the *df*). This issue will be a major motivation for mixed-effects models (chapter 8).

3.5.2.2 Example: Type II error from nonindependence

Consider checking whether there is an effect of voicing on vowel duration for data from subject 14 in the `neutralization` data—which is one of the *t*-tests in the multiple comparisons example in section 3.3.6.1. Within each subject observations come in voiced/voiceless pairs—thus, the two observations for each pair are not independent. What effect does taking

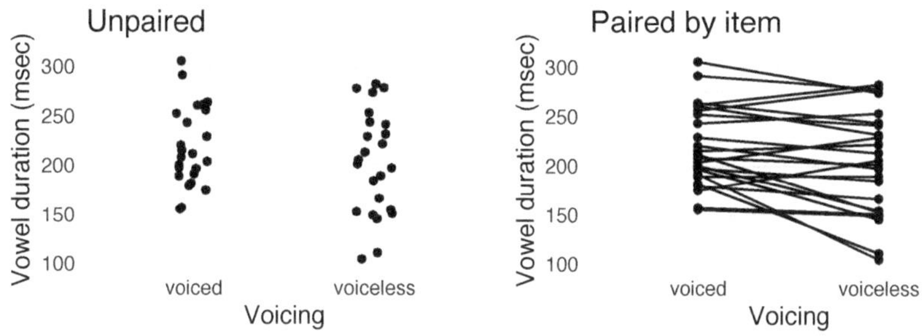

Figure 3.10
Distribution of vowel duration as a function of following consonant voicing for speaker 14 observations from the `neutralization` data without (left) and with (right) observations paired by `item_pair`.

this nonindependence into account have on our conclusion? Figure 3.10 plots `vowel_dur` as a function of `voicing` for subject 14, in two ways.

If we assume the VOICED and VOICELESS observations are independent (left), the effect of voicing looks very weak—small group difference relative to within-group variability. But if we know that VOICED and VOICELESS observations come in pairs (right), the pattern is clearer: VOICED>VOICELESS for 17 pairs (71%), suggesting a clear effect of voicing. These different intuitions are borne out by a two-sample t-test (false independence assumption) versus a paired t-test (independence assumption satisfied), whose results are summarized in table 3.2.

```
filter(neutralization, subject == 14) %>%
  t.test(vowel_dur ~ voicing, data = .)
filter(neut_paired, subject == 14) %>%
  t.test(.$voiced, .$voiceless, paired = TRUE, data = .)
```

Table 3.2
Results of two-sample t-tests corresponding to the left and right panels of figure 3.10

Data	Estimate	p	t	df	95% CI
Unpaired	15.93	0.25	1.17	43.08	[−11.5, 43.3]
Paired	15.93	0.02	2.59	23.00	[3.2, 28.7]

The estimated voiced/voiceless difference (= effect size) is the same in each test (16 msec), but it is only reliably different from 0 in the second test. Thus, assuming subject 14 does in reality have a positive voiced/voiceless duration difference, the fact that observations from the same item are not independent would result in a type II error (and a conservative p-value).

Until we cover mixed-effects regression, we will get around the fact that most linguistic data involve grouping structure (and hence t-tests/linear/logistic regressions will have nonindependent errors) in one of three ways:

1. Using datasets *without grouping structure*, either because this is how the raw data looks (e.g., studies of a language's lexicon such as `regularity`), or because observations have been preaveraged to remove grouping structure (such as `english`).
2. *Taking a subset* of a dataset that removes grouping structure (e.g., `transitions_sub` in chapter 2).
3. Committing *pseudoreplication*: analyzing datasets with grouping structure (e.g., `neutralization`), using methods that assume independence of errors (e.g., *t*-tests, linear regression), with the understanding that our results are more likely to be wrong in some sense (type I/II/M/S errors).

3.6 Further Reading

References on power, effect size, and related issues include Kline (2013), Cohen (1988, 1992), Hallahan and Rosenthal (1996), Cumming (2012), Kelley and Preacher (2012), Pek and Flora (2018) (who give concrete examples for reporting effect size), as well as the APA and ASA resources cited in chapter 2. Kline and Cumming are part of the "new statistics" movement to emphasize estimation and precision over traditional NHST (*p*-values); Cumming (2014) gives an overview. For linguistic data, power and effect size are discussed by Kirby and Sonderegger (2018b, 2018a), Vasishth and Nicenboim (2016), and Winter (2019, chapter 9).

Type M and type S errors are discussed by Gelman and Carlin (2014) and more technically by Gelman and Tuerlinckx (2000) and Lu, Qiu, and Deng (2019). In linguistics, Vasishth and Nicenboim (2016) introduce type M and S errors, which have been used in a growing number of psycholinguistic and phonetic studies (e.g., Nicenboim, Roettger, and Vasishth 2018; Vasishth, Mertzen, et al. 2018), including design analysis for the German IN case by Kirby and Sonderegger (2018a).

Exercises

3.1. Write code to calculate the effect sizes (*d*) for `prosodic_boundary`, `accent_type`, and `gender` shown in section 3.2.3.2

3.2. As noted in box 3.4, most *t*-tests carried out in this chapter on the `neutralization` data involve pseudoreplication, which is a violation of independence assumptions.

(a) Examine how the data are organized by `subject`, `item_pair`, and `voicing` (e.g., by cross-tabulating using `xtabs`). How and why are observations nonindependent in this dataset?

(b) Find two examples of hypothesis tests in this chapter where independence assumptions are violated (outside of section 3.5.2) and explain why they are violated in each case.

(c) Find the one example in this chapter (outside of section 3.5.2) where independence assumptions are not violated and explain why not.

4 Linear Regression 1

This chapter introduces regression in general (section 4.2), and then discusses linear regression for one predictor (section 4.3) and several predictors (sections 4.4 and 4.5), focusing on how to fit and interpret basic models.

4.1 Preliminaries

4.1.1 Packages

This chapter assumes that you have loaded packages from the previous chapters, in addition to broom, a tidyverse package for working with fitted models, and sjPlot for plotting model predictions (Robinson, Hayes, and Couch 2021; Lüdecke 2021):

```
library(languageR)
library(tidyverse)
library(patchwork)
library(broom)
```

4.1.2 Data

This chapter also assumes that you have loaded the `neutralization` dataset, as in chapter 3:

```
neutralization <- read.csv("data/neutralization_rmld.csv",
  stringsAsFactors = TRUE) %>%
  filter(!is.na(prosodic_boundary))
```

A row from `neutralization` with missing data for a variable we will use (`prosodic_boundary`) has also been dropped.

The `english` dataset

We'll make use of the `english` dataset, which is part of languageR and is described in detail on its help page (`?english`). This psycholinguistic dataset comes from the English Lexicon Project (Balota et al. 2007), which obtained measures of lexical access for a large subset of the English lexicon, across many speakers of two age groups (`AgeSubject`: levels OLD, YOUNG), to understand what properties of words affect how quickly they are perceived and produced. The `english` dataset contains a subset of the English Lexicon Project data ($n = 2284$ words, all nouns and verbs), with two rows per word ($n = 4568$) corresponding to the two age groups, and columns including

- Lexical access measures, averaged across all subjects from the same age group:
  - Reaction time (RT) for lexical decision task (`RTlexdec`: msec, log-transformed), where participants judge whether a visually presented string of letters is a word (e.g., "flip") or not (e.g., "flirp").
  - RT for word naming task (`RTnaming`: msec, log-transformed), where participants produce a visually presented word as quickly as possible.
  - Proportion of subjects who accepted this word in the lexical decision task (`CorrectLexdec`).
- Word frequency (`WrittenFrequency`: log-transformed)
- Word familiarity (`Familiarity`): subjective ratings.
- Number of morphologically related words (`FamilySize`: log-transformed)

We change the `AgeSubject` factor so the YOUNG level comes before OLD, which makes more conceptual sense:

```
english <- mutate(english, AgeSubject = relevel(AgeSubject, "young"))
```

Section 4.5.3 discusses "releveling" further.

4.2 Regression: General Introduction

Regression analysis is "a conceptually simple method for investigating functional relationships among variables" (Chatterjee and Hadi 2012, 1). The variable to be explained, written y, is called the *response* (or "dependent variable"), often written y. The explanatory variables, written $x_1, x_2, \ldots$, are called *predictors* (or "independent variables").

Regression modeling can have one of several objectives (Faraway 2015, section 1.3; Shmueli 2010):

1. *Prediction*: of future values of y, given values of the predictors.
2. *Explanation*, or *estimation*: of the effect of predictors on the response, or the relationship between the two.
3. *Description*: of the relationship between the predictors and the response, as in an exploratory study.

For (2) usually we would like to infer causal relationships (x_1 affects y, not vice versa), but this is often not possible, hence "effect" (causal) versus "relationship" (causality uncertain). *Estimation* refers to the actual values of parameters characterizing relationships (e.g., regression coefficients). *Description* is rarely the end goal in a study, but regressions can be part of exploratory data analysis, and probably many analyses of linguistic data are a mixture of exploratory and confirmatory (even when we think they are just confirmatory), as discussed in section 1.3.2.

The goal of the analysis matters because it can affect what methodology is best, as we will see. The important distinction is between (1) and (2)—whether we prioritize predictive accuracy or parameter estimation, which can lead to different models.

Usually in language sciences explanation is the goal or the goal is not explicitly stated. So we will often use language that implies the "explanation/estimation goal".

As a running example, we consider a regression analysis modeling a word's lexical decision RT (or just "RT": `RTlexdec`) for the `english` dataset, as a function of predictors (e.g., age of subject, word frequency). The goal of the analysis could be

- Prediction: predict `RTlexdec` as accurately as possible given a new word of English.
- Explanation: determine how much `RTlexdec` increases as a result of `AgeSubject` changing (causal: we wait 20 years), or how much it differs between YOUNG and OLD subjects (noncausal).
- Description: generate hypotheses about which predictors are most likely to affect RT, to be tested on a new lexical decision dataset for English speakers (of another dialect, say), or as part of planning an English Lexicon Project–like study for a new language where resources are much more limited and only a few predictors can be measured.

Box 4.1
Broader Context: What Do Regressions Tell Us?

While *estimation* can be precisely defined (estimation of parameters, in the terms discussed in chapter 2), what *explanation* actually means is tricky. Regression has a directional interpretation (predictors affect the response), and typically we would like to think of the model in causal terms. But arguing for causality in regression models is hard, especially in observational data (e.g., from linguistic corpora). It requires the tools of *causal inference*, which are beyond the scope of this book (see, e.g., Faraway 2015, section 5.2; McElreath 2020; Gelman and Hill 2007, chapter 9). Even in well-designed controlled experiments, where only one or two things vary, a causal interpretation relies on many assumptions, only some of which can be checked. The difficulty of causal interpretation doesn't detract from the power of statistical analyses to give insights into research questions. But in general we must just always assume that we are fitting correlational models, and we must bear the "correlation is not causation" adage in mind when interpreting our results.

Some of these issues are discussed for linguistic typology by Roberts and Winters (2013).

4.2.1 Linear Models: Terminology

The relationship between variables is captured by a regression *model*:

$$y = f(x_1, x_2, ..., x_k) + \varepsilon.$$

In this model, y is a (random) variable approximated by a function of the k predictors, and the difference between the model and reality is called the *error* (ε).

Throughout much of this book we will be dealing with *linear models*, where the model can be written in this form (Harrell 2015, 33):

$$C(y|x_1, \ldots, x_k) = \beta_0 + \beta_1 x_1 + \beta_2 x_2 + \cdots + \beta_k x_k. \tag{4.1}$$

Here, C is some function of y, which differs depending on the type of regression model. The β's are called *regression coefficients*. This turns out to be a very general class of model, which can be applied to a wide range of phenomena. Some important kinds of models that don't appear at first glance to fit equation (4.1), such as regression as a function of polynomials (like $y = \beta_0 + \beta_1 x + \beta 21 x^2$: see section 5.5.1), can be written in this form.

Two broad types of regression are considered in this book. First is *linear regression*, where the response (y) is a continuous variable—such as modeling RT (`RTlexdec`) as a function of word frequency (`WrittenFrequency`) for the `english` dataset. In this case, the function C is just the identity plus an error term, and the regression model is

$$y = \beta_0 + \beta_1 x_1 + \beta_2 x_2 + \cdots + \beta_k x_k + \varepsilon$$
$$\varepsilon \sim N(0, \sigma) \qquad (4.2)$$

Chapter 6 will consider *logistic regression*, where the response is binary: 0 or 1. An example would be modeling whether a Dutch verb is regular or not (`Regularity`) as a function of which auxiliary it takes (`Auxiliary`) and its frequency (`WrittenFrequency`), in the `regularity` dataset. In this case C predicts the probability that response is 1, given the values of the predictors.

Regression with just one predictor ($k = 1$) is called *simple*, and regression with multiple predictors ($k > 1$) is called *multiple*. The two examples just given would be examples of "simple linear regression" and "multiple logistic regression".

Predictors can be *continuous*, such as word frequency, or *categorical* (also called *factors*) such as participant age group or verb auxiliary.

Box 4.2
Broader Context: What about ANOVAs?

Certain special cases of linear models that are common go by their own names:

- *Analysis of variance* (ANOVA) models continuous y using categorical predictors. ANOVA is used to analyze variability between groups.
- *Analysis of covariance* (ANCOVA) models continuous y with a mix of categorical and continuous predictors. ANCOVA is used to model variability between groups (the categorical predictors) while controlling for 'covariates' (the continuous predictors).

This book does not cover these cases, but they (especially ANOVAs) are widely used in language research, and you may have seen them before. Introductions for linguists include Rietveld and Van Hout (2005), Levshina (2015, chapter 8), and Johnson (2008, chapter 4). Classical statistics training in behavioral sciences emphasizes ANOVAs, as part of an emphasis on data from factorial experiments (where differences between groups or conditions are of primary interest)—so if you have taken a statistics course, chances are it covered ANOVAs. Given that quantitative studies in linguistics take on many forms besides factorial designs, it is useful to just think of ANOVAs as a special case of linear regression (see, e.g., Cohen 1968; Field, Miles, and Field 2012, section 10.2.3; Vasishth and Broe 2011, section 5.4). Once you understand linear regression well, understanding ANOVA analyses is relatively straightforward.

4.2.2 Steps and Assumptions of Regression Analysis

Following Chatterjee and Hadi (2012, section 1.4), regression analyses have these broad steps:

1. Statement of the (scientific) problem
2. Selection of possible variables

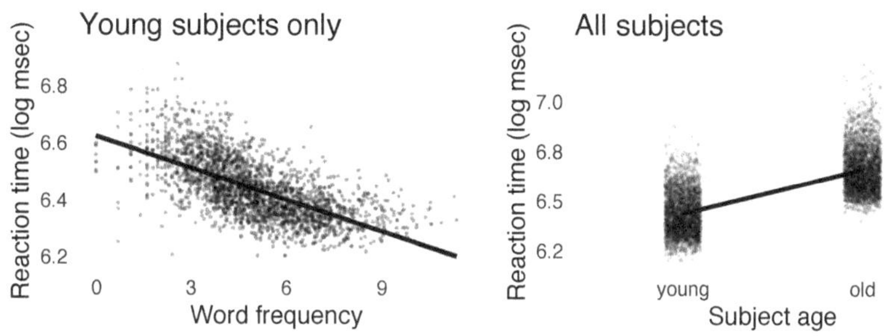

Figure 4.1
Scatterplot of `RTlexdec` as a function of `WrittenFrequency` (left) and `AgeSubject` (right), for the `english` dataset. Lines show least-squares lines of best fit, which for the right plot is the line between group means.

3. Data collection/processing
4. Model specification
5. Model fitting
6. Model criticism and validation
7. Model selection
8. Using the chosen model to address the problem

For our running `english` example (section 4.2), steps (1)–(4) might be:

- Question: How do word frequency and subject age affect lexical decision time?
- Response=`RTLexdec`, predictors=`WrittenFrequency`, `AgeSubject`, any control predictors.
- Load these variables from the `english` dataset; do any processing and exploratory analysis to find outliers or problematic cases to be excluded.
- Linear regression with y=`RTlexdec`, x_1=`WrittenFrequency`, x_2=`AgeSubject`.

In this book we generally assume that steps (1)–(3) are given (you know what question you want to address, with a dataset you already have), and we are usually brief on (8) (which would be the focus of a publication).

In this chapter, we focus on model specification and fitting (steps 4 and 5) for simple linear regression (section 4.3), then multiple linear regression (sections 4.4, 4.5). For now, we are assuming that the assumptions of linear regression are met (6), and which predictors go in the model has been fixed (7); we discuss these steps in chapter 5.

While posing the problem (1) and drawing conclusions about it based on the final model (8) must be done before and after anything else, the order of steps (2)–(7) are not fixed, because **regression analysis is an iterative process**. New models may be built based on model criticism, predictors may be included or excluded due to model selection or finding a mistake in data processing during model validation, and so on.

4.3 Simple Linear Regression

We first consider the simplest application of simple linear regression, to model a relationship between two continuous variables. As an example, consider the relationship between

WrittenFrequency (x) and RTlexdec (y) for the subset of the english data from YOUNG speakers, shown in figure 4.1 (left).

4.3.1 Continuous Predictor

The mathematical form of simple linear regression, written in terms of individual observations, is:

$$y_i = \beta_0 + \beta_1 x_i + \varepsilon_i$$
$$\varepsilon_i \sim N(0, \sigma), \quad \text{for } i = 1, \ldots, n \tag{4.3}$$

In these equations, β_0 and β_1 are *coefficients*. Conventionally β_0 is called the *intercept* and any other coefficient is called a *slope*.

x_i and y_i are the value of the predictor and the response for the ith observation—so the relationship between y and x is just a line, and deviations from this line (the *error*: ε_i) are normally distributed and independent.

This is our first linear model of one (random) variable (y) as a function of another (x). The same model, written in terms of these variables rather than individual observations, is:

$$y = \beta_0 + \beta_1 x + \varepsilon$$
$$\varepsilon \sim N(0, \sigma) \tag{4.4}$$

That is, we write variables in our model (predictors or responses) using Roman letters, and refer to actual values they take on by adding a subscript. (See section 1.3.3 for notation conventions.)

For example, for the english example, the first few observations are:

```
select(english, WrittenFrequency, RTlexdec)
##       WrittenFrequency RTlexdec
## 1                 3.91      6.5
## 2                 4.52      6.4
## 3                 6.51      6.3
## 4                 5.02      6.4
...
```

So $x_2 = 4.5$ and $y_2 = 6.4$ are the values of the predictor (x) and the response (y) for the second observation, and so on.

4.3.2 Parameter Estimation

To get a line of best fit we need to estimate β_0 and β_1, which are population values, using the sample (section 2.2.1). These estimates are written $\hat{\beta}_0$ and $\hat{\beta}_1$. Given these estimates, we can plug into equation (4.3) to get *fitted values* for each observation, written $\hat{y}_i$:

$$\hat{y}_i = \hat{\beta}_0 + \hat{\beta}_1 x_i, \quad \text{for } i = 1, \ldots, n. \tag{4.5}$$

Note that there are no errors in this equation—ε_i are again population values, which we can't observe. Our estimates of the errors, given an estimated line of best fit, are the *residuals*, written $\hat{\varepsilon}_i$:

$$\varepsilon_i = y_i - \hat{y}_i, \quad \text{for } i = 1, \ldots, n.$$

The most common way to estimate coefficient values is using the *least-squares estimates*: the values that minimize the difference between the observed and expected values (their magnitudes: $|\hat{\varepsilon}_i|$), by minimizing the sum of the squared errors (ε_i^2).

Example The `lm()` function fits linear regressions, with the regression formula specified in a particular format (box 4.3). This code fits a simple linear regression for our example (y = reaction time, x = word frequency) for YOUNG speakers in the `english` dataset:

```
english_young <- filter(english, AgeSubject == "young")
## SLR: Simple Linear Regression
slr_mod_1 <- lm(RTlexdec ~ WrittenFrequency, data = english_young)
```

The estimated coefficients for this model are:

```
coefficients(slr_mod_1)
##      (Intercept) WrittenFrequency
##            6.626           -0.037
```

The interpretations of these coefficients are:

- $\hat{\beta}_0$ (6.6): the predicted `RTlexdec` value when `WrittenFrequency` is 0.
- $\hat{\beta}_1$ (–0.037): the predicted change in `RTlexdec` when `WrittenFrequency` is changed by 1 (a *unit change*).

Substituting these values into equation (4.5) gives the regression line. The prediction for observation i is:

$$\text{Predicted } \texttt{RTlexdec}_i = 6.6 - 0.037 \cdot \texttt{WrittenFrequency}_i.$$

Box 4.3
Practical Note: Formula Notation

The first argument of `lm()` is a *formula* in R format: the tilde (~) separates the left- and right-hand sides. For regression models in R, usually the response goes on the left-hand side and predictors on the right-hand side. Formula notation is used idiosyncratically in different types of R functions, but in regression models it makes sense: the formula looks similar to the model written in terms of variables (e.g., equation (4.4)), with the conventions that the intercept and error terms aren't written. The tilde makes sense because a regression model specifies a probability distribution for y. With some abuse of notation, we could write the simple-linear regression formula for our example as

$$\underline{\texttt{RTlexdec} \sim} N(\beta_0 + \beta_1 \underline{\texttt{WrittenFrequency}}, \sigma).$$

The underlined parts make up the model's formula in R format.

R model formulas automatically include an intercept term, which you can also include explicitly as `1`. So another way to write the `slr_mod_1` formula would be:

$$\texttt{RTlexdec} \sim \texttt{1 + WrittenFrequency},$$

which would correspond more closely to the distribution equation:

$$\underline{\texttt{RTlexdec}} \sim N(\beta_0 \cdot \underline{1} + \beta_1 \cdot \underline{\texttt{WrittenFrequency}}, \sigma).$$

4.3.3 Hypothesis Testing

In addition to estimating the regression coefficients, we would like measures of uncertainty and precision in our estimates. The least-squares estimates $\hat{\beta}_0$ and $\hat{\beta}_1$ are random variables that follow a sampling distribution, analogously to the sample mean (section 2.2.2). $\hat{\beta}_0$ and $\hat{\beta}_1$ are normally distributed (Chatterjee and Hadi 2012, section 2.6), and so both have standard deviations, which we approximate by standard errors (SEs) calculated using the sample: $SE(\hat{\beta}_0)$ and $SE(\hat{\beta}_1)$ (box 4.5).

By the same logic as for a sample mean (section 2.4.3), we can then use the estimates and their SEs to carry out (one-sample) t-tests and obtain confidence intervals (CIs).

For example, for β_1, the null hypothesis is of no relationship between x and y ($H_0 : \beta_1 = 0$), and the t-test has $n-2$ degrees of freedom (intuitively, because two estimates are being made using n observations). The resulting p-value tells us how surprised are we to get a slope this far from 0, under H_0.

Because this is a t-test, we can also compute t-based 95% CIs (section 2.3.4) for β_1. Similar logic applies for β_0.

In R, the simplest way to print all this information for a fitted model is to apply the `summary()` function. For our example:

```
summary(slr_mod_1)
##
## Call:
## lm(formula = RTlexdec ~ WrittenFrequency, data = english_young)
##
## Residuals:
##      Min       1Q   Median       3Q      Max
## -0.3466 -0.0552 -0.0055  0.0517  0.3488
##
## Coefficients:
##                   Estimate Std. Error t value Pr(>|t|)
## (Intercept)       6.625556   0.004943  1340.3   <2e-16
## WrittenFrequency -0.037107   0.000924   -40.1   <2e-16
##
## Residual standard error: 0.081 on 2282 degrees of freedom
## Multiple R-squared:  0.414,Adjusted R-squared:  0.414
## F-statistic: 1.61e+03 on 1 and 2282 DF,  p-value: <2e-16
```

The kind of table reported after `Coefficients` is called the *model table* (or "regression table", or "coefficient table").

However `summary()` also gives a lot of other output summarizing the model that we often don't need, such as the quantiles of residuals. It also doesn't give the CIs, which we could obtain using:

```
confint(slr_mod_1)
##                       2.5 % 97.5 %
## (Intercept)           6.616  6.635
## WrittenFrequency -0.039 -0.035
```

For example, the 95% CI for the `WrittenFrequency` slope is [–0.039, –0.035]. Neither CI includes 0, which is consistent with the very low p-values in the model table.

> **Box 4.4**
> **Practical Note: Stargazing**
>
> The default output of `summary()` for a regression model (e.g., running `summary(slr_mod_1)` yourself) shows "significance codes" that are commonly used to indicate whether p is below conventional thresholds: * for $p<0.05$, ** for $p<0.01$, and so on. These codes are not shown in this book (by setting `options(show.signif.stars = FALSE)`) to discourage *stargazing*: interpreting only terms with p-values below a certain threshold as important, and models with more stars as better.

A useful tidyverse function for seeing just information about coefficients, in a format suitable for further processing, is `tidy()` (from the broom package). `tidy()` is very general; it usually works whenever `summary` would. For our example:

```
tidy(slr_mod_1, conf.int = TRUE)
##                  term estimate std.error statistic   p.value conf.low
## 1         (Intercept)    6.626   0.00494      1340  0.0e+00    6.616
## 2 WrittenFrequency   -0.037   0.00092       -40 3.9e-267   -0.039
##     conf.high
## 1       6.635
## 2      -0.035
```

> **Box 4.5**
> **Broader Context: Simple Linear Regression—Math Details and Intuitions**
>
> You don't need to know the formulas used for coefficient estimation and hypothesis testing, for any regression (e.g., chapter 2). But the mathematical details can be useful for developing intuitions, which apply to more complex models, where we can't easily write down formulas—these are italicized in this box.
>
> The least-squares estimates of β_0 and β_1 for simple linear regression can be written as:
>
> $$\hat{\beta}_1 = r_{xy}\frac{\hat{\sigma}_x}{\hat{\sigma}_y}, \quad \hat{\beta}_0 = \bar{y} - \hat{\beta}_1\bar{x},$$
>
> where $\bar{x}$ is the average of the x_i, $\bar{y}$ is the average of the y_i, r_{xy} is the (Pearson's) correlation of x and y (equation (3.2)), and $\hat{\sigma}_x$ and $\hat{\sigma}_y$ are estimates of the SDs of x and y. (Equation (2.4), but with n instead of $n-1$ in the denominator.)
>
> From the equation for $\hat{\beta}_0$, we see that the line of best fit passes through the "average value" of the data, $(\bar{x}, \bar{y})$. One consequence is that the *least-squares estimates are easily thrown off by outliers*—observations which are very far from $\bar{x}$ or $\bar{y}$.
>
> The equation for $\hat{\beta}_1$ shows that *the slope of the regression line is essentially the correlation* between x and y (which lies between -1 and 1)—and is adjusted for the scales of x and y.

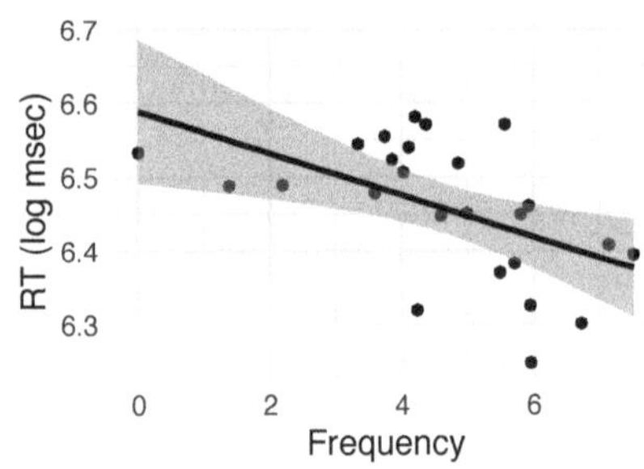

Figure 4.2
Reaction time versus word frequency (as in figure 4.1 left), for a subset of 25 points from young speakers in the `english` dataset. Line and shading are least-squares line of best fit and 95% CI.

The SEs for both estimates are related to σ, the degree of variability in the data, for which an unbiased estimator is

$$\hat{\sigma} = \sqrt{\frac{\sum_{i=1}^{n} \hat{\varepsilon}_i^2}{n-2}}. \tag{4.6}$$

The numerator sums the error across observations, and $n-2$ is the degrees of freedom. So $\hat{\sigma}^2$ measures "average error" (a.k.a. *mean-squared error*), and $\hat{\sigma}$ can be thought of as a measure of error that is "leftover": the *residual standard error*, which is reported by `summary()` for a linear regression model.

The SEs of the coefficients turn out to be:

$$SE(\hat{\beta}_0) = \frac{\hat{\sigma}}{\sqrt{n}} \sqrt{1 + \left(\frac{\bar{x}}{\hat{\sigma}_x}\right)^2}, \quad SE(\hat{\beta}_1) = \frac{\hat{\sigma}}{\sqrt{n}} \frac{1}{\hat{\sigma}_x}.$$

We can make a few observations—focusing on β_1 (we usually don't care about the intercept):

- Uncertainty in coefficient estimates increases with the noisiness of the data ($\hat{\sigma}$).
- Uncertainty decreases as the square root of sample size increases (box 2.4).
- Uncertainty in the effect of a predictor x (here, $\hat{\beta}_1$) decreases the wider the range of x in the sample.

So *to get more precise results about how x affects y, we can collect more data, less noisy data, or data from a wider range of x.* "More precise results" directly translates into lower p-values/narrower CIs.

Although the equations for multiple linear regression (and beyond) get more complicated, the basic (italicized) points always hold.

4.3.4 Example: Small Subset

It is a little silly to compute p-values and 95% CIs for the full `english` dataset for young speakers, given how much data there are—the line of best fit has a tiny CI. Let's look at a line of best fit for a subset of just $n = 25$ points (figure 4.2), corresponding to this simple linear regression:

```
set.seed(2903)
young_sample <- english_young %>% sample_n(25)
```

The shading around the line is the 95% confidence interval, and "can we reject H_0?" is equivalent to asking, "can a line with 0 slope cross the shaded area through the range of x

(and going through $(\bar{x}, \bar{y})$)?" Although the exact line is uncertain, we can be confident that its slope is negative, as reflected by the p-value and 95% CI for `WrittenFrequency` in the coefficient table:

```
tidy(slr_mod_2, conf.int = TRUE) %>%
  select(term, statistic, p.value, conf.low, conf.high)
##                term statistic p.value conf.low conf.high
## 1       (Intercept)       140 3.1e-35    6.492    6.6864
## 2 WrittenFrequency         -3 7.0e-03   -0.048   -0.0085
```

4.3.5 Goodness of Fit

Figures 4.1 (left) and 4.2 plot the lines of best fit from the models we have discussed, on top of empirical data. We often want a metric quantifying how well a model fits the data—the *goodness of fit*.

For simple linear regression, we can derive such a metric as follows (e.g., Chatterjee and Hadi 2012, section 2.9). For each observation, the amount of variation (deviation from the grand mean) can be partitioned into pieces explained and not explained by the model:

$$\underbrace{(y_i - \bar{y})}_{\text{deviation from mean}} = \underbrace{(\hat{y}_i - \bar{y})}_{\text{deviation explained by model}} + \underbrace{(y_i - \hat{y}_i)}_{\text{remaining deviation ('residual')}} . \tag{4.7}$$

Summing each piece across all observations, we define three quantities:

$$TSS = \sum_{i=1}^{n} (y_i - \bar{y})^2 \qquad \textit{total sum of squares} \tag{4.8}$$

$$ESS = \sum_{i=1}^{n} (\hat{y}_i - \bar{y})^2 \qquad \textit{explained sum of squares} \tag{4.9}$$

$$RSS = \sum_{i=1}^{n} (y_i - \hat{y}_i)^2 \qquad \textit{residual sum of squares} \tag{4.10}$$

It is a remarkable property of linear regression that the same relationship that holds for individual observations (equation (4.7)) holds across the whole sample:

$$TSS = ESS + RSS.$$

This relationship can be used to give a measure of goodness of fit: how much of the total variance in the data (TSS) is explained by the model (ESS):

$$R^2 = \frac{ESS}{TSS}. \tag{4.11}$$

For simple linear regression, R^2 is (a) the *proportion of total variability in y explained* by x. R^2 lies between 0 (no variance explained) and 1 (all variance explained). It turns out R^2 is also the square of: (b) Pearson's *correlation between predictor and response* (r_{xy}^2; see equation (3.2)); and (c) The *correlation between observed and fitted values* ($\text{cor}(y, \hat{y})$).

Interpretations (a)–(c) are useful in different settings, and hold across all linear regression models. R^2 is printed at the bottom of the `summary()` output for a linear regression, or it can be extracted directly:

```
summary(slr_mod_2)$r.squared
## [1] 0.28
```

4.3.6 Categorical Predictor

Simple linear regression easily extends to the case of a binary predictor x (a two-level factor).

For example, we can model the relationship between `AgeSubject` (x) and `RTlexdec` (y) for the `english` data, shown in figure 4.1 (right). We now have

- $x_i = 0$: if `AgeSubject` = YOUNG
- $x_i = 1$: if `AgeSubject` = OLD

Everything else is the same as for the case where x is continuous (section 4.3.1), including the regression equation for observation i (equation (4.3)):

$$y_i = \beta_0 + \beta_1 x_i + \varepsilon_i.$$

Only the interpretation of the coefficients differs:

- β_0: mean `RTlexdec` when `AgeSubject`=YOUNG (because $x_i = 0$)
- β_1: difference in means of `RTlexdec` between OLD and YOUNG

Hypothesis tests, p-values, CIs, and goodness of fit work the same as for a continuous predictor (section 4.3.3).

We would fit this model as follows in R:

```
slr_mod_3 <- lm(RTlexdec ~ AgeSubject, data = english)
tidy(slr_mod_3)
##              term estimate std.error statistic p.value
## 1     (Intercept)     6.44    0.0023      2771       0
## 2 AgeSubjectold       0.22    0.0033        67       0
```

The model suggests that YOUNG speakers have significantly longer RT than OLD speakers do: $\hat{\beta} = 0.222$, $SE(\hat{\beta}) = 0.0033$, $t(4566) = 67.5$, $p < 0.001$.

Box 4.6
Broader Context: Simple Linear Regression Equivalences

Conceptually, simple linear regression with a binary predictor does the same thing as a two-sample t-test: tests the difference between two groups in the value of a continuous variable. In fact, the two are equivalent, as we can see by carrying out the equivalent t-test to model `slr_mod_3`:

```
t.test(RTlexdec ~ AgeSubject, data = english, var.equal = TRUE)
...
## data:  RTlexdec by AgeSubject
## t = -67, df = 4566, p-value <2e-16
## alternative hypothesis: true difference in means between
## group young and group old is not equal to 0
## 95 percent confidence interval:
##  -0.23 -0.22
...
```

(The `var.equal` option forces the t-test to assume equal variances in both groups, which is an assumption of linear regression we'll return to in section 5.3.3.)

The results (t, p, df, CIs) are identical to the regression results. (Except that t/CIs have reversed sign, because `t.test()` ignores our releveling of the AGESUBJECT factor, which doesn't matter for this point.) So, a t-test can be thought of as a special case of simple linear regression.

Similarly, the hypothesis test of interest for simple linear regression with a continuous predictor (for the slope) does conceptually the same thing as Pearson's correlation: it tests whether there is an association between x and y. The two are almost equivalent, as we can see by carrying out a hypothesis test for Pearson's r for the data used for the section 4.3.4 example, testing whether it is significantly different from 0:

```
cor.test(young_sample$WrittenFrequency, young_sample$RTlexdec)
...
## data:  young_sample$WrittenFrequency and young_sample$RTlexdec
## t = -3, df = 23, p-value = 0.007
## alternative hypothesis: true correlation is not equal to 0
## 95 percent confidence interval:
##  -0.76 -0.16
## sample estimates:
##   cor
## -0.53
```

t, p, and r are exactly the same as for `slr_mod_2` (shown in sections 4.3.4 and 4.3.5)—so a correlation test and a simple linear regression are equivalent ways of testing for an association between (continuous) x and y. The correlation test can be thought of as a subset of the regression model: the regression gives the same information, but it also lets us predict y as a function of x.

4.4 Multiple Linear Regression

In *multiple linear regression*, we use a linear model to predict a continuous response with k predictors $k > 1$. Analogously to equation (4.4) for simple linear regression:

$$y = \beta_0 + \beta_1 x_1 + \beta_2 x_2 + \cdots + \beta_k x_k + \varepsilon.$$
$$\varepsilon \sim N(0, \sigma). \tag{4.12}$$

Each predictor x_i can be continuous or categorical, and there are now k coefficients to be estimated. Least-squares estimation generalizes from single to multiple linear regression (e.g., Chatterjee and Hadi 2012, chapter 3), giving

- Estimates of the coefficients ($\hat{\beta}_1, \ldots, \hat{\beta}_k$);
- SEs, p-values, and 95% CIs for each coefficient (using t-tests with $n-k-1$ degrees of freedom).

4.4.1 Example

For the `english` data, let's model (lexical decision) RT as a function of word frequency and participant age. In this model, the response y is `RTlexdec`. The two predictors are `WrittenFrequency` (x_1) and `AgeSubject` (x_2: OLD = 0, YOUNG = 1).

Because there are two predictors ($k=2$), the regression equation for observation i is:

$$y_i = \beta_0 + \beta_1 x_{i1} + \beta_2 x_{i2} + \varepsilon_i, \tag{4.13}$$

where x_{ij} means "the jth predictor for the ith observation."

To fit the model we again use `lm()`:

```
mlr_mod_1 <- lm(RTlexdec ~ WrittenFrequency + AgeSubject,
  data = english)
```

We can summarize its results using `summary()` and `confint()` (or just `tidy()`):

```
summary(mlr_mod_1)
...
##
## Coefficients:
##                   Estimate Std. Error t value Pr(>|t|)
## (Intercept)       6.625071   0.003979  1664.9   <2e-16
## WrittenFrequency -0.037010   0.000703   -52.6   <2e-16
## AgeSubjectold     0.221721   0.002593    85.5   <2e-16
##
## Residual standard error: 0.088 on 4565 degrees of freedom
## Multiple R-squared:  0.688,Adjusted R-squared:  0.688
## F-statistic: 5.04e+03 on 2 and 4565 DF,  p-value: <2e-16
```

```
confint(mlr_mod_1)
##                   2.5 % 97.5 %
## (Intercept)       6.617  6.633
## WrittenFrequency -0.038 -0.036
## AgeSubjectold     0.217  0.227
```

This model tells us that the least-squares solution for the regression line is

$$\texttt{RTlexdec} = \underbrace{6.6}_{\hat{\beta}_0} + \underbrace{-0.037}_{\hat{\beta}_1} \cdot \texttt{WrittenFrequency} + \underbrace{0.22}_{\hat{\beta}_2} \cdot \texttt{AgeSubject}.$$

For example, for observation 3, where `WrittenFrequency` = 6.5 and `AgeSubject` = YOUNG, the model predicts:

$$\begin{aligned}\texttt{RTlexdec} &= 6.6 + (-0.037 \cdot 6.5) + (0.22 \cdot 0) \\ &= 6.4.\end{aligned}$$

You can make such model predictions automatically using the `predict()` function, which works for different kinds of regression models. For example, the predicted value for the third observation is

```
predict(mlr_mod_1, english[3, ])
##   3
## 6.4
```

Or we can predict for new observations:

```
## Four new observations:
expand.grid(AgeSubject = c("young", "old"),
  WrittenFrequency = c(5, 6)) %>%
## Add model predictions for each observation
  mutate(prediction = predict(mlr_mod_1, newdata = .))
##   AgeSubject WrittenFrequency prediction
## 1      young                5        6.4
## 2        old                5        6.7
## 3      young                6        6.4
## 4        old                6        6.6
```

`expand.grid()`, or analogous functions in tidyr (see `?expand`), makes a dataframe with all combinations of listed variables.

4.4.2 Interpretation of Coefficients

In a multiple linear regression, the interpretation of the coefficients is

- Intercept (β_0): predicted value of y when all predictors are held at 0 ($x_1 = 0, x_2 = 0, \ldots$).
- Slope k (e.g., β_1): predicted change in y for a unit change in the kth predictor, when other predictors are held constant.

In the previous example, YOUNG speakers are predicted to have RTs 0.22 faster than OLD speakers, when word frequency is held constant.

4.4.3 Goodness of Fit Metrics and Significance

The *residual standard error* (RSE) is one measure of goodness of fit: it is denoted $\hat{\sigma}$ (equation (4.6)), the model's estimate of the variance of the error term, so it measures how much variance in y is leftover in the units of the predictor. RSE is an "unstandardized effect size" (section 3.2.2), so its advantage is interpretability in terms of the actual predictor value (which is useful if the units are meaningful), and the disadvantage is that interpreting it requires understanding what values on the scale of the predictor mean.

The *multiple* R^2 is just R^2 defined exactly as for simple linear regression. The formulas for sums of squares (equations (4.8–4.10)) work regardless of the number of predictors, so we can again use this equation to quantify the "proportion of variance explained" in y by all predictors together:

$$R^2 = \frac{ESS}{TSS}.$$

Multiple R^2 is most widely used goodness-of-fit metric and has the advantages of a *standardized effect size* (section 3.2.2): it doesn't depend on the problem domain and has a straightforward interpretation. It has the important disadvantage that it always increases as more predictors are added to a model. Adding predictors increases "explained variance" (ESS), even when a predictor in reality has no effect on the response ($\beta_i = 0$).

This issue is addressed by an alternative to R^2, *adjusted* R^2, which is defined as

$$R_a^2 = 1 - \frac{ESS/(n-p-1)}{TSS/(n-1)}.$$

This expression is a ratio comparing the amount of variance explained by two models: the "full" model (with p predictors) and the "baseline" model (with just the intercept). Each model's sum of squares is scaled by its degrees of freedom, which is a measure of how much variance is explained given the number of predictors. This property makes adjusted R^2 more appropriate than R^2 for comparing different models (section 5.9). The disadvantage of adjusted R^2 is that it no longer has the nice interpretation as the "percentage of variation explained."

The `summary()` output for a linear regression reports all three of these measures (as in the section 4.4.1 example): RSE, multiple R^2, and adjusted R^2.

Example Consider a simple linear regression of the effect of `voicing` on vowel duration in the `neutralization` data used in the last chapter. To fit and summarize this model (output not shown):

```
mlr_mod_2 <- lm(vowel_dur ~ voicing, data = neutralization)
summary(mlr_mod_2)
```

We know that the `voicing` effect is small (same as the t-test result in section 3.2.2: 9.1 msec, Cohen's $d = 0.21$)—the research question underlying the study is whether there is an effect at all ("incomplete neutralization")! But the goodness-of-fit measures give us some extra ways to think about this, besides the effect's size and p-value. Let's extract these measures using another handy tidyverse function, `glance()` (from the broom package), which extracts model summary numbers—all of which we have seen, or will see in later chapters:

```
glance(mlr_mod_2)
##   r.squared adj.r.squared sigma statistic p.value df logLik  AIC
## 1     0.011        0.0094    43       8.1  0.0045  1  -3885 7775
##    BIC deviance df.residual nobs
## 1 7789  1402790         747  749
```

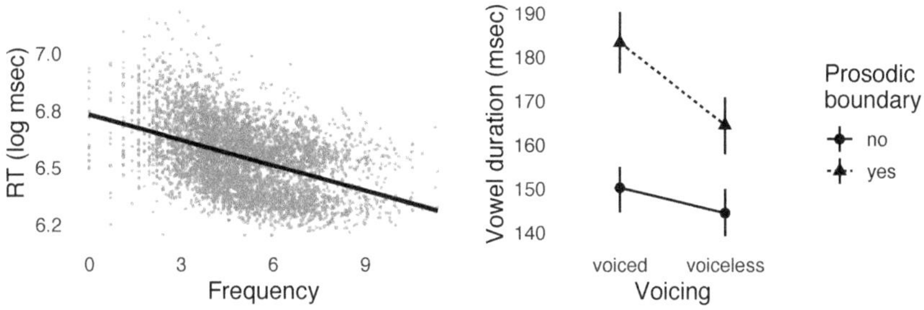

Figure 4.3
Empirical summaries of effects of two predictors on a response. *Left* (`english` data): Line of best fit for word frequency effect on reaction time, for each subject age group. (95% CIs are included, but are invisible due to high sample size.) *Right* (`neutralization` data): Mean and 95% bootstrapped CI for each combination of consonant voicing and prosodic boundary presence.

Focusing on the first three columns:

- Multiple R^2 (`r.squared`): `voicing` explains just 1% of variance in vowel duration
- Adjusted R^2 (`adj.r.squared`): `voicing` is only expected to improve *prediction* of vowel duration (for new observations) by 0.9%, relative to a model where we just always predict the mean value.
- RSE (`sigma`): this value is large in practical terms, for this kind of data (section 3.2.3.2). It is also almost the same as the SD of the original variable (`vowel_dur`: $s = 44$), indicating that little variance is explained.

4.5 Interactions

The multiple regression models considered so far assume that each predictor affects the response independently. For example, the multiple regression example (`mlr_mod_1`) assumes that the slope of the frequency effect on RT is the same for OLD and YOUNG subjects. This looks like it might be approximately true, in that both groups of speakers show a similarly negative slope (figure 4.3, left).

Figure 4.3 (right) shows an example from the `neutralization` data where the independence assumption isn't necessarily true: vowel duration (y: `vowel_dur`) as a function of consonant voicing (x_1: `voicing`—of primary interest) and presence of a prosodic boundary (x_2: `prosodic_boundary`).

Just from this empirical plot, it looks like the effect of voicing on vowel duration may be higher for words with a following prosodic boundary. In this case, we say there is an *interaction* between voicing and prosodic boundary: the value of one predictor modulates the effect of the other. This interaction is modeled by adding an extra coefficient to regression equation (4.13), for the product of x_1 and x_2:

$$y = \beta_0 + \beta_1 x_1 + \beta_2 x_2 + \beta_3 x_1 x_2 + \varepsilon$$

(where $\varepsilon \sim N(0, \sigma)$). In a model with interaction terms, the noninteraction terms are called *main effects*.

Note that the middle terms can be rewritten as

$$\beta_1 x_1 + \beta_2 x_2 + \beta_3 x_1 x_2 = \beta_2 x_2 + \underline{(\beta_1 + \beta_3 x_2) x_1}.$$

The underlined term shows that the interaction coefficient (β_3) modulates the slope of x_1: depending on the value of x_2, x_1 has a different effect on y. Equivalently, the interaction coefficient can be interpreted as how much the effect of x_2 changes depending on x_1:

$$\beta_1 x_1 + \beta_2 x_2 + \beta_3 x_1 x_2 = \beta_1 x_1 + \underline{(\beta_2 + \beta_3 x_1) x_2}.$$

Whichever interpretation makes more sense for one's research questions can be chosen. For the `neutralization` example, because `voicing` (x_1) is of primary interest, it makes sense to think of the interaction as asking "how does prosodic position modulate the effect of voicing?"

4.5.1 Example

Let's fit the model previously described. x_1 and x_2 are coded as:

- `voicing`: VOICED = 0, VOICELESS = 1
- `prosodic_boundary`: NO = 0, YES = 1

```
mlr_mod_3 <- lm(vowel_dur ~ prosodic_boundary + voicing +
    voicing:prosodic_boundary, data = neutralization)
```

In an R model formula, `x1:x2` means "interaction between `x1` and `x2`." The notation `x1*x2`, which expands automatically to `x1 + x2 + x:x2`, means "this interaction and all lower-order terms" (box 4.7). So the model could also be written

```
lm(vowel_dur ~ voicing * prosodic_boundary, data = neutralization)
```

Of interest is the `prosodic_boundaryyes:voicingvoiceless` row in the model table:

```
tidy(mlr_mod_3) %>% select(term, estimate, p.value)
##                                    term estimate  p.value
## 1                           (Intercept)    150.3 1.1e-280
## 2                     prosodic_boundaryyes     33.1  3.5e-12
## 3                      voicingvoiceless     -5.7  1.2e-01
## 4 prosodic_boundaryyes:voicingvoiceless    -13.0  4.7e-02
```

The negative value of the interaction coefficient means "the difference between $x_1 = 1$ and $x_1 = 0$ (here, VOICELESS-VOICED) is smaller when $x_2 = 1$ (here, YES prosodic boundary)." Put in more intuitive terms, the difference between VOICED and VOICELESS stops is predicted to be 13 msec larger when there is a following prosodic boundary; this difference is statistically significant at the $\alpha = 0.05$ level ($p = 0.047$).

Just for exposition, let's work out the model's prediction for an observation with `prosodic_boundary` = NO and `voicing` = VOICELESS:

$$\begin{aligned}\hat{y} &= \beta_0 + \beta_1 x_1 + \beta_2 x_2 + \beta_3 x1 x_2 \\ &= 150 + -5.7 \cdot 1 + 33 \cdot 0 + -13 \cdot 0 \cdot 1 \\ &= 145 \text{ msec.}\end{aligned}$$

To make such predictions automatically we would use `predict` (Exercise 4.5).

Box 4.7
Broader Context: Lower-Order Terms

R automatically adds lower-order terms (`x1`, `x2`) when an interaction (`x1*x2`) is added to a model because an interaction term does not typically make sense otherwise (Baguley 2012, section 14.3.1). With the lower-order terms, `x1:x2` has a "moderator" interpretation: How does `x1` change the effect of `x2`? Without one or both lower-order terms, `x1:x2` doesn't have a straightforward interpretation. By the same logic, including a three-way `x1*x2*x3` interaction in a model entails including all lower-order interactions as well (`x1:x2`, `x1:x3`, `x2:x3`, `x1`, `x2`, `x3`), and so on. With these terms included, the `x1:x2:x3` coefficient has a clear (if complex) interpretation: how does `x3` change the degree by which `x2` modifies the effect of `x1`? The number of lower-order terms is one reason why four-way or higher interactions are rare in practice, and it's best to interpret three-way interactions using additional tools (introduced in section 7.4) rather than examining coefficients.

This logic is sometimes called the *hierarchy principle*: including an interaction term in a model entails including all the lower-order terms.

4.5.2 Visualizing Interactions

Interpreting interaction terms directly from a model table is possible but confusing. Making plots is indispensable for interpreting interactions. And more generally, visualizing the predictions made by a model is often the easiest way to understand it.

4.5.2.1 Option 1: Plotting model predictions

Visualizing a model's predictions is a topic we'll return to several times in this book; each time demonstrating different methods. Good packages exist for many common cases of making model predictions—including the effects and sjPlot/ggeffects packages (Fox and Weisberg 2019; Lüdecke 2021, 2018) as of this writing—so this will often use such packages rather than making model predictions from scratch.[1]

Two useful kinds of model prediction plots are

- *Partial effect plots*: these show the model's prediction as one predictor is changed, holding others constant.
- *Interaction plots*: these show the model's predictions as two or more predictors are changed.

1. It is also useful to know how to make predictions yourself from scratch, as preexisting packages make various assumptions and don't always give exactly what you want, but we'll return to this later (e.g., sections 4.4.1, 8.3.4).

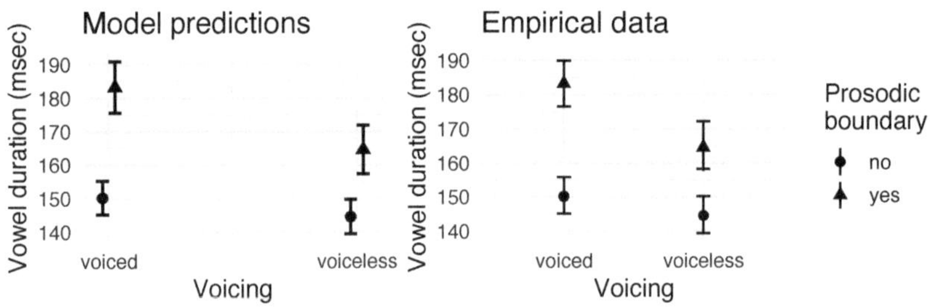

Figure 4.4
Two plots for interpretation of the `voicing:prosodic_boundary` interaction. *Left*: Model predictions with 95% CIs for model `mlr_mod_3`. *Right*: Empirical means and 95% CIs.

For example, we can use `plot_model()` from sjPlot package to make partial effect plots for the two main effects from model `mlr_mod_3` of the `neutralization` data (plots not shown):

```
plot_model(mlr_mod_3, type = "pred", terms = c("voicing"))
plot_model(mlr_mod_3, type = "pred", terms = c("prosodic_boundary"))
```

The same function can be used to plot the predicted `voicing:prosodic_boundary` interaction (figure 4.4, left):

```
plot_model(mlr_mod_3,
  type = "pred",
  terms = c("voicing", "prosodic_boundary"),
  title = "Model predictions"
)
```

4.5.2.2 Option 2: Empirical summaries

Another way to visualize an interaction between variables is with relevant plots of the empirical data, summarizing the data's distribution as two or more predictors vary. For example, for the `voicing:prosodic_boundary` interaction, we could make the plot shown in figure 4.4 (right), showing the means and 95% CIs for each combination of voicing and prosodic boundary in the `neutralization data`. (Remember that this book does not show ggplot2 code in text, but all plotting code is in the code file for this chapter.)

It is very useful to know how to make such empirical plots for for understanding data generally. But for current purposes, what's important is that this kind of empirical interaction plot often gives a reasonable approximation of the model's predictions. In the examples in figure 4.4, they are nearly identical. Although you should take empirical plots less seriously than model prediction plots (box 4.8), empirical plots are often easier to make, and serve as useful sanity checks.

Box 4.8
Practical Note: Model Predictions or Empirical Plots?

When interpreting a regression model or writing it up for publication, is it better to show model predictions or empirical plots?

At first glance the answer seems obvious: model prediction plots are clearly better for summarizing your model, because empirical plots don't show the actual results of the model. In particular, empirical plots don't do the main thing we are fitting a statistical model for: controlling for other predictors. It is very easy to make empirical plots that suggest an effect is present, when it is in fact not significantly different from null when even the simplest statistical analysis is applied.

The counter-argument appeals to common sense: empirical plots serve as a sanity check. They are more intuitive, and we would like to see a model's predictions "come out" in empirical data to have confidence that something isn't wrong (e.g., a data processing bug, using the wrong model). If we make the appropriate empirical plot and see qualitative disagreement with model predictions, it is very often a sign that something is wrong. Especially as you fit more complex models, it is easy to get a significant "result" that is in fact an artifact of something about the data (e.g., a few influential points resulting from annotation error) that would emerge in careful exploratory plots.

My personal opinion is that model prediction plots are 'better' (similarly to Gries 2021, section 5.2.6.3), but making and cross-checking both kinds of plots are important when writing up results—whichever one goes in the actual paper. (It is also fine to include both kinds of plots, if you have space.)

4.5.2.3 Why the difference matters

In general, although examining empirical plots is good practice, model prediction plots are necessary to actually interpret your model, and the two kinds of plots can lead to different conclusions. This is an important point because it is common to advocate "just plotting the data" as primary over statistical modeling.

As a simple example where the distinction matters, suppose we add a single term to the previous model: `accent_type`, which is 0 if the word has prenuclear accent and 1 if it has nuclear accent. It would make sense to control for this variable, because it is highly correlated with `prosodic_boundary` and the two are conceptually related. To fit the model:

```
## Exclude very infrequent accent_type (n=11) for this example
neut_sub <- filter(neutralization, accent_type != "deaccented")

## Relevel factor so that accented (nuclear) > unaccented (prenuclear)
neut_sub <- mutate(neut_sub, accent_type = relevel(accent_type,
  "prenuclear"))
mlr_bad_mod_1 <- lm(vowel_dur ~ prosodic_boundary * voicing +
    accent_type, data = neut_sub)
```

From the coefficient table alone, it looks like we would make a similar conclusion about the interaction as in the previous model:

```
mlr_bad_mod_1 %>%
  tidy(conf.int = TRUE) %>%
  select(term, estimate, p.value)
##                                  term estimate  p.value
## 1                         (Intercept)      150 7.1e-271
```

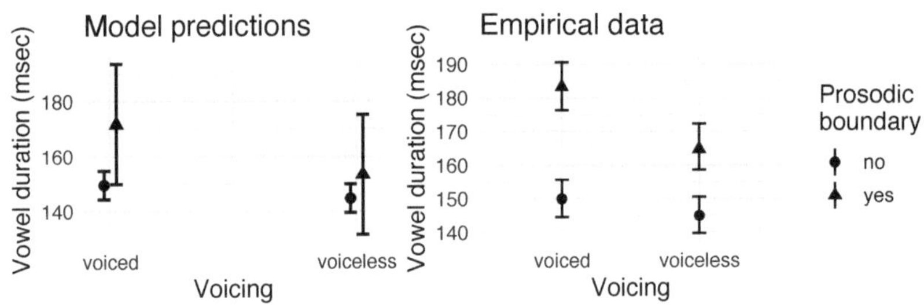

Figure 4.5
Two plots for interpretation of the `voicing:prosodic_boundary` interaction in model `mlr_bad_mod_1`, which includes `accent_type`. *Left*: Model predictions with 95% CIs. *Right*: Empirical means and 95% CIs.

```
## 2                    prosodic_boundaryyes       22  4.7e-02
## 3                        voicingvoiceless       -5  1.8e-01
## 4                      accent_typenuclear       12  2.7e-01
## 5 prosodic_boundaryyes:voicingvoiceless      -14  3.9e-02
```

The same is true if we just looked at the empirical plot corresponding to this interaction (figure 4.5, right). However, making the model prediction plot for the interaction (figure 4.5, left) shows a very different picture from the previous model—there seems to be huge uncertainty about predicted vowel duration for observations that aren't before a prosodic boundary. And just based on the errorbars, it seems odd that the `voicing:prosodic_boundary` interaction is statistically significant. These oddities should spur us to investigate further. What is going here results from very high correlation between `accent_type` and `prosodic_boundary`, and points to a real issue with this data that our model didn't take into account. (It's not possible in this data to disentangle the effects of accent type and prosodic boundary in these data.)

The point is that if we examined the empirical summary alone, we would have missed important information from the fitted model, which is crucial for our research question.

4.5.3 Releveling Factors for Model Interpretability

R automatically puts the levels of factors into alphabetical order when loading data. It is often useful to change this ordering, so that the resulting regression coefficients have interpretations that are more intuitive—like changing `AgeSubject` so YOUNG comes before OLD (section 4.1.2)—or correspond better to research questions the data are being used to address.

For example, for the `neutralization` example modeling the interaction between `voicing` and `prosodic_boundary` (section 4.5.1), by default the `voicing` factor orders VOICED before VOICELESS (alphabetical). Thus

- The `voicing` coefficient means "VOICELESS–VOICED."
- The `voicing:prosodic_boundary` coefficient means "change in (VOICELESS – VOICED) depending on presence of a prosodic boundary."

(The lowest level is the *base level*, here VOICED.)

But we expect the response (vowel_dur) to be higher for VOICED than VOICELESS, and the VOICED–VOICELESS difference (the amount of "incomplete neutralization": IN) is of primary interest for these data. So it makes more sense for the base level to be VOICELESS. The interpretations of the two coefficients would now be

- Amount of IN
- How much IN increases due to the presence of a prosodic boundary

We would expect both coefficients to be positive.

We can reorder factor levels using relevel() and refit model mlr_mod_3 from section 4.5.1 using update(), which is a useful function for refitting a model with only certain pieces changed:

```
neutralization <- neutralization %>%
  mutate(voicing = relevel(voicing, "voiceless"))

# Refit model using new 'neutralization' dataframe
# . ~ . means "same model formula"
mlr_mod_4 <- update(mlr_mod_3, . ~ ., data = neutralization)
```

The coefficient estimates are now:

```
tidy(mlr_mod_4) %>% select(term, estimate)
##                                  term estimate
## 1                         (Intercept)    144.6
## 2                 prosodic_boundaryyes     20.1
## 3                       voicingvoiced      5.7
## 4 prosodic_boundaryyes:voicingvoiced     13.0
```

Comparing to the coefficients of mlr_mod_3, the magnitudes of the voicing and voicing:prosodic_boundary coefficients are the same, but are now positive instead of negative, as expected.

4.6 Reporting a Linear Regression Model

There are two parts to reporting a multiple linear regression model in a write-up: individual coefficients, and the whole model. The general principles are similar to reporting hypothesis tests and parameter estimates (section 2.7).

4.6.1 Coefficients

Each regression coefficient corresponds to a parameter estimate and a hypothesis test, so you minimally report sufficient information to interpret these. The most common minimums are the following:

- (a) Regression *coefficient estimate* $\hat{\beta}$.
- (b) The coefficient's *standard error* $SE(\hat{\beta})$.
- (c) *Test statistic* value, and
- (d) corresponding *p-value*.

Note that $\hat{\beta}$ is an effect size/parameter estimate and $SE(\hat{\beta})$ is its corresponding measure of variability (items (4) and (6) from section 2.7). The *df* for each coefficient's *t*-test is the same $(n-p-1)$, so it is usually just reported once or not at all (it can be inferred from the sample size and number of terms in the regression).

It is strongly recommended to report *all* coefficients for a multiple regression model, in a table. A full table (which could be in an appendix or an online repository) is needed for your write-up to meet the APA manual standard of "sufficient information to help the reader fully understand the analyses conducted" (quoted in section 2.7). This table could be just the output of `summary` or `tidy`, but formatted more nicely (example 1 below). Individual coefficients can be highlighted in the text (example 2 below).

Besides (a)–(d), it is also nice to give the following (by the same logic as in section 2.7):

- (e) *Visualizations*—especially for interactions and effects of primary interest—either using model prediction or empirical plots.
- (f) *Confidence intervals* for the coefficients.
- (g) Basic *descriptive statistics* (e.g., mean and SD of y for each level of factor x; correlation of y with continuous x).

It is neither possible nor effective to report all of (a)–(g) for every effect in a paper, or even effects of primary interest. In a very short paper you may not even be able to report more than (a)–(d) for the effects that directly address your research questions. Exactly how you report regression models is a matter of writing and context, determined both by what is important to convey given the details of your study, and by personal style. For example, if the important effects for your research questions are all interactions, you probably need visualizations to unpack these effects; if only main effects are of interest, you may not need visualizations. If the null hypothesis is obviously wrong for your effect of interest, and/or you are following recommendations to deemphasize statistical significance in statistical reporting, you could report 95% CIs but not *p*-values.

As with other kinds of writing, a good way to develop your voice is using relevant papers as models and seeking guidance from colleagues or written sources that you trust.

4.6.2 Model Summaries

It is common to report some quantitative summaries of the model: its goodness of fit (R^2, adjusted-R^2, $\hat{\sigma}$, and associated *df*; see section 4.4.3), measures of model likelihood or information (*F*-test, AIC, BIC, log-likelihood, deviance; see section 5.9), or number of parameters (p) and observations ($n = df + p$). Some of these are reported by `summary()`, and most are reported by `glance()`.

Whichever summaries are used, they are reported at the bottom of the regression table. There is no standard set to report. In linguistics (including my own work), sometimes no summaries are reported, which probably isn't a good option. At a minimum we should probably report R^2 and $\hat{\sigma}$ and its residual *df* (as complementary measures of goodness of fit), as well as *n* (which can also be given in the text), as in Example 1.[2] As for individual coefficients, different summaries make sense depending on your study and the analysis being carried out.

Example 1 A possible regression table for the model in section 4.5.1:

Coefficient	$\hat{\beta}$	$SE(\hat{\beta})$	t	p
Intercept	150.25	2.57	58.49	<0.001
Prosodic boundary	33.06	4.67	7.07	<0.001
Voicing	–5.69	3.67	–1.55	0.12
PB:Voicing	–12.97	6.53	–1.99	0.047

$R^2 = 0.095$, Residual SE = 42 ($df = 745$), $n = 748$.

Example 2 Possible in-text reports of coefficients for the model in section 4.4.1:

Words with higher frequencies had significantly decreased reaction time ($\hat{\beta} = -0.037$, $SE(\hat{\beta}) = 0.0007$, $t(4565) = -53$, $p < 0.001$). Old speakers (mean RT: 6.7) had significantly longer reaction times than young speakers (mean RT: 6.4): $\hat{\beta} = 0.22$, $SE(\hat{\beta}) = 0.0026$, $t(4565) = 86$, $p < 0.001$.

Simple linear regression Reporting a simple regression model is a special case of reporting a multiple regression model where there is only one coefficient. In this case a regression table may be overkill and just reporting in text can make sense, especially if space is limited.

4.7 Further Reading

The topics in this chapter are covered in introductions to regression modeling, including Faraway (2015, chapters 1–3), Chatterjee and Hadi (2012, chapters 1–3), Gelman and Hill (2007, chapter 3), or to general statistics (e.g., Navarro 2016, chapter 15; Field, Miles, and Field 2012, chapter 7). For linguistic data specifically, Winter (2019, chapters 4, 6, 11) and Gries (2021, chapter 5) give particularly extensive introductions from a complementary perspective; most sources listed in the preface give briefer introductions (e.g., Baayen 2008, sections 6.1–6.2).

Exercises

4.1. For model `slr_mod_1` fitted in section 4.3.2:

(a) What is the predicted `RTlexdec` when `WrittenFrequency` = 5? 10?

(b) Why doesn't it make sense to predict `RTlexdec` for `WrittenFrequency` = –1?

4.2. In the coefficient table `summary()` outputs by default for a linear regression model (as in section 4.3.3), the column for the *p*-value says "`Pr(>|t|)`." Why is the *p*-value written this way?

2. Here I have just suggested the most common summaries used by sources cited in this chapter (e.g., Faraway 2015; Chatterjee and Hadi 2012, `summary()`, `tidy()`).

4.3. In section 4.3.5, R^2 for model `slr_mod_2` is shown at the bottom of the model summary. Verify that you get the same value by calculating the correlation between x and y, and between observed and fitted values of y. (Hint: `?fitted`)

4.4. Figure 4.3 (left) suggests that the effects of subject age and word frequency on lexical decision RT are essentially independent, for the `english` data. Let's now ask whether this is also the case for word naming RT, `RTnaming`.

(a) Make a plot like figure 4.3 (left), but with `RTnaming` rather than `RTlexdec` on the y-axis. Does it look like the effect of `WrittenFrequency` differs between OLD and YOUNG speakers? If so, does the amount of change seem (visually) large or small?

(b) Now fit a multiple regression model with response `RTnaming` and an interaction between `WrittenFrequency` and `AgeSubject`. Is there a significant interaction between the two predictors? What is its interpretation, in words?

(c) Calculate the slope of `WrittenFrequency` predicted for OLD and YOUNG speakers by this model. How much steeper, as a percentage, is the slope of `WrittenFrequency` for old speakers compared to young speakers?

(d) How does this percentage compare to your judgement from the visualization in part (a)? If there is a discrepancy, can you explain why it comes about?

4.5. For model `mlr_mod_2` fitted in section 4.4.3, calculate the values shown in figure 4.4 (left) by hand. That is, use R functions to make a dataframe with each combination of x_1 and x_2, then to get a model prediction and 95% CIs for each row. (Hint: `?predict.lm`)

4.6. In sections 4.5.1 and 4.5.3, the same regression model is fit for the `neutralization` data with the base level of the `voicing` factor changed. The text explains why the `voicing` and `voicing:prosodic_boundary` estimates for the two models are the same except with the opposite sign. However, the other coefficients (the intercept, PROSODIC_BOUNDARY) have different values.

(a) Why are the `prosodic_boundary` coefficients different in the two models? What is the interpretation of this coefficient in each model?

(b) Same question, this time for the intercept (the `(Intercept)` coefficient).

5 Linear Regression 2

In the previous chapter, we fit linear regressions; we assumed that all assumptions of linear regression were met and the set of variables to include in the model was fixed. In this chapter we will first discuss validation of a fitted model: the assumptions of linear regression, how to detect and address them when they are violated (sections 5.2–5.7), and assessing overfitting or underfitting to the data (section 5.8). We will then turn to methods for comparing different models (section 5.9), including selecting which predictors to include (section 5.10), culminating in different possible methods for building up a model.

5.1 Preliminaries

5.1.1 Packages

This chapter assumes that you have loaded packages from previous chapters, as well as the arm, car, performance, and MuMIn packages (Gelman and Su 2021; Fox and Weisberg 2019; Lüdecke et al. 2021; Barton 2020), which contain useful functionality for working with regression models and assessing their performance.

```
library(tidyverse)
library(broom)
library(languageR)
library(car)
library(arm)
library(performance)
library(MuMIn)

## Ensures that `rescale', a function we will use,
## is not the version from the `scales' package.
rescale <- arm::rescale
```

5.1.2 Data

The `vot` dataset

We also assume that you have loaded the `vot` dataset:

```
vot <- read.csv("data/vot_rmld.csv")
```

This dataset, analyzed in Sonderegger, Bane, and Graff (2017) and described in more detail in appendix A, section A.2, contains positive voice onset times (VOTs: column VOT, in msec) measured for word-initial stop consonants (p/t/k/b/d/g) in spontaneous speech from a corpus of speakers of different British English dialects. VOT is the primary acoustic cue in English signaling whether a stop is phonologically VOICED (p/t/k) or VOICELESS (b/d/g) (column VOICING), but VOT also varies as a function of many other properties of the word, context, or speaker, including:

- `speaking_rate` (in syllables/second) of the surrounding phrase
- `place` (levels ALVEOLAR, LABIAL, VELAR): the consonant's place of articulation
- `cons_cluster` (levels NO, YES): whether the stop is in a consonant cluster (e.g., "tick," "trick")
- `foll_high_vowel` (levels NO, YES): whether the following vowel is high or nonhigh (e.g., "tease," "tack")
- `gender` of the speaker (levels F, M)
- `log_celex_freq`: the word's frequency (log-transformed)

We will consider different research questions in several analyses of these data.

Observations from the same speaker (`speaker`) or word (`word`) are not independent. To create a plausibly independent dataset that we can analyze using linear regression (as discussed in section 3.5.2.2), we take a subset `vot_michael` of data consisting of observations from a single speaker (MICHAEL) and one observation per word:

```
vot_michael <- filter(vot, speaker == "michael") %>%
  ## First use of each unique word
  filter(!duplicated(word)) %>%
  droplevels()
```

This dataframe contains $n = 593$ observations. In addition, because we have not covered multilevel factors yet, we simplify the `place` column to have just two levels (LABIAL, NON_LABIAL):

```
vot_michael <- mutate(vot_michael,
  place = recode_factor(place,
    alveolar = "non_labial", velar = "non_labial"
  )
)
```

5.2 Linear Regression Assumptions

Any regression model makes a range of assumptions about the data, the form of the relationship between predictors and the response, and so on. Following Faraway (2015, chapters 6–7), it is useful to think of these in terms of concrete problems that could apply to a given model:

- Problems with the *errors* (e.g., nonnormality; section 5.3)
- Problems with the *model* (e.g., nonlinearity; section 5.4)
- Problems with *predictors* (e.g., collinearity; section 5.6)
- Problems with *observations* (e.g., outliers; section 5.7)

In each case it is important to know how to check for the problem and, if the problem exists, to determine how serious it is and what can be done to address it.

Closely related topics are assessing how well the model fits the data (goodness of fit; see section 4.3.5) and how well it generalizes to unseen data. These can be thought of as *problems with fit* (section 5.8): either not fitting the dataset well or not generalizing well to unseen data. Fitting observed and unseen data well are not assumptions of linear regression models, but they are obvious desiderata for a model.

All these topics are part of *model validation*—the iterative process of diagnosing and (possibly) addressing problems, and (possibly) changing the data and model as a result, to arrive at a statistical model that we can trust.[1] Only then should we interpret our model to address research questions or use it to predict new data. This process can feel overwhelming because there are so many things that could be wrong with a model, but the point is not to give a "correct" model. As Faraway (2016, 14) puts it: "It is virtually impossible to verify that a given model is exactly correct. The purpose of the diagnostics is more to *check whether the model is not grossly wrong*" (emphasis mine).

There is an important additional assumption for every regression model "that the data at hand are actually relevant for the question of interest" (97). There are no diagnostics to check this.

Visual methods Visualization is crucial for checking model assumptions and for data analysis in general. A famous example illustrating this is *Anscombe's quartet* (Anscombe 1973; `anscombe` in R), a set of four small datasets ($n = 11$) of (x, y) pairs that seem similar just from numerical summary statistics. The (sample) mean and SD for x and y, as well as the least-squares regression line (and thus the correlation between x and y: $r_{xy} = 0.82$), are the same for each dataset.

Yet it is immediately clear from plotting the datasets (figure 5.1) that they show qualitatively different patterns. Dataset 1 is what we instinctively imagine given the summary statistics: a line of best fit, with each observation about evenly well-predicted, and no one observation having much more influence than any other. But these assumptions are violated for datasets 2, 3, and 4 (respectively), each of which corresponds to a different problem we'd like to detect through model validation: nonlinearity, outliers, and overly influential observations.

With more than one predictor it is not possible to check regression assumptions by just plotting the data, but other visual methods (such as residual plots) are available. Although

1. Some authors (e.g., Baayen 2008) distinguish between "model criticism" (checking and correcting regression assumptions) and "model validation" (assessing goodness of fit and overfitting). This book uses "validation" for simplicity.

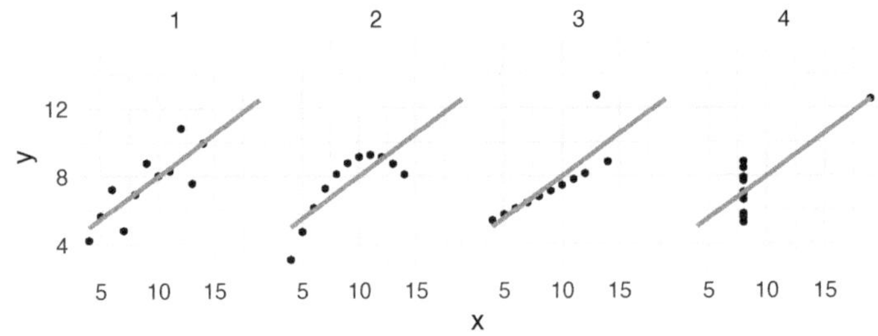

Figure 5.1
Anscombe's quartet.

there are usually numeric "tests" available as well to check regression assumptions, in general visual methods are more effective (e.g., Faraway 2015, 75; Chatterjee and Hadi 2012, 85; Zuur, Ieno, and Elphick 2010).

5.3 Problems with the Errors

Any linear regression model, such as equation (4.12), contains a normally distributed *error* term (ε_i for the *i*th observation). There are several associated assumptions made by the model.

5.3.1 Independence of Errors

A crucial assumption is that these ε_i are independent: knowing the error for one observation shouldn't tell you anything about the error for another observation.

Unfortunately, violations of this assumption are endemic in linguistic data, and ignoring this assumption in data analysis ("pseudoreplication") can cause serious problems—as we discussed with respect to *t*-tests in section 3.5.2. Violations of the independence assumption are often anticonservative: CIs will be too narrow and *p*-values too small if the lack of independence of errors is not taken into account by the model. The examples in section 3.5.2, which carry over directly to linear regression models,[2] show the effects of the most common kind of nonindependence in linguistic data: grouping structure (e.g., by subject and item). This is arguably the most important assumption underlying statistical analysis of linguistic data more generally and a primary motivation for using regression models that can take grouping structure into account, which we will turn to in chapter 8. Nonindependent errors can also arise due to *autocorrelation* among observations that are somehow near each other, usually in time or space (box 5.1).

In practice it is usually hard to detect problems with independence via diagnostic plots. You have to think about the structure of your data to know whether the independence assumptions hold and how to transform or subset it to satisfy them (see discussion in section 3.5.2.2).

2. Because they are analyzed with *t*-tests (section 4.4).

Box 5.1
Broader Context: Autocorrelation

Autocorrelation is fairly common in raw linguistic data. For example, nearby observations in time will be correlated in acoustic measurements of speech (pitch or formant tracks) or time series from behavioral experiments (e.g., eye-tracking data, self-paced reading). In data from dialectology or neurological studies (where the output is a map of brain activity), observations spatially near each other may be correlated.

For data where each observation is a time series (like a pitch or eye-tracker track), it is common in linguistics to deal with temporal autocorrelation by collapsing to a single number, by averaging over a region of interest (e.g., 100–200 msec from stimulus onset) or taking a single point (e.g., 33% through a vowel). But unlike grouping structure, it is sometimes possible to account for temporal or spatial structure by simply including predictors for time/space (e.g., a linear or polynomial effect of `time`) in the regression model.

If correlated errors still exist, you can use regression models that take autocorrelation into account by explicitly modeling the errors, such as "generalized least squares" or "time series" models (most commonly autoregressive moving average: ARMA). Modeling autocorrelated errors is beyond the scope of this book, but some places to start are Faraway (2015, chapter 8) and Zuur, Ieno, and Smith (2007, chapters 16–19). For linguistic data, recent work deals with autocorrelated errors using more complex methods (e.g., Baayen et al. 2018; Wieling 2018; Grieve 2017 for psycholinguistics, phonetics, dialectology).

5.3.2 Normality of Errors

The next major assumption is that the errors ε_i are normally distributed with mean 0 and a fixed variance. This assumption is impossible to check directly, because we never observe the true errors ε_i, only the residuals $\hat{\varepsilon}_i$. In general, we check assumptions about errors by examining the distribution of the residuals. If the normality of errors assumption holds, then the residuals will be normally distributed with mean 0 and fixed variance. So if they are not, we know the assumption does not hold. (If the results are normally distributed, it's not a guarantee that the assumption holds, but we hope for the best.)

Box 5.2
Broader Context: Standardized/Studentized Residuals

Technically there is a flaw in the preceding reasoning: the residuals of a linear regression are not normally distributed (even if the errors are), because some observations will be more influential than others in determining the fitted responses $\hat{y}_i$ when fitting the least-squares estimates of the regression coefficients. This can be seen in a simple linear regression line of best fit (e.g., figure 4.2): the CI is wider for points further from the means of x and y, because these points are more influential, causing the variance of their residuals to increase. Thus the variance of the residuals is not constant, so they don't follow a normal distribution. It is possible to transform the residuals in a way that accounts for the different influence of different observations to "Studentized" or "standardized" residuals and then use these transformed residuals in diagnostic plots. This is recommended by some regression texts (e.g., Chatterjee and Hadi 2012, section 4.3), whereas others (e.g., Faraway 2015, section 6.1) think it is sufficient to just use plots based on untransformed residuals. This book usually uses untransformed residuals, but it sometimes uses standardized residuals when this is the default of an R function or when standardizing is particularly important (defining outliers; see section 5.7.2).

5.3.2.1 Example

Consider a first model of the `vot_michael` data, where we predict VOT as a function of three predictors expected to have strong effects (section 5.1.2): `voicing`, `place`, `cons_cluster`, `speaking_rate`:

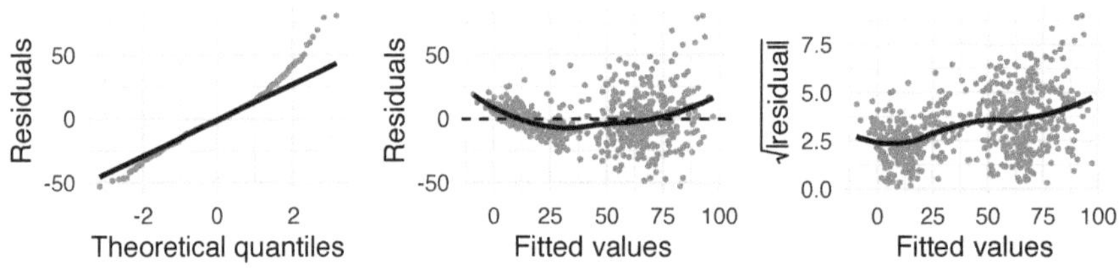

Figure 5.2
Three residual diagnostic plots for `vot_mod_1`: Q-Q plot of residuals (*left*), fitted value-residual plot (*middle*), scale-location plot (*right*)

```
vot_mod_1 <- lm(vot ~ voicing + place + cons_cluster + speaking_rate,
  data = vot_michael)
```

We can extract the residuals using `residuals(vot_mod_1)` or the tidy `augment()` function (from `broom`), which adds several useful columns of information for diagnostic plots to the original dataframe, including the fitted values ($\hat{y}_i$) and residuals ($\hat{\varepsilon}_i$):

```
augment(vot_mod_1) %>% head(n=3)
##    vot   voicing      place cons_cluster speaking_rate .fitted
## 1   45 voiceless non_labial           no           5.4      60
## 2   56 voiceless non_labial           no           5.0      62
## 3   39 voiceless non_labial           no           5.3      60
##    .resid    .hat .sigma .cooksd .std.resid
## 1   -14.7 0.0062     18 0.00086      -0.83
## 2    -5.8 0.0052     18 0.00011      -0.33
## 3   -21.0 0.0058     18 0.00167      -1.19
```

The augmented dataframe can then be used to make diagnostic plots, such as a *Q-Q plot* of residuals to check normality (figure 5.2, left). The residuals are somewhat nonnormally distributed (histogram in figure 5.3, left). The Q-Q plot suggests that this is primarily due to observations with the very highest and lowest residuals. In the *fitted values-residuals plot* (figure 5.2, left), we see that the points responsible for nonnormality tend to be those with the highest and lowest values of the response (`vot`), which are also responsible for the nonnormal distribution of the response (figure 5.3, right). This illustrates the most common source of nonnormal residuals: a highly nonnormal distribution of the predictor or response.

The italicized diagnostic plots are among those made automatically by `plot` applied to a linear regression model in R (e.g., `plot(vot_mod_1)`).

Nonnormality of residuals is a common violation of regression assumptions. In practice it often doesn't matter much, at least if our goal is estimation of the regression coefficients (as opposed to prediction), especially for large samples, where the central limit theorem kicks in (Faraway 2015, section 6.4; Gelman and Hill 2007, 46).

However, it is always a good idea to examine the distribution of residuals because nonnormality of residuals can signal other issues with the data, such as the presence of outliers or the predictor or response being on the wrong scale. Nonnormality of residuals can often

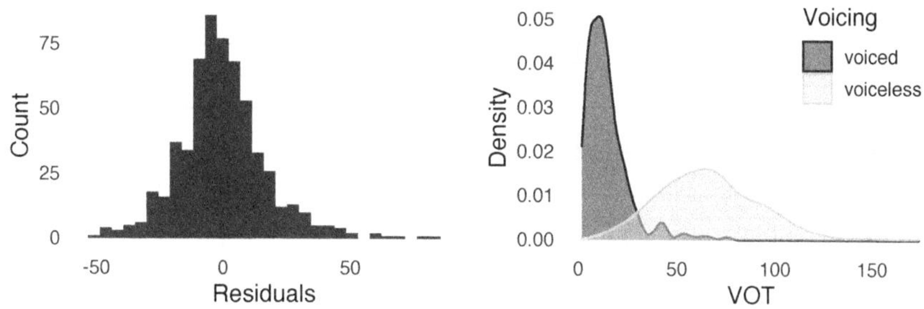

Figure 5.3
Distributions of residuals of `vot_mod_1` (*left*), and of VOT in the `vot_michael` dataset (*right*) for VOICED and VOICELESS consonants.

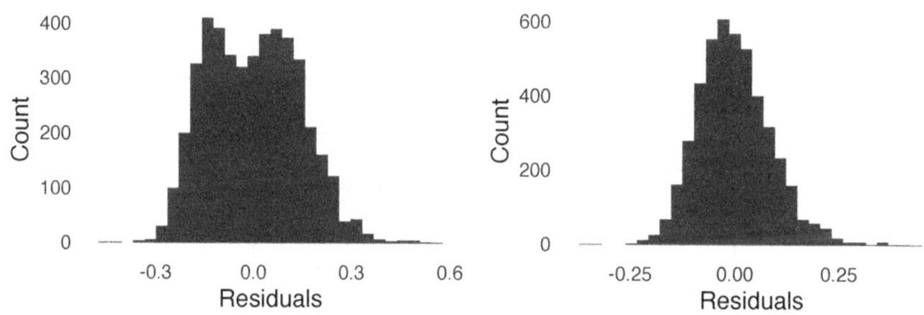

Figure 5.4
Distributions of residuals of `english_mod_1` (*left*) and `english_mod_2` (*right*).

be dealt with by transforming the predictor or response to have a more normal distribution. Section 5.5 discusses these points further.

Nonnormality of residuals can also signal other errors, such as that an important predictor is missing.

5.3.2.2 Example

Consider two linear regressions of reaction time (RT: variable `RTlexdec`) for the `english` data: as a function of word frequency (`WrittenFrequency`) alone and as a function of word frequency and subject age (`AgeSubject`).

```
english_mod_1 <- lm(RTlexdec ~ WrittenFrequency, data = english)
## Add one predictor to model formula
english_mod_2 <- update(english_mod_1, . ~ . + AgeSubject)
```

The residuals of the first model are not normal (figure 5.4, left), primarily because they are bimodal: the two modes correspond to values of `AgeSubject` (left, right—OLD, YOUNG: Exercise 5.1). The residuals of the second model (figure 5.4, right) look closer to normality; they are no longer bimodal.

This example shows one reason that examining the residual distribution is useful. If we didn't already know what the missing predictor was, then the nonnormality of the residual distribution would give us a way to look for an explanatory variable. (Look at observations in each mode of the distribution to see what they have in common.)

5.3.3 Constancy of Variance

Linear regressions assume not just that the errors ε_i are normally distributed, but also that the variance of the normal distribution is the same across all values of the predictor.

For example, in our model of VOT as a function of three predictors (`vot_mod_1`; see section 5.3.2.1), it is assumed that the amount of (remaining) variability in VOT is similar for voiced and voiceless stops, for prevocalic voiced stops and preconsonant voiceless stops (`voicing` = VOICED and `cons_cluster` = NO vs. `voicing` = VOICELESS and `cons_cluster` = YES), and so on.

Constancy of variance can be examined with plots of the fitted values or a predictor (x-axis) versus the residuals (y-axis). If constancy of variance holds, then the residuals are uncorrelated with predictor values and with the fitted values, and there should be a constant spread (variance) of residual values (y-axis) for each fitted or predictor value (x-axis). Thus, it is common to make diagnostic plots of residuals versus fitted values and versus predictors. The desired pattern is a flat line that has the same variance for different x-axis values.

In the VOT example, a fitted value-residual plot (figure 5.2, middle) shows nonconstant variance: higher values of $\hat{y}$ show higher variance in the residuals. Sometimes the relationship is examined with a "scale-location plot," where fitted values are plotted against $\sqrt{|\varepsilon_i|}$ (figure 5.2, right). For normally distributed errors, how vertically spread out the points are ("scale" or "spread") should be the same for all values of $\hat{y}$ or any predictor, and the average spread should not change; we want to see a horizontal line (Faraway 2015, section 6.1.1).

Nonconstant variances arise frequently in some kinds of linguistic data, such as count data (e.g., from lexical statistics, typology, dialectology), which is one motivation for analyzing them using more complex models than linear regression (e.g., Poisson regression; see Baayen [2008, 35]; Winter [2019, chapter 13]; Faraway [2016, chapter 5]).

In general, estimates of least-squares coefficients in the presence of nonconstant variance are unbiased, but SEs will be under- or overestimated (Chatterjee and Hadi 2012, section 6.4). This means that CIs will be too narrow/wide and p-values too low/high.

An important reason to check for constant variance is that it can suggest changes that should be made to the data and model. This is because one can often correct for nonconstant variance by transforming the response or predictors, including a "missing" interaction between predictors or nonlinear effect of a predictor, and so on. (chapter 4, Faraway 2015, chapter 9, Zuur, Ieno, and Smith 2007, section 5.1). For example, we will see in section 5.5.2.2 that log-transforming the response will resolve the nonconstant variance in our VOT model.

5.4 Problems with the Model

Recall that linear regression is a type of linear model (section 4.2.1), where the relationship between y and k predictors is

$$E(y) = \beta_0 + \beta_1 x_1 + \beta_2 x_2 + \cdots + \beta_p x_k \qquad (5.1)$$

Here, $E(y)$ is the *expectation* of y—its average value across many trials.

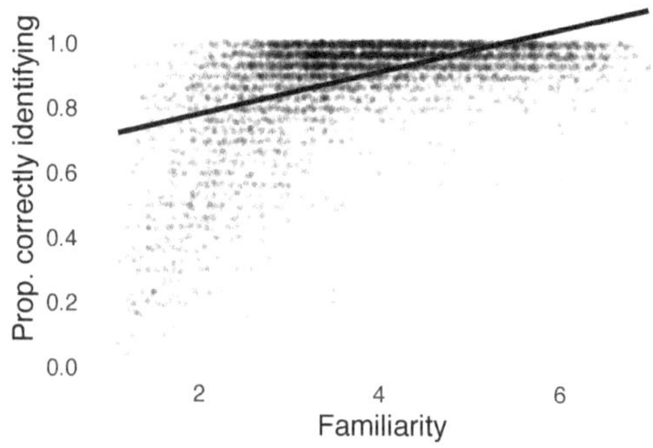

Figure 5.5
An inappropriate linear regression for the `english` data, which predicts impossible *y* values (proportion of subjects who correctly identified a word can be at most 1).

This is the basic relationship assumed between the response and predictors in linear regression (for an "average observation"), abstracting away from the error. Previously we considered problems related to the error term. This section describes problems related to equation (5.1) not being a good model for the data.

5.4.1 Bounded or Discrete *y*

If you are modeling data where *y* is constrained to be between lower and/or upper bounds, fitting a linear regression to the data doesn't make conceptual sense a priori, because equation (5.1) can give nonsensical values of *y* (outside the bounds) for new predictor values.

For example, suppose we'd like to predict the proportion of subjects who correctly identified each word (*y*: `CorrectLexDec/30`) for the `english` data. A linear regression model wouldn't be appropriate here because *y* can only be between 0 and 1, but this model would allow *y* to take on values outside [0, 1] for some predictor values. For example, figure 5.5 shows the relationship between *y* and a word's subjective familiarity (*x*: `Familiarity`). A simple linear regression of *y* on *x* predicts $y > 1$ even for values of *x* in the data the model is fitted to. It makes sense that *x* would strongly affect *y* (more familiar words are more often identified correctly), but a linear regression is not an appropriate model of the relationship.

A more common case is where the fitted model predicts valid *y* values for data the model was fitted on, but also it would predict invalid *y* values for new data (with predictor values not in the training data).

This kind of problem can be detected by just thinking about the data or by making exploratory plots of *y* against possible predictors. Sometimes it's possible to get around the issue by transforming *y* (an example is presented in section 5.5.2.2); sometimes a different method is needed (logistic regression, in this example).

By the same logic, it isn't technically valid to model *y*, which can only take on a finite set of values (e.g., a 1–7 Likert scale), as opposed to a continuous variable (e.g., VOT), with a linear regression, because *y* is bounded and the linear regression model predicts that *y* can lie between the discrete values. The latter problem rapidly diminishes as the number of levels increases. It is a reasonable approximation to model *y* as continuous the more

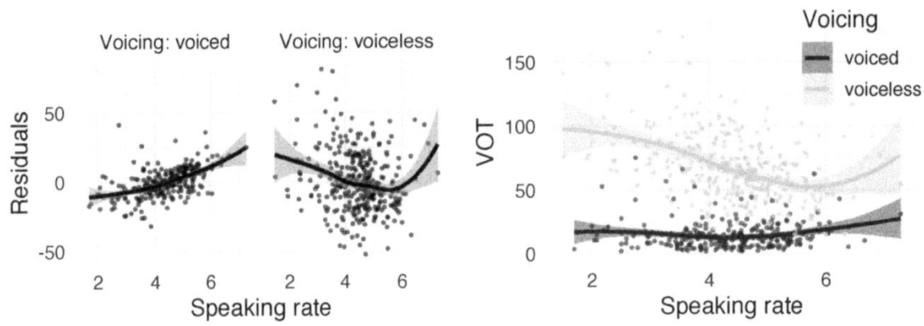

Figure 5.6
Residuals for `vot_mod_1` as a function of predictors `speaking_rate` and `voicing` (*left*), and `vot` as a function of these two predictors for the `vot_michael` data (*right*), with nonlinear smoothers (LOESS) and 95% CIs.

the distribution of y looks continuous: the number of levels is large (10+, say), and most levels are actually used (Gelman and Hill 2007, section 6.5), for example, responses are not clustered near the highest and lowest levels. Otherwise a method that treats y as an ordinal should be used (Liddell and Kruschke 2018).

5.4.2 Linearity

A primary assumption of a linear regression model is that the relationship between the response and the predictors is indeed linear: in the right-hand side of equation (4.12), the predictors can be each multiplied by a constant and added together to get their effect on y. While seemingly obvious, this assumption is very important because modeling a curve as a line can lead to serious errors.

Nonlinearity in a single predictor The linearity assumption can be a concern for any continuous predictor x_i: intuitively, its effect on y may not look like a line. Such nonlinearities often show up in diagnostic plots of one or more x_i versus residuals. For example, figure 5.6 (left) plots the distribution of residuals for `vot_mod_1`, as a function of two predictors—`speaking_rate` and `voicing`. The nonlinear trends in each panel suggest there may be a nonlinear effect of `speaking_rate`, while the differing trends in the two panels suggest a possible omitted `speaking_rate:voicing` interaction. The existence of nonlinearity and an interaction with `voicing` is also suggested by the empirical plot in figure 5.6 (right). You can confirm (Exercise 5.2) that after adding these terms to the model, the residual plot looks better.

A closely related technique is to make *added-variable* plots that isolate the effect of each x_i after all other predictors are controlled for (e.g., Faraway 2015, section 6.3; Fox and Weisberg 2019, section 8.2).

More simply, linearity can be partially checked by making plots of y as a function of each (continuous) predictor x_i, which is good to do before statistical analysis in any case. Because you'll also typically want to examine relationships between the predictors themselves as well to assess collinearity (section 5.6.1), it is efficient to make *pairwise plots* for each pair of a set of variables. It helps to include a nonlinear smoother—a smooth curve interpolated based on nearby observations (see section 7.5.1)—in each plot. If a line is not compatible

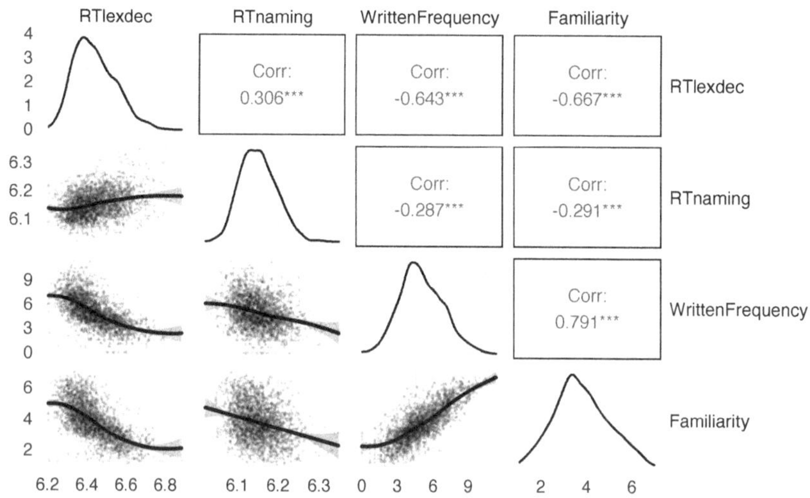

Figure 5.7
Pairwise plots of several variables from the `english` dataset, with nonlinear (generalized additive model) smoothers and 95% CIs (lines/shading), and Pearson correlation for each pair. Diagonal panels are empirical distributions of each variable.

with the 95% CI of this curve (meaning you could draw a straight line within the CI), any effect of x_i on y may not be linear.

For example, figure 5.7 shows pairwise plots between two possible responses (`RTlexdec`, `RTnaming`) and two possible predictors (`WrittenFrequency`, `Familiarity`) for the `english` data for YOUNG speakers. The relationship between `RTnaming` and each predictor looks plausibly linear, while the relationship between `RTlexdec` and each predictor looks clearly nonlinear: for lower and higher frequency words, the decrease in RT "levels off."

Box 5.3
Broader Context: Missing Interactions as Nonlinearity

In addition to nonlinear effects of continuous predictors, the linearity assumption also covers another common case: whenever the effect of a predictor x_1 depends on the value of x_2—that is, there is an interaction—omitting the interaction term is a linearity violation. To see why, recall that the models with and without an interaction term would be

$$y = \beta_0 + \beta_1 x_1 + \beta_2 x_2,$$

$$y = \beta_0 + \beta_1 x_1 + \beta_2 x_2 + \beta_3 x_1 \cdot x_2.$$

The existence of an interaction means that the second model is the correct one, with $\beta_3 \neq 0$, so x_1 and x_2 do not linearly predict y because the first model is incorrect. Defining $x_3 = x_1 x_2$, the second model satisfies linearity because y is a linear function of x_1, x_2, and x_3.

The existence of missing interactions can be assessed with exploratory plots (of y as a function of each pair of predictors) or by variable selection methods (section 5.10). How much to look for "missing" interactions is a tricky subject in practice, because the number of possible interactions increases exponentially with the number of predictors, but every new term considered for inclusion in a regression model carries potential benefits (section 5.8.1) and risks (spurious effects).

5.5 Transformations

Let's take a detour from problems to discuss a type of solution. Transformations of the response or predictors (y, the x_i) can be useful to address violations of model assumptions or for interpretability of model coefficients.

5.5.1 Nonlinear Effects: Polynomials

The most obvious example is for the kind of nonlinear effect of a predictor just discussed. This kind of nonlinear relationship can be included in a linear regression model, but it has to be written as a sum of linear terms. For the `RTlexdec~WrittenFrequency` relationship, for example, it turns out the nonlinear relationship is decently approximated by a cubic polynomial.

So the relationship can be captured in a linear regression with three predictors by setting x_1, x_2, and x_3 to be frequency, frequency^2, and frequency^3. We could fit this model as follows:

```
lm(RTlexdec ~ WrittenFrequency + I(WrittenFrequency^2) +
    I(WrittenFrequency^3), english_young)
```

(Here, the `I(x^2)` notation is used because `x^2` would be interpreted as `x + x:x`; see section 4.5.1.)

We will cover nonlinear effects of predictors in detail in section 7.5. The point here is that a model of `RTlexdec` with just `WrittenFrequency` as a predictor would violate the linearity assumption, while a model with squared and cubed terms might not.

5.5.2 Transforming to Normality

Normality of the distribution of the response and predictors (y and the x_i) is not an assumption of linear regression. This is a common misconception.

However, there is still good reason to be circumspect if y or (continuous) x_i are not normally distributed, because this can often lead to violations of regression assumptions. This is why it is often useful to transform y or x_i to normality—to fix various problems covered so far (normality of errors, constant variance, linearity, etc.). Transforming one or more variables to normality makes it less likely that a regression assumption will be violated, but doing so also changes the interpretation of the transformed variable(s), which may make it harder to interpret the model's results.

For linguistic data, logarithmic transformations are often useful when working with skewed distributions, because many common variables are roughly *log-normally* distributed, meaning the log-transformed variable is normally distributed. Some examples are lexical statistics (e.g., word frequency), reaction times, or duration measures in phonetics (e.g., VOT, vowel duration). For example, figure 5.8 shows the distribution of word frequencies in the `english` dataset before log-transforming (∼log-normal) and after transforming (∼normal).

Other transformations besides log are also used: reaction times are sometimes inverse or inverse-log-transformed (1/RT, log(1/RT)), and durations are sometimes square root-transformed.

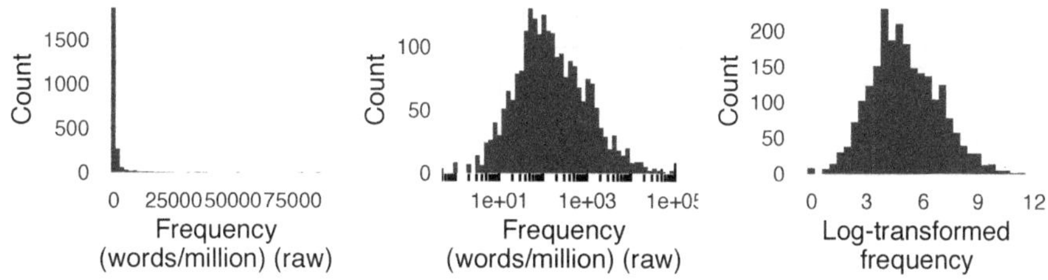

Figure 5.8
Distribution of `WrittenFrequency` for words in the `english` dataset, before (*left*) and after log-transforming frequency, with x-axis in log (*center*) and original (*right*) scale.

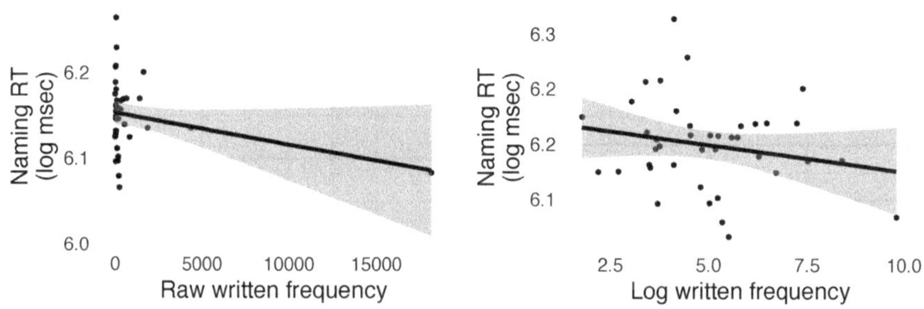

Figure 5.9
`RTnaming` as a function of untransformed word frequency (`WrittenFrequency_raw`; *left*) and log-transformed word frequency (`WrittenFrequency`; *right*), with line of best fit and 95% CI.

For visualization of transformed variables, it is often more interpretable to make plots showing untransformed values on a transformed scale, as in figure 5.8 (center) where word frequencies are shown in words-per-million, but the x-axis is log-transformed.

5.5.2.1 Example: Transforming a predictor

Consider a small ($n = 40$) subset of the `english` data, which we will use in a few examples:

```
english_young <- filter(english, AgeSubject == "young")
set.seed(2900)
english_40 <- sample_n(english_young, 40)
```

This subset simulates a small-scale version of the `english` experiment where only 40 words have been examined for YOUNG speakers. It is easier to see the effects of problems that can affect linear regressions when using smaller samples.

Consider a simple linear regression of `RTnaming` as a function of raw written frequency, before and after log-transforming:

```
english_40$WrittenFrequency_raw <- exp(english_40$WrittenFrequency)
slr_mod_4 <- lm(RTnaming ~ WrittenFrequency_raw, data = english_40)
slr_mod_5 <- lm(RTnaming ~ WrittenFrequency, data = english_40)
```

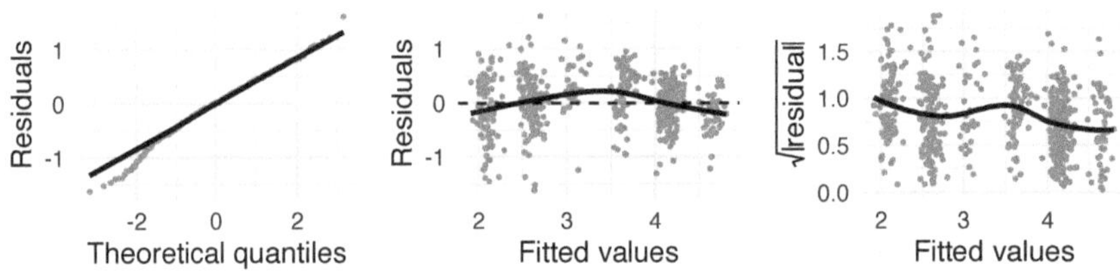

Figure 5.10
Diagnostic plots for `vot_mod_2`: Q-Q plot of residuals (*left*), fitted value-residual plot (*middle*), scale-location plot (*right*).

The first regression corresponds to the line of best fit in figure 5.9 (left). It is clear that fit is poor, and any effect of (raw) frequency on RT may be nonlinear. Making other diagnostic plots (e.g., `plot(slr_mod_4)`) will show a host of other problems (e.g., nonconstant variance). All these issues are resolved by log-transforming frequency, so that when x = `WrittenFrequency`, the effect of x on y looks closer to linear (figure 5.9, right) and so on.

5.5.2.2 Example: Transforming the response

The VOT model fitted in section 5.3.2.1 has a number of problems, including nonnormal residuals and nonconstant variance of residuals. In addition, there is a conceptual problem with the model because y is bounded—VOT can only be positive, by definition (section 5.1.2)—but in a model where y is `vot` nothing stops negative y from being predicted.

There are several reasons to think these issues can be resolved by log-transforming y (or square root-transforming):

1. A model predicting $y = \log(\text{VOT})$ would allow y to be positive or negative but can only predict VOT > 0, addressing the conceptual problem.
2. y itself is not normally distributed within VOICED or VOICELESS stops (figure 5.3), and log-transforming often brings right-skewed distributions closer to normal.
3. The Q-Q plot of residuals (figure 5.2, left) implies a skewed distribution, which can result from a skewed distribution of y.
4. The "fan" pattern in the fitted-residuals plot (figure 5.2, center), where variance increases as a function of y, often is fixed by log- or square-root–transforming y.

Let's refit the model with y = `log_vot`:

```
## Change the response in the model formula
vot_mod_2 <- update(vot_mod_1, log_vot ~ .)
```

The diagnostic plots now look better (compare figure 5.10 to figure 5.2): the Q-Q plot suggests a residual distribution closer to normality, and the fitted-value and scale-location plots suggest roughly constant variance.[3]

3. There is some evidence for nonconstant variance that should be investigated further (especially around $\hat{y} = 3$).

5.5.3 Interpretation Issues

The interpretation of every coefficient in a regression model changes when y is transformed. For example, for the log-transformation, the effects of predictors on y now multiply rather than add (Faraway 2016, section 9.1). To see why, consider a simple linear regressions of $\log(y)$ on x:

$$\log(y) = \beta_0 + \beta_1 x + \varepsilon.$$

Exponentiating both sides gives

$$y = \exp \beta_0 \cdot (\exp \beta_1)^x \cdot \exp \varepsilon.$$

In terms of y, each nonintercept coefficient β in this model has the interpretation "exp(β) is how much y is multiplied by for a unit change in x, holding other predictors constant." This means that increasing x by 1 will multiply y by a number (e^{β}), increasing x by 2 will multiply y by the square of that number, and so on. When there are multiple predictors, the right-hand side of the model now has one term ($(\exp \beta_i)^{x_i}$) per predictor, which multiply together to predict y.[4]

For our example, consider just the estimated `voicing` coefficient for our VOT models `vot_mod_1` (response: `vot`) and `vot_mod_2` (response: `log_vot`):

```
coefficients(vot_mod_1)[["voicingvoiceless"]]
## [1] 50
coefficients(vot_mod_2)[["voicingvoiceless"]]
## [1] 1.6
```

The interpretations of these coefficients are

- "VOICELESS stops have VOT 50 msec longer than VOICED stops,"
- "VOICELESS stops have VOT 4.8 ($= e^{1.6}$) times higher than VOICED stops."

The `vot_mod_2` result is less interpretable, because linguists have almost always thought about VOT in terms of msec and msec differences, not proportions. A 50 msec difference is immediately interpretable to a phonetician (it's a big effect), while a 4.8 times increase is not. So even though `vot_mod_2` is better in terms of satisfying regression assumptions, it is worse in terms of interpretability. This is often the situation when y is transformed.

5.5.4 Centering and Scaling Predictors

An important aspect of interpreting regression model results is being able to interpret and compare coefficient values. This is not the case for our models so far because of how the predictors are coded.

To see the issue, consider a model of lexical decision time as a function of three predictors for the `english` data: word frequency, the length of the word in letters, and subject age:

4. Interpretation also changes when a continuous x_i is transformed, but only for the coefficients involving that predictor. For example, if we use $\log(x_i)$ instead of x_i as a predictor in a model, the effect of x_i on y is now multiplicative. The coefficient β_i now means not "how much y changes for a unit change in x_i," but "how much y changes for a unit change in $\log(x_i)$," which is equivalent to "when x_i is multiplied by $\sim$1.7." (Because $e - 1 \approx 1.7$.)

```
# Add to model containing WrittenFrequency and AgeSubject
english_mod_3 <- update(english_mod_2, . ~ . + LengthInLetters)
tidy(english_mod_3)
##                term estimate std.error statistic p.value
## 1       (Intercept)   6.6076    0.0080     826.5   0.000
## 2 WrittenFrequency  -0.0369    0.0007     -52.4   0.000
## 3     AgeSubjectold   0.2217    0.0026      85.6   0.000
## 4   LengthInLetters   0.0039    0.0015       2.5   0.012
```

Two issues make the coefficients difficult to interpret. First, the *intercept value* is not interesting or meaningful. The intercept here means "predicted RT when all predictors are 0," that is "for YOUNG subjects, when `WrittenFrequency` = 0 and `LengthInLetters` = 0." But zero frequency is not an interesting value (it means an extremely low frequency word), and zero length is impossible. Even without the latter issue (suppose `LengthInLetters` = 0 actually meant "one letter long"), "very low frequency and very short length" describes a word that is not representative of English—shorter words tend to have higher frequencies (in every language)—so the intercept describes a very unusual word. Second, the *coefficient values aren't comparable*, because the predictors are on different scales. A unit change in `AgeSubject` covers 100% of the data (because 0, 1 = YOUNG, OLD), whereas a unit change in `WrittenFrequency` only covers part of the data (range: 0–11). Comparable coefficient values are valuable to assess effect size (section 3.2.3.2).

We can address these two issues by transforming predictors before fitting the model.[5] There are a few common ways to "standardize" predictors to address, and it is worth being aware of different options because they are all widely used (box 5.4). For clarity of presentation we will discuss just a single option that we will use going forward, following Gelman and Hill (2007, section 4.2):[6]

1. Continuous predictors are centered (subtract the mean) and divided by two SDs (2σ).
2. Binary factors are transformed to 0 and 1, then centered.

This results in a factor with equal numbers of observations for levels A and B (a *balanced* factor) being coded as –0.5 and 0.5; a factor with 25% A and 75% B observations being coded as –0.75 and 0.25, and so on.

Because all predictors have been centered, the intercept is now interpretable as the predicted y at the average value of each predictor, which tends to be more meaningful than setting all predictors to 0. Dividing by 2σ makes the effect sizes of coefficients for continuous and binary predictors comparable: as long as the data are not too unbalanced, a unit change now corresponds to a change of 2σ for any predictor.[7]

This book often uses *rescaling* to refer to the particular standardization in (1) and (2), following the usage of Gelman and Hill (2007) and the `rescale()` function in the arm

5. Alternatively one can transform the coefficents after fitting the model to get standardized coefficients.

6. For more in-depth discussion of the advantages of centering/standardizing variables, and how it affects model interpretation, see, e.g., Schielzeth (2010), Cohen et al. (2002), and Dalal and Zickar (2012).

7. This is because $\sigma \approx 0.5$ for a Bernoulli-distributed variable with $p \approx 0.5$; see box 2.3. For (very) unbalanced data, $\sigma \ll 0.5$, so the coefficients for binary predictors still mean "difference between A and B" but their sizes are not comparable with coefficients for continuous predictors.

package, which can be used to do (1) and (2) automatically. However, this is not standard terminology, and it is always important when writing up results to say exactly how you standardized variables, for example, "continuous variables were standardized by centering and dividing by two SDs."

Box 5.4
Practical Note: Standardizing/Normalizing/Scaling

Standardizing (or "normalizing") predictors is commonly done in a few different ways.

For continuous predictors, it is common to *z-score* (center, then divide by 1σ). *z*-scoring makes interpretation of coefficients for continuous predictors more intuitive, while dividing by 2σ makes the interpretation of binary predictors more intuitive ("difference between levels A and B") and the effect sizes of binary and continuous predictors comparable.

For binary predictors, a few options are common:

1. Leave as 0 and 1 (no centering—R default).
2. Change to –0.5 and 0.5 (centering, for balanced data).
3. (1) then subtract mean (centering).
4. (3) then divide by 2σ (centering and scaling).

The advantage of (4) is that the coefficients for binary predictors are truly comparable with those of continuous predictors, even for unbalanced data; the disadvantage is that the coefficients no longer are interpretable as "difference between levels A and B" (as for (1)–(3)). We will discuss different ways of coding categorical variables as numeric predictors ("contrast coding") further in chapter 7.

After rescaling predictors, the interpretations of the coefficients are more intuitive:

1. Intercept (β_0): the predicted y when all predictors are held at their mean values.
2. Main effect coefficient (β_i) for continuous x_i: the predicted change in y when x_i is changed by 2σ, with other predictors held at their mean values.
3. Main effect coefficient (β_j) for binary x_j: the predicted difference in y between the levels of x_j, with other predictors held at their mean values.

Holding other predictors at their mean values in this case (all x_i are continuous or binary) is the simplest example of *marginalizing*—summarizing the effect of one predictor, averaged across values of other predictors in some way—which we will discuss further for logistic regression (sections 6.7.3, 9.8.3), where this method is particularly useful.

Technically in (2) and (3) we could think of other predictors being "held constant" at any value. It is useful to use the wording in (2) and (3), where predictors are held constant at mean values, because it carries over to when interactions are included in the model: the main effect β_i is then interpretable as the effect of x_i marginalizing over other predictors, even in the presence of interactions between x_i and other predictors.

Example As a concrete example of what rescaling does to coefficient interpretations, let's refit the `english_mod_3` model with rescaled predictors.

```
english_2 <- mutate(english,
  WrittenFrequency = rescale(WrittenFrequency),
  LengthInLetters  = rescale(LengthInLetters),
  AgeSubject       = rescale(AgeSubject)
)
english_mod_4 <- update(english_mod_3, data = english_2)
```

The coefficient tables for the two models are

```
# Raw predictors
tidy(english_mod_3)
##                 term estimate std.error statistic p.value
## 1        (Intercept)   6.6076    0.0080     826.5   0.000
## 2 WrittenFrequency  -0.0369    0.0007     -52.4   0.000
## 3      AgeSubjectold   0.2217    0.0026      85.6   0.000
## 4   LengthInLetters   0.0039    0.0015       2.5   0.012
# Rescaled predictors
tidy(english_mod_4)
##                 term estimate std.error statistic p.value
## 1        (Intercept)   6.5501    0.0013    5055.0   0.000
## 2 WrittenFrequency  -0.1360    0.0026     -52.4   0.000
## 3         AgeSubject   0.2217    0.0026      85.6   0.000
## 4   LengthInLetters   0.0065    0.0026       2.5   0.012
```

The interpretations of coefficients in the two models are

Term	Raw	Rescaled
`(Intercept)`	RT for a YOUNG speaker and a word with `WrittenFrequency` = 0 and `LengthInLetters` = 0	RT for an "average" speaker (across OLD and YOUNG) and "average" word (mean `WrittenFrequency` and `LengthInLetters` values)
`WrittenFrequency`	RT difference when `WrittenFrequency` increased by 1	RT difference for a 2σ increase in `WrittenFrequency`
`LengthInLetters`	Similar to `WrittenFrequency`	
`AgeSubject`	RT difference between OLD and YOUNG speakers	same

Note that the `AgeSubject` coefficient is the same in the two models. Although `AgeSubject` is a two-level factor in one model and a centered numeric variable in the other model, the regression coefficient captures the difference between the two levels (OLD and YOUNG speakers) in each case.

We can read off the relative importance of the three predictors from the magnitudes of the coefficients in the model using rescaled predictors: `AgeSubject` > `WrittenFrequency` > `LengthInLetters`.

5.6 Problems with Predictors

5.6.1 Nonindependent Predictors

A crucial assumption of linear regression is that the predictors are *linearly independent*, meaning it isn't possible to write one predictor as a linear function of the others. If you can do so, it's impossible to disentangle the effect of different predictors on the response. For example, temperatures in Fahrenheit and Celsius are related by a linear function (F = 9/5C + 32), so F and C are linearly dependent. Another way of thinking of linear dependence is that in a linear regression where you predict one predictor as a function of the others, R^2 would be 1. Linear dependence of predictors will either give a model error or a weird model output if you fit a linear regression in R, because the math to find least-squares estimates of coefficients doesn't work out if there is linear dependence: there is no longer a unique optimal solution. (For example, if the slope of F and C in the optimal model were $\beta_1 = 2$ and $\beta_2 = 0$, then a model using slopes of 0 and 10/9 would also be optimal.)

Using a set of linearly dependent predictors happens surprisingly often, by accident. For example, when modeling VOT we might want to include these predictors, all of which have been found in laboratory studies to affect VOT:

1. Speaking rate
2. Number of syllables in the word
3. Word duration

Furthermore, we might log-transform each predictor, because all will have right-skewed distributions. However, if speaking rate is defined as "syllables per second," then the predictors are linearly dependent:

$$\log(\text{speaking rate}) = \log(\#\text{ syllables}) - \log(\text{word duration}).$$

Box 5.5
Broader Context: Singularities

You may encounter the terms *singular* and *singularity* in R output; this usually means that some variable can be perfectly predicted from others, and this may be a problem. The source of these terms is not important, but if you know some linear algebra, then you know a set of predictors is linearly dependent when one column of the *design matrix* X (where each column is the values of predictor i, across all observations) can be predicted from the others. The matrix X^TX, which is inverted to find the least-squares regression coefficient estimates, then has no inverse, for which the technical term is *singular*; alternatively, there is a *singularity*.

To see what linear dependence looks like in practice, let's simulate perfect dependence among predictors for the `english_40` data. We define a new predictor `new_var`, which is the sum of `WrittenFrequency` and `Familiarity`, and fit a linear regression of lexical decision time (`RTlexdec`) as a function of these three predictors:

```
bad_mod <- english_40 %>%
  # . notation: use output of mutate as the 'data' model is fitted to
  mutate(new_var = WrittenFrequency + Familiarity) %>%
  lm(RTlexdec ~ WrittenFrequency + Familiarity + new_var, data = .)
```

```
summary(bad_mod)
...
## Coefficients: (1 not defined because of singularities)
##                   Estimate Std. Error t value Pr(>|t|)
## (Intercept)         6.7017     0.0468  143.31   <2e-16
## WrittenFrequency   -0.0234     0.0134   -1.76    0.087
## Familiarity        -0.0418     0.0206   -2.03    0.049
## new_var                 NA         NA      NA       NA
...
```

The *NA*s in the row for new_var indicate there is something wrong. The "1 not defined because of singularities" message means the predictors are linearly dependent (box 5.5). We're getting this message because we have asked the model to use three predictors, but they only contain as much information as two independent predictors.

5.6.2 Collinearity

Full linear dependence of predictors is usually a sign that something is conceptually wrong with your data or model structure and is relatively easy to detect (e.g., the NA rows in the preceding model).

However, it is very common for there to be *partial* dependence between predictors: that is, $0 < R^2 < 1$ when you regress one predictor on the others. This is called *multicollinearity*, or just *collinearity*.

There can be high collinearity among several predictors, where one predictor is largely predictable as a function of several others. For example, in the previous VOT example, speaking rate (predictor 1) will be more predictable from number of syllables and word duration (predictors 2 and 3) than any single predictor would be from any other. (For the vot_michael data, R^2 for pairs of (1)–(3) are 0.43–0.5, whereas $R^2 = 0.75$ for predicting (1) from (2) and (3).)

Collinearity is ubiquitous in linguistic data and can significantly affect the estimates and interpretations of regression coefficients, particularly when collinearity is "high" (using the diagnostics in section 5.6.3). But collinearity is not a violation of the assumptions of linear regression, so whether it is a problem is debatable, as will be discussed in section 5.6.3. Let's first demonstrate the effects of very high and high collinearity through a couple examples.

5.6.2.1 Example: Very high collinearity

Consider the same situation as that of the linear dependence example, except now we define new_var to have very high correlation ($R^2 = 0.99$) with WrittenFrequency + Familiarity, rather than perfect correlation. We fit a model with and without new_var:

```
## Without new_var
good_mod <- lm(RTlexdec ~ WrittenFrequency + Familiarity,
  data = english_40)

set.seed(2000)
## With new_var
english_40_bad <- mutate(english_40,
```

```
  ep = rnorm(nrow(english_40), mean = 0, sd = 0.01),
  new_var = (WrittenFrequency + Familiarity) * (1 + ep)
)
bad_mod_2 <- update(good_mod, . ~ . + new_var, data = english_40_bad)
```

Now compare the output for these models:

```
summary(good_mod)
...
## Coefficients:
##                   Estimate Std. Error t value Pr(>|t|)
## (Intercept)         6.7017     0.0468  143.31   <2e-16
## WrittenFrequency   -0.0234     0.0134   -1.76    0.087
## Familiarity        -0.0418     0.0206   -2.03    0.049
##
## Residual standard error: 0.082 on 37 degrees of freedom
## Multiple R-squared:  0.508,Adjusted R-squared:  0.481
## F-statistic: 19.1 on 2 and 37 DF,  p-value: 2e-06
```

```
summary(bad_mod_2)
...
## Coefficients:
##                   Estimate Std. Error t value Pr(>|t|)
## (Intercept)         6.7199     0.0486  138.31   <2e-16
## WrittenFrequency    0.1644     0.1493    1.10     0.28
## Familiarity         0.1379     0.1438    0.96     0.34
## new_var            -0.1866     0.1478   -1.26     0.21
##
## Residual standard error: 0.081 on 36 degrees of freedom
## Multiple R-squared:  0.529,Adjusted R-squared:  0.49
## F-statistic: 13.5 on 3 and 36 DF,  p-value: 4.74e-06
```

If we assume that the coefficient estimates for the first model are relatively close to reality (which you can check by redoing the model on the full `english` dataset), the results of the second model are bizarre. The effects of `WrittenFrequency` and `Familiarity` are in the opposite direction from common-sense expectations (e.g., "higher frequency words should be identified faster"), and neither effect has a low p-value. Furthermore, the model seems "unstable" in the sense that adding a single predictor drastically changes our qualitative conclusions.

Note that nothing in the output of `bad_mod_2` (alone) indicates a problem, despite there being near-perfect linear dependence among predictors. This is part of what's dangerous about collinearity; because it's not a violation of model assumptions, it can easily go unnoticed.

5.6.2.2 Example: High collinearity

Suppose that the `english_40` data were from a pilot study where we are interested in the effect of word familiarity (`Familiarity`) on lexical decision RT (`RTlexdec`)

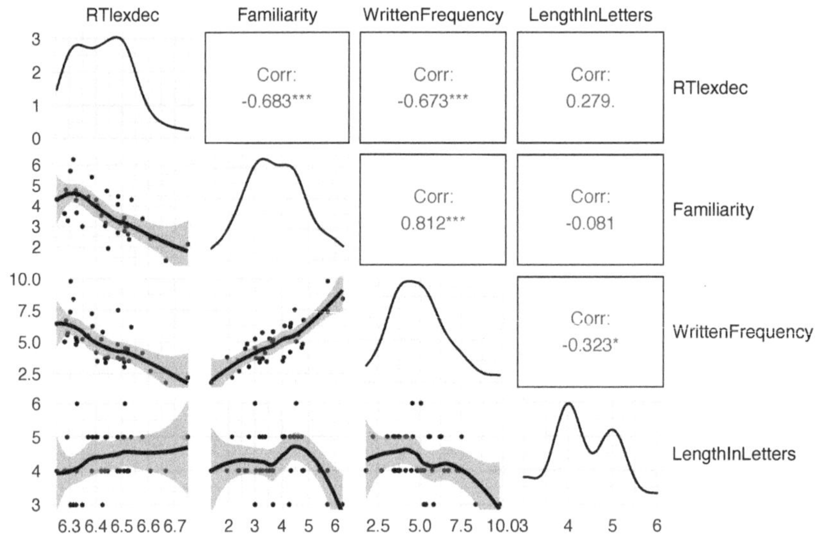

Figure 5.11
Pairwise plots of several variables from the `english_40` dataset, with nonlinear (generalized additive model) smoothers and 95% CIs (lines/shading), and Pearson correlation for each pair. Diagonal panels are empirical distributions of each variable.

by young speakers. We would like to control for a word's `WrittenFrequency` and `LengthInLetters`, because both are expected to correlate with `Familiarity` and to affect RT. For this example, let $\alpha = 0.01$ be the significance threshold.

Consider the correlations between these variables (figure 5.11). As expected, both frequency and familiarity are negatively correlated with RT, and positively correlated with each other. Correlations with word length look weaker than anticipated.

Fitting a simple linear regression of just familiarity on RT would give a significant negative effect:

```
lm(RTlexdec ~ Familiarity, data = english_40) %>% summary()
...
## Coefficients:
##              Estimate Std. Error t value Pr(>|t|)
## (Intercept)   6.6955     0.0479  139.82  < 2e-16
## Familiarity  -0.0712     0.0123   -5.77  1.2e-06
##
## Residual standard error: 0.084 on 38 degrees of freedom
## Multiple R-squared:  0.467,Adjusted R-squared:  0.453
## F-statistic: 33.3 on 1 and 38 DF,  p-value: 1.17e-06
```

Now consider a model of RT as a function of all three predictors:

```
english_mod_5 <- lm(RTlexdec ~ Familiarity + WrittenFrequency +
    LengthInLetters, data = english_40)
summary(english_mod_5)
```

```
...
## Coefficients:
##                   Estimate Std. Error t value Pr(>|t|)
## (Intercept)         6.5861     0.0990   66.53   <2e-16
## Familiarity        -0.0511     0.0216   -2.37    0.023
## WrittenFrequency  -0.0149     0.0147   -1.01    0.320
## LengthInLetters    0.0250     0.0189    1.32    0.195
##
## Residual standard error: 0.081 on 36 degrees of freedom
## Multiple R-squared:  0.531,Adjusted R-squared:  0.492
## F-statistic: 13.6 on 3 and 36 DF,  p-value: 4.42e-06
```

This model finds that none one of the three variables significantly affect RT at the $\alpha = 0.01$ level. `Familiarity` in particular now does not have a significant effect.

Comparing the two models, we can note a couple surprising things. First, although the coefficient estimates for `Familiarity` are consistent (each is in within $2 \cdot$ SE of the other), the SE of the coefficient roughly doubles in the second model. This is what is primarily behind the `Familiarity` effect "losing" significance. Second, almost half of the variance in `RTlexdec` is explained in the second model ($R^2 = 0.53$), despite no predictor having a low p-value.

This kind of situation is a *credit assignment problem*: we can tell that some combination of predictors together affects the response, but not whether an individual predictor does, after controlling for other predictors.

To get a better sense of what this means, consider the *correlation matrix* of the coefficient estimates, which describes how correlated the estimates of different coefficients are:

```
## Transform variance-covariance matrix to correlation matrix:
vcov(english_mod_5) %>% cov2cor()
##                   (Intercept) Familiarity WrittenFrequency
## (Intercept)              1.00        0.07            -0.42
## Familiarity              0.07        1.00            -0.83
## WrittenFrequency        -0.42       -0.83             1.00
## LengthInLetters         -0.88       -0.33             0.44
##                   LengthInLetters
## (Intercept)                 -0.88
## Familiarity                 -0.33
## WrittenFrequency             0.44
## LengthInLetters              1.00
```

For example, the coefficient estimate for `Familiarity` is highly correlated with the estimate for `WrittenFrequency` ($r = -0.83$) and moderately correlated with the estimate for `LengthInLetters` ($r = -0.33$). Figure 5.12 shows 99% confidence ellipses for each of these two pairs of coefficients. The model is confident that the `Familiarity` and `WrittenFrequency` coefficients lie somewhere roughly along a line; a higher estimate for one implies a lower estimate for the other. In contrast, the estimates for the `Familiarity` and `LengthInLetters` coefficients are weakly related. The left plot illustrates the credit

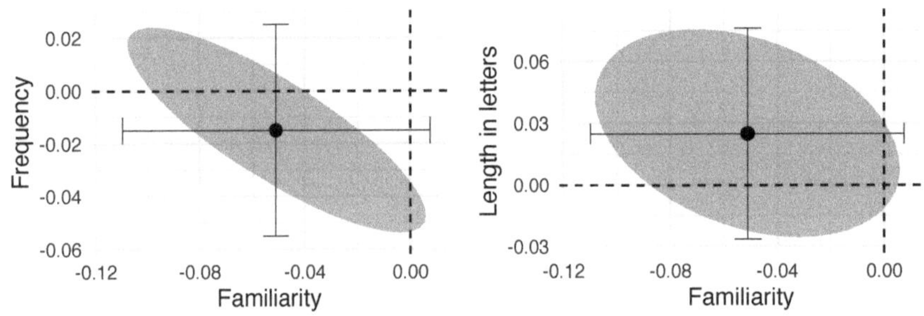

Figure 5.12
These are 99% confidence ellipses for pairs of coefficient estimates in model `english_mod_5`. Errorbars are 99% CIs for point estimates of coefficients. Dotted lines are 0.

assignment problem: the 99% confidence ellipse is consistent with either coefficient individually being 0 (overlaps with the dotted lines), but not with both coefficients being 0 (no overlap with the origin), which is possible because the estimates are highly correlated.

5.6.3 Effects of Collinearity and Diagnostics

High collinearity can have effects that seem undesirable, each of which we have observed in the examples thus far (e.g., Chatterjee and Hadi 2012, section 9.4):

- *Unstable coefficients*: large changes in values of the $\hat{\beta}_i$ when predictors/data points are added/dropped.
- *Nonsensical coefficients*: signs of $\hat{\beta}_i$ don't conform to prior expectations.
- *Unexpected nonsignificance*: values of $\hat{\beta}_i$ for predictors expected to be important (e.g., from exploratory plots) have large SEs and high p-values, or there are no "significant" coefficients despite R^2 well above 0.

Any of these can be thought of as a warning sign that there may be substantial collinearity in your data. They all follow from the fact that when data are highly collinear, a linear relationship between predictors almost holds, so there are many regression coefficient estimates that give models almost as good as the least-squares estimates. A common situation with high collinearity is where several closely related predictors are each correlated with the response (as in the credit assignment problem). In general, collinearity does not affect the actual values of coefficient estimates, just their SEs.

However, should we even be worried about these issues? Is (high) collinearity a problem at all? There are two philosophies here (box 5.6); the first one ("yes") has been influential in statistics-for-linguistics, while I am personally partial to the "no" philosophy. But this ultimately comes down to a choice between different analysis strategies with different risks.

One must also distinguish between *essential collinearity*, where predictors are inherently related, and *nonessential collinearity*, which can be eliminated by centering all predictor variables (Cohen et al. 2002, section 7.2). Only essential collinearity is potentially a problem. In addition, there are common situations where terms are collinear by construction, so collinearity can't be a problem: nonlinear effects (we expect `x` and `x^2` to be correlated) or interactions (`x:y` will usually be correlated with `x` and with `y`). There are other good reasons in these cases to consider transformations that would reduce collinearity (e.g., interactions are more interpretable for centered predictors: section 7.2; Dalal and Zickar 2012), but this is a side effect rather than the goal.

Whichever perspective on collinearity one takes, it is important to be able to diagnose collinearity, to decide to what extent it could affect inference for each predictor (mostly SEs, and thus p-values).

The qualitative warning signs we've discussed are not good diagnostics, because they are neither necessary nor sufficient to detect how much collinearity is present and whether it may be harmful (Belsley, Kuh, and Welsch 2004, 92). The degree of collinearity can be quantified in a couple common ways.

Box 5.6
Broader Context: Is Collinearity a Problem?

Yes Collinearity is often treated as a problem (e.g., Chatterjee and Hadi 2012; Zuur, Ieno, and Smith 2007; Belsley, Kuh, and Welsch 2004, Faraway 2015, section 7.3; and for language sciences Baayen 2008; Levshina 2015; Tomaschek, Hendrix, and Baayen 2018), because it can cause the issues discussed in the text; it can also slow down or foil model fitting (historically a major issue, less so with modern statistical software); and increased SEs will increase the likelihood of type II errors.

Therefore, collinearity should be dealt with somehow. Techniques include removing predictors, defining new predictors with the effects of others "residualized" out, using dimensionality reduction to combine collinear predictors into a single predictor giving equivalent information (e.g., "principal components"), or using a regression technique such as "ridge regression," which shrinks coefficient values (Harrell 2015, section 4.7, Tomaschek, Hendrix, and Baayen 2018) .

No The other perspective is "there is no general sense in which collinearity is a problem" (Morrissey and Ruxton 2018, 1), because collinearity is a property of the data, not a violation of model assumptions (e.g., Gelman and Hill 2007; O'Brien 2007; and for linguistics Levy 2012; Vanhove 2021). The issues listed previously reflect a lack of information in your data and how hard it is to detect effects of variables of interest using this dataset. As such, collinearity may help explain the results of a model (e.g., high collinearity suggests a type II error is likely), but this is not a problem in itself—in fact, it can provide useful information. Therefore, you should either fit the model with your original set of predictors (which acknowledges the lack of information), or collect more data (which raises power, decreasing the risk of a type II error). In the previous example, it makes intuitive sense that it is harder to detect a "real" effect of `Familiarity` when `WrittenFrequency` is added as a predictor, because of the correlation between `Familiarity` and `WrittenFrequency`.

Methods to deal with collinearity can have unintended consequences on interpretability or lead to biased coefficients. Residualizing one predictor, the most commonly used strategy to deal with collinearity in linguistics over the past ~15 years (e.g., in my work: Carlson, Sonderegger, and Bane 2014), is generally not a good idea, because it complicates the interpretation of the regression coefficients in unintended ways (York 2012; Wurm and Fisicaro 2014). Dimensionality reduction can result in predictors whose coefficients are hard to interpret. Most importantly, dropping predictors will in general bias the coefficient estimates of remaining predictors, as does ridge regression, potentially resulting in type I errors. These methods may still all make sense in a given case (except residualizing), but you should be aware that they are not without issues.

What to do? Whether to deal with collinearity or not is an *analysis choice*, which often comes down to type I versus type II error risk (section 3.3.5). (This assumes you have already minimized "nonessential collinearity" which is uncontroversial.) More generally, dealing with collinearity is often convenient from the perspective of actually answering one's research questions, but doing so can require more complex interpretation and/or caveats in reporting results. Not dealing with collinearity is technically more correct, especially if we prefer avoiding type I errors, but can lead to less useful conclusions about research questions. (Usually one would like to say more than "more data are needed.")

If you work with linguistic data where collinearity among theoretically important predictors is an important issue—such as disentangling frequency and probability effects on language production—it is worth reading a full treatment of these issues, which are subtle but not difficult to understand. Morrissey and Ruxton (2018) is particularly readable.

5.6.3.1 Condition number

First, we can quantify the answer to "how far is the whole set of predictors from being linearly independent?" by using the *condition number* κ, which is calculated for the design matrix (where each column = values of one predictor, across the dataset). κ characterizes how sensitive the coefficients of a linear regression are to changes in the predictors (lower sensitivity is better). $\kappa > 30$ and $\kappa < 5$ are common cutoffs for "high" and "no" collinearity, but like all quantitative diagnostic measures what counts as "bad" is primarily an empirical question. (See Belsley, Kuh, and Welsch 2004, chapter 3 for details.)

For example, we can apply `collin.fnc()` from languageR to the matrices of predictors for the previous examples to calculate their condition numbers:[8]

```
# 'Very high' collinearity example
X1 <- english_40_bad %>% select(WrittenFrequency, Familiarity, new_var)
collin.fnc(X1)$cnumber
## [1] 254
# 'High' collinearity example
X2 <- english_40 %>% select(Familiarity, WrittenFrequency, LengthInLetters)
collin.fnc(X2)$cnumber
## [1] 21
```

The "very high collinearity" example has $\kappa = 254$, which indicates extreme collinearity. The "high collinearity" example has $\kappa = 21$, indicating moderate collinearity (by the cutoffs discussed).

5.6.3.2 Variance inflation factors

Another measure of collinearity uses both the predictors and the response (y). Because the main problem caused by collinearity is to increase SEs for each predictor x_i, an intuitive measure of "how much does collinearity increase SE?" is a *variance inflation factor* (VIF): the ratio of (sample) variance for x_i in the multiple regression ($y \sim$ all predictors) to a simple regression ($y \sim x_i$). This turns out to be just

$$\text{VIF}_i = \frac{1}{1 - R_i^2},$$

where R_i^2 is R^2 for a regression of x_i on all other predictors. VIF > 10, meaning at least one coefficient has variance inflated at least 10-fold, is a common (if arbitrary) diagnostic for collinearity that "may be causing problems" (Chatterjee and Hadi 2012, 238).

A common way to "do something" about collinearity if this choice is made (box 5.6) is to use VIF or κ cutoffs, typically to remove or combine predictors. This is a controversial

8. This implementation first adds the intercept to the design matrix and scales each predictor (divided by one SD), as recommended by Belsley, Kuh, and Welsch (2004, 157).

practice, not least because neither measure accounts for sample size (for high enough sample size, even highly correlated effects can be precisely estimated), but also this practice is very common, so it's good to know about.

VIFs can be calculated using `vif()` in the car package. For the previous examples:

```
vif(bad_mod_2)
## WrittenFrequency       Familiarity             new_var
##              374               146                 888
vif(english_mod_5)
##      Familiarity WrittenFrequency   LengthInLetters
##              3.3              3.7               1.3
```

By the VIF = 10 cutoff, we would conclude that there is potentially problematic collinearity in `bad_mod_2`, with the variance of `new_var` particularly inflated (by over 800 times), but that the degree of collinearity in `english_mod_5` is acceptable (maximum VIF = 3.7).

Box 5.7
Broader Context: Measurement Error

A further assumption of most regression models is that all predictor values—x_{11}, x_{52}, and so on—are measured without error. This is often not the case; for example in the `vot` dataset, speaking rate is based on an automatic algorithm with some margin of error; word frequencies were determined based on a finite corpus; and so on. The effects of measurement error are complex, but as a rule if measurement error **is** taken into account for x_i, its effect in the fitted model will be weaker (β_i closer to 0). Thus, usually our regression coefficients are biased away from 0 (meaning anticonservative p-values, etc.), and we should keep this in mind in interpreting a model. Effects of measurement error are small as long as measurement errors are small relative to the residual error (ε_i). If you know that there is nontrivial measurement error for an important predictor in your model, by this metric, it's worth reading more (e.g., Faraway 2015, section 7.1; Kline 2013, 140–142).

5.7 Problems with Observations

The main linear regression assumption about observations is that they are roughly equally **influential**. This assumption is important, whether our goal is estimation or prediction—both abstract away from our finite sample. If certain observations are much more influential than others, they skew the regression results to reflect not the population, but the particular sample we happened to draw.

We are most interested in diagnosing observations with high *influence*. These tend to be observations that have extreme predictor values, called high *leverage*, or observations that the model fits particularly badly (∼large residuals), called *outliers*. It is useful to identify outliers or high leverage points for two reasons. First, they could be influential, and for many regression models (e.g., mixed-effects models) it is easy to find outliers or compute leverage, but hard (computationally intensive) to compute influence. Second, they are worth examining for their own sake, as they often indicate a problem with the data or model. What to do about influential observations (or high leverage/outlier observations) when they don't reflect errors is more subjective. Often, some points are inherently more influential than

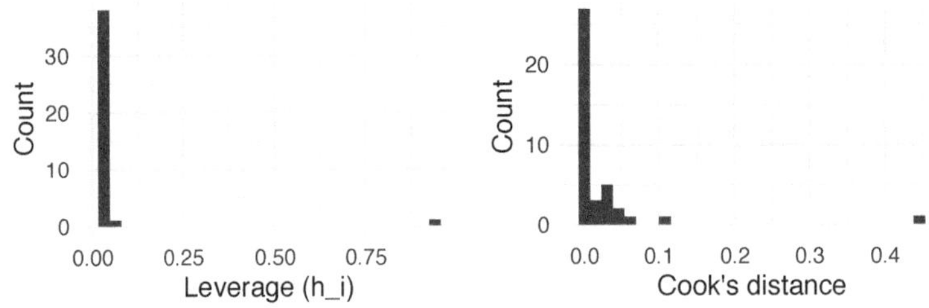

Figure 5.13
Distribution of leverage and influence measures for model `slr_mod_4`.

others (say, a couple participants with behavior very different from others), and then we will need to decide how to proceed in fitting and interpreting the model.

The presence of highly influential observations can lead to either type I or type II errors: the influential observations might be responsible for a spurious result, or they might obscure a pattern that would be clear if they were excluded.

5.7.1 Leverage

The leverage of observation i, written h_i, is how far out in predictor space the observation is from an "average" observation (each predictor = its mean value), on a scale from 0 to 1.[9] These are also called "hat values" (hence the h), and they can be calculated in R using `hatvalues()` or just as a column resulting from `augment()` (see example in section 5.3.2.1).

For example, consider model `slr_mod_4` (from section 5.5.2.1), which corresponds to the relationship in figure 5.9 between raw frequency (x) and naming latency (y) in the `english_40` data. Plotting a histogram of leverage (h_i) values shows that there are two extreme observations (figure 5.13, left).

These are the observations with raw frequency greater than 5000. It is visually clear that these observations will have much greater influence than other observations; hence the very large CIs for this frequency range.

This is a simple example of a common cause of high leverage: a nonnormally distributed predictor. When this is the case, often transforming the predictor to normality results in more equal leverage across observations.

5.7.2 Outliers

There are two aspects to defining outliers: choosing a metric (outlier in what space?) and a criterion for outlier-hood in that space.

5.7.2.1 Defining outliers

Informally, researchers often define outliers as points with an unusual y value, sometimes using a cutoff like "3 SDs away from the mean"—so, the metric is "z_i, the standardized value of the yth observation." This can be a good first pass, especially to check for observations

9. Formally, h_i is the Mahalanobis distance of observation i from the average observation, using the covariance matrix of the predictors (Faraway 2015, 84).

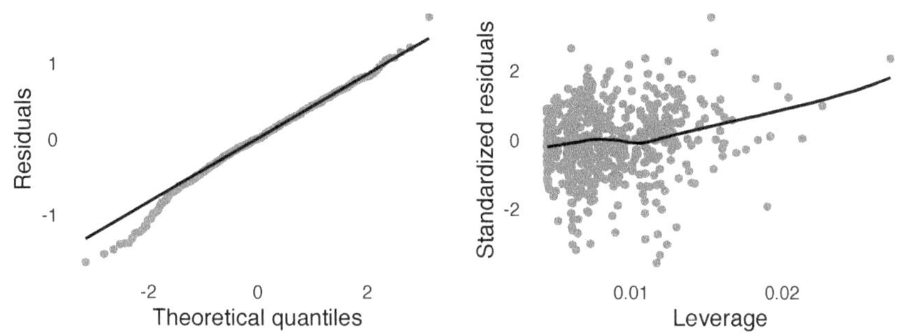

Figure 5.14
Diagnostic plots for vot_mod_2: Q-Q plot of residuals (*left*) and residual-leverage plot (*right*).

with gross problems (e.g., annotation errors). But because the distribution of y does not control for any predictors (which is a primary motivation for doing a regression), it makes more sense to define outliers in terms of how well the model fits observations, and doing so requires a metric for comparing $\hat{y}$ and y.

The (absolute value of the) residual, $|\hat{\varepsilon}_i| = |\hat{y}_i - y_i|$, seems like a natural metric, but it has two issues: it is not on an interpretable scale (we would like "three" to mean "3 SDs from the mean"), and it does not account for the fact that observations further from the mean have higher variance (box 5.2), that is, lower leverage (h_i). Defining the *standardized residuals* r_i addresses both issues

$$r_i = \frac{\hat{\varepsilon}_i}{\hat{\sigma}\sqrt{1-h_i}},$$

where $\hat{\sigma}$ is the residual standard error (equation (4.6)), the estimate for "how much variability is left" in the data.

Outliers are points that have unusual values of r_i by some criterion: either visual inspection or a formal test, such as the Bonferroni outlier test (`outlierTest()` in the car package; e.g., Faraway [2015, 6.2.2]). As always, visual inspection is preferred to quantitative diagnostics for model validation, but sometimes visual inspection is impractical, such as if you are running dozens of models. If you do use a quantitative diagnostic for outlier detection, it is better to use a test than a cutoff ("any observation with $|r_i| > 3$ is an outlier").

So to summarize: r_i is a better metric than z_i, and for deciding outlier-hood, visual inspection is better than a formal test, which is better than a strict cutoff.

5.7.2.2 Example

The standardized residuals can be calculated using `augment()` (the `.std.resid` column) or `rstandard()`. Standardized residuals are often used in a *residual-leverage* diagnostic plot (one of those made automatically when `plot()` is applied to a linear regression model), which shows h_i (x-axis) versus r_i (y-axis). Points that have high values on either x or y may have high influence and should be flagged for further inspection. Often it is useful to determine *why* they are outliers—why is the prediction unusual given the predictor values?

Figure 5.14 shows a Q-Q plot of standardized residuals and a residual-leverage plot for model `vot_mod_2`. The Q-Q plot suggests one possible outlier with high r_i and maybe some with low r_i. The highest and lowest-r_i observations are

```
vot_mod2_sorted <- augment(vot_mod_2) %>%
  arrange(.std.resid)
tail(vot_mod2_sorted, n = 1)
##   log_vot voicing  place cons_cluster speaking_rate .fitted .resid
## 1     4.3  voiced labial          yes           2.7     2.7    1.6
##      .hat .sigma .cooksd .std.resid
## 1 0.015   0.46   0.039        3.6
head(vot_mod2_sorted, n = 1)
##   log_vot voicing  place cons_cluster speaking_rate .fitted .resid
## 1       1  voiced labial          yes           3.6     2.6   -1.6
##      .hat .sigma .cooksd .std.resid
## 1 0.012   0.46    0.03       -3.5
```

The first one turns out to be the voiced stop observation (`voicing` = VOICED) with the highest VOT (75 msec; compare with figure 5.3, right). r_i is high because the observation has a very high VOT for a voiced stop. The second one is the lowest-VOT observation among observations with a consonant cluster (`cons_cluster` = YES), which tends to dramatically raise VOT. Both points bear further inspection; perhaps they are annotation errors.

5.7.3 Influence

There are different ways to quantify how influential an observation is for a model. Most commonly used is *Cook's distance* (CD), C_i, which measures how much the regression coefficients change when the model is fitted with all data except observation i. This turns out to be (Chatterjee and Hadi 2012, section 4.9)

$$C_i = \frac{r_i^2}{k+1} \frac{h_i}{1-h_i}, \tag{5.2}$$

where k is the number of predictors. On the right-hand side of this equation, the first term is high for points with high outlier-ness (the squared standardized residual), and the second term is high for points with high leverage (h_i closer to 1), so CD will be high for observations that are outliers or have high leverage. Observations with high CD are highly influential and should be flagged and examined. Again, it is better to decide what CD is "high" by visual inspection than by a numerical cutoff; 1 and $4/n$ are commonly used cutoffs.

Equation (5.2) captures the intuition that highly influential points tend to be either outliers or have high leverage, but neither is a necessary or sufficient condition for high influence. This intuition also holds for other measures of influence, such as DFBETA (`dfbeta()` in R), which quantifies the influence of each observation on a particular coefficient.

Examples CD can be calculated using `cooks.distance()` or `augment()` (the `.cooksd` column).

For the `slr_mod_4` model, we saw that there are high-leverage points, but the distribution of CD (figure 5.13, right) suggests that there are no highly influential points—those with very different CD from other points, or with CD > 1. This is because the high-leverage points are consistent with the general trend (higher $x \Rightarrow$ lower y). Similarly, it turns out there are no clear high-influence points for the `vot_mod_2` model even though there were outliers (Exercise 5.3).

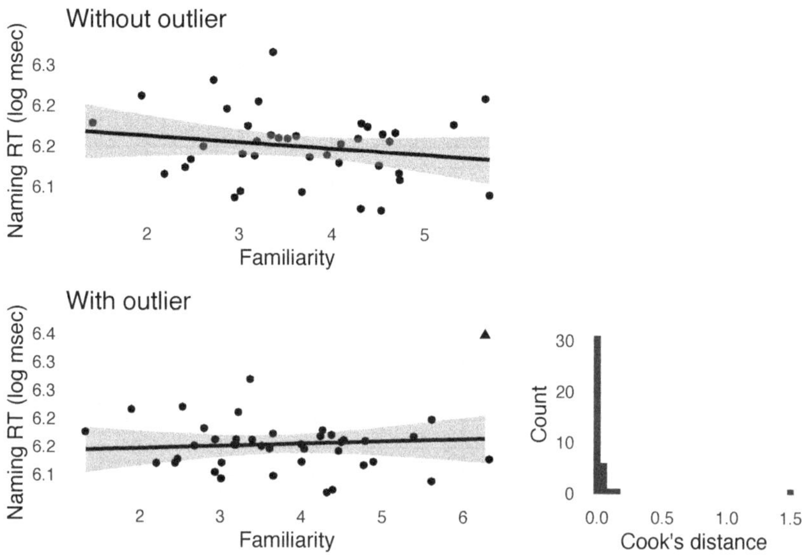

Figure 5.15
Relationship between word familiarity and naming RT for the `english_40` data, with line of best fit and 95% CI, before (*top left*) and after (*bottom left*) changing one observation (the triangle point). Histogram (*right*) shows distribution of influence measures for the latter.

To get an intuition for what it means for observations to have high influence, let's tweak the `english_40` data (from section 5.5.2.1) so that the point with highest `Familiarity` has high `RTnaming`:

```
# Tweak data to illustrate point
english_40_new <- english_40
english_40_new[english_40_new$Familiarity > 6, ]$RTnaming <- 6.35
```

Figure 5.15 shows the relationship between these variables predicted by a simple linear regression without and with this point (left/center panels). The point with high `RTnaming` has much higher influence than the others, as measured by CD (right panel), with the effect that removing this point changes the direction of the `Familiarity` effect, and thus our qualitative conclusion from the model. If we were writing up an analysis of this dataset it would be particularly important to flag and discuss this point.

5.7.4 What to Do about Influential (and Other) Observations

Compared to finding influential (or high-leverage/outlier) observations, "deciding what to do about them can be difficult" (Faraway 2015, 83). There are different philosophies, and what makes sense to do depends on one's dataset and research questions. Still, a few general points can be made.

At a minimum, during data analysis:

1. You should look for high-influence/outlier/high-leverage observations (often just called "outliers"). Just examining these observations often reveals gross errors, such as a participant who always gave the same response or data coding errors that cannot be corrected; these data can be discarded. Examination could also reveal issues that will improve your analysis, such as a predictor that should be transformed (e.g., in the preceding example).
2. Such observations that are not "errors" should be flagged and examined. You might try excluding them to see what effect this has on the analysis.
3. Report how influential/high-leverage/outlier points were dealt with, what points were discarded, and ideally whether excluding nonerror influential points (if step (2) is taken) would affect your analysis.

Note that (3) does not necessarily mean detailed reporting of the model validation itself (e.g., plots of CD). These could make sense for an appendix but are not usually reported in the text of a write-up. A short summary suffices.

By far the most common method to deal with influential (or high-leverage/outlier) points, including in linguistic research (e.g., my own work: Stuart-Smith et al. 2019), is simply to exclude them automatically (e.g., all points with standardized residual >2.5). This is not a good method (Faraway 2015, 89) because it can eliminate observations that are actually the most meaningful, but it is better than nothing, especially for cases where manual inspection of influential points is infeasible. However, refitting a model after automated outlier exclusion is a nice way to check that your qualitative conclusions are not just due to influential points (Baayen 2008, 279).

Another option for dealing with outliers is not to exclude them at all (even if they are clear errors), but to model them explicitly, by using a method that allows for outliers (e.g., "robust regression"; see Faraway [2015, section 8.4]).

5.8 Trade-Offs between Models

Before discussing the problem of choosing between different regression models, we need a high-level understanding of consequences of using one model or another. In building any statistical model there are closely related trade-offs: underfitting versus overfitting and bias versus variance.

We start by assuming the goal of model-building is estimation, as is often the case for linguistic data, to introduce the bias-variance trade-off. We then turn to underfitting and overfitting, which is easier to think about assuming our goal is prediction. (However, both trade-offs hold whether the goal is estimation or prediction.)

5.8.1 The Bias/Variance Trade-Off

If our goal in building a model is coefficient estimates, different possible models trade off between two problems, corresponding to too few and too many predictors (Chatterjee and Hadi 2012, section 11.2): high bias and high variance (defined in section 3.4.1). Suppose for simplicity that we are particularly interested in estimating the effect of x_1, corresponding to coefficient β_1, and have to decide whether to include a term for x_2; the general points hold for the more general case (multiple effects of interest, multiple candidate predictors to add).

Bias If x_2 does in reality affect the response y ($\beta_2 \neq 0$), and x_2 isn't completely orthogonal to x_1, the bias in our estimate of β_1 will increase when x_2 is omitted: on average, our estimate will be off by more. This is called *omitted variable bias* and makes intuitive sense: if x_1 and x_2 affect y but only x_1 is included in a model, some of what looks like an effect of x_1 could in reality be an effect of x_2 (or the effect of x_1 could be stronger once x_2 is controlled for).

Variance On the other hand, if x_2 does not in reality affect y ($\beta_2 = 0$) or is orthogonal to x_1, and we do not omit x_2, the degree of variability in our estimate of β_1 across runs of the experiment (its *variance*) will increase. This follows similar logic to our discussion of collinearity, where we saw that adding a highly collinear predictor x_2 increases the SE of the effect of x_1, because the model is "less sure" what variance in y to attribute to x_1 versus x_2. (Unless x_1 and x_2 are completely orthogonal.)

Bias-variance trade-off Because we usually don't know whether x_2 does have an effect, the choice of whether to add x_2 to the model risks either increasing bias or variance, which is a specific instantiation of the *bias-variance trade-off*. Intuitively, too many predictors lead to overfitting to noise, but any method to control overfitting (e.g., dropping some predictors or using robust regression) introduces bias. A more quantitative way to think of the bias-variance trade-off is by thinking of the goal of estimation as minimizing *mean-squared error*: how off our estimate $\hat{\beta}$ of a parameter (whose true value is) β is, on average, $E(\beta - \hat{\beta})^2$. It turns out that (Hastie, Tibshirani, and Friedman 2009, section 7.3)

$$E(\beta - \hat{\beta})^2 = \text{Variance}(\hat{\beta} \mid \beta) + \text{Bias}(\hat{\beta} \mid \beta)^2$$

That is, the mean-squared error is the sum of the variance and the square of bias. We want an estimator (in our case, a linear regression model with certain predictors) to minimize both variance and bias, hence reducing error; in reality different estimators tend to trade off. Our goal is to find the sweet spot between bias and variance.

As a concrete example, thus far, our VOT models assume that `speaking_rate` affects VOICED and VOICELESS stops equally, but in fact this is probably wrong (figure 5.6, right): speaking rate affects voiceless stops much more, which we could account for by adding a SPEAKING_RATE:VOICING term to the model. Without this term, our estimate for "`speaking_rate` effect for voiced stops" was biased away from 0, and our estimate for "`speaking_rate` effect for voiceless stops" was biased toward 0. On the other hand, there are many possible predictors whose effects might interact with `voicing` and aren't orthogonal to speech rate (e.g., word frequency); the more interactions we include with `voicing` for these terms, the more the SE of our `speaking_rate` effects will go up, and the more variable the coefficient values would be if we repeat the experiment. The "best" model is one where the added terms, such as `speaking_rate:voicing` (which from previous work we are pretty sure has a large effect), decrease variance more than the square of the bias the *not*-added terms introduce.

5.8.2 Overfitting and Underfitting

A closely related trade-off when terms are added/removed is between a model that doesn't fit the observed data well (*underfitting*) and a model that doesn't generalize well to unobserved data (*overfitting*); this is easiest to get an intuition for if our modeling goal is prediction. We are used to thinking about underfitting; this is just goodness of fit on the dataset used to fit the model, captured for example by R^2. For overfitting it's helpful to see an example.

Example Consider the `vot_michael` data. Suppose we had just 50 observations, which we simulate by choosing them randomly; this is the "training" data, and the rest of `vot_michael` is the "test" data:

```
set.seed(100)
inds <- sample(1:nrow(vot_michael), 70)
vot_train <- vot_michael[inds, ]
vot_test <- vot_michael[-inds, ]
```

Now consider three different models of VOT fitted to the test data, each obtained by adding more predictors to the previous model:

- Model 1: `voicing` only ($k = 1$)
- Model 2: add five other plausible predictors, based on previous work ($k = 6$)
- Model 3: add all two-way interactions ($k = 16$)

```
ou_mod_1 <- lm(log_vot ~ voicing, data = vot_train)
ou_mod_2 <- update(ou_mod_1, . ~ (voicing + speaking_rate +
    foll_high_vowel + cons_cluster + log_corpus_freq + stress))
ou_mod_3 <- update(ou_mod_1, . ~ (voicing + speaking_rate +
    foll_high_vowel + cons_cluster + log_corpus_freq + stress)^2)
## show formula, to see how many terms this is
formula(ou_mod_3)
## log_vot ~ voicing + speaking_rate + foll_high_vowel + cons_cluster +
##       log_corpus_freq + stress + voicing:speaking_rate +
##       voicing:foll_high_vowel + voicing:cons_cluster +
##       voicing:log_corpus_freq + voicing:stress +
##       speaking_rate:foll_high_vowel + speaking_rate:cons_cluster +
##       speaking_rate:log_corpus_freq + speaking_rate:stress +
##       foll_high_vowel:cons_cluster + foll_high_vowel:log_corpus_freq +
##       foll_high_vowel:stress + cons_cluster:log_corpus_freq +
##       cons_cluster:stress + log_corpus_freq:stress
```

Now consider the correlation between predicted and observed values of y—that is, R^2—on the training and test data (code not shown):

	Train R^2	Test R^2
Model 1	0.79	0.67
Model 2	0.82	0.72
Model 3	0.86	0.72

For the training data, R^2 is highest for the most complicated model (3), as we expect: adding more predictors can only increase the proportion of variance explained. A too-simple model such as model 1 underfits.[10]

But on the test data, as predictors are added, R^2 goes up and then down: it is model 2 that has highest R^2 on data not used to fit the model. Model 3 has the lowest R^2 because it overfits to noise in the training data.

At the extremes (e.g., models 1 and 3), we can see why overfitting and underfitting are bad. We would like the model that trades off best between them and that would predict best on unseen data. But we don't typically have unseen data, so we have to somehow find this sweet spot using just the data we have.

Assessing overfitting Different criteria covered elsewhere (adjusted R^2, F-test, AIC, BIC; see sections 4.4.3, 5.9.1, 5.9.2) try to balance over- and underfitting when comparing different models, but they are not suitable for assessing the degree to which a *single* model overfits.

A useful rule of thumb is that a linear regression model fitted to n observations can contain about $n/15$ predictors before overfitting becomes a concern (Harrell 2015, section 4.4). This rule of thumb is rough—other factors, especially predictor distributions, affect whether overfitting occurs—but it is helpful for critically reading papers or when a better method is computationally intractable.

There are various more rigorous methods to assess overfitting, called "internal validation" or "external validation" depending on whether you check for overfitting using held-out data (section 5.3). In general external validation is better as long as the test set is large enough and you can truly keep it unseen; otherwise performance on the test set isn't a good metric. For regression modeling in scientific research (as well as machine learning for smaller datasets), this option is often unrealistic because data are too expensive to give up a substantial fraction and it is hard to keep it truly unseen.

Here are some common internal methods, ranked in order of worst to best (section 5.3). These methods are described in terms of R^2, but they apply to any goodness-of-fit measure.

- Simply reporting a measure of goodness of fit (e.g., R^2) on the data used to fit the model. This is the most common in linguistic research.
- *Data-splitting*, where the data are split into training and test sets, then the model is fit and validated using these subdatasets in some fashion, before refitting on the full dataset. For example, a method used in variationist sociolinguistics to avoid overfitting is to split the dataset in half and fit a model with all candidate terms to each half (e.g., Labov 1994). The final model, fitted to the full dataset, only contains terms which were significant in both submodels.
- *Cross-validation*, where the data are split into k groups. For each group, generate predictions using a model trained on the other $k-1$ groups. Compute R^2 using the full set of predictions.

10. In this case we know model 1 underfits, by construction. Much previous work suggests `voicing` is not the only thing that affects VOT.

- *Bootstrapping*, where we resample from the dataset many (hundreds) of times to make new datasets; then refit the model to each dataset and calculate R^2 on the original dataset. This gives a distribution of R^2 values, which lets us know how much our observed R^2 values (on the original dataset) should be adjusted downward for overfitting.

For linear and logistic regression models, there is excellent functionality in the rms package, discussed for linguistic data by Baayen (2008, section 6.2.4), which makes the bootstrapping option easy. (Other packages, such as caret, have better functionality but are a bit harder to use.) Section 6.8.4 gives a practical example for a case where overfitting is a real concern.

For our example (code not shown), R^2 adjusted for overfitting using a bootstrap method (from `rms`) ends up being 0.78, 0.79, and 0.63 (models 1, 2, 3), suggesting that model 2 trades off the best between underfitting and overfitting. This is not the same thing as "is the best model" (box 5.8), but it's important nonetheless.

Box 5.8
Broader Context: How Much Should We Care about Overfitting?

A thornier issue is how important avoiding overfitting is when building regression models for scientific understanding. If we are *primarily* interested in prediction—building a model that predicts well on unseen data—overfitting is a primary concern. This is the usual setting in machine learning and artificial intelligence. But commonly in language sciences, prediction is not our (only) goal; often estimation/explanation is more important (section 4.2). From the estimation perspective, overfitting does matter but may not be as much of a concern (Harrell 2015, section 4.12.2). Biased estimates may be a larger concern than variable ones, if we are most concerned about type I errors. There are often important qualitative desiderata for a regression model: to test specific hypotheses and to be interpretable in order to give insight into scientific questions. Both considerations can suggest including certain predictors in a model, whether they may lead to overfitting (Baguley 2012, section 12.4.1). Also, the concept of "unseen" data may not make sense. Sometimes our dataset is all the data (for example, all closed-class words of English), and sometimes (e.g., in observational studies or when studying highly endangered languages), we can't afford to actually "hold out" data, which is expensive to collect and includes crucial cases for our scientific questions. Put simply, regression models developed to optimize prediction (held-out test data, cross-validation to choose parameters, and so on) would not necessarily be optimized for insight into the scientific questions that motivate us.

There are two corresponding sorts of high-level philosophies about regression modeling that you may encounter (Shmueli 2010), and it is useful to be aware that they exist because they lead to rather different recommendations, neither of which is "right." The first is that overfitting is not a primary issue, unless the researcher's primary goal is prediction. This viewpoint underlies many regression textbooks, such as Chatterjee and Hadi (2012) and Faraway (2015), where overfitting is barely mentioned. The second is that the primary goal of modeling is prediction on unseen data, and even if your scientific interest is in estimation, prediction is a well-defined goal, optimizing which often leads to a model that is also optimized for estimation/explanation. This view is held by others such as Harrell (2015), Pinheiro and Bates (2000), and McElreath (2020), who have been influential in statistics-for-linguists.

5.9 Model Comparison

In realistic data analysis we usually do not know in advance the exact form of the regression model—which predictors will be used and what terms will be included—so it is necessary to compare several possible models of the same data. Choosing between models is called *variable selection* (or *model selection*). To perform variable selection, we need a way to perform *model comparison*: comparing two or more candidate models to assess which one is better, in the sense of how well it balances fit and parsimony (number of predictors), which is very similar to balancing bias/variance or overfitting/underfitting.

We wish to compare models of the same dataset, with the same response (y) and different sets of predictors (the x_i). Model comparison techniques differ on whether they can compare "nested" models only or both nested and "nonnested" models.

5.9.1 Nested Model Comparison

Two models are *nested* if one is a subset of the other, in terms of the set of predictors included; they are called the *full* (or *superset*) and *reduced* (or *subset*) models.

A very common case is where the models differ in a single predictor, such as these two nested models predicting `RTlexdec` for the `english` dataset:

$$M_1 \; : \; \texttt{RTlexdec} = \beta_0 + \beta_1 \cdot \texttt{WrittenFrequency} + \varepsilon$$

$$M_2 \; : \; \texttt{RTlexdec} = \beta_0 + \beta_1 \cdot \texttt{WrittenFrequency} + \beta_2 \cdot \texttt{Familiarity} + \varepsilon$$

To compare these two models, we wish to perform a test of the null hypothesis that $\beta_2 = 0$. More generally, the two nested models can contain any number of predictors:

- M_1: response y, predictors $x_1, ..., x_q$
- M_2: response y, predictors $x_1, ..., x_k$ (where $q < k$)

We would like to compare these models by testing the null hypothesis that all the additional predictors in the superset model have no effect:

$$H_0 \; : \; \beta_{k+1} = \beta_{k+2} = \cdots = \beta_q = 0. \tag{5.3}$$

We do so using the RSS for each model (equation (4.10)), RSS_1 and RSS_2, which measures how well the model fits the data (lower is better). The RSS of a regression model can be divided by $n-k-1$—the "*df* of the model," where k is the number of predictors— to measure how well the model fits the data, given the sample size and number of predictors.

In addition, the difference in RSS values between the two nested models can be scaled to give a measure of how much RSS has gone down, relative to what's expected given the dropped predictors:

$$(RSS_1 - RSS_2)/(k-q).$$

Thus, the following test statistic measures how much the unexplained variance is reduced in the reduced model, with respect to the full model:

$$F = \frac{(RSS_1 - RSS_2)/(k-q)}{RSS_2/(n-k-1)}.$$

Intuitively, F divides how much effect dropping the $k-q$ extra predictors has on explained variance. RSS_2 will always be smaller than RSS_1, because adding predictors to a model always improves fit. The hypothesis test checks whether the superset model fits the data significantly better than the subset model, given the added complexity.

This test statistic follows an F distribution (under the null hypothesis; see equation (5.3)) with $k-q$, $n-k-1$ degrees of freedom, written as $F(k-q, n-k-1)$ (e.g., Chatterjee and

Hadi 2012, section 3.9). We can thus perform hypothesis testing using an *F-test*, which may be familiar if you have seen ANOVAs, and intuitively checks whether "explained variance" (significantly) increases relative to "unexplained variance."

For instance, for the `english_40` example, $k-q=1$ and $n-k-1=40-2-1=37$. So an F-test would be used, with test statistic $F(1,37)$.

Another common case of nested model comparison appears at the bottom of the `summary()` of every linear regression, which tests whether the k predictors together significantly affect the response. This is a comparison of the model with k predictors to a model containing only the intercept. This F-test can be thought of as the "significance" of the regression, so it is reported next to measures of goodness of fit (mean-squared error, adjusted/raw R^2), which give complementary information ("effect size"). This F-test is usually unsurprising because "significantly different from a model containing just the intercept" is a very low bar, but in some circumstances it can be interesting.

For example, consider model `english_mod_5` from section 5.6.2.2, where no single predictor had a low p-value despite the predictors together explaining a lot of variance (R^2):

```
summary(english_mod_5)
...
## Residual standard error: 0.081 on 36 degrees of freedom
## Multiple R-squared:  0.531,Adjusted R-squared:  0.492
## F-statistic: 13.6 on 3 and 36 DF,  p-value: 4.42e-06
```

The F-test confirms that the $k=3$ predictors together significantly reduce unexplained variance.

5.9.1.1 Effect size

The preceding F-test only addresses the significance of the difference between two nested linear regressions, but does not give an effect size, which is important to report with any hypothesis test (section 2.7).

An intuitive effect size in this case is the fraction by which the unexplained variance, $1-R^2$, increases when the extra predictors are dropped from the full model. If we write R^2 for models M_1 and M_2 as R_1^2 and R_2^2, this measure is

$$f^2 = \frac{R_2^2 - R_1^2}{1 - R_2^2}. \tag{5.4}$$

We'll call this *Cohen's f* (one of a family of measures with this name). A rule of thumb is that f^2 of 0.01, 0.06, and 0.16 are small/medium/large (Cohen 1992), keeping in mind the caveats on rules of thumb (section 3.2.3.1).

There is not an R function to calculate f^2 for this exact case, but it's simple to write one ourselves (using the `r2()` function from performance):

```
## m1, m2: full model, reduced model
cohens_f2 <- function(m1, m2) {
  as.numeric((r2(m2)$R2 - r2(m1)$R2) / (1 - r2(m2)$R2))
}
```

Also commonly used are variants of η^2, but these are less intuitive.

5.9.1.2 Examples

Suppose for the purposes of these examples that $\alpha = 0.01$. To do model comparisons in R you fit nested models and then conduct the F-tests using `anova()`. For the first `english` example:

```
m1 <- lm(RTlexdec ~ WrittenFrequency, english)
m2 <- update(m1, . ~ . + Familiarity)
anova(m1, m2)
## Analysis of Variance Table
##
## Model 1: RTlexdec ~ WrittenFrequency
## Model 2: RTlexdec ~ WrittenFrequency + Familiarity
##   Res.Df  RSS Df Sum of Sq   F Pr(>F)
## 1   4566 91.2
## 2   4565 88.2  1      3.03 157 <2e-16
```

This model comparison suggests that adding `Familiarity` to `m1` significantly reduces its unexplained variance. Cohen's f^2 is 0.034, suggesting this is a small effect.

```
cohens_f2(m1, m2)
## [1] 0.034
```

This kind of model comparison, to check whether dropping or adding a single term to a regression model is justified, is extremely common in practice.

As an example comparing two models differing in multiple terms, let's compare the first and second models from our overfitting example (section 5.8.2), which used $q = 1$ and $k = 6$ predictors to explain log(VOT):

```
anova(ou_mod_1, ou_mod_2)
...
##   Res.Df  RSS Df Sum of Sq    F Pr(>F)
## 1     68 14.6
## 2     63 12.2  5      2.37 2.44  0.044
```

Adding the five predictors does not significantly improve the model (at $\alpha = 0.01$)—even though there are individual predictors (like `cons_cluster`) that would significantly improve the model if added in isolation, and the effect size of adding these predictors is large ($f^2 = 0.19$).

This kind of comparison is sometimes done for testing whether a group of conceptually related predictors together improve a model. Another common case is testing whether a factor with three or more levels improves a model (section 7.3).

It is useful to know that `anova()` can be used for nested model comparison of three or more models as well (example in section 5.10.3.1), for example, `anova(ou_mod_1,` ou_mod_2, ou_mod_3), in which case ou_mod_2 is compared to ou_mod_1 and ou_mod_3 to ou_mod_2.

Box 5.9
Practical Notes: F-tests and Nested Model Comparison

In linguistics and psychology, the F-test is sometimes reported without comment or *model comparison* used as shorthand, in the case of nested model comparison via an F-test to compare two linear regressions.

To report a model comparison, you follow the same guidelines as for any NHST test (section 2.7), reporting at minimum *df*, test statistic, p-value, and effect size. For instance, for the second of the preceding examples:

> "These five additional predictors expected to affect VOT based on prior research did not significantly improve the model of log(VOT) at the $\alpha = 0.01$ level ($F(5, 63) = 2.4$, $p = 0.044$), though the effect size of adding these predictors was large ($f^2 = 0.19$)."

In current practice it is common to report model comparisons without an effect size.

5.9.2 Nonnested Model Comparison

In nonnested model comparison, one model isn't a subset of the other. For example, these three models of RTlexdec for the english_40 data are not nested, because each uses a different set of predictors:

```
nn_m1 <- lm(RTlexdec ~ WrittenFrequency + Familiarity, english_40)
nn_m2 <- lm(RTlexdec ~ WrittenFrequency + LengthInLetters, english_40)
nn_m3 <- lm(RTlexdec ~ Familiarity + LengthInLetters, english_40)
```

Nonnested models can no longer be compared using sums of squares (via an F-statistic). Instead, a very different approach is used: compare models using a metric that captures how well they meet two conflicting goals we saw in section 5.8: fit (fitting the data as well as possible) and parsimony (have as few predictors as possible). We have seen one such metric already: adjusted R^2 (section 4.4.3), which measures how much variance is explained given the number of predictors.

More commonly used are *information criteria*, which combine model likelihood (L: higher is better) and number of predictors (k: lower is better) into a single value. The most widely used criteria are the *Akaike information criterion* (AIC):

$$AIC = 2k - 2\log(L), \tag{5.5}$$

and the *Bayesian information criterion* (BIC):

$$BIC = k\log(n) - 2\log(L), \tag{5.6}$$

where n is the number of observations.

"Corrected AIC" (AICc; use `AICc()` in the MuMIn package) is a variant of AIC, which is recommended for small enough sample size ($n/df < 40$; see Burnham and Anderson [2002]):

$$AICc = 2k\left[1 + \frac{k+1}{n-k-1}\right] - 2\log(L). \tag{5.7}$$

Written this way, we see that AICc is AIC increased by a penalty term that goes to 0 as sample size increases.[11]

To apply an information criterion you calculate its value for each model in a set of candidate models and pick the model with the lowest value. Information criteria can be used to compare both nested and nonnested models.

Example Consider the three candidate models used in the overfitting example (section 5.8.2). We can use AIC, AICc, or BIC to compare the models:

```
AIC(ou_mod_1, ou_mod_2, ou_mod_3)
##          df AIC
## ou_mod_1  3  95
## ou_mod_2  8  93
## ou_mod_3 22 104
```

```
AICc(ou_mod_1, ou_mod_2, ou_mod_3)
##          df AICc
## ou_mod_1  3   95
## ou_mod_2  8   95
## ou_mod_3 22  125
```

```
BIC(ou_mod_1, ou_mod_2, ou_mod_3)
##          df BIC
## ou_mod_1  3 102
## ou_mod_2  8 111
## ou_mod_3 22 153
```

AICc is a better choice than AIC here because the sample size is small for these models; using either criterion we would choose `ou_mod_2` (with 7 predictors). Using BIC we would choose `ou_mod_1` (with 1 predictor). This illustrates a couple of points: different model selection criteria don't necessarily give the same answer, and BIC tends to choose more parsimonious models than AIC does, because it penalizes additional parameters more. (AICc lies between the two.)

These are nested models, but we could do the same thing to compare nonnested models, such as `nn_m1`, `nn_m2`, and `nn_m3` which were defined previously. We would choose model 3 whether AIC, AICc, or BIC is used (not shown).

Box 5.10
Broader Context: Which Information Criterion to Use?

There is a vast literature on information criteria and their use for model selection (e.g., Burnham and Anderson 2002; McElreath 2020, chapter 6). Besides AIC/AICc/BIC some common ones are the deviance information criterion (DIC) and the widely-applicable information criterion (WAIC). Which one is better is a

11. $n-k-1$ in this equation is df for a linear regression.

complex and much-debated issue, which depends on your modeling goals and the type of data you're modeling. It is also an important issue because AIC(c) and BIC often give different results; this is not an analytical choice where we can say "if the choice you make affects the conclusions you can draw, beware." Things are even less clear for more complex regression models (e.g., mixed-effects models; see section 8.6.2). If model selection is important for your analysis you should read up on what method makes sense.

At least for non-mixed-effects models, a crude rule of thumb is that AIC is better when the goal is prediction and BIC is better when the goal is estimation (Shmueli 2010; Faraway 2015, section 10.3). The intuition is that AIC will tend to keep terms in that help in prediction even while increasing the variance of "large" effects you'd like to estimate, but BIC will remove such terms, leading to much more precise (lower-variance) estimates. (But BIC might eliminate terms that you want to estimate given your research questions, showing a limitation of this rule of thumb.) Another rule of thumb is that AIC is better when the true model contains effects of progressively smaller magnitude ("tapering"), but BIC is better when the true model contains no or a few large effects, and all other effects are 0 (Burnham and Anderson 2002). The tapering scenario sounds more plausible for corpus data, but I am not sure about other kinds of linguistic data.

5.10 Variable Selection

Now that we have seen some methods to compare models with different sets of predictors, we can turn to variable selection: how to decide what predictors to keep in a given model.

There is no single best way to do variable selection out of context; each method has pros and cons, and what method to use depends on the goals of your study: both in terms of your research questions, and whether your objective is prediction, explanation/estimation, or description (section 4.2).

A priori, your research questions will help determine which predictors you consider for your model and, depending on your modeling philosophy, may dictate that you leave certain predictors in the model no matter what. When your objective is an explanation, as is often the case for linguistic studies, typically it does not make sense to consider dropping terms that directly test your research questions, regardless of significance, because these effect estimates are effectively why you are doing the analysis (Baguley 2012, section 12.4.1).

But, even given your research questions, there are very different methods for variable selection, and it is important to know that these methods can result in different final models that are appropriate for different goals. Many papers that report variable selection simply say what they did without justification. It is up to the reader to think, what might the final model be if a different variable selection method were used? We already saw one example: using AIC for model selection tends to lead to larger models than using BIC does.

There are three *components* of any variable selection method:

1. How models are *compared* (e.g., F-test on nested models, AIC).
2. How the *quality* of a single model is evaluated.
3. How it's decided which of a set of models to *select*, based on components 1 and 2.

We have already seen a few possible methods for variable selection, by comparing models containing different sets of variables (component 1):

- Comparison using F-tests (nested linear regression models only)
- Comparison of information criterion values (e.g., AIC, BIC)
- Comparison of adjusted-R^2 values

In each example we implicitly used "choose automatically" for component 3.

We'll consider a couple more methods here. The first, stepwise model selection, is widely used but problematic. The second is an example of "holistic" variable selection methods that take theory and domain knowledge into account. This type of method is not currently in wide use in linguistic data analysis, but in my opinion should be the default.

5.10.1 Stepwise Variable Selection

The most common type of variable selection procedure, *stepwise* variable selection, also uses choose automatically for component 3 (how do we decide which model to prefer). Using this method we decide whether to add or drop terms based on a model comparison procedure (Component 1): AIC, BIC, F test with $p < 0.05$, etc.

Stepwise model selection can be conducted "forward" (start with an intercept-only model and add terms), "backward" (start with a full model and drop terms), or both.

A common option is that used by the `step()` function in R by default, stepwise backward selection using AIC:

1. Start with the complete regression model (all possible predictors) and obtain AIC.
2. At each step, remove the predictor that would decrease AIC the most.
3. Repeat steps 1 and 2 until the model has lower AIC than any of the models resulting from removing a predictor.

Stepwise model selection is very popular, because it is easy to do using statistical software, doesn't require much thought, and feels objective. But it has serious drawbacks, as we illustrate through an example. Box 5.11 gives more context.

Example: 34 potential predictors One basic property a variable selection method should have is consistency: if we apply the method to two similar datasets, we should get similar results. In this example we split the `english` data into two random halves, then fit a full model of lexical decision time (`RTlexdec`) with 34 potential predictors to each half and apply stepwise backward selection. This procedure simulates repeating the experiment twice that generated the `english` dataset.

The 34 predictors are simply most of the columns of the `english` data, all presumably included in the dataset because they might help explain RT.[12]

```
## Split the English data in half, randomly
set.seed(2903)
english_half_1 <- english %>% sample_n(nrow(english) / 2)
english_half_2 <- english[!(row.names(english) %in%
    row.names(english_half_1)), ]

step_mod_1 <- lm(RTlexdec ~
WrittenFrequency + Familiarity + AgeSubject +
  FamilySize + DerivationalEntropy + WrittenSpokenFrequencyRatio +
  WordCategory + InflectionalEntropy + NumberSimplexSynsets +
```

12. Note that this is not an unreasonable number of predictors, because we could have up to about 150 predictors before worrying about overfitting (by the divide-by-15 rule).

```
  NumberComplexSynsets + LengthInLetters + Ncount +
  MeanBigramFrequency + FrequencyInitialDiphone +
  FrequencyInitialDiphoneWord + FrequencyInitialDiphoneSyllable +
  Obstruent + Voice + CV + Frication + CorrectLexdec +
  ConspelV + ConspelN + ConphonV + ConphonN +
  ConfriendsV + ConfriendsN + ConffV + ConffN +
  ConfbV + ConfbN, english_half_1)

step_mod_2 <- update(step_mod_1, data = english_half_2)
vars <- names(step_mod_1$coefficients)

step_mod_1 <- step(step_mod_1, trace = 0)
step_mod_2 <- step(step_mod_2, trace = 0)
```

To compare the two resulting models you should examine their summaries (not shown):

```
summary(step_mod_1)
summary(step_mod_2)
```

Here we do some basic comparisons. Consider the set of variables selected by each model:

```
vars_1 <- tidy(step_mod_1) %>% .$term
vars_2 <- tidy(step_mod_2) %>% .$term

## Variables selected by both
intersect(vars_1, vars_2)
##  [1] "(Intercept)"             "WrittenFrequency"
##  [3] "Familiarity"             "AgeSubjectold"
##  [5] "FamilySize"              "InflectionalEntropy"
##  [7] "NumberComplexSynsets"    "MeanBigramFrequency"
##  [9] "FrequencyInitialDiphone" "Voicevoiceless"
## [11] "CorrectLexdec"           "ConfbV"
## Variables selected in one model but not the other
union(setdiff(vars_1, vars_2), setdiff(vars_2, vars_1))
##  [1] "DerivationalEntropy"
##  [2] "FrequencyInitialDiphoneWord"
##  [3] "FrequencyInitialDiphoneSyllable"
##  [4] "Fricationfrication"
##  [5] "Fricationlong"
##  [6] "Fricationshort"
##  [7] "ConspelV"
##  [8] "ConfriendsV"
##  [9] "ConffV"
## [10] "WrittenSpokenFrequencyRatio"
## [11] "CVV"
## [12] "ConspelN"
## [13] "ConphonV"
## [14] "ConfriendsN"
```

While the models agree on 29 potential predictors (13 included, 16 excluded), they differ on 5 others. This is worrying given that the main goal of variable selection is to end up with a set of predictors. Furthermore, 2 of these 5 predictors have $p < 0.01$ and two have $p \approx 0.05$, so if we used *p*-values to decide which terms are "important," we would make different qualitative conclusions from the two models.

Now consider the terms that both models have "kept." Ideally the two models would give similar results for the same term. We consider just the coefficient value ($\hat{\beta}$) and *p*-value for exposition. Comparing the coefficient tables should show that while the broad qualitative conclusions are largely the same (coefficients have the same sign and have the same significance using a $\alpha = 0.05$ criterion), their actual coefficient values and *p*-values differ greatly. To quantify this, we compute the ratio of coefficient values (`coeff_rat`) and the log of the ratio of *p*-values for the two models (`log_p_rat`):

```
tidy(step_mod_1) %>%
  left_join(tidy(step_mod_2), by = "term") %>%
  na.omit() %>%
  mutate(
    coeff_rat = estimate.x / estimate.y,
    log_p_rat = log(p.value.x) - log(p.value.y)
  ) %>%
  select(term, coeff_rat, log_p_rat)
##                        term coeff_rat log_p_rat
## 1               (Intercept)      1.01       NaN
## 2          WrittenFrequency      1.01     -7.89
## 3               Familiarity      1.13    -11.71
## 4             AgeSubjectold      1.05       NaN
## 5                FamilySize     -0.69      4.11
## 6       InflectionalEntropy      0.43      9.86
## 7       NumberComplexSynsets      0.79      4.27
## 8       MeanBigramFrequency      1.04     -2.08
## 9  FrequencyInitialDiphone      0.88      0.33
## 10           Voicevoiceless      2.70     -4.60
## 11            CorrectLexdec      1.49   -129.25
## 12                   ConfbV     -0.34      2.45
```

If the models gave the same result, `coeff_rat` would be near 1 and `log_p_rat` would be near 0, but this is far from the case. Our estimated effect for `MeanBigramFrequency` differs by 20% between the two models, the *p*-value for `CV` is 0.06 versus 0.01, and so on.

In sum, stepwise variable selection can easily give unreliable models that don't meet even the basic standard of consistency.

Box 5.11
Broader Context: Why Stepwise Is Unwise

Statistical texts generally cover stepwise variable selection because it is so widely used, while strongly recommending against it. Harrell (2015, 67) is representative: "[stepwise variable selection] ... violates every principle of statistical estimation and hypothesis testing."

Harrell then lists eight major problems with stepwise variable selection, including that every aspect of the resulting model (coefficient estimates, *p*-values, SEs) is biased, usually anticonservatively (i.e., *p*-values are too low, CIs are too narrow), meaning that one easily finds spurious effects. Essentially, automatically dropping terms with high p-values leads to lowered p-values for remaining terms. Stepwise procedures can also (to a lesser extent) miss true effects.

But, in general, stepwise model selection is dangerous, and some statisticians recommend it not be used at all. It can be helpful as a tool in some settings, such as selecting among a huge set of predictors as a first pass, but should never be used as the only tool for variable selection. I tend to agree with Winter (2019, 277): "Simply put, linguists should not use stepwise regression."

5.10.2 Holistic Approaches—Gelman and Hill (2007)

Better than automated variable selection are holistic approaches where you combine quantitative tests with thinking about your data and problem setting. An example is the approach suggested by Gelman and Hill (2007, 69), which takes common sense into account in evaluating model quality (component 2) in addition to quantitative model comparison (component 1):

1. Include all predictors that, *for substantive reasons*, are expected to be important, for example, because they are part of the study's design or are covariates known from previous work to have large effects on the response.[13]
2. For predictors with large effects, consider interactions as well.
3. Now decide whether to exclude predictors, one at a time:
    - Not significant, coefficient has expected sign: consider leaving in.
    - Not significant, wrong sign: remove.
    - Significant, wrong sign: think hard about whether something's wrong.
    - Significant, right sign: keep in.

This approach requires you to have a sense of what the "right" and "wrong" signs of different coefficients are, ideally for substantive reasons (previous work suggests an effect direction). It is also possible to decide what "right" and "wrong" are based on exploratory plots, but if your sense of right/wrong for a given model coefficient is based *solely* on exploratory data analysis, remember that you may simply be modeling this dataset well (goal = description/exploratory analysis) rather than obtaining results that generalize to unseen data (goal = prediction or explanation).

5.10.3 Example: Building a VOT Model

Let's consider the models we would build under the different strategies we've discussed, to address a concrete research question for the `vot_michael` data: *what factors modulate the voicing contrast*: is the VOICED/VOICELESS VOT difference larger in slower speech, for higher-frequency words, and so on? In terms of variables in this dataset, this means "What interactions are there with `voicing`?" We model $y = \log(\text{VOT})$ for the reasons discussed in section 5.5.2.2.

13. For example, any phonetic study of vowel duration where the research questions are based on effects of predictors x_1 and x_2 should include these predictors (regardless of significance), as well as (minimally) measures of speech rate and vowel height because these have large effects on vowel duration.

The variables we consider as possible modulators of `voicing` for this exercise are `speaking_rate`, `foll_high_vowel`, `cons_cluster`, `place`, and `log_corpus_freq`. Only one of these, speaking rate, is certain to modulate `voicing` from previous work.

We now consider the three different model selection strategies introduced. Let's first rescale all predictors, as defined in section 5.5.4, so that their coefficients are comparable as effect sizes:

```
vot_michael <- vot_michael %>%
  mutate(
    speaking_rate = rescale(speaking_rate),
    log_corpus_freq = rescale(log_corpus_freq),
    ## Categorical variables -> 0/1-valued
    voicing = rescale(voicing),
    cons_cluster = rescale(cons_cluster),
    foll_high_vowel = rescale(foll_high_vowel),
    place = rescale(place)
  )
```

In the following, to keep things simpler, we do not consider any three-way interaction terms.

5.10.3.1 Model comparison

Let's consider these three candidate models:

```
mc_mod_1 <- lm(log_vot ~ voicing * speaking_rate + foll_high_vowel +
    cons_cluster + log_corpus_freq + place, data = vot_michael)
mc_mod_2 <- update(mc_mod_1, . ~ . + voicing:(foll_high_vowel +
    cons_cluster + log_corpus_freq + place))
mc_mod_3 <- update(mc_mod_2, . ~ . + (foll_high_vowel +
    cons_cluster + log_corpus_freq + place)^2)
```

These are all reasonable candidate models a priori:

1. Only speaking rate modulates voicing.
2. All variables modulate voicing.
3. All variables modulate voicing once we control for all possible two-way interactions.

Taking $\alpha = 0.05$ we would choose model 3:

```
anova(mc_mod_1, mc_mod_2, mc_mod_3)
...
##      foll_high_vowel:place + cons_cluster:log_corpus_freq +
##      cons_cluster:place + log_corpus_freq:place
##    Res.Df RSS Df Sum of Sq     F Pr(>F)
## 1     585 117
## 2     581 104  4     13.19 18.66  2e-14
## 3     575 102  6      2.31  2.18  0.044
```

In this model, the interactions with VOICING are

```
tidy(mc_mod_3) %>%
  # Regular expression to grab "voicing...:" terms
  filter(str_detect(term, "voicing.*:"))
##                       term estimate std.error statistic p.value
## 1     voicing:speaking_rate   -0.327     0.071      -4.6 5.6e-06
## 2 voicing:foll_high_vowel    0.082     0.078       1.1 2.9e-01
## 3      voicing:cons_cluster   -0.345     0.084      -4.1 4.6e-05
## 4 voicing:log_corpus_freq   -0.147     0.076      -1.9 5.3e-02
## 5             voicing:place    0.546     0.074       7.4 6.3e-13
```

Using $p < 0.05$ as a filter, we would conclude that place, cons_cluster, and speaking_rate modulate voicing most strongly (highest effect size).

5.10.3.2 Stepwise variable selection

Now we obtain a model by applying stepwise backward model selection to the largest model (mc_mod_3):

```
# Remove 'trace=0' to see details about terms dropped
step_mod <- step(mc_mod_3, trace = 0)
```

In this model, the interactions with VOICING are

```
tidy(step_mod) %>%
  filter(str_detect(term, "voicing.*:"))
##                       term estimate std.error statistic p.value
## 1     voicing:speaking_rate    -0.33     0.071      -4.7 3.5e-06
## 2      voicing:cons_cluster    -0.33     0.083      -4.0 6.6e-05
## 3 voicing:log_corpus_freq     -0.13     0.074      -1.8 7.4e-02
## 4             voicing:place     0.54     0.074       7.4 5.8e-13
```

If we take "X:voicing remains in the stepped-down model" to mean that "X modulates voicing" for our research question, we would conclude that the same three variables modulate voicing as we did using model comparison, as well as log_corpus_freq. The frequency effect is "weaker" than the others, with a smaller effect size and $p > \alpha$.

5.10.3.3 Gelman and Hill's method

This method requires more domain-specific knowledge, which we'll walk through; see appendix A, section A.2 for more information.

We begin with model mc_mod_1, where all predictors are included as main effects—they are all expected to affect VOT for substantive reasons—as well as the interaction with voicing expected from previous work. The expected effect directions are

- speaking_rate: negative effect for VOICELESS stops and a very small effect for VOICED stops
- log_corpus_freq: negative effect
- cons_cluster, foll_high_vowel, place: positive effects expected

The fitted model is

```
gh_mod_1 <- mc_mod_1
tidy(gh_mod_1)
##                     term estimate std.error statistic  p.value
## 1            (Intercept)    3.375     0.018     183.6  0.0e+00
## 2                voicing    1.569     0.038      41.3 1.8e-175
## 3          speaking_rate   -0.147     0.037      -3.9  9.9e-05
## 4         foll_high_vowel    0.116     0.040       2.9  4.1e-03
## 5           cons_cluster    0.454     0.043      10.5  9.2e-24
## 6         log_corpus_freq   -0.075     0.038      -2.0  4.9e-02
## 7                  place   -0.513     0.038     -13.4  5.2e-36
## 8 voicing:speaking_rate   -0.351     0.074      -4.7  2.8e-06
```

The `cons_cluster`, `foll_high_vowel`, `voicing`, and `place` terms have much larger effect sizes than other main effects (remembering that the effect sizes are comparable because we have standardized predictors), so we consider their two-way interactions. Our expected directions for these interactions would be

- `foll_high_vowel:place`: *positive*, because nonlabial voiceless stops in English are often aspirated before high vowels (e.g., Nearey and Rochet 1994), which would make the high/nonhigh difference larger when `voicing` is 1 (VOICELESS).
- `foll_high_vowel:cons_cluster`, `cons_cluster:place`: *unclear*.
- `voicing:cons_cluster`: *unclear*.
- `voicing:foll_high_vowel`: *positive*, because some voiceless stops are aspirated before high vowels in English.
- `voicing:place`: *negative*, because no interaction would be expected in VOT itself, which implies a larger `place` effect for VOICED stops in log(VOT).[14]

We do not consider three-way interactions, where the large `voicing:speaking_rate` interaction would be interacted with other terms with large effect sizes, because we want to keep the model interpretable for this example and because we would not have clear predicted directions for these three-way interactions.

We could add in the interactions one at a time, but for brevity let's add them all at once:

```
gh_mod_2 <- update(gh_mod_1, . ~ . +
    (voicing + foll_high_vowel + cons_cluster + place)^2)
tidy(gh_mod_2)
##                              term estimate std.error statistic
## 1                     (Intercept)    3.406     0.018    190.70
## 2                         voicing    1.559     0.036     42.98
## 3                   speaking_rate   -0.160     0.035     -4.50
## 4                 foll_high_vowel    0.105     0.039      2.72
## 5                    cons_cluster    0.451     0.041     10.88
## 6                 log_corpus_freq   -0.080     0.036     -2.20
## 7                           place   -0.493     0.037    -13.45
## 8           voicing:speaking_rate   -0.344     0.071     -4.88
```

14. This is my guess, from sparse discussion of the place effect differing between voiced and voiceless stops in previous work. If we assume that it doesn't differ, this means a larger difference in log(VOT) for voiced stops.

```
## 9         voicing:foll_high_vowel   0.082   0.078    1.05
## 10          voicing:cons_cluster  -0.314   0.083   -3.79
## 11                 voicing:place   0.566   0.073    7.75
## 12 foll_high_vowel:cons_cluster  -0.086   0.088   -0.98
## 13         foll_high_vowel:place   0.167   0.078    2.13
## 14            cons_cluster:place  -0.071   0.083   -0.85
##     p.value
## 1   0.0e+00
## 2  2.7e-182
## 3   8.3e-06
## 4   6.7e-03
## 5   3.2e-25
## 6   2.8e-02
## 7   4.6e-36
## 8   1.4e-06
## 9   2.9e-01
## 10  1.7e-04
## 11  4.1e-14
## 12  3.3e-01
## 13  3.3e-02
## 14  4.0e-01
```

We now move on to the step where we consider each term based on expected sign and its significance ($p < 0.05$):

- Most terms have expected sign—leave them in.
- `voicing:foll_high_vowel`: wrong sign and nonsignificant—remove.
- `cons_cluster:place`, `foll_high_vowel:cons_cluster`: not significant and expected sign unclear—remove.
- `voicing:cons_cluster`: significant, and expected sign unclear. Because interactions with `voicing` are of primary interest, it is prudent to keep this term in the model, in case it's controlling for something that might otherwise give spurious interactions of other variables with `voicing`.

The final model is

```
gh_mod_3 <- update(gh_mod_2, . ~ . - voicing:foll_high_vowel -
    cons_cluster:place - foll_high_vowel:cons_cluster)
tidy(gh_mod_3)
##                      term estimate std.error statistic  p.value
## 1             (Intercept)    3.402     0.018     192.6  0.0e+00
## 2                 voicing    1.556     0.036      43.1 2.0e-183
## 3           speaking_rate   -0.160     0.035      -4.5  7.2e-06
## 4         foll_high_vowel    0.103     0.038       2.7  7.2e-03
## 5            cons_cluster    0.452     0.041      10.9  1.8e-25
## 6         log_corpus_freq   -0.078     0.036      -2.2  3.1e-02
## 7                   place   -0.488     0.036     -13.4  4.6e-36
## 8   voicing:speaking_rate   -0.343     0.070      -4.9  1.3e-06
## 9    voicing:cons_cluster   -0.304     0.083      -3.7  2.5e-04
## 10          voicing:place    0.558     0.072       7.7  5.7e-14
```

```
## 11 foll_high_vowel:place    0.162      0.078        2.1  3.7e-02
```

We would conclude that `place`, `cons_cluster`, and `speaking_rate` modulate the `voicing` effect, with the `place` effect stronger than the others.

This conclusion differs from the stepwise method in not concluding that word frequency modulates `voicing`, rather than concluding that it has a weak but real effect. Whether this difference matters is a matter for interpretation in the context of our study.

Box 5.12
Practical Note: A Fallible Master Recipe

Chapters 4 and 5 have contained a lot of information, and it seems natural to sum up by giving you a suggested recipe for building a multiple linear regression model. But it should also be clear by now that there is no single recipe; what steps your analysis should contain depends on your research questions, data, and modeling philosophy.

Nonetheless, there is a list of steps you should always consider in regression analysis, and they could be thought of as a "master recipe" if you would like somewhere to start. What follows is a possible master recipe, which I have given in instructions for linear regression class assignments.

1. Preliminaries
   - State the problem (what are the goals of this analysis?).
   - Select the response(s) and relevant predictors.
   - Determine continuous variables: center and scale, possibly transform to normality.
   - Determine categorical variables: center/choose coding scheme.
2. Perform exploratory data analysis
   - Examine exploratory plots and summary statistics relevant for analysis goals.
   - Identify potential outliers.
   - Determine (if applicable) potential interactions and (if applicable) examine by-subject and by-item plots.
3. Choose models to fit, based on steps 1 and 2.
4. Fit candidate model(s).
5. Model validation
   - Assess linearity assumption.
   - Examine distribution of (standardized) residuals.
   - Examine plots of residuals against fitted values and predictors.
   - Look for influential, outlier, and high-leverage points.
   - Check for high collinearity and possibly address if high.
   - Check for overfitting (optional).
6. Revise. Based on step 5, possibly
   - Exclude some data.
   - Transform the predictor and/or response.
   - Then refit the model (step 4).
7. Iterate. Repeat steps 4-6 until model validates well.

5.11 Further Reading

These discussions of regression diagnostics and model validation draw on Chatterjee and Hadi (2012, chapters 4, 6, 9), Faraway (2015, chapters 6, 7, 9), and Harrell (2015, chapters 2, 4, 5). Other valuable sources are Fox and Weisberg (2019, chapter 8), Gelman and

Hill (2007, chapters 4, 5); for linguistic data Baayen (2008, section 6.2), Gries (2021, section 5.6), and Winter (2019, sections 4.4, 6.3) are most extensive. Zuur, Ieno, and Elphick (2010) summarizes all these issues and gives a protocol for addressing them through "data exploration"; Zuur, Ieno, and Smith (2007) gives more detail. Both are part of the rich literature on data analysis for ecologists, which tends to be readable, thorough, and easily portable to linguistic data.

Many sources cover the bias/variance and overfitting/underfitting trade-offs, such as McElreath (2020, chapter 6) and Hastie, Tibshirani, and Friedman (2009, chapter 7), and model comparison/variable selection (e.g., Chatterjee and Hadi 2012, chapter 11; Faraway 2015, chapter 10). Winter (2019, chapter 16) and Gries (2021, section 5.5) discuss model selection for linguistic data. All sources emphasize the role of hypotheses and domain knowledge.

Exercises

5.1. Section 5.3.2.2 claims that the residuals of model `english_mod_1` are bimodal because the `AgeSubject` predictor has not been included in the model. Make a plot that confirms this.

5.2. Section 5.4.2 suggests that the residual plots in figure 5.6 look better once two missing nonlinearities—an interaction and a nonlinear `speaking_rate` effect—are included in `vot_mod_1`. Verify that this is the case.

- **(a)** Refit `vot_mod_1` such that the effect of `speaking_rate` is allowed to be quadratic and to interact with `voicing`.
- **(b)** Examine the coefficient table for the resulting model. Can you interpret what each row involving `speaking_rate` means? (There should be four such rows.) Which row(s) confirm that there is a nonlinear `speaking_rate` effect in this data and that VOICED and VOICELESS stops show different rate effects?
- **(c)** Make the same residual plot as in figure 5.6 (left), but for the new model. (The code for the original plot is in the code file for this chapter if you need it.) In what ways does the new plot look better than the original plot? What remaining issue with residuals is indicated by the new plot?

5.3. Verify this statement from section 5.7.3: "There are arguably no highly influential points for model `vot_mod_2`, even though there were outliers." You should use notions of "outliers" and "influential" as they were defined in the text.

6 Categorical Data Analysis and Logistic Regression

So far we have mostly considered data analysis where the outcome is a continuous variable, such reaction time or vowel duration. In this chapter we'll turn to *categorical data analysis*, where the outcome is a categorical variable (sections 6.2 and 6.3), focusing on *logistic regression*, which predicts a binary outcome (sections 6.4–6.9).

6.1 Preliminaries

6.1.1 Packages

This chapter assumes that you have loaded several packages from previous chapters, as well as the ModelMetrics and margins in packages (Hunt 2020; Leeper 2021):

```
library(tidyverse)
library(broom)
library(languageR)
library(arm)
library(sjPlot)
library(performance)
library(ggeffects)
library(ModelMetrics)
library(margins)
library(rms)

## ensures that `rescale' is not the version from the `scales' package
rescale <- arm::rescale
```

6.1.2 Data

The `diatones` dataset

We also assume that you have loaded the `diatones` dataset:

```
diatones <- read.csv("data/diatones_rmld.csv", stringsAsFactors = TRUE)
```

The English lexicon contains many pairs of nouns and verbs that are spelled the same, such as *protest* and *decay*. Some of these pairs are pronounced identically and some differ only in the placement of primary stress (e.g., initial-/final-stressed 'protest' = noun/verb). The placement of stress for some words has changed over time, typically by the noun changing from final-stressed to initial-stressed. Pairs with initial stress on the noun and final stress on the verb are called *diatones*.

This dataset, analyzed in Sonderegger (2010) and described in more detail there, contains information about 130 pairs of disyllabic (two-syllable) noun/verb pairs, all of which had final-stressed noun and verb in British English in 1700 (e.g., 'protest' noun/verb were pronounced 'proTEST'). For a subset of noun/verb pairs (column `stress_shifted`), the noun's stress stress shifted by 2005 (e.g., 'PROtest' noun). (We will call noun/verb pairs "words" from here on for simplicity.)

Several variables capture each word's *frequency* and aspects of its phonological *structure*, which might condition stress shift based on previous work:

- `frequency`: the word's frequency (log-transformed, from a large corpus)
- `syll1_coda` (levels NO, YES): whether the first syllable ends in a vowel or a consonant (e.g., 'debate', 'disdain')
- `syll2_coda` (levels O, C, CC, CCC): whether the second syllable ends in a vowel or in one to three consonants
- `syll2_coda_td` (levels NO, YES): whether the second syllable ends in /t/ or /d/

Frequency and structure are representative of two broad sources of pronunciation change (roughly, how a word is used vs. how it is represented in the mind), so it is of interest whether a word's likelihood of shifting stress is predicted by each, and their relative contributions.

For simplicity, in this chapter we will code all the "structure" factors as numeric, while renaming the original variables:

```
diatones <- diatones %>% mutate(
  syll1_coda_orig = syll1_coda,
  syll2_coda_orig = syll2_coda,
  syll2_td_orig = syll2_td,
  ## Change no/yes -> 0/1
  syll1_coda = ifelse(syll1_coda == "no", 0, 1),
  ## Change '0'/'C'/'CC'/'CCC' -> 0/1/2/3
  syll2_coda = str_count(syll2_coda_orig, "C"),
  syll2_td = ifelse(syll2_td == "no", 0, 1),
)
```

Coding `syll1_coda` and `syll2_td` as 0 and 1 is the same as we have previously done for two-level factors (e.g., section 5.5.4). Coding `syll2_coda` as numeric makes some sense because its levels have a clear order (number of consonants), and we haven't covered multilevel factors yet.

Figure 6.1 shows the proportion of words with shifted stress as a function of each variable.

Box 6.1
Practical Note: Empirical Plots for Binomial Data

Empirical plots of the effect of a predictor x for binomial data (where $y = 0, 1$) should take into account that what is being plotted are (estimated) probabilities, not averages of a response y that happens to only take on

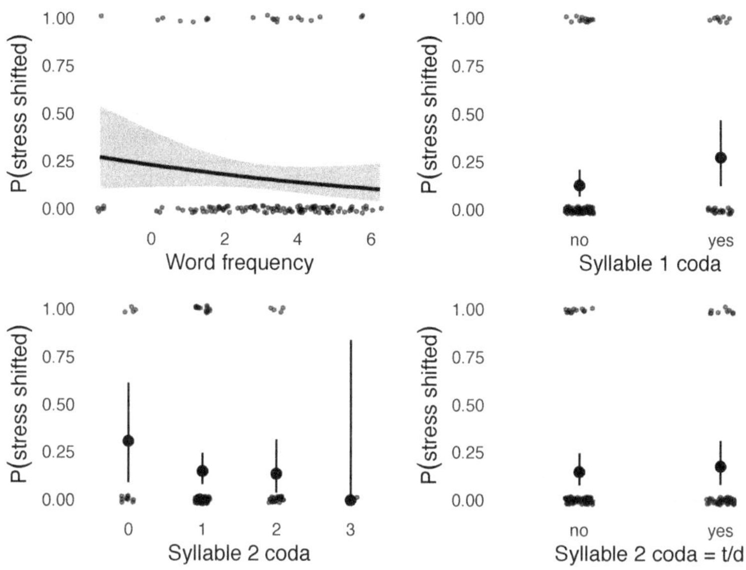

Figure 6.1
Proportion of words with shifted stress as a function of each predictor in the `diatones` dataset. Points are individual observations (at $y=0, 1$); dots/errorbars are means and exact 95% CIs; and the line/shading are predictions and 95% CIs from simple logistic regression.

values 0 and 1. This reflects the structure of logistic regressions, which will model the probability that $y=1$ rather than y itself.

For continuous x, a simple logistic regression should be used to calculate the empirical effect (with 95% CIs), which will look like part of an S-shaped curve. This is better than just showing the distributions of x for $y=0$ and $y=1$ (e.g., using boxplots), which is common, but doesn't respect the fact that x is a predictor.

For categorical x, the empirical effect is just the proportion p of cases with $y=1$ for each value of x, but the 95% CIs should be calculated using an exact method (e.g., from `binom.test()`), which can be very different from bootstrapped CIs, especially for small sample size or p near 0 and 1.

Figure 6.1 shows examples for continuous (top left) and categorical (other panels) predictors; figure 7.11 shows more examples. The code files show how to make such plots.

This topic seems technical, but it comes up a lot in practice and raises a more general point: empirical plots should summarize data in a way that matches the statistical models to be used, as much as possible.

6.2 Categorical Data Analysis

This section reviews some basics of categorical data analysis. Section 6.10 gives some places to read more if this is your first exposure to categorical data analysis.

6.2.1 Contingency Tables

Contingency tables—or *cross-tabulation*, hence the R function `xtabs()`—show the number of observations for each combination of values of categorical variables. A 2×2

contingency table in particular shows the number of observations for each combination of two categorical variables, x_1 and x_2, each of which has two levels.

For example, for the `diatones` data, let's consider the variables capturing whether the first syllable has a coda (`syll1_coda`) and whether stress shifted (`stress_shifted`), each of which has two levels. (In the dataframe these are coded as numeric variables with values 0 and 1, which is equivalent.) A contingency table showing the number of observations for each combination of these variables is

```
diatones_xtab <- xtabs(~ syll1_coda + stress_shifted, data = diatones)
diatones_xtab
##             stress_shifted
## syll1_coda  0  1
##           0 88 13
##           1 21  8
```

It looks like stress shift isn't independent of coda presence: `stress_shifted` = 1 is much more likely relative to `stress_shifted` = 0 when `syll1_coda` = 1.

Let's formally test this hypothesis of nonindependence by using the null hypothesis that `stress_shifted` and `syll1_coda` occur independently in this data. Under this null hypothesis, the expected count in a cell with $x_1 = a$ and $x_2 = b$, given a total of n observations, is

$$P(x_1 = a) \cdot P(x_2 = b) \cdot n.$$

The two probabilities can be estimated by just summing over the relevant row or column of the contingency table and dividing by n. For example, for the preceding contingency table, we can calculate the expected count when `syll1_coda` = 1 and `stress_shifted` = 0, given that $n = 130$:

$$\begin{aligned} P(\texttt{syll1\_coda}=1) \cdot P(\texttt{stress\_shifted}=0) \cdot n &= \frac{(21+8)}{130}\frac{(88+21)}{130} \cdot 130 \\ &= 24.3. \end{aligned}$$

6.2.2 The Chi-Squared Test

The preceding logic applies to all four cells of a 2×2 contingency table, giving two numbers for the ith cell: the *observed* count O_i and the *expected* count E_i (under the null hypothesis of independence). Let m be the number of cells in the table; here $m = 4$.

Pearson's test statistic X^2 measures how much the observed and expected values differ across the whole contingency table:

$$X^2 = \sum_{i=1}^{m} \frac{(O_i - E_i)^2}{E_i}. \tag{6.1}$$

X^2 is 0 when the observed and expected values are exactly the same, and it increases the more the observed and expected values differ.

Pearson's test statistic for a 2×2 table approximately follows a χ^2 distribution (pronounced "chi-squared") with one *df*, denoted $\chi^2(1)$. Thus, we can test the null hypothesis that x_1 and x_2 are independent by comparing the value of X^2 to the $\chi^2(1)$ distribution; this is a *chi-squared test* (or χ^2 test), one of the most widely used hypothesis tests.

The same methodology applies for testing independence of two categorical variables with any number of levels. For x_1 and x_2 with p and q levels, X^2 is calculated using equation (6.1), $m = pq$, and the null hypothesis tested using a $\chi^2(df)$ distribution, where $df = (p-1)(q-1)$. For example, a 2×2 contingency table has $df = 1$ and a 3×4 contingency table has $df = 6$.

The chi-squared test relies on the approximation that X^2 follows a chi-squared distribution (the "χ^2 approximation"), which is only valid when the expected count in each cell is "big enough"—otherwise, p-values are anticonservative. A common rule of thumb is five (expected) observations per cell, although for tables larger than 2×2 about 20% of cells can have expected counts of 1–5 (Agresti 2007, section 5.3.4).

Examples `chisq.test()` applied to a contingency table carries out a chi-squared test. To test whether `syll1_coda` and `stress_shifted` are associated for the previous contingency table:

```
diatones_xtab_test <- chisq.test(diatones_xtab)
diatones_xtab_test
##
##  Pearson's Chi-squared test with Yates' continuity
##  correction
##
## data:  diatones_xtab
## X-squared = 3, df = 1, p-value = 0.1
```

Whether a word's first syllable has a coda is not significantly associated (at the $\alpha = 0.05$ level) with whether its stress shifted.

Executing this code gives a warning message:

```
## Warning in chisq.test(diatones_xtab): Chi-squared approximation may be
incorrect
```

This message refers to the expected counts rule of thumb, as one cell in the contingency table has an expected count below 5:

```
diatones_xtab_test$expected
##            stress_shifted
## syll1_coda  0    1
##           0 85 16.3
##           1 24  4.7
```

Box 6.2
Broader Context: Whither Chi-Squared Tests?

It is common to have fewer than five observations in one or more cells in a contingency table, making "chi-squared tests" in the usual sense of the term (p-value calculated assuming that X^2 follows a chi-squared distribution) frequently inappropriate. Widespread use of the chi-squared test is largely a holdover from when it was computationally difficult to compute "exact" tests (which don't require a large-sample approximation to compute p-values), and there is little reason to use the χ^2 approximation today.

For example, you can run a chi-squared test for the `diatones_xtab_test` example without assuming the χ^2 approximation by using the `simulate.p.value = TRUE` argument to `chisq.test()`, which gets rid of the error.

It is still worth knowing about the chi-squared test and its limitations because it is still frequently used, and in older literature is very widely used. For example, if you are reading a paper where the crucial result relies on a chi-squared test with $p = 0.02$ for a 2×2 contingency table with five observations in one cell, you should be suspicious. If the contingency table has 20+ observations per cell, you shouldn't be.

6.2.3 Fisher's Exact Test

Given that the assumptions of a chi-squared test are often not met in practice (box 6.2), an alternative that is a good default is *Fisher's exact test*, which as an "exact test" does not place any assumptions on counts per cell. Fisher's exact test asks a slightly different question from a chi-squared test: given the row and column totals in the contingency table, how likely would an arrangement of the data at least this extreme be if x_1 and x_2 are independent? In practice, both tests can be thought of as testing the independence of two categorical variables.

For example, let us test whether a verb's `Regularity` (two levels) and its `Auxiliary` (three levels) are independent, for the Dutch verb `regularity` data (section 2.2.3). The contingency table is

```
regularity_xtab <- xtabs(~ Auxiliary + Regularity, regularity)
regularity_xtab
##           Regularity
## Auxiliary irregular regular
##    hebben         108     469
##    zijn            12       8
##    zijnheb         39      64
```

Fisher's exact test asks, for example: if we have 159 irregular verbs, assuming independence, how likely would we be to have $\geq$108 *hebben*, $\leq$12 *zijn*, $\leq$39 *zijnheb*, and so on? In R

```
fisher.test(regularity_xtab)
##
##  Fisher's Exact Test for Count Data
##
## data:  regularity_xtab
## p-value = 2e-07
## alternative hypothesis: two.sided
```

As expected, whether a verb is irregular and which `Auxiliary` it takes are significantly associated ($\alpha = 0.05$).

6.2.4 Effect Sizes and Reporting

As for continuous y, it is essential to report effect sizes for categorical data analysis, and many options are available (Kline 2013, chapter 6). For associations between two categorical variables x_1 and x_2, an effect size that can be used for any $p \times q$ contingency table (and hence chi-squared test) is *Cramer's V*:

$$V = \sqrt{\frac{X^2}{df \cdot n}},$$

where df is the minimum of $p-1$ and $q-1$. Cramer's V can be thought of as X^2 (equation (6.1)) scaled so its maximum value is 1 (172). So $V=0$ and $V=1$ mean "no association" and "perfect association," respectively. For a 2×2 contingency table ($df=1$), V is the same as $|r|$, the absolute value of Pearson's correlation between x_1 and x_2 (coded as 0 and 1 for the two levels). So small/medium/large values of V are the same as for correlations (section 3.2.3.1), divided by $\sqrt{df}$.

For the `diatones_xtab` and `regularity_xtab` tables, $V=0.17$ and 0.22 with $df=1$; both are "small" effects (exercise 6.1).

It is also very common for 2×2 tables to just use an odds ratio (section 6.3.3) as an effect size measure, especially when reporting Fisher's exact test.

Box 6.3
Practical Note: Interpreting and Reporting Tests of Association

When the null hypothesis is rejected, one must still interpret what the association means. This is straightforward for a 2×2 table (e.g., "stress shifting is more likely when there is a coda in syllable 1"), but harder for larger contingency tables. It can be useful to examine the *observed/expected ratios* O_i/E_i for intuition, which give a sense of how often combinations of the two variables co-occur relative to what is expected by chance. O/E ratios can be can be any number >0, with O/E < 1 meaning the variables "co-occur less often than expected," O/E $\approx$ 1 meaning they co-occur "about as often as expected," and O/E > 1 meaning they co-occur "more often than expected." O/E is commonly used to describe some kinds of linguistic data, such as co-occurrence of sounds in phonology (e.g., Frisch, Pierrehumbert, and Broe 2004).

For the `regularity` example:

```
with(chisq.test(regularity_xtab), observed / expected)
##            Regularity
## Auxiliary irregular regular
##   hebben        0.82    1.05
##   zijn          2.64    0.52
##   zijnheb       1.67    0.80
```

Verbs are much more likely to be irregular for the auxiliary ZIJN than for the auxiliary HEBBEN, with verbs that can take either auxiliary falling in between.

A test of association is reported using the same principles as for other hypothesis tests (section 2.7), bearing in mind that Fisher's exact test doesn't have a test statistic or df.

A long report for the `diatones` chi-squared test might say:

> Although a higher proportion of verbs with first-syllable codas shifted stress (28%) than verbs without codas (13%), the association was not significant at the $\alpha = 0.05$ level by a chi-squared test (Cramer's $V = 0.17$, $X^2 = 3.6$, exact $p = 0.083$).

A short report for the `regularity` Fisher's exact test might say:

> There was a small (Cramer's $V = 0.22$) and significant association between verb regularity and auxiliary class (Fisher's exact test: $p < 0.001$), primarily reflecting that verb irregularity is more likely for verbs that take auxiliary ZIJN.

6.3 Odds and Odds Ratios

Intuitively, in logistic regression we will predict the probability that some event happens (e.g., stress shift) as a function of predictors, given data consisting of observations of whether the event happened or not ($y = 0$ or 1).

It would be tempting to just apply linear regression to this task, but this would not work. We can't use y as the response because we want to predict a probability (and y only takes on the values 0 and 1). We also can't do a linear regression using a probability as a response, because probabilities are bounded by 0 and 1 and linear regression assumes that y isn't bounded.

To predict probabilities using a regression model, we need a way to think of probabilities on a continuous, unbounded scale and a way to estimate these probabilities, such that the sample statistic is normally distributed. The latter is necessary so that we can apply all the statistical inference machinery we have used so far to do things like conduct hypothesis tests.

The answer turns out to be to use log-odds, for which we must first discuss odds.

6.3.1 Odds

Odds are a way of thinking about probability, in terms of how likely an event is to happen versus not happen. For example, if we say "the odds are 3:1 that it will be sunny tomorrow" (pronounced "three-to-one"), it means the probability of sun is three times the probability of not-sun.

The odds corresponding to a probability p is

$$\text{odds}(p) = \frac{p}{1-p}.$$

For example, if the probability that it is sunny tomorrow is 0.75, then the odds of it being sunny tomorrow are 3 ($= 0.75/(1 - 0.75)$). The odds of it not being sunny tomorrow are 1/3 ($= 0.25/(1 - 0.25)$).

Figure 6.2 (left) shows odds as a function of probability. Odds are more intuitive than probabilities, but they cannot be predicted in our regression model because they are always positive.

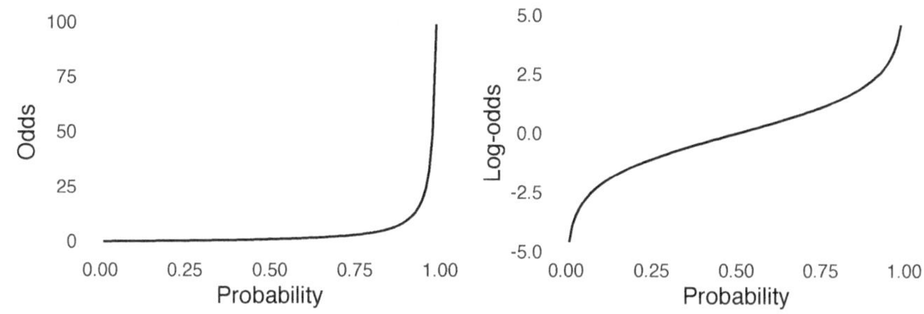

Figure 6.2
Odds (*left*) and log-odds (*right*) as a function of probability (from 0.01 to 0.99).

6.3.2 Log-Odds

The *log-odds* corresponding to a probability p are

$$\text{log-odds}(p) = \log \frac{p}{1-p}.$$

Figure 6.2 (right) shows log-odds as a function of probability.

> **Box 6.4**
> **Broader Context: Logarithm/Exponentials Refresher**
>
> Categorical data analysis and (especially) interpreting logistic regressions involve a lot of thinking in logarithms and exponentials. If you haven't seen these for a while, some important facts are
>
> - Logarithms transform division to subtraction: $\log(x/y) = \log(x) - \log(y)$
> - Logarithms transform multiplication to addition: $\log(xy) = \log(x) + \log(y)$
> - Exponentials are the inverse of logarithms: if $y = \log(x)$, $x = e^y$.
>
> Note that we use "log(x)" to refer to the *natural logarithm*: the power to which e (the "base") would need to be raised to get x, where $e = 2.718...$ is Euler's number.
>
> You may have seen "log" used to mean something else (usually logarithm with base 10) or "ln" used to mean natural logarithm.

Log-odds meet both desiderata as the response for a regression: they range from $-\infty$ to ∞, and they turn out to have normally distributed sampling statistics (section 6.3.4). Log-odds have the drawback of being unintuitive, compared to odds or probability, but with practice you can think in terms of log-odds.

For example, here are some correspondences of p, odds (approximated), and log-odds:

p	Odds	Log-odds
0.27	1:2.7	−1
0.50	1:1	0
0.73	2.7:1	1
0.88	7.4:1	2
0.95	20.1:1	3

This example illustrates a few intuitions about log-odds, which you can see in figure 6.2 (right):

- *Probability space is contracted* in log-odds, as we approach 0 or 1: a change of 1 in log-odds corresponds to a smaller change in p.
- *Log-odds of 3/–3 is big*: log-odds of –3 to 3 encompasses 90% of probabilities ($p \approx 0.05, 0.95$).
- *Log-odds of 5/–5 is huge*: $p \approx 0.99, 0.01$.

6.3.3 Odds Ratios

Let us return to the `diatones` data example where we looked at the association between `syll1_coda` and `stress_shifted` (section 6.2.2), but now let's think of it as "how does the likelihood of stress shifting depend on the presence of a coda?" That is, the response y and predictor x are

- y = `stress_shifted` (1 = yes, 0 = no)
- x = `syll1_coda` (1 = yes, 0 = no)

The proportion of cases in each cell for our data are

```
diatones_ptab <- prop.table(diatones_xtab)
diatones_ptab
##              stress_shifted
## syll1_coda       0     1
##            0 0.677 0.100
##            1 0.162 0.062
```

The `prop.table()` function calculates a table of proportions corresponding to a contingency table. For example, 0.1 of observations have $x=0$ and $y=1$.

We would like a measure of how much more likely $y=1$ (stress shifting) is when $x=1$ than when $x=0$, for which we calculate the odds of stress shifting in each case, then take their ratio:

```
## Odds(y=1 | x=0)
odds_y1x0 <- diatones_ptab[1, 2] / diatones_ptab[1, 1]
## Odds(y=1 | x=1)
odds_y1x1 <- diatones_ptab[2, 2] / diatones_ptab[2, 1]

## odds ratio
odds_y1x1 / odds_y1x0
## [1] 2.6
```

So the odds of stress shifting are about 2.5 times higher when the first syllable has a coda relative to when it does not.

To define this *odds ratio* (OR) more generally, suppose that two binary random variables x and y co-occur with these (population) probabilities:

$$\begin{array}{c|cc} & y=1 & y=0 \\ \hline x=1 & p_{11} & p_{10} \\ x=0 & p_{01} & p_{00} \end{array} \tag{6.2}$$

The odds that $y=1$ are then p_{11}/p_{10} when $x=1$ and p_{01}/p_{00} when $x=0$. The OR, as a population statistic, is defined as

$$\text{OR} = \frac{p_{11}/p_{10}}{p_{01}/p_{00}}.$$

The OR describes how much the occurrence of one (binary) event depends on another (binary) event and is often used as an intuitive measure, as in "your odds of developing lung cancer are halved if you quit smoking." Note that ORs are interpreted multiplicatively. It makes sense to say "A raises the odds of B by a factor of 2," but not "A raises the odds of B by 2."

ORs are in part a way of thinking about changes in probability. For the `diatones` example, our estimates of the probability of stress shifting in each case are

$$P(\texttt{stress\_shifted} = 1 | \texttt{syll1\_coda=1}) \approx 0.13,$$

$$P(\texttt{stress\_shifted} = 1 | \texttt{syll1\_coda=0}) \approx 0.28. \tag{6.3}$$

Thus, the change in probability depending on whether a coda is present is 0.15. However, ORs are not equivalent to changes in probability, because with ORs, what a change from p to $p+\Delta p$ corresponds to depends on the probability (p) you start at. You can see this in figure 6.2 (right): the further p is from 0.5 (x-axis), the bigger the change in log-odds (y-axis) corresponding to a change of one "square" in p ($\Delta p=0.125$).

For example, for a change in probability from 0.5 to 0.65, the odds increase about 1.8-fold, while for a change in probability from 0.8 to 0.95 the odds increase 4.8-fold.

To summarize, odds correspond to probabilities, but changes in odds don't straightforwardly correspond to changes in probabilities.

6.3.4 Log-Odds: Sample and Population

To estimate the population value of the odds ratio, we need a sample statistic to estimate it from observed counts for each x/y combination. Using the same notation as in equation (6.2), let n_{01} be the number of observations where $x=0$ and $y=1$, and so on. To estimate the OR, we estimate each probability from the observed counts (as we did in equation (6.3)), then calculate the odds for each case. For example, for $x=1$:

$$P(y=1|x=1) \approx \frac{n_{11}}{n_{11}+n_{10}}, \quad P(y=0|x=1) \approx \frac{n_{10}}{n_{11}+n_{10}}$$

$$\Rightarrow \text{odds}(y=1|x=1) = \frac{n_{11}}{n_{10}}.$$

Similarly estimating odds($y=1|x=0$), then taking the logarithm of odds($y=1|x=1$)/odds($y=1|x=0$), gives a sample statistic for the change in log odds:

$$\hat{L} = \log\left[\frac{n_{11}/n_{10}}{n_{01}/n_{00}}\right].$$

It turns out that $\hat{L}$ is (approximately) normally distributed with mean $\log(OR)$, where OR is the population value of the OR (Agresti 2007, section 2.3.3). Thus, if we want to do inferential statistics on the effect of binary x on binary y, it makes sense to work with log-odds, rather than probabilities or odds, estimates of which are not normally distributed.

6.3.5 Calculation and Interpretation

The function from probability to log-odds is called *logit* ($\text{logit}(p)$), and the function from log-odds to probability is called *inverse logit* ($\text{logit}^{-1}(x)$), or the "logistic function" ($1/(1+e^{-x})$).

Both can be calculated using `qlogis()` and `plogis()` in base R, or the more intuitively named `logit()` and `invlogit()` in the arm package.

Note that adding in the log-odds scale corresponds to multiplying in the odds scale, by a power of e (box 6.4). So an increase of x in log-odds corresponds to multiplying the odds by e^x.

For instance, for the `diatones` example, the (estimated) odds of stress shifting for words without and with a first syllable coda are 0.15 and 0.38 (`odds_y1x0`, `odds_y1x1`). The change in log-odds is

```
log(odds_y1x1) - log(odds_y1x0)
## [1] 0.95
```

This change corresponds to multiplying the odds by $e^{0.95}$ (= 2.6, the OR).

6.4 Simple Logistic Regression

Analogously to "simple linear regression" (section 4.3) for continuous y, in *simple logistic regression* we predict a binary y from a single predictor x, which may be continuous or discrete.

In this setting our sample consists of n observations, $(x_1, y_1), ..., (x_n, y_n)$, where each y_i is 0 or 1. We model the log-odds of $y=1$ for the ith observation as a function of the single predictor:

$$\text{logit}[p(y_i = 1)] = \beta_0 + \beta_1 x_i, \quad \text{for } i = 1, \dots, n. \tag{6.4}$$

This is the model in "logit space," where the right-hand side of the equation looks the same as for simple linear regression (equation (4.3)) except for the lack of an error term, while the left-hand side of the equation looks similar to simple linear regression except for the logit function. (These differences are explored further in box 6.6.)

The same model can be written in "probability space":

$$P(y_i = 1) = \text{logit}^{-1}(\beta_0 + \beta_1 x_i) = \frac{1}{1 + e^{-(\beta_0 + \beta_1 x_i)}}, \quad \text{for } i = 1, \dots, n.$$

The model in equation (6.4) predicts the log-odds of $y = 1$ for the ith observation to be

$$\text{logit}(\hat{p}_i) = \hat{\beta}_0 + \hat{\beta}_1 x_i.$$

Table 6.1
Interpretation of coefficients of a simple logistic regression

	Intercept interpretation	Slope interpretation
Log-odds	$\hat{\beta}_0$: predicted log-odds of $y = 1$ when $x = 0$	Predicted change in log-odds of $y = 1$ for a unit change in x
Odds	$e^{\hat{\beta}_0}$: predicted odds of $y = 1$ when $x = 0$	$e^{\hat{\beta}_1}$: predicted amount odds of $y = 1$ is multiplied by for a unit change in x
Probability	$\text{logit}^{-1}(\hat{\beta}_0)$: probability of $y = 1$ when $x = 0$	$\hat{\beta}_1$ interpretation in probability depends on the value of x

The fitted regression coefficients, $\hat{\beta}_0$ and $\hat{\beta}_1$, can be interpreted in terms of log-odds, odds, or probability (table 6.1).

We will discuss estimation of these coefficients and the usual associated measures (SEs, CIs, p-values), after we work through some examples of fitted models for intuition.

6.4.1 Examples

Logistic regressions are fit using the `glm()` function with the argument `family = "binomial"`, because they are one kind of "generalized linear model" (box 6.5).

Let us fit a simple logistic regression, which predicts whether a verb is regular as a function of its frequency, for the `regularity` dataset. In this case y is `Regularity` (0/1 = IRREGULAR/REGULAR) and x is `WrittenFrequency`.

This code fits this model after rescaling `WrittenFrequency` (center and divide by two SDs) for interpretability:

```
regularity <- mutate(regularity,
  written_frequency = rescale(WrittenFrequency))

logreg_mod_1 <- glm(Regularity ~ written_frequency,
  data = regularity, family = "binomial")
```

As for linear regression, we can see the model's results using either `summary()`, which is more standard but includes a lot of output besides the coefficient table of primary interest, or `tidy()` (from the broom package).

```
tidy(logreg_mod_1)
##                  term estimate std.error statistic p.value
## 1         (Intercept)      1.5      0.11      13.6 4.7e-42
## 2 written_frequency     -2.2      0.23      -9.6 9.5e-22
```

Box 6.5
Broader Context: Logistic Regression as a GLM

Logistic regression is one type of *generalized linear model* (GLM), which is a family of models that look like linear regression, but with different choices for each part of equation (6.4) which make it look different from the equation for linear regression (equation (4.3)), which is the *link function* (here, logit) tying the right-hand side of the equation (the *linear predictor*) to the probability *distribution* over y. (In a linear regression, the link function is just the identity, and y is normally distributed.) GLMs let us model many different types of data just using (almost) the same tools as for linear regression; in R this is the `glm()` function.

Logistic regression assumes a binary response y and uses the logit link function. Because the logit link is the most commonly used for binomially distributed data (of which a binary response is a special case, with $n = 1$), specifying `family = "binomial"` in `glm` gives a logit link by default (it's the same as writing `family = binomial(link = "logit")`). Logistic regression can be characterized either as the probability of a binary outcome for each observation or as modeling the number of "hits" for a Bernoulli distribution over n outcomes, for each set of observations that share the same predictor values. Both are possible using `glm()` with `family = "binomial"`, but we only cover the binary outcome option.

Logistic regression is the only type of GLM covered in this book and is the GLM most commonly used for linguistic data. But other types of GLM linear model are very useful and cover many other types of data. Some others that are used in linguistics are

- *Poisson regression* is used to model count data, where y counts how many times an event occurs, such as the number of times a word occurs in a corpus or the number of times a participant gestures during a conversation (Winter and Bürkner 2021).
- *Multinomial regression* is a generalization of logistic regression, where the response is one of k discrete outcomes, such as different types of syntactic construction or qualitatively different phonetic realizations for the same meaning/sound (Levshina 2015, chapter 12; Gries 2021, section 5.4.1).
- *Ordinal (logistic) regression* is a special case of multinomial regression used to model a response that can take on several levels that have an intuitive "order," such as ratings on a discrete scale in psycholinguistic experiments (Baayen 2008, section 6.3; Liddell and Kruschke 2018).

Other kinds of GLM are a good fit for different kinds of linguistic data, but they are less widely used in linguistics, including gamma regression (for skewed continuous y) or log-normal regression (for right-skewed data where y must be positive).

There is no comprehensive introduction to GLMs for linguistic data, but different sources cover the GLM framework generally or individual model types (in addition to the references cited in this box, see also Gries 2021, sections 5.3–5.4; Coupé 2018; Winter 2019, chapter 13). Faraway (2016, chapters 1–9) is a general introduction to GLMs using R.

6.4.2 Interpreting Coefficients

We can interpret the model's coefficients in log-odds or probability. The log-odds of regularity for a word with average frequency (`written_frequency` = 0) is 1.5, and increasing `written_frequency` by 1 (i.e., increasing frequency by two SDs) decreases the log-odds of regularity by 2.2.

Turning to probability, the probability of regularity for the "average word" is 0.81 (`invlogit(1.5)`). Figure 6.3 shows the predicted probability *P*(`Regularity`) as a function of `written_frequency`, which shows the characteristic s-shaped logistic curve. This plot illustrates why it is not straightforward to describe the effect of a predictor as a change in probability: the slope in probability space (literally, the slope of the s-shaped curve) changes depending on where on the curve you are.

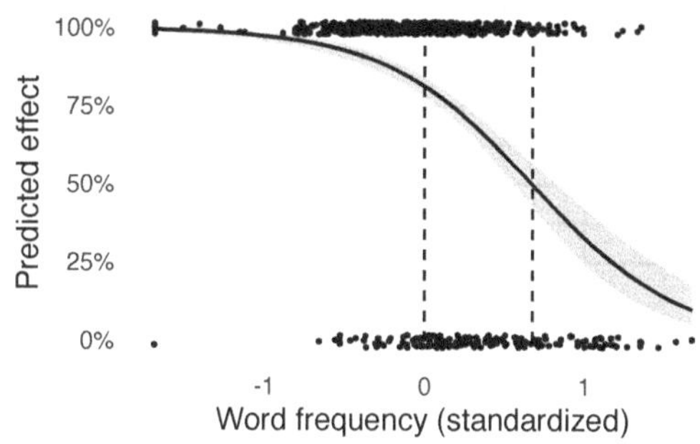

Figure 6.3
Predicted probability of verb regularity as a function of `written_frequency` for model `logreg_mod_1`, with 95% CIs (shading). Dotted lines are at the mean of `written_frequency` (*left*) and the maximum slope (*right*).

Nonetheless it is more intuitive to interpret logistic regression coefficients as changes in probability, so this is often done holding the predictor at (1) its average value or (2) the value that maximizes the slope (that is, where predicted probability = 0.5).

Let p_0 denote the predicted probability when all predictors are held at average values ($p_0 = \text{logit}^{-1}(\beta_0)$). The slopes at points (1) and (2) are

1. $\beta_1 \cdot p_0(1-p_0)$
2. $\beta_1/4$

It is also useful to know how to calculate a probability-space interpretation automatically. One option is the "average marginal effect" (AME), using functionality from the margins package, which will be described in section 9.8.3:

```
margins(logreg_mod_1)
##  written_frequency
##            -0.3151
```

This is the slope of `written_frequency` (in probability), averaging over the dataset.

For the current example, (1) and (2) correspond to the left and right dotted lines in figure 6.3 and have values –0.33 and –0.55.

The probability of verb regularity is thus predicted to decrease by 0.33 or 0.55 as x is changed by two SDs, which is a large effect. Visually this corresponds to the s-shaped curve spanning most of probability space (0–1) for the observed range of x.

As an example with categorical x, let's now fit a logistic regression corresponding to the `diatones` example (section 6.3.3) to see whether stress shift is more likely for words with a first-syllable coda. In this case y is `stress_shifted` and x is `syll1_coda` (0/1 = no coda/coda).

The regression model is still equation (6.4), where the intercept β_0 is the log-odds of stress shifting for a word without a coda and the slope β_1 is the change in log-odds of stress shifting between words with and without a coda.

To fit this model and see its coefficients:

```
logreg_mod_2 <- glm(stress_shifted ~ syll1_coda,
  data = diatones, family = "binomial")
summary(logreg_mod_2)
...
## Coefficients:
##               Estimate Std. Error z value Pr(>|z|)
## (Intercept)    -1.912      0.297   -6.44  1.2e-10
## syll1_coda      0.947      0.511    1.85    0.064
...
```

The predicted probabilities of stress shifting for `syll1_coda` = 0 and 1 words are

```
logreg_mod_2 %>%
  predict(list(syll1_coda = c(0, 1)), type = "response")
##    1    2
## 0.13 0.28
```

We can compare to the observed proportion of stress-shifted words with and without a coda:

```
diatones %>%
  group_by(syll1_coda) %>%
  summarise(prop_shifted = sum(stress_shifted) / n())
##   syll1_coda prop_shifted
## 1          0         0.13
## 2          1         0.28
```

The model matches these proportions exactly, because predicting these two numbers is all a model with one categorical predictor has to do.

Box 6.6
Broader Context: Logistic/Linear Regression Differences, Log-Odds versus Probability Space

There are two important differences between logistic regression and linear regression.

First, the response and what is being modeled are not the same thing as each other: in logistic regression, we observe y, which takes on values 0 or 1, but we model the expected value of y: the probability that $y = 1$. In linear regression y is the response variable, which is also what the regression is modeling.

The most important consequence is that "logistic regressions are inherently more difficult than linear regressions to interpret" (Gelman and Hill 2007, 101), because the model is only linear in a scale we're not interested in (log-odds) and is *nonlinear* in the actual scale of interest (probabilities). So when interpreting logistic regressions (e.g., plotting model predictions, interpreting model coefficients) we must always choose between (1) a more interpretable model (almost identical to linear regression) on an unintuitive scale and (2) an intuitive scale (probabilities) but a less-interpretable model. The same issue exists with any GLM. The tradition in language sciences prefers option (1): just report logistic regression without much discussion of results in probability space, which assumes that results in log-odds are meaningful. Interestingly, researchers in some other fields prefer option (2) (e.g., political science, sociology, ecology: chapter 5; Long and Freese 2006, chapter 4; McElreath 2020, section 9.2): it is assumed that only results on the probability scale are of interest, so a nonlinear model's results must be interpreted, which in turn requires more subtle understanding of fitted models, and care in interpreting their results. For example, a significant interaction effect in log-odds is neither necessary nor sufficient for a meaningful interaction in *probability*

(Berry, DeMeritt, and Esarey 2010). Option (2) seems like a reasonable default to consider for linguistic data as well, but we focus on option (1) given current practice.

The second difference is the presence of an error term. In linear regression, there is an error term ε_i, capturing the difference between the fitted value ($\hat{y}_i$) and the actual value (y_i) for each observation, along with the linear predictor ($\beta_0 + \beta_1 x_{i1} + \cdots + \beta_p x_{ip}$). In logistic regression, there is no error term: the right-hand side of equation (6.4) predicts $P(y_i = 1)$ using just the linear predictor.

Why? As explained in box 2.3, for a Bernoulli random variable y, the expectation (p) and variance ($p(1-p)$) only depend on one unknown parameter, p (the probability of a "success"). So unlike for linear regression, where the mean (μ) and variance (σ) in our model of the response can be tweaked independently to better fit the data, in logistic regression these two things are not independent. By fitting the mean (probability that $y = 1$), we have already determined the variance.

6.5 Inference for Logistic Regression

6.5.1 Parameter Estimation: Maximum Likelihood

For linear regression, it was possible to estimate the regression coefficients by least squares (section 4.3.2): minimizing the difference between the model's prediction and the observed data (squared residuals), across all points. For least squares, the regression coefficients can be solved for by plugging the predictor and response values for all observations into an equation.

For logistic regression, there are no residuals (box 6.6), so we can't use a least-squares method. Instead, we use a method based on the *likelihood*: the probability of the observed data as a function of values of the regression coefficients. A *maximum likelihood* (ML) method finds the values of regression coefficients that maximize this probability.

For simple logistic regression, for any β_0 and β_1, the likelihood of the data is

$$L(\beta_0, \beta_1) = \prod_{i=1}^{n} p_i^{y_i}(1-p_i)^{1-y_i}, \tag{6.5}$$

where p_i is the predicted probability for the ith observation:

$$p_i = \text{logit}^{-1}(\beta_0 + \beta_1 x_i), \quad i = 1, \ldots, n.$$

We choose $\hat{\beta}_0, \hat{\beta}_1$ that maximize the likelihood; these are the ML estimates. There is no equation to determine $\hat{\beta}_0, \hat{\beta}_1$, which must instead be estimated numerically, using an estimation algorithm. The `glm()` function uses the Fisher scoring algorithm by default.

6.5.2 Uncertainty

There are two approaches to quantifying uncertainty in the coefficients, and thus obtaining SEs, CIs, and p-values: Wald tests and likelihood methods. It is worth knowing that both methods exist, because the former is used by default in R in many settings (e.g., mixed models), and the latter is typically recommended for use whenever possible.

6.5.2.1 Wald test approach

This approach assumes that coefficient estimates are normally distributed, which is true when estimating log-odds for large enough samples. Thus, a SE can be obtained for each coefficient, and the test statistic of the form mean/SE is (approximately) normally distributed (Agresti 2003, section 5.5.2). For example for β_1, the test statistic is

$$z = \frac{\hat{\beta}_1}{SE(\hat{\beta}_1)}. \tag{6.6}$$

As for any normally distributed test statistic, we can perform a *Wald test* of the null hypothesis that $\hat{\beta}_1 = 0$, in other words, that there is no relationship between $P(y=1)$ and x. From this test we obtain a p-value and CI for the slope $\hat{\beta}_1$. (The same is true for the intercept $\hat{\beta}_0$.) This is the method used in `glm()` to calculate the SEs, z-values, and p-values shown in the model summary (e.g., `logreg_mod_2`). The corresponding CIs (which are not the default) can be calculated using `confint.default()`:

```
confint.default(logreg_mod_2)
##                2.5 % 97.5 %
## (Intercept) -2.495   -1.3
## syll1_coda  -0.054    1.9
```

6.5.2.2 Likelihood approach

The likelihood approach calculates p-values by dropping terms from the model using a *likelihood-ratio test* (LRT), analogously to how model comparison was performed for linear regression models, where the full model was compared to the intercept-only model, and an F-test was used to test whether the change in sum of squares was significant, given the added predictors in the full model (section 5.9.1). Logistic regression models don't have sums of squares because they don't have residuals. Instead, we compare the likelihood (equation (6.5)) of the two models: Is the change in likelihood between M_0 and M_1 significant?

Likelihood methods typically use log-transformed likelihood to avoid numerical issues when computing with very small numbers. It is also customary to multiply by –2 and work with the *deviance* of a model M (Agresti 2007, section 3.4.3):

$$D = -2 \log L(M).$$

Thus, a better model (higher likelihood) has lower deviance. To get some intuition, we can substitute in for $L(M)$ from equation (6.5):

$$D = -2 \sum_{i=1}^{n} \log \hat{p}_i^{y_i} (1 - \hat{p}_i)^{1-y_i} \tag{6.7}$$

$$= \sum_{i=1}^{n} 2\{y_i \log[1/\hat{p}_i] + [1 - y_i] \log[1/(1 - \hat{p}_i)]\} \tag{6.8}$$

So the deviance sums up an error from each observation (the part in brackets in equation (6.8)). For $y = 1$ observations, this error is 0 when $p = 1$ (perfect prediction) and larger the further p is from 1. For $y = 0$ observations, this error is 0 when $p = 0$ (perfect prediction) and larger the further p is from 0.

Returning to model comparison, the difference in deviance (ΔD) between models with (M_1) and without (M_0) a single predictor x is related to the log of their *likelihood ratio* (LR):

$$\Delta D = -2 \log \frac{L(M_0)}{L(M_1)}. \quad (6.9)$$

For large enough sample size, ΔD follows a $\chi^2(1)$ distribution if the slope of x is 0. Thus, we can use ΔD as a test statistic for the null hypothesis that $\beta_1 = 0$. This test is a kind of LRT, and gives a p-value.

The corresponding CI uses a *profile method*: with all coefficients except β_1 held at the values that maximize likelihood, find the values of β_1 such that an LRT dropping x would reject the null hypothesis.

For example, we can calculate an LRT-based p-value and 95% CI for the `syll1_coda` slope in model `logreg_mod_2`, using the useful function `drop1()` for excluding a single term from a model:

```
drop1(logreg_mod_2, "syll1_coda", test = "Chisq")
...
##              Df Deviance AIC  LRT Pr(>Chi)
## <none>               112 116
## syll1_coda  1        115 117 3.26    0.071
```

```
confint(logreg_mod_2)["syll1_coda", ]
##  2.5 % 97.5 %
## -0.085  1.940
```

Note that this method gives a slightly larger p-value/larger CI than the Wald-test-based method:

```
##   method     p confint_width
## 1   Wald 0.064             2
## 2     LR 0.071             2
```

The LR method is more conservative in general. in

Box 6.7
Practical Note: Does It Matter How We Get p-Values?

The likelihood and Wald-test-based methods will give very similar results for large samples. The Wald test can be calculated instantaneously, but it relies on a large-sample approximation, which is poor even for medium-sized samples ($n < 100$ is sometimes given) or very large $|\beta|$. For these reasons, Agresti (2003, section 5.2.1) recommends always using an LRT to assess coefficient significances. But an LRT takes longer because it requires fitting two models. Similarly, a likelihood-based CI is preferable to a Wald-based CI, but it takes longer to compute because profile methods are slow. For the example here speed isn't an issue (small dataset/simple model), but in more realistic cases it quickly becomes an issue.

The R defaults for logistic regression reflect this tension. By default `summary()` shows Wald-based p-values, whereas `confint()` shows likelihood-based CIs.

> This kind of situation comes up often: for most regression models there are different ways of calculating the "same thing" (especially for hypothesis testing), which tend to vary from "easier/faster but less accurate" to "harder/slower but very accurate."

6.6 Goodness of Fit

We would like a measure of goodness of fit for logistic regression models: How well does the model predict y? For linear regression, we used R^2 to quantify how similar the model's predictions ($\hat{y}_i$) were to the observations (y_i), relative to a baseline model where the model's prediction is always the grand mean: R^2 was simply the correlation between observed and predicted values.

However, for logistic regression we cannot directly compare model predictions (p_i: [log-odds of] probabilities) to observations (y_i: 0 or 1), so we need a different measure of goodness of fit. Let's discuss a few options.

6.6.1 Classification-Based Measures

The simplest way to assess goodness of fit comes from thinking of the model as a classifier that predicts y, as follows:

$$\hat{y}_i = \begin{cases} 1 & \text{if } \hat{p}_i > k \\ 0 & \text{if } \hat{p}_i \leq k, \end{cases}$$

where k is some threshold, typically $k = 0.5$.

That is, we just predict $y = 1$ if the predicted log-odds are positive and $y = 0$ if the predicted log-odds are negative. We can then define *classification accuracy* as the percentage of observations where the predicted and observed values are the same ($\hat{y}_i = y_i$). Classification accuracy needs to be compared to a baseline, usually "how often would we classify correctly if we chose the most common case?"

Although quantities like this we can calculate for ourselves with simple commands, it is good to start using packages for evaluating the quality of regression models that will generalize to more complex models. In this case we use `performance_pcp()` from the performance package, which calculates the expected classification accuracy on unseen data (using threshold $k = 0.5$), which is what we really care about, rather than just calculating the classifier's performance on this data (see `?performance_pcp`; Herron 1999, 90–92).[1] For the `regularity` model:

```
performance_pcp(logreg_mod_1)
## # Percentage of Correct Predictions from Logistic Regression Model
##
##   Full model: 71.19% [67.84% - 74.55%]
##   Null model: 64.89% [61.35% - 68.43%]
##
```

1. Expected classification accuracy is the sum of $\hat{p}_i$ for $y = 1$ observations and $1 - \hat{p}_i$ for $y = 0$ observations, divided by n. Correct predictions count more the closer $\hat{p}_i$ is to 0 or 1.

...

The classification accuracy of the model with its 95% CI is shown first (full model), followed by the baseline accuracy (null model, which consists of an intercept only). The model's improvement over baseline is a measure of effect size and seems substantial (~6%). However the 95% CIs overlap, suggesting the model does not give "significantly" better accuracy than the baseline.

Although widely used for linguistic (and other) data in classification tasks, accuracy alone is not a good metric for classifier performance, because it hides important information, especially when class sizes are very different. Accuracy doesn't distinguish between false positives and false negatives (i.e., type I/II errors; see section 3.3); accuracy can be very high just because one class is more frequent. So when reporting performance of a classifier it is better to report two or more values that better summarize the trade-off between false positives and false negatives, such as sensitivity and specificity (true positive and true negative rates, for a given threshold k), or *area under the ROC curve* (AUC), which summarizes performance as k is varied, trading off true positives for true negatives.[2]

We will report AUC by default. AUC lies between 0.5 and 1, where the baseline model has AUC = 0.5. AUC can be calculated using `auc()` in the ModelMetrics package, for example, for the `regularity` model:

```
auc(logreg_mod_1)
## [1] 0.77
```

This value of AUC suggests reasonably good fit even taking the skew of observed values (REGULAR verbs more common) into account.

Box 6.8
Broader Context: AUC Intuition

In the logistic regression context, where probabilities are being predicted, it may be more intuitive to think of AUC (rather than a classification metric) as related to the (rank) correlation between predicted probabilities and observed responses. This is called *Somers' D_{xy}* (Harrell 2015, section 10.8), and it is widely reported in linguistics (e.g., Baayen 2008). As a correlation, D_{xy} ranges between −1 and 1; AUC for a logistic regression model is just D_{xy} scaled to lie between 0 and 1 ($AUC = D_{xy}/2 - 0.5$). Because we can always make a correlation lie between 0 and 1 by just switching the $y=0$ and $y=1$ labels, it is usually assumed that D_{xy} lies between 0 and 1; hence AUC lies between 0.5 and 1. (AUC <0.5 would mean predicting worse than randomly guessing, but for a regression model used as a classifier this shouldn't happen, because just setting all nonintercept terms to 0 already gives AUC = 0.5.)

2. *ROC* stands for "receiving operator characteristic," but sometimes it is just used as an acronym (like "laser" is).

6.6.2 Pseudo-R^2 Measures

Pseudo-R^2 measures of goodness of fit quantify the degree of variance accounted for by the model, which was one interpretation of R^2 for linear regressions. These pseudo measures replace variance by likelihood to define R^2, because there is no variance for a logistic regression.

The most popular pseudo-R^2 is *Nagelkerke's* R^2 (e.g., Faraway 2016, section 2.6). The measure captures the relative likelihoods of the fitted model ($\hat{L}$) and a baseline model (including only the intercept: $\hat{L}_0$), scaled so that the measure lies between 0 and 1:

$$R_n^2 = \frac{1-(\hat{L}_0/\hat{L})^{2/n}}{1-\hat{L}_0^{2/n}}.$$

R_n^2 can be calculated using `r2_nagelkerke()` in the performance package (which also computes other pseudo-R^2 measures, such as Cox-Snell, McFadden, and Tjur). For the `regularity` example:

```
r2_nagelkerke(logreg_mod_1)
## Nagelkerke's R2
##              0.24
```

Because no pseudo-R^2 measure has all the properties of R^2 for linear regression, interpreting their values is not intuitive, and they are not recommended as a goodness-of-fit measure except as a way of comparing different logistic regression models. Nonetheless R_n^2 is widely reported in linguistics.

6.7 Multiple Logistic Regression

Generalizing from one predictor to multiple predictors is similar for logistic regression as for linear regression (section 4.4). We assume there are k predictors ($x_1, ..., x_k$) of a binary response y, where each x_i can be continuous or categorical. The logistic regression model is now:

$$\text{logit}[P(y=1)] = \beta_0 + \beta_1 x_1 + \cdots + \beta_p x_k. \tag{6.10}$$

Written in terms of individual observations, the log-odds of $y=1$ for the ith observation would be (using the notation x_{ij} as in section 4.4.1):

$$\text{logit}[P(y_i=1)] = \beta_0 + \beta_1 x_{i1} + \cdots + \beta_p x_{ik}.$$

We can rewrite the model in equation (6.10) in terms of odds by exponentiating each side:

$$\text{odds}(y=1) = \frac{P(y=1)}{P(y=0)} \tag{6.11}$$

$$= e^{\beta_0} e^{\beta_1 x_1} \ldots e^{\beta_p x_p}. \tag{6.12}$$

If we redefine the coefficients as $\alpha_i = e^{\beta_i}$, which we do to interpret them in terms of odds, this becomes

$$\text{odds}(y=1) = \alpha_0 \alpha_1^{x_1} \cdots \alpha_p^{x_p}.$$

That is, each predictor's effect on the odds is multiplicative. A predictor that has no effect corresponds to multiplying by 1 ($\alpha_i = 1$), whereas predictors with "negative" effects decrease the odds ($\alpha_i < 1$) and predictors with "positive" effects increase the odds ($\alpha_i > 1$).

For example, suppose the predictor x_1 is a two-level factor that can take on values 0 and 1, and $\alpha_1 = 2$. When $x_1 = 1$, the odds are multiplied by 2 ($\alpha_1^{x_1} = 2^1$) relative to when $x_0 = 0$ ($\alpha_1^{x_1} = 2^0$). If x_1 were a continuous predictor, then every increase of 1 in x_1 would correspond to multiplying the odds by 2.

6.7.1 Likelihood-Ratio Test: General Case

The Wald test and likelihood approaches (section 6.5.2) generalize to the multiple regression case, for calculating SEs, CIs, and p-values for coefficients of a multiple logistic regression. The likelihood is now:

$$L(\beta_0, \beta_1, \ldots, \beta_k) = \prod_{i=1}^{n} p_i^{y_i}(1-p_i)^{1-y_i}, \tag{6.13}$$

where p_i is defined in equation (6.10).

The LRT also generalizes: the difference in deviance ΔD between models M_0 (subset) and M_1 (superset) is still defined as equation (6.9), but now the models have k and q predictors:

$$M_0 : \text{predictors } x_1, \ldots, x_k, \tag{6.14}$$

$$M_1 : \text{predictors } x_1, \ldots, x_k, x_{k+1}, \ldots, x_q. \tag{6.15}$$

ΔD approximately follows a $\chi^2(q-k)$ distribution, under the null hypothesis that the additional terms in M_1 have no effect ($\beta_{k+1} = \beta_{k+2} = \cdots = \beta_q = 0$; see Agresti 2007, section 3.4.3). Thus, we can use an LRT to test whether adding these predictors significantly improves model likelihood. Examples are shown below (section 6.7.2) for two common applications: assessing whether a model improves at all on the baseline and whether adding additional terms is justified.

Effect size An effect size should be reported with any hypothesis test, but it is unclear whether there is a standard effect size to report for this kind of LRT, which is very common in practice. As for the linear regression case (section 5.9.1.1), we will just use one measure that is intuitive rather than discussing different options.

Let $odds_0$ and $odds_1$ be the accuracy of the two models (as defined previously, with threshold 0.5), transformed to odds. The OR, $odds_1/odds_0$, can then be used as an effect size. What counts as small/medium/large ORs is unclear, because this depends on the base rate (Chen, Cohen, and Chen 2010), but a rough rule of thumb is that OR < 1.5 is small and OR > 4 is large.

There is not an R function for this exact case, but we can write one using the performance functionality used to compute model accuracy:

```
# m1: reduced model
# m2: full model
oddsratio_lr <- function(m1, m2) {
  p2 <- performance_pcp(m2)$pcp_model
  p1 <- performance_pcp(m1)$pcp_model
  return((p2 / (1-p2)) / (p1 / (1-p1)))
}
```

6.7.2 Example

In this example we model whether stress shifts for the `diatones` data as a function of word `frequency` and syllable structure (captured by `syll1_coda`, `syll2_coda`, and `syll2_td`). We first fit a simple model with only main effect terms (no interactions), as an example of multiple logistic regression, interpreting coefficients, and goodness of fit. We then consider adding interaction terms, which leads to the first model from Sonderegger (2010).

6.7.2.1 Model 1

We first standardize all predictors, for interpretability (section 5.5.4):

```
diatones <- mutate(diatones,
  syll1_coda = rescale(syll1_coda_orig),
  syll2_td = rescale(syll2_td_orig),
  syll2_coda = rescale(syll2_coda),
  frequency = rescale(frequency)
)
```

Because the first two predictors are two-level factors, they have been transformed to 0 and 1 and then centered; the second two predictors are numeric, so they have been centered and divided by two SDs. The coefficients will mean

- `syll1_coda`: predicted change (in log-odds of stress shifting) between when a first-syllable coda is/isn't present (similar for `syll2_td`).
- `syll2_coda`: predicted change when the number of second-syllable coda consonants is increased by two SDs (similar for `frequency`). In this case, an increase of two SDs corresponds to an increase of 1.2 consonants, so the coefficient is a bit more than the predicted change in log-odds for each added consonant.

To fit the model and view its coefficient table:

```
mlogreg_mod_1 <- glm(stress_shifted ~ syll2_coda + syll2_td +
    frequency + syll1_coda, data = diatones, family = "binomial")
mlogreg_mod_1 %>% tidy()
##          term estimate std.error statistic p.value
## 1 (Intercept)    -1.83      0.28      -6.6 3.4e-11
## 2  syll2_coda    -1.31      0.64      -2.1 4.0e-02
## 3    syll2_td     0.84      0.58       1.5 1.5e-01
## 4   frequency    -0.61      0.49      -1.2 2.2e-01
```

```
## 5  syll1_coda      1.24       0.55         2.2 2.5e-02
```

The effects capture the empirical trends shown in figure 6.1. The main patterns are that words with a first syllable coda or without a second syllable coda seem more likely to shift stress. In addition higher frequency words may be slightly less likely to shift stress.

Because the predictors are standardized, we can interpret their values as relative effect sizes: the `syll2_coda` and `syll1_coda` effects are stronger than the others, and `frequency` has a smaller effect than the other predictors (which index syllable structure).

Let's interpret a couple coefficient estimates from the model, in terms of log-odds and probability.

The intercept is $\hat{\beta}_0 = -1.8$, which corresponds to probability of 0.14 (inverse-logit of −1.8). Thus the predicted probability of stress shifting is 14%, for an "average word" (all predictors held at mean values).

The `syll1_coda` coefficient is 1.2, meaning the odds of stress shifting are predicted to be 3.5 times higher for words where the first syllable has a coda.

The `frequency` coefficient is smaller: −0.61, meaning the odds of stress shifting are predicted to be about 50% lower for words with frequency two SDs higher ($e^{-0.61} = 0.54$).

Each term β_i of a multiple logistic regression can also be interpreted in terms of probability, similarly to simple logistic regression (section 6.4.2): either as the slope with all predictors held at 0, or at the maximum possible slope (when probability = 0.5), or as the average marginal effect. Exercise 6.2 asks you to do this for model 1.

We can evaluate the model using an LRT and AUC, as we have already discussed. To test how much likelihood improves from a null model:

```
mlogreg_mod_1_baseline <- update(mlogreg_mod_1, . ~ 1)
anova(mlogreg_mod_1, mlogreg_mod_1_baseline, test = "Chisq")
...
##   Resid. Df Resid. Dev Df Deviance Pr(>Chi)
## 1       125        105
## 2       129        115 -4    -9.58    0.048
```

Note that the dfs of this comparison is $k - q = 4$, because there are four predictors in the full model. The four terms significantly contribute to overall model likelihood ($\alpha = 0.05$), but the p-value is high. Intuitively the null model already predicts quite well (because ∼80% of words don't shift stress), so there is a high bar for improving it. Indeed, the effect size (as an odds ratio) of adding the four terms is small (OR < 1.5):

```
oddsratio_lr(mlogreg_mod_1_baseline, mlogreg_mod_1)
## [1] 1.1
```

The AUC value is 0.71 (`auc(mlogreg_mod_1)`), suggesting the model has some explanatory power to predict whether stress shifts (baseline would be AUC = 0.5).

6.7.2.2 Model 2

We now add all the frequency-structure interaction terms, as these are of interest for the research questions for the diatones study. To fit and summarize the model:

```
mlogreg_mod_2 <- update(mlogreg_mod_1, . ~ . +
    (syll2_td + syll1_coda + syll2_coda):frequency)
tidy(mlogreg_mod_2)
##                      term estimate std.error statistic p.value
## 1             (Intercept)    -2.23      0.37     -6.08 1.2e-09
## 2              syll2_coda    -1.94      0.79     -2.44 1.5e-02
## 3                syll2_td     1.80      0.73      2.46 1.4e-02
## 4               frequency    -0.77      0.57     -1.35 1.8e-01
## 5              syll1_coda     1.87      0.71      2.66 7.9e-03
## 6      syll2_td:frequency     2.87      1.23      2.33 2.0e-02
## 7   frequency:syll1_coda     4.14      1.53      2.71 6.8e-03
## 8  syll2_coda:frequency    -1.56      1.59     -0.98 3.3e-01
```

We can verify using LRTs that the `syll2_coda:frequency` term does not significantly contribute, but other terms do. For example (output not shown):

```
mlogreg_mod_3 <- update(mlogreg_mod_2, . ~ . - syll2_coda:frequency)
## Test the syll2_coda:frequency interaction
anova(mlogreg_mod_2, mlogreg_mod_3, test = "Chisq")
```

Following the Gelman and Hill model selection guidelines (section 5.10.2), because we have no expectations about the directions of the interaction effects, we will keep those that significantly contribute to model likelihood and drop those that do not: the updated model is `mlogreg_mod_3`. The coefficients for this final model are

```
tidy(mlogreg_mod_3)
##                      term estimate std.error statistic p.value
## 1             (Intercept)    -2.17      0.35      -6.3 3.8e-10
## 2              syll2_coda    -1.76      0.74      -2.4 1.7e-02
## 3                syll2_td     1.66      0.70       2.4 1.7e-02
## 4               frequency    -0.59      0.52      -1.1 2.5e-01
## 5              syll1_coda     1.77      0.68       2.6 8.7e-03
## 6      syll2_td:frequency     2.43      1.13       2.1 3.2e-02
## 7   frequency:syll1_coda     3.79      1.46       2.6 9.6e-03
```

We will call this model 2. The predicted effects of "structure" predictors remain the same as in model 1: stress shift is predicted to be more likely for words with a first-syllable coda, or a second syllable ending in /t/ or /d/ (averaging over `frequency`), or without a second-syllable coda.

Otherwise, interpreting the model's results just from the coefficient table is unintuitive. To go further—in particular interpreting the frequency effect of theoretical interest—we must consider the interaction terms. This will serve as an example of how to visualize effects for logistic regression.

6.7.3 Visualizing Effects for Logistic Regression

Model prediction plots are especially useful for interpreting the results of logistic regression models containing multiple predictors, because in *probability* space, the effect of each predictor depends on the value others are held at. In addition, prediction plots are always helpful for understanding models containing interaction terms. We can exemplify this using the `mlogreg_mod_3` model.

It is useful when making prediction plots to refit with all two-level factors coded as factors, so that plotting functions treat these variables as categorical rather than continuous:

```
mlogreg_mod_3_fact <- update(mlogreg_mod_3, . ~ syll2_coda +
  (syll2_td_orig + syll1_coda_orig) * frequency)
```

We have previously used sjPlot to plot predictions of linear regression models (section 4.5.2; figure 6.3). Functions in this package like `plot_model()` (1) compute model predictions using functions from the `ggeffects` package, then (2) plot them. The computation step (1) is the hard part, because various choices must be made, such as whether/how to marginalize over predictors not plotted.

For the plotting step (2), you can either work with automatically generated plots, using `plot_model()` or a similar function (e.g., `allEffects()` in the effects package), or work directly with the model predictions to make customized plots.

For the `mlogreg_mod_3` model, we first visualize the main effect of each predictor using `plot_model`. For example, the code for the `syll1_coda` partial-effect plot is

```
plot_model(mlogreg_mod_3_fact, "eff",
  terms = "syll1_coda_orig", title = "") +
  ylab("Pred. probability") + xlab("Syll 1 coda")
```

Figure 6.4 shows the "marginal effect" of each predictor: other predictors are held at an *average* value, which is the mean for continuous predictors and the weighted average (using the proportion of observations at each level) for factors. This is the "marginal effect at the mean" from box 6.9.

Box 6.9
Broader Context: Marginal Effects

Marginal effects are a very useful tool for interpreting regression results: they quantify the effect of varying one or more predictors, as others are held constant or averaged over in some way that is often of substantive interest. Marginal effects are widely used in other fields (especially social sciences: e.g., Long and Freese 2006) and deserve to be more widely used in linguistics. It is useful to define some terminology to connect with existing resources, especially existing R packages such as emmeans, ggeffects, and margins.

Suppose we are just referring to one predictor x and an outcome y. Consider the slope of x when other predictors are held constant at some values—how much y changes per unit change in x (formally, the partial derivative of y with respect to x). There are three intuitive quantities we could calculate. The *average marginal effect* (AME) of x is the "average effect", which is the slope of x averaged over every observation in the dataset. The *marginal effect at the mean* is the slope of x for an "average observation", which is the slope at the mean values of all covariates. The coefficients of main effects when predictors are standardized can be thought of as either marginal effects at the mean ("all other predictors held at 0") or AMEs. Finally,

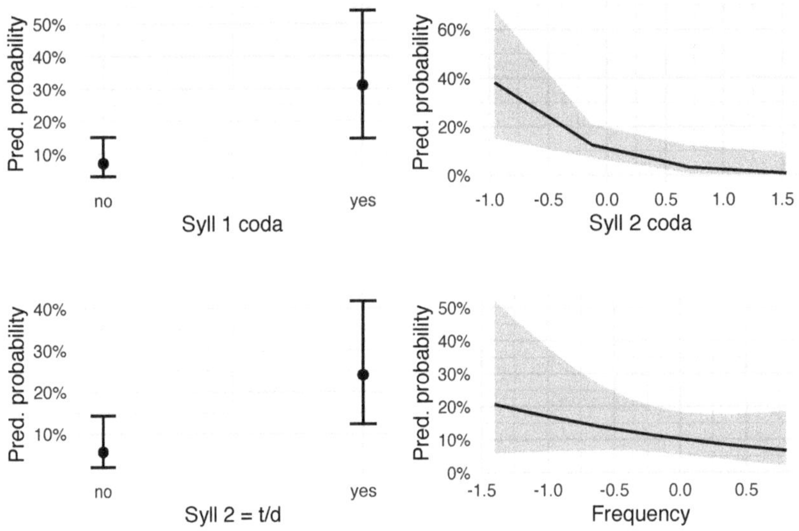

Figure 6.4
Partial effect of each predictor in model `mlogreg_mod_3`, marginalizing over other predictors (at mean values).

> the *marginal effect at representative values* are the slopes of x at theoretically interesting values of other predictors.
>
> The vignettes for the margins package (especially Leeper 2018, section 1) are a good place to read more.

To visualize the `frequency:syll1_coda` and `frequency:syll2_td` interactions, we will use `ggeffect` directly to see how it works:

```
int1_eff <- ggeffect(mlogreg_mod_3_fact,
  terms = c("frequency", "syll1_coda_orig")
)
int2_eff <- ggeffect(mlogreg_mod_3_fact,
  terms = c("frequency", "syll2_td_orig")
)
```

This function computes dataframes consisting of predicted probabilities, with SEs and 95% CIs, as the predictions to be plotted are varied, with others marginalized over.

For example, part of one dataframe is

```
head(int2_eff)
## # Predicted probabilities of stress_shifted
##
## # syll2_td_orig = no
##
## frequency | Predicted |        95% CI
```

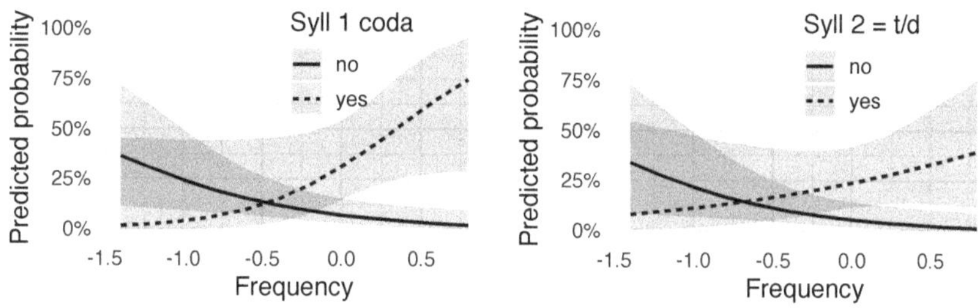

Figure 6.5
Interaction plots for each interaction in model `mlogreg_mod_3`, marginalizing over other predictors (at mean values).

```
## -------------------------------------
##      -1.40 |        0.34 | [0.09, 0.73]
##      -1.20 |        0.27 | [0.08, 0.62]
##      -1.00 |        0.22 | [0.07, 0.49]
##
## # syll2_td_orig = yes
##
## frequency | Predicted |         95% CI
## -------------------------------------
##      -1.40 |        0.08 | [0.01, 0.54]
##      -1.20 |        0.10 | [0.01, 0.51]
##      -1.00 |        0.11 | [0.02, 0.48]
```

Dataframes produced by `ggeffects` are automatically plotted using `ggplot()` (e.g., `plot(int2_eff)`), which allows us to change the appearance of the plots just like any other ggplot. The code to make figure 6.5, which plots the two interactions, is in the code file.

These plots help us interpret the interactions: for words without a first-syllable coda or where the second syllable does not end in t/d, stress shift is predicted to be more likely for lower-frequency words. Otherwise, this frequency effect is weaker and may even change direction (stress shift more likely for higher frequency words).

Together with the main effects visualized in figure 6.4, the overall interpretation is that stress shift is more likely for words with certain syllable structures (first syllable coda, lack of second syllable coda, second-syllable t/d coda), which is in line with the empirical trends (figure 6.1). There is no overall effect of frequency. Instead, the frequency effect depends on the word's syllable structure. Essentially, there is only a clear (and negative) frequency effect for words whose syllable structure does *not* promote stress shift.

The modulation of the frequency effect by syllable structure, as well as the effect of each structure predictor, is statistically significant, but with large CIs, as reflected in the large CIs in the prediction plots.

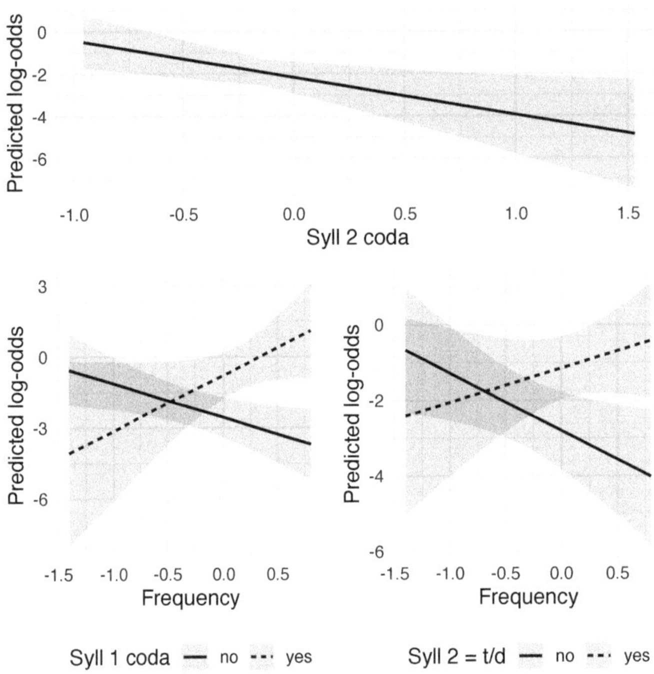

Figure 6.6
Partial-effect and interaction plots for model `mlogreg_mod_3`, showing predictions in log-odds marginalizing over other predictors, corresponding to figure 6.4 (top-right) and figure 6.5.

Box 6.10
Practical Note: Probability versus Log-Odds for Interpreting Predictions

Interpreting logistic regressions in probability versus log-odds has pros and cons (box 6.6), which carry over to plotting predictions. We can use the prediction dataframes constructed in this section to instead plot predictions in log-odds by just logit-transforming the predictions and CIs before plotting. Figure 6.6 shows some of these plots.

Some observations about the model are clearer from the log-odds plots. It is clearer from these plots how to interpret the interaction terms: each interaction only lets us be confident that a particular syllable structure modulates the frequency effect, not what the frequency effect is for any particular syllable structure. Examining the model prediction plots, it looks like the predicted frequency effect is negative for some syllable structures (the CIs for the frequency effect when `syll1_coda` = NO or `syll2_td` = NO can't contain a horizontal line), but it may not be positive for some syllable structures. We will see in section 7.4.1 how to make this observation more precise.

More generally, both predicted probability and predicted log-odds plots are useful for understanding logistic regression results. It is tempting to always just examine probabilities, as they are more intuitive, but often the model's interpretation is clearer from log-odds plots.

6.8 Model Validation

Chapter 5 discussed model validation for linear regression in detail: the iterative process of diagnosing and addressing problems in the data and model to arrive at a model we have confidence in. Because model validation addresses broad problems that are common to most regression models, most model assumptions and diagnostic tools are not qualitatively different for logistic regression. For example, the "linearity" assumption is now that the predictors are linearly related to the log-odds that $y=1$, rather than to y itself (as for linear regression). There are some aspects of model validation that differ from linear regression, which we will highlight.

We will discuss aspects of model validation for logistic regression using model 2, in the same order as we did for linear regression in sections 5.3–5.8. Because of the overlap with linear regression, our discussion is briefer for logistic regression (see section 6.10, Further Reading).

As for linear regressions, `augment()` in broom is a quick way to add additional information that is useful for model validation to the dataframe, including predicted values (in log-odds: `.fitted`), residuals (`.resid`), and Cook's distance (`.cooksd`), which will be discussed in the following sections.

```
mlogreg_mod_3_df <- augment(mlogreg_mod_3, data = diatones)
```

6.8.1 Problems with the Errors

6.8.1.1 Residuals and residual plots

Similarly to linear regression, residual plots are useful for diagnosing issues with the data or model, by looking for patterns in the errors the model makes. For logistic regression there is no error term in the model, so we must define residuals that preserve some intuitive properties of linear regression residuals. There are several options that preserve different properties (e.g., Agresti 2007, section 3.4); here we'll just discuss the most commonly used option.[3]

Deviance residuals intuitively capture each observation's contribution to the model's error on the data, analogously to the residual sum of squares for linear regression. Recall from equation (6.8) that deviance can be thought of as a sum of errors from each observation. We can define these errors in a way that makes deviance look like the residual sum of squares ($RSS=\sum_{i=1}^{n} e_i^2$):

$$D=\sum_{i=1}^{n} 2\cdot\{y_i \log[1/\hat{p}_i]+[1-y_i]\log[1/(1-\hat{p}_i)]\} \tag{6.16}$$

$$=\sum_{i=1}^{n} r_i^2. \tag{6.17}$$

3. Gelman and Hill (2007, section 5.6) illustrates using raw residuals for diagnostic plots.

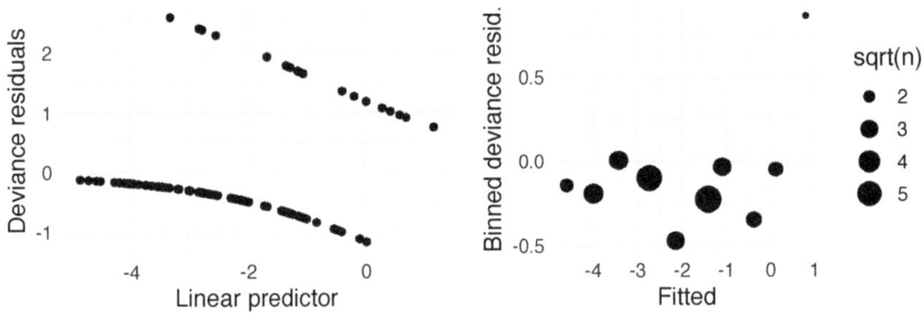

Figure 6.7
Fitted-residual (*left*) and fitted-binned-residual plots (*right*; with 10 bins, size scaled by number of observations) for model `mlogreg_mod_3`.

Written in this way, it is natural to define the *deviance residuals* as

$$d_i = \text{sign}(y_i - \hat{p}_i) r_i.$$

These residuals are positive for observations where $y = 1$ (because the predicted probability is always <1) and negative for $y = 0$ observations. These residuals, which are generated by `augment()` (or `residuals()`) by default, can be used to make diagnostic plots, analogously to linear regression: fitted-residual plots, plots of each predictor versus residuals, and so on.

Figure 6.7 shows two such plots for model 2. The left panel is a fitted-residual plot: two lines corresponding to 0 and 1 observations. This plot is not very useful, but could show gross outliers (there are none here).

It is more helpful to use *binned residuals*: the average residual over observations within a certain range of expected values. That is, you chop the plot in the left panel into vertical bins and average the residuals value within each bin to get a single y-axis number for each slice. The right panel shows the fitted-residual plot for model 2, using 10 bins.

This plot now looks more like a fitted-residual plot for a linear regression. An important difference from linear regression is that logistic regression residuals are not assumed to follow a normal (or any) distribution, so it is not strictly necessary to check whether the residuals show constant variance, are normally distributed, or have mean of 0. Nonetheless, residual plots remain useful for diagnosing qualitative problems with the data or model (e.g., nonlinearity, outliers), similarly to linear regression.

For example, in the fitted-binned-residual plot, we expect the residuals to have roughly constant mean and variance (regardless of whether they are normally distributed): qualitatively they are like the ideal fitted/residual plot for a linear regression. A trend in the mean could indicate a nonlinear predictor effect, for example. This plot doesn't suggest any major issue.

6.8.1.2 Independence

Logistic regression assumes that observations are independent: knowing the predicted log-odds for one observation should not give you any information about the predicted log-odds of another observation, after accounting for their predictor values. As for linear regression, violations of independence are the most important issue to address for logistic regression

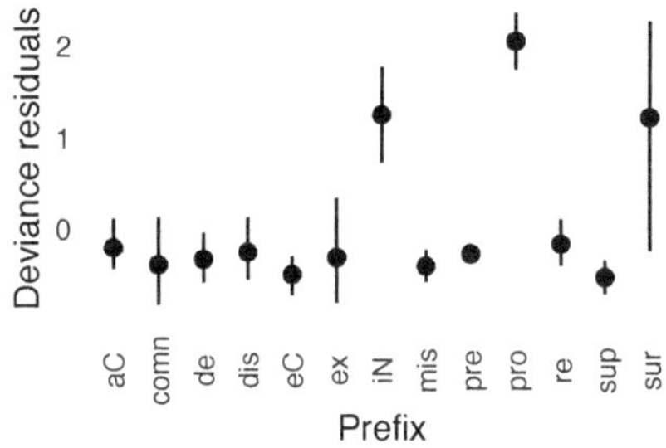

Figure 6.8
Average deviance residuals for model `mlogreg_mod_3` by word `prefix` (errorbars are 95% CIs).

and the hardest to diagnose (same as section 5.3.1), because they require thinking about the structure of your data for possible sources of nonindependence.

For the `diatones` data, there is a natural possibility to check: that words sharing the same prefix (e.g., 'survey', 'surprise') are not independent. We can check this by examining the distribution of residuals for different values of `prefix` (figure 6.8).

It looks like different prefixes mostly have similar residual distributions, but a few have much higher residuals than others—especially the prefixes IN and PRO. We will return to this issue later (section 9.2.1), where we will check whether fitting a model that accounts for this source of nonindependence changes the qualitative conclusions we draw from the model.

6.8.2 Problems with the Model and Predictors

6.8.2.1 Linearity, transformations, collinearity

Possible problems with the model, and how these can be addressed by nonlinear effects or transformations (sections 5.4 and 5.5), are broadly similar to linear regression. Most importantly, *linearity* remains a crucial assumption for logistic regression models (no missing nonlinear effects or interactions), which can be checked using similar methods to linear regression such as plotting (binned) residuals as a function of a continuous predictor or having two or more predictors. It is no longer possible to transform the response for a logistic regression, because its values must remain 0 and 1. But all motivations for transforming predictors—especially centering and scaling to improve interpretability of regression coefficients and to make their magnitudes comparable—are just as important as for linear regression. As for linear regression (section 5.6), the predictors in logistic regression are assumed to not be linearly dependent, and the presence of collinearity has similar effects on regression coefficients.

Exercise 6.4 asks you to assess possible nonlinearity in the effect of `frequency` and assess the degree of collinearity for model 2.

6.8.2.2 Separation and quasi-separation

A new potential problem with predictors for logistic regression is *separation*: when a subset of predictor values is associated only with $y = 1$ or only with $y = 0$.

As an example, first note that for some parts of the `diatones` data no words have shifted stress, such as words with second syllable coda = CCC:

```
xtabs(~ syll2_coda_orig + stress_shifted, diatones)
##                stress_shifted
## syll2_coda_orig  0  1
##             0    9  4
##             C   73 13
##             CC  25  4
##             CCC  2  0
```

Fitting a simple model to this data, predicting stress shift just as a function of `syll1_coda_orig` and `syll2_coda_orig`, gives an odd result:

```
glm(stress_shifted ~ syll1_coda_orig + syll2_coda_orig,
  data = diatones, family = "binomial"
) %>% tidy()
##                   term estimate std.error statistic p.value
## 1          (Intercept)    -0.81      0.60     -1.35   0.177
## 2 syll1_coda_origyes      1.24      0.55      2.27   0.023
## 3    syll2_coda_origC    -1.32      0.71     -1.86   0.063
## 4   syll2_coda_origCC    -1.43      0.85     -1.69   0.091
## 5 syll2_coda_origCCC   -16.48   1635.11     -0.01   0.992
```

The `syll2_coda_origCCC` coefficient has a very high estimate and SE, which means that some observations (those with second syllable coda = CCC) will have a predicted probability very near 0. This is *complete separation*. To fit this data, the model has chosen coefficients that are very far from 0 so that the predicted probability is effectively 0 or 1 for some observations. The problem with this is similar to one for linear dependence: the model is unstable in the sense that there is no longer a unique set of "best" coefficients. Choosing a `syll2_coda_origCCC` estimate that is half or twice as large will make effectively the same model predictions, hence the SEs are huge.

Like linear dependence, *complete separation* is fairly easy to spot, and it indicates a problem with your data or model that should be fixed.[4] Trickier is *quasi-complete separation*, which is analogous to very high collinearity: if the predicted probability for some set of predictor values is very close to 0 or 1, some coefficients will have high values/SEs because a range of coefficient values give similar model predictions (Baguley 2012, section 17.4.3). The resulting issues are similar to high multicollinearity, and it is similarly debatable whether quasi-complete separation is a "problem."

Complete/quasi-complete separation is essentially a case of overfitting: it is unlikely that the true probabilities are 0 or 1, but the more predictors are added to a model, the more likely it is that some combinations of predictors will have only $y = 0$ or only $y = 1$ observations. Thus, possible solutions are those applicable to overfitting: collecting more data, using a simpler model (e.g., combining factor levels, in this case CC and CCC), or using a variant of logistic regression that penalizes large coefficients.

---

4. For more extreme cases, R throws a warning: `glm.fit: fitted probabilities numerically 0 or 1 occurred`.

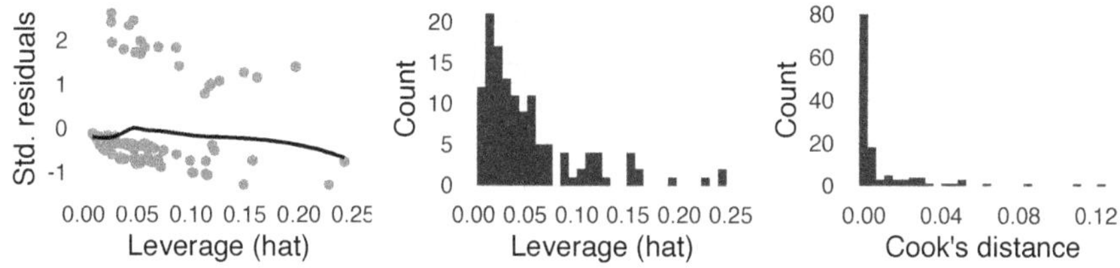

Figure 6.9
Diagnostic plots for unusual observations for model `mlogreg_mod_3`: leverage-residual plot (*left*), histograms of leverage and Cook's distance (*center, right*).

See Agresti (2015, section 5.4.2), Baguley (2012, section 17.4.3), and Levshina (2015, 273) for further discussion of (quasi-)separation and possible solutions.

6.8.3 Problems with Observations

For the same reasons as for linear regression, we are interested in identifying observations that have particularly extreme predictor values, for which the model predicts particularly badly, or that have high influence on coefficient values (section 5.7). These are quantified by defining measures of outlier-ness (standardized residual), leverage (hat values), and CD, similarly to linear regression; previously we added these columns to `mlogreg_mod_3_df` using `augment()`.

Again, visual inspection or quantitative diagnostics can be used to decide which points are "influential" and so on. For example, `outlierTest()` in the car package suggests no clear outliers.

```
outlierTest(mlogreg_mod_3)
## No Studentized residuals with Bonferroni p < 0.05
## Largest |rstudent|:
##    rstudent unadjusted p-value Bonferroni p
## 50      2.7             0.0064         0.83
```

This can also be assessed by a leverage-residual plot (figure 6.9, left). This plot suggests a few possible outliers.

Histograms of hat values and Cook's distance (figure 6.9 center, right) don't suggest any points that have unusual leverage or influence. The most influential points are a couple words with prefix RE:

```
filter(mlogreg_mod_3_df, .cooksd > 0.1)
##       word syll1_coda syll2_coda syll2_td frequency prefix
## 1 rebound       -0.22         0.7     0.62     -0.65     re
## 2 relapse       -0.22         0.7    -0.38     -0.47     re
...
```

These words are also the potential outliers in the leverage-residual plot; they are influential because they have shifted stress while the model predicts a low probability of stress shift.

We might flag these points to see what effect they have on the model. It turns out that the model has higher predictive power (AUC) and all qualitative results are strengthened when these points are excluded (exercise 6.4). So we can be confident that the model's results aren't due to a few points. This should be reported in a write-up of the model.

6.8.4 Example: Assessing Overfitting

In section 5.8.2 we discussed different internal validation methods for assessing overfitting (for linear regression), without a practical example. Because the `diatones` dataset is fairly small relative to the number of predictors, this is a good case to show one way to assess whether a model overfits and whether overfitting affects the conclusion of the analysis.

The discussion there carries over to logistic regression, with the rough rule of thumb that overfitting may be a concern for a model with more than $n/15$ predictors–where n is now the number of observations with the *less common* outcome (Harrell 2015, section 4.4). For the `diatones` data, $n = 21$ (words with shifted stress), so $n/15 = 1.4$, and our best model (model 2) has six more predictors than this, so overfitting is a possibility.

At a high level, we will be

1. Assessing the degree of overfitting (with a bootstrapping method)
2. Refitting the model using a method robust to overfitting
3. Comparing the results of the original and new model

We use functionality from the rms package for (1)–(3), which is described in more detail by Baayen (2008, section 6.2.4). There are different ways to do (1), (2), and (3), but the general idea is the same.

We first refit the model using the logistic regression function from rms:

```
rms_mod_1 <- lrm(formula(mlogreg_mod_3), data = diatones,
  x = T, y = T)
```

This model has an identical coefficient table to model 2 fitted with `glm()` (not shown). Its goodness of fit, as an AUC value, can be extracted from an `lrm` model as follows:

```
## The C-index is the same as AUC for a logistic regression
rms_mod_1$stats[["C"]]
## [1] 0.78
```

We assess overfitting with the `validate()` function, which applies a bootstrapping method (Harrell 2015, section 5.3.5) that refits the model on many resampled datasets, each time recalculating goodness of fit using the new model on the full dataset and the resampled dataset—the difference between these is the "optimism," or how much the goodness-of-fit measure is inflated due to overfitting.

```
val <- rms::validate(rms_mod_1, B = 100)
val
##           index.orig training  test optimism index.corrected   n
## Dxy            0.553    0.608 0.508    0.100           0.453 100
```

```
...
```

In this case, the AUC measure has optimism (estimated over 100 runs) of 0.05 (the D_{xy} value divided by 2), so the corrected AUC is 0.73 (0.5 + 0.45/2). That's a big drop (recall that 0.5 is baseline), so there is some overfitting, as expected.

To assess whether it matters that the model is overfitted with respect to our research questions, we need to refit the model, correcting for overfitting, and compare to our original results. Specifying the `penalty` argument to `lrm()` fits the model using *penalized maximum likelihood*, where a term is added to the likelihood that penalizes coefficients further from 0. This has the effect of shrinking coefficients toward 0.

```
# Find 'optimal' penalty: maximizes modified AIC
pt_result <- pentrace(rms_mod_1, seq(0, 2.0, by = 0.02))
## Refit model with this penalty
rms_mod_2 <- update(rms_mod_1, penalty = pt_result$penality)
```

```
rms_mod_2
...
##                          Coef    S.E.   Wald Z Pr(>|Z|)
##  Intercept              -2.1700 0.3466 -6.26  <0.0001
##  syll2_coda             -1.7646 0.7400 -2.38  0.0171
##  syll2_td                1.6612 0.6968  2.38  0.0171
##  frequency              -0.5914 0.5190 -1.14  0.2545
##  syll1_coda              1.7714 0.6756  2.62  0.0087
##  syll2_td * frequency    2.4293 1.1332  2.14  0.0321
##  frequency * syll1_coda  3.7858 1.4609  2.59  0.0096
##
```

Compare to the unpenalized model:

```
rms_mod_1
...
##
##                          Coef    S.E.   Wald Z Pr(>|Z|)
##  Intercept              -2.1700 0.3466 -6.26  <0.0001
##  syll2_coda             -1.7646 0.7400 -2.38  0.0171
##  syll2_td                1.6612 0.6968  2.38  0.0171
##  frequency              -0.5914 0.5190 -1.14  0.2545
##  syll1_coda              1.7714 0.6756  2.62  0.0087
##  syll2_td * frequency    2.4293 1.1332  2.14  0.0321
##  frequency * syll1_coda  3.7858 1.4609  2.59  0.0096
##
```

Every (nonintercept) coefficient in the penalized model is closer to 0, and has a z-value closer to 0 (= a higher p-value). However, the *qualitative* results are almost the same: the coefficients have the same signs, the same relative effect sizes, and relatively similar p-values (e.g., on the same side of $\alpha = 0.05$ or 0.01). We don't expect any of our qualitative conclusions about the model to change, as can be verified by making the same prediction plots as in section 6.7.3: although all CIs are wider and all effects a bit "flatter," the qualitative patterns are identical. Thus, (1) our original model was overfitted, but (2) overfitting wasn't severe enough to affect the conclusions we draw from the analysis. At this point we could either report the penalized model (as in Sonderegger 2010) or report the original model, noting (1) and (2). This choice is an instance of the bias-variance tradeoff (section 5.8.1): coefficients from the penalized model are more biased than the original model (by design), but have lower variance.

Two important takeaways: (1) it is important to evaluate overfitting when sample size is small relative to the model's complexity, but (2) an overfitted model is not necessarily a problem if the conclusions drawn in your analysis are robust.

6.9 Reporting and Summarizing

6.9.1 Reporting a Logistic Regression Model

Reporting a logistic regression model in a write-up is generally similar to reporting a linear regression model: the guidelines and rationale in section 4.6 for reporting individual coefficients and the whole model hold, with some adjustments.

6.9.1.1 Coefficients

For each regression coefficient you report at a minimum the coefficient estimate, its SE, the test statistic value (whether z for a Wald test or χ^2 and *df* for a likelihood-based test; see section 6.5.2), and corresponding p-value. It is best to report all model coefficients in a table (see the following example), in addition to possible in-text reporting.

As for linear regression, it is useful to also give visualizations, CIs, and basic descriptive statistics, but what is appropriate will depend on context and space. Model prediction plots are especially important for interpreting logistic regressions, as discussed in section 6.7.3.

6.9.1.2 Model summaries

It is common to report some quantitative measures of the model under the regression table: its goodness of fit (e.g., AUC, D_{xy}, Nagelkerke R^2, accuracy), measures of likelihood/information (deviance of the full and baseline model with associated *df*, AIC or BIC, log-likelihood), or the number of parameters and observations. These and other measures are reported by `summary()`, `glance()`, or packages discussed in section 6.6.

There is no standard set of measures to report; the minimum seems to be one or more measures of goodness of fit (such as AUC or D_{xy}, not accuracy), number of observations, and one or more measures of model likelihood versus the baseline (e.g., deviance of full and null models, difference in deviance), all of which give important information about the model and from which other information of interest can be determined.

6.9.1.3 Example

This is one possible regression table for model 2 (`mlr_mod_3`), following Faraway (2016) (and adding AUC):

Coefficient	$\hat{\beta}$	$SE(\hat{\beta})$	z	p
Intercept	−2.17	0.35	−6.26	<0.001
Syllable 2 coda	−1.76	0.74	−2.38	0.017
Syllable 2 t/d	1.66	0.70	2.38	0.017
Frequency	−0.59	0.52	−1.14	0.25
Syllable 1 coda	1.77	0.68	2.62	0.0087
Syllable 2 t/d:Frequency	2.43	1.13	2.14	0.032
Syllable 1 coda:Frequency	3.79	1.46	2.59	0.0096

$n = 130$, AUC = 0.78, $p = 7$, deviance = 94, null deviance = 115 (difference = 21).

6.9.2 Variable Importance in More Complex Models

It is often of interest to compare the relative importance of different predictors for a model. We have discussed how to do this in two ways, so far only for relatively simple linear and logistic regression models, which contain only continuous predictors or two-level factors (no interactions, factors with three or more levels, or nonlinear effects). The *effect size* of different predictors can be read off from the model coefficients (when predictors are standardized). The *significance* of different predictors can be compared via their p-values.

These methods don't generalize to more complex models, where we want to ask the question: "what is the importance of predictor x?" For example, given the motivation of the `diatones` study, it is of interest for model 1 to know how important each predictor is overall—`frequency` and each structure predictor—including both the main effect and interactions including this predictor. Let's discuss a couple ways to do this, each defining "importance" differently.

6.9.2.1 Model comparison

The contribution of each predictor can be assessed by comparing the full model with a model where the predictor and all its interaction terms are dropped.

The "significance" of each comparison can be assessed by an appropriate hypothesis test. For example, an LRT for `syll1_coda` would exclude the `syll1_coda` and `syll1_coda:frequency` terms:

```
mlogreg_mod_3_nosyll1_coda <- update(mlogreg_mod_3, . ~ . -
  frequency:syll1_coda - syll1_coda)
anova(mlogreg_mod_3, mlogreg_mod_3_nosyll1_coda, test = "Chisq")
...
##   Resid. Df Resid. Dev Df Deviance Pr(>Chi)
## 1       123       93.7
## 2       125      108.3 -2    -14.6  0.00067
```

Similar tests would be conducted for `syll2_coda`, `syll2_td`, and `frequency`. As a shortcut, we can apply `anova()` to the model refitted using `lrm`. (This applies Wald tests rather than LRTs, but the idea is the same.)

```
anova(rms_mod_1, india=FALSE)
##                Wald Statistics          Response: stress_shifted
```

```
##
## Factor                                                    Chi-Square
## syll2_coda                                                 5.7
## syll2_td  (Factor+Higher Order Factors)                    8.2
## frequency  (Factor+Higher Order Factors)                  11.2
## syll1_coda  (Factor+Higher Order Factors)                 13.3
## syll2_td * frequency  (Factor+Higher Order Factors)        4.6
## frequency * syll1_coda  (Factor+Higher Order Factors)      6.7
## TOTAL INTERACTION                                          9.5
## TOTAL                                                     15.6
## d.f. P
## 1    0.0171
## 2    0.0163
## 3    0.0109
## 2    0.0013
## 1    0.0321
## 1    0.0096
## 2    0.0087
## 6    0.0162
```

(The "india" option excludes some rows we don't need.) Here, the `syll1_coda` row reports the hypothesis test when the `syll1_coda` and `frequency * syll1_coda` terms are excluded and so on.[5] The p-values for the first four rows are metrics of the *significance* of each variable's contribution. This gives the following order of predictor importance (exercise 6.5):

`syll1_coda` > `frequency` > `syll2_td` > `syll2_coda`

That is, the p-value is lowest for `syll1_coda` and so on.

Alternatively, an appropriate *effect size* can be calculated for each model comparison (Cohen's f for linear regression, the odds ratio for logistic regression; see equation (5.4), section 6.7.1) and used as a metric of each predictor's importance. For example, for `syll1_coda` the odds ratio would be

```
oddsratio_lr(mlogreg_mod_3_nosyll1_coda, mlogreg_mod_3)
## [1] 1.2
```

The ORs turn out to give the same order of predictor importance (exercise 6.5), but in general the conclusions from effect size and significance could be different.

This order makes intuitive sense, looking at the partial-effect plots for this model (figure 6.6). The first syllable coda clearly has a large effect. Frequency also has a large effect, if we are thinking of its *absolute effect*, which is when we add together its positive and negative effects for different values of structure variables.

5. This method also works for linear regression models, by refitting the model using the `ols()` function from rms.

6.9.2.2 Average marginal effects

Another approach is based on how much a change in the predictor actually affects the response, with other predictors held constant—the *marginal effect* (box 6.9). The AME is the mean of the marginal effect calculated for each observation in the dataset, and it gives a sense of the average effect of the predictor.[6]

The `margins` package (Leeper 2021, 2018) computes AMEs, including uncertainties, for a wide range of models. When applied to our `mlogreg_mod_3`, the `margins` function from this package gives the AME for each predictor, as well as measures of uncertainty:

```
ame <- margins(mlogreg_mod_3)
summary(ame)
##       factor     AME     SE       z      p   lower   upper
##    frequency -0.0145 0.0565 -0.2567 0.7974 -0.1252  0.0962
##   syll1_coda  0.1559 0.0744  2.0962 0.0361  0.0101  0.3017
##   syll2_coda -0.1962 0.0789 -2.4877 0.0129 -0.3508 -0.0416
##     syll2_td  0.1584 0.0730  2.1705 0.0300  0.0154  0.3014
```

Each AME is in probability space. So, for example, a unit change in frequency leads to a decrease of –0.014 in probability, with 95% CI (–0.13, 0.096).

We can use the absolute values of the AMEs as a measure of variable importance:

`syll2_coda` > `syll1_coda` > `syll2_td` > `frequency`

This is very different from the previous ordering: `frequency` now has the smallest effect, while the three structure predictors have comparable importance (taking into account their overlapping CIs). Frequency now has the smallest effect because *on average* it has a *near-zero effect*, averaging across the cases where its effect is positive and negative (figure 6.6, center and right panels). This ordering is what we'd expect based on the empirical plots for each predictor in figure 6.1 where other predictors are effectively averaged over.

Which option makes more sense for assessing variable importance (hypothesis testing or AMEs) will depend on context. For the `diatones` data, one research question was how important a word's frequency is in determining stress shift. Rephrasing this question as "how much does frequency help predict stress shift?", the hypothesis testing method makes more sense.

Box 6.11
Broader Context: Average Marginal Effects as Standardized Regression Coefficients

The order of predictor importance we got using AMEs (`syll2_coda` > `syll1_coda` > `syll2_td` > `frequency`) is the same as what we would get from comparing the absolute value of the main effect coefficients for model `mlogreg_mod_3`, which is another way to assess predictor importance in models where predictors are standardized by "rescaling" (section 5.5.4).

In fact, calculating the AMEs in log-odds space, we see that the AMEs are the same as the main effect coefficients for this model:

6. Sonderegger (2010) quantifies the effect of each predictor for models of the `diatones` data using the very similar "average predictive difference" method of Gelman and Hill (2007, section 21.4).

```
margins(mlogreg_mod_3, type = "link")
##   syll2_coda syll2_td frequency syll1_coda
##       -1.765    1.661   -0.5914      1.771
coefficients(mlogreg_mod_3)[2:5]
## syll2_coda   syll2_td  frequency syll1_coda
##      -1.76       1.66      -0.59       1.77
```

This is because we have standardized all variables. One motivation for standardizing was so that each main effect coefficient could be interpreted as an average effect, averaging across other predictors.

Calculating AMEs is still useful for a model where the predictors aren't standardized or for a more complex model (e.g., with nonlinear effects) where standardizing isn't possible. It is also good to know that AMEs are a way to calculate the average effect of predictors for *any* model, regardless of how you've coded the predictors.

6.10 Further Reading

Many sources cover topics in this chapter, including detailed treatments of logistic regression. I have relied primarily on Agresti (2007, chapters 1–5), Chatterjee and Hadi (2012, chapter 12), Faraway (2016, chapters 1, 6), and Gelman and Hill (2007, chapters 4, 5). Chatterjee and Hadi (2012) and Faraway (2016) are particularly good on model validation. Agresti (2007) is an accessible resource for categorical data analysis generally, and Agresti (2003) is a more technical/comprehensive version.

For language scientists specifically, there are resources focusing on specific subfields where logistic regression has been used the longest—variationist sociolinguistics and corpus linguistics (Tagliamonte 2006; Johnson 2008, chapter 5; Speelman 2014; Gries 2021, section 5.3)—as well as more general treatments (Winter 2019, chapter 12; Baayen 2008, section 6.3.1; Levshina 2015, chapter 13). Jaeger (2008) introduces logistic regression in comparison to ANOVAs.

Exercises

6.1. Calculate Cramer's V for the `diatones` and `regularity` examples in section 6.2.4. Hint: you can use `chisq.test()` with argument `correct = FALSE` to compute χ^2.

6.2. For the model `mlogreg_mod_1` in section 6.7.2.1, we interpreted the `syll1_coda` and `frequency` coefficients in terms of odds. What is the interpretation of each coefficient in terms of probability, calculated in one of the ways discussed in the text (e.g., AMEs)?

6.3. The model `mlogreg_mod_1` will predict the log-odds of stress shifting by the year 2005 for any English diatone. Suppose that "disfug" were an English diatone with average `frequency`, and `syll1_coda` = `yes`, `syll2_coda` = `C`, `syll2_td` = `no`. What is the predicted probability of stress shifting by 2005 for this word? (Hint: to use `predict()`, you'll need to determine what each predictor's value is for the standardized versions used to fit the model.)

6.4. Additional model validation for `mlogreg_mod_2`:

(a) Assess whether the assumption of a linear effect of `frequency` is appropriate, using one or more diagnostic plots.

(b) Assess the degree of collinearity in the data this model is fitted to, using diagnostics from section 5.6.3.

(c) In section 6.8.3 we saw there are some potentially influential points for this model. Check what happens to AUC and to the qualitative conclusions we can make from the model when these points are excluded. Is the model robust to these observations?

6.5. Carry out the full set of model comparisons and effect size computations referred to in section 6.9.2, resulting in four significances and effect sizes, and verify that they give the same order of importance for the four predictors.

7 Practical Regression Topics

Thus far we have only fitted models with relatively simple structure: categorical predictors ("factors") with two levels, linear effects of continuous predictors, and two-way interactions between such predictors. This chapter introduces tools for fitting and interpreting regression models that include factors with more than two levels (sections 7.2 and 7.3: contrast coding, omnibus and post hoc tests), more complex interactions (section 7.4), or nonlinear effects (section 7.5).

7.1 Preliminaries

7.1.1 Packages

This discussion assumes that you have loaded several packages from previous chapters, as well as the splines and emmeans packages (Lenth 2021).

```
library(tidyverse)
library(broom)
library(arm)
library(languageR)
library(performance)
library(splines)
library(emmeans)
```

7.1.2 Data

This discussion also assumes you have reloaded the `vot_michael` dataset from section 5.1.2, but now we'll be keeping `place` as a three-level factor, so this dataset can be used to introduce multilevel factors.

```
vot <- read.csv("data/vot_rmld.csv", stringsAsFactors = TRUE)
# One speaker, first use of each unique word
vot_michael <- filter(vot, speaker == "michael") %>%
  filter(!duplicated(word)) %>%
  mutate(place = fct_relevel(place, "labial")) %>%
  droplevels()
```

We also recoded the levels of PLACE to go in the theoretically expected order LABIAL < ALVEOLAR < VELAR (see appendix A, section A.2).

We will also need the young speakers subset of the `english` data (section 4.3.2):

```
english_young <- filter(english, AgeSubject == "young")
```

The `french_cdi_24` dataset

This discussion also assumes that you have loaded the French CDI dataset:

```
french_cdi_24 <- read_csv("data/french_cdi_24_rmld.csv")
```

This dataset comes from Wordbank, an open database of children's vocabulary growth that aggregates data from thousands of parent-report questionnaires on lexical acquisition—the MacArthur–Bates Communicative Development Inventory (CDI)—across many languages (Frank et al. 2021). Data were extracted to create a small subset: 664 words spoken by Quebec French–learning children aged 24 months.[1]

Each row corresponds to one word, from a single child (`data_id`). The column `produces` is 1 if the parent reports the child can produce this word (column `definition`) and 0 otherwise; this is a proxy for whether the child "knows" the word. Of interest is how a word's likelihood of being known is predicted by the child's age and gender, and properties of the word such as its `lexical_class` (levels: FUNCTION_WORDS, VERBS, ADJECTIVES, NOUNS, OTHER). Broadly speaking, children are thought to acquire nouns earlier than predicate types (verbs/adjectives) and predicates earlier than function words. This is commonly summarized as "noun bias" (nouns acquired earliest); Frank et al. (2021, chapter 11) argue from the Wordbank data that "function-word bias" (function words acquired latest) is an even more robust generalization across languages.

For this chapter, we use data from a single child (`data_id` 140275), so that observations are independent. Of interest is how `lexical_class` affects the probability the child can produce this word; figure 7.1 (right) shows the empirical pattern. For simplicity we exclude OTHER words and reorder `lexical_class` to the theoretically expected order:

```
french_cdi_24 <- filter(french_cdi_24, data_id == 140275) %>%
  filter(lexical_class != "other") %>%
  mutate(lexical_class = fct_relevel(
    lexical_class, "function_words",
    "verbs", "adjectives", "nouns"
  )) %>%
  droplevels()
```

1. Wordbank is at http://wordbank.stanford.edu/. To make this dataset, item and instrument-level responses were downloaded for all available "WS" forms in Quebec French from Wordbank (Frank et al. 2017; Boudreault et al. 2007; Trudeau and Sutton 2011) in September 2020, then restricted to words (gestures excluded), and combined with lexical measures (frequency, etc.) from Braginsky et al. (2019). I thank Mika Braginsky for their help.

7.2 Multilevel Factors: Contrast Coding

As running examples of a factor x with $k>2$ levels in regression models, we use the effect of `place` on VOT ($k=3$) in the `vot_michael` data, and the effect of `lexical_class` on word production ($k=4$) in the `french_cdi_24` data.

There are two broad approaches to understand the question: "How does x affect the response?"

1. *Contrast coding* (section 7.2): A factor with k levels corresponds to $k-1$ predictors in a regression model, the *contrasts*. By examining the model's results for each contrast, we address $k-1$ questions about how the factor affects the response.
2. *Omnibus/post hoc tests* (section 7.3): First we ask, "do levels [significantly] differ at all?" (using an "omnibus test"); if so, we apply additional hypothesis tests to dig into which levels (significantly) differ (using "post hoc tests"), correcting for multiple comparisons.

The contrast coding approach is more relevant for addressing research questions using regression models, but the omnibus/post hoc test approach is more common in current practice (especially for ANOVA analyses). The line between these approaches is blurry, and it is often useful to combine methods for a more flexible approach (e.g., interpreting interactions; see section 7.4), But it's useful to keep them conceptually separate to understand how they are used in practice in many papers (box 7.1).

Box 7.1
Broader Context: Terminology for Multilevel Factors

The terminology for multilevel factors in regression modeling (at least in linguistics and psychology), and the divide into two approaches described, come from the most common use case historically: a classic experiment testing for differences between groups, analyzed using ANOVAs, with the statistical analysis decided before data are collected. This makes the terminology confusing if you have not had a traditional "psych stats" training or don't often analyze data from this use case. But the terminology is so widely used that it is useful to know the context (e.g., Field, Miles, and Field 2012, section 10.4).

Suppose you are doing a classic experiment: there are k groups, and you want to know whether the groups differ in the response y. (Note that the "groups" could be a control and several treatments, four groups differing in two between-subject manipulations (a 2×2 design), for instance) An example would be a phonetic study of how `place` affects `vot` for a new language. You must decide how to do the statistical analysis before collecting data, and your analysis can ask $k-1$ independent questions about how groups differ. There are two ways you could analyze the data.

First, you could prespecify a single question—do groups differ at all?—which you test with an ANOVA (or a model comparison, in the regression setting). If they do, you carry out additional hypothesis tests (e.g., t-tests, in the ANOVA setting) to see exactly how groups differ—typically by "pairwise" comparisons (every possible pair of levels: LABIAL/ALVEOLAR, etc.), adjusting p-values for multiple comparisons. These additional tests are post hoc, because you did not specify them in advance to test specific research hypotheses.

Alternatively, you could define the $k-1$ comparisons you are going to test, to address specific research hypotheses (e.g., "do LABIAL and ALVEOLAR differ?", "do ALVEOLAR and VELAR differ?"), before you collect the data; these are *planned comparisons* (or "planned contrasts"). You carry out the corresponding hypothesis tests (in one multiple regression or with $k-1$ t-tests), without correcting for multiple comparisons, and report their results.

This all makes sense until you move outside the assumptions of this "classic" setting: that every test is either planned or post hoc, and that this lines up with whether a test is targeted to address research questions or is agnostic to them. In a more flexible data analysis workflow with both exploratory and confirmatory analysis, which is common in practice (section 1.3.2) both for observational and experimental data, the line between planned and post hoc is not so clear and is analogous to the division between EDA and CDA. We might choose a contrast coding scheme not to address research questions, but because it helps with model interpretation (e.g., contrasts are *centered*, or reflect the empirical pattern observed in the data). Are these planned or post hoc comparisons? And do we need to correct for multiple comparisons if the contrasts weren't planned? And so on.

My opinion is that researchers should be aware of the general issues, rather than following strict rules: how k-level categorical predictors are coded into $k-1$ numeric predictors, hypothesis testing in this setting, the logic of correcting for multiple comparisons, and especially "researcher degrees of freedom"—the more ways you analyze your data, the less your p-values mean (Simmons, Nelson, and Simonsohn 2011; Roettger 2019).

7.2.1 Background

For exposition of contrasts, which describe comparisons between levels of a factor, it is useful to introduce this notation (Bolker 2019a):

$$\mu_i = \text{the model-predicted value for level } i.$$

We also use μ_i to mean the value's approximation from the empirical data (the average of y for observations in level i). For example, for `vot`, μ_1 and μ_2 are the average VOT for LABIAL and VELAR stops.

We also need to define the average value of y for a factor, for example, "the average of `place`." This is often called the "grand mean," and it can be defined in two ways (e.g., Cohen et al. 2002, section 8.3.1):

1. *Weighted (grand) mean* is the average of y across the dataset.
2. *Unweighted (grand) mean* is the average of the averages of y for each level of the factor (i.e., the average of μ_1, μ_2...).

For the `vot` example (y is `vot`), these two means are

```
vot_means <- vot_michael %>%
  group_by(place) %>%
  summarise(m = mean(vot), n = n()) %>%
  summarise(weighted = weighted.mean(m, n), unweighted = mean(m))
vot_means
##   weighted unweighted
## 1       44         45
```

For *balanced* data, where there are an equal number of observations per factor level, the weighted and unweighted means are the same, and it is fine to say just "grand mean." But for unbalanced data (which most linguistic data are), the weighted mean will be skewed toward levels that have more observations. Often it is useful to consider the unweighted mean, which estimates what the grand mean *would* be if the data were balanced. The terminology is confusing, and it is common to use "grand mean" to mean either weighted or unweighted mean or both (when the meaning is clear).

7.2.2 Contrast Coding: Introduction

In a regression model, a factor x with k levels corresponds to $k-1$ numeric predictors, also called *contrasts*, which capture hypotheses/questions about how the response differs between groups. There are different ways of mapping k levels into $k-1$ predictors (*contrast coding schemes*), corresponding to different sets of hypotheses. Importantly, choosing different contrast coding schemes for x results in *equivalent models*, which make the same predictions, but whose coefficients have different interpretations.

For example, for the predictor `place` (levels = BILABIAL, ALVEOLAR, VELAR), the pairs of questions might be

- What is the difference between BILABIAL and ALVEOLAR? Between BILABIAL and VELAR?
- Or: What is the difference between BILABIAL and non-BILABIAL stops? Between ALVEOLAR and VELAR?

Only $k-1$ contrasts (here, two contrasts) can be included in a model in order for these predictors to not be linearly dependent—the intercept is the kth degree of freedom. As a result, contrasts never correspond to y at a certain level of x (e.g., "`vot` when `place` = BILABIAL"): they always correspond to differences between levels. This is a frequent point of confusion.

Choosing a contrast coding scheme for x fixes the interpretation of the regression coefficients for x (in terms of levels of x, such as, "`alveolar` – `bilabial`"), and the intercept term. Thus, how to interpret the regression coefficients corresponding to a factor depends on the coding scheme, so some understanding of contrast coding is essential for interpreting regression models.

Different schemes are used in practice, with two broad motivations:

1. *Matching model structure to research questions*, by choosing contrasts that test focused hypotheses
2. *Making practical considerations*, such as model interpretability, minimizing unnecessary collinearity, and ease of fitting (for more complex models)

The first motivation is more important, but it is also case-specific; this is the focus of most introductions to contrast coding (e.g., Schad et al. 2020; Cohen et al. 2002, chapter 8).[2] Choosing the right contrasts greatly improves the model's utility for addressing research questions and raises statistical power by requiring fewer hypothesis tests. We focus on (1) only when discussing constructing a custom coding scheme, but in practice many research questions can be mapped into the commonly used schemes (Helmert, etc.).

Our discussion focuses more on (2), because these considerations apply to *any* model containing categorical variables (you always have to choose a contrast coding scheme), and in practice the coding scheme used greatly affects model fitting and interpretability for

---

2. Cohen et al. (2002, 303) advises that "researchers should choose the coding system that provides information that most directly addresses their substantive research questions."

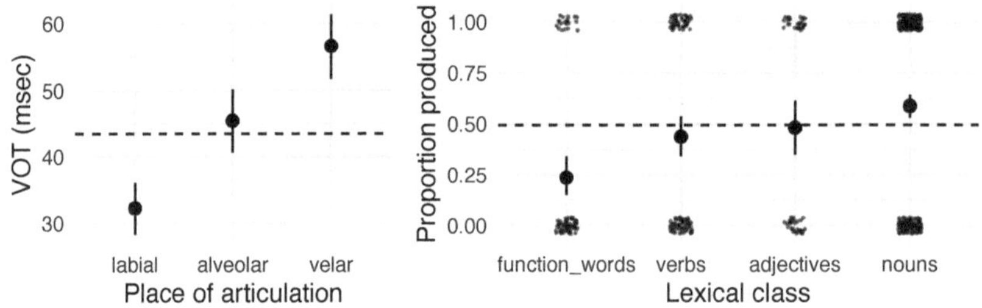

Figure 7.1
Left: `vot` as a function of `place` (means with bootstrapped 95% CIs) for the `vot_michael` data. *Right*: Proportion (with exact 95% CIs) of words produced (`produces`) as a function of `lexical_class` for the `french_cdi_24` data. Dotted lines are averages across all observations.

more complex models (e.g., mixed-effects models) so it is good to become familiar with the issues using simpler models.

First, this discussion exemplifies a few commonly used "standard" coding schemes and then shows how to construct a custom coding scheme, using the empirical cases shown in figure 7.1 (for the `vot_michael`, `french_cdi_24` data).

7.2.3 Treatment Coding

The default in R is *treatment coding*, also known as *dummy coding*. In treatment coding of a factor with k levels, the intercept corresponds to level 1, the first contrast corresponds to level 2 minus level 1, the second contrast to level 3 minus level 1, and so on. The idea is that level 1 is the "control," and each subsequent level is a "treatment," which is compared to the control.

For example, consider a linear regression of `vot` as just a function of `place` for the `vot_michael` data:

```
vot_cc_mod_1 <- lm(vot ~ place, data = vot_michael)
tidy(vot_cc_mod_1)
##              term estimate std.error statistic p.value
## 1     (Intercept)       32       2.1      15.7 1.9e-46
## 2 placealveolar         13       3.2       4.1 4.2e-05
## 3      placevelar       24       3.2       7.7 4.9e-14
```

The `place` predictor by default is mapped to two contrasts—the `placealveolar` and `placevelar` rows. Each one is a numeric predictor, whose coefficient therefore has an estimate and SE (and has a test statistic and p-value from a t-test).

To verify the interpretation of these contrasts (and the intercept), consider the mean `vot` for each level of `place` in these data:

```
vot_michael_place_summ <- vot_michael %>%
  group_by(place) %>% summarise(mean_vot = mean(vot)) %>%
  column_to_rownames("place")
vot_michael_place_summ
```

```
##           mean_vot
## labial          32
## alveolar        46
## velar           57
```

The model's coefficient estimates are the same as the following:

1. The mean for LABIAL stops (`(Intercept)` coefficient)
2. The difference in means for ALVEOLAR and LABIAL stops (coefficient `placealveolar`: $13 = 46 - 32$)
3. The difference in means for VELAR and LABIAL stops (coefficient `placevelar`)

7.2.4 Implementation

In R, contrasts are specified in a *contrast matrix* showing the mapping from the k levels to the $k-1$ numeric predictors (the contrasts). For the previous example:

```
contrasts(vot_michael$place)
##           alveolar velar
## labial           0     0
## alveolar         1     0
## velar            0     1
```

There are two numeric predictors: The first is 1 for ALVEOLAR observations and 0 otherwise. The second is 1 for VELAR observations and 0 otherwise. There is always an implicit additional column of 1s, because the intercept is 1 for every observation.[3]

Here are the first few rows of the *design matrix*, which is what a model containing a categorical variable uses under the hood—a matrix where each column is the values of a single predictor across all observations and each row is one observation:

```
model.matrix(vot_cc_mod_1)
##      (Intercept) placealveolar placevelar
## 1              1             0          1
## 2              1             0          1
## 3              1             0          1
## 4              1             0          1
...
```

Compare that to the first few rows of the actual data we inputted to the model:

```
model.frame(vot_cc_mod_1)
##         vot    place
## 1      45.0    velar
## 2      56.0    velar
## 3      39.3    velar
```

3. By default R gives the kth contrast the name of the $k+1$st level. This is confusing, but standard—often contrasts are just reported as LEVEL-NAME in papers, with the convention that treatment coding is being used so the contrast means "difference between this level and level 1."

...

The first observation has `place` = VELAR. This is mapped into three predictors: the intercept and two numeric predictors whose values are 0 and 1 (the LABIAL row of the contrast matrix).

Note that for a *two-level factor* ($k = 2$), using treatment coding is equivalent to converting the factor to a numeric variable with values 0 and 1 (box 7.2).

> **Box 7.2**
> **Practical Note: Two-Level Factors**
>
> Examples in this chapter use factors with three or more levels ($k > 2$). In chapters 4–6 we included any two-level factor x in models by first converting to numeric predictors—and optionally standardizing (section 5.5.4). What happens if x is left as a two-level factor?
>
> Depending on the coding scheme chosen, x corresponds to a single contrast whose interpretation is "the difference between level 2 and level 1,"—possibly multiplied by a constant. (In R, treatment/Helmert/sum contrasts correspond to 1/0.5/–1 times ($\mu_1 - \mu_2$), which is confusing.) The intercept corresponds to the different meanings of 0 under different standardizations of a binary predictor (box 5.4): LEVEL 1 (treatment coding = no standardization), the average of LEVEL 1 and LEVEL 2 (Helmert, sum coding: set 0/1 to –0.5/0.5), or their weighted average (weighted Helmert: transform to 0 and 1, then subtract the mean, as in our "rescaling").

7.2.5 Centered and Orthogonal Contrasts

Treatment coding is the most traditional coding scheme for a categorical variable x, and easiest to understand, but it lacks two important properties.

First, the contrasts are not *centered*: because each column of the contrast matrix is just 0s and 1s, each contrast won't average to 0 across the dataset. A coding scheme is "centered" when the entries in each column of the contrast matrix sum to 0. (This means the contrast will have mean 0 for "balanced" data, where there are an equal number of observations for each level.) When this is the case, the intercept is interpretable as the (unweighted) grand mean across the dataset, for example, "average of `vot` across LABIAL, ALVEOLAR, and VELAR." This is in line with the main motivation for centering for predictors discussed previously (section 5.5.4): interpretability of the intercept and of the main effects in the presence of interactions. In dummy coding, the intercept is LEVEL 1, which is often not of direct interest.

Second, the contrasts are not *orthogonal*: the contrasts are correlated across the dataset, meaning that knowing the value of one contrast gives you information about others. For the VOT example, if I tell you that the first contrast is 1 for an observation, you already know the value of the second contrast (it must be 0, because the observation can only be ALVEOLAR if the first contrast is 1). Nonorthogonal contrasts implies there will be collinearity between predictors for no good reason ("nonessential collinearity"; see section 5.6.3), because the contrasts encode x using $k-1$ dimensions, and we could choose these dimensions to be independent. Formally, contrasts are orthogonal (or "independent") when the correlation between any two columns of the contrast matrix is 0.[4] In this case, any pair of contrasts is uncorrelated across the dataset for balanced data. (Unbalanced data are addressed in section 7.2.9.)

These two properties—being centered and orthogonal—are the "practical reasons" to choose a contrast coding scheme (section 7.2.2). Because treatment coding lacks both properties, it is not recommended to use this coding scheme without good reason (i.e., level 1 is a control group of direct interest, to which others are compared). Instead it is best to use a coding scheme where contrasts are centered and ideally orthogonal.

7.2.6 Interpretations of Contrasts

The preceding discussion stated that the contrast matrix for treatment coding of a factor with 3 levels (the output of `contrasts(vot_michael$place)`) is equivalent to the following interpretation of the intercept and contrasts, in terms of factor levels:

$$\begin{aligned} \beta_0 &= \mu_1, \\ \beta_1 &= \mu_2 - \mu_1, \\ \beta_2 &= \mu_3 - \mu_1. \end{aligned} \tag{7.1}$$

This is not obvious: in general the interpretation of the contrasts does not follow from the contrast matrix.[5] It's not so important to understand why (box 7.3), but it is very useful to know how to go back and forth between the contrast matrix and the *interpretation matrix*, where each row shows how to interpret one contrast (or intercept) in terms of group means.

To do this, we add a column of 1s to the contrast matrix, then take its inverse (e.g., using `solve()`). For the VOT example:

```
cMat <- solve(cbind(1, contrasts(vot_michael$place))) %>% fractions()
cMat
##          labial alveolar velar
##           1      0        0
## alveolar -1      1        0
## velar    -1      0        1
```

(The `fractions()` function from the MASS package is just to clean up the output.) In this matrix, the first row corresponds to the interpretation of the intercept and subsequent rows to the interpretation of the contrasts; each column is one factor level. Let's add some labeling to be clearer:

```
rownames(cMat) <- cbind("Intercept", "contrast1", "contrast2")
cMat
##           labial alveolar velar
## Intercept  1      0        0
```

4. In linear algebra this condition is more commonly stated as "when the dot product of any two columns is 0" (see Schad et al. 2020 for details).

5. For orthogonal contrasts the columns of the contrast matrix do correspond to the contrasts' interpretations, up to a constant.

```
## contrast1 -1        1        0
## contrast2 -1        0        1
```

The interpretation of each row is

- Intercept: LABIAL ($1 \cdot$ LABIAL $+ 0 \cdot$ ALVEOLAR $+ 0 \cdot$ VELAR)
- Contrast 1: ALVEOLAR $-$ LABIAL
- Contrast 2: VELAR $-$ LABIAL

You can get a similar matrix for any contrast matrix `x` using `solve(cbind(1, x))`.

More usefully, you can go in the opposite direction: specify the interpretation you want the intercept and contrasts to have, then take the inverse of that matrix to determine the contrast coding scheme that captures this interpretation. This *inverse trick* is especially useful for defining custom contrast coding schemes.

Box 7.3
Broader Context: Contrasts ↔ Interpretations

Bolker (2019a)'s online notes (or Schad et al. 2020), summarized here, explain clearly why the inverse trick works. The basic idea requires some linear algebra. We would like to know the matrix C that describes what the regression coefficients (the intercept + the $k-1$ contrasts) mean in terms of group means:

$$C \begin{pmatrix} \mu_1 \\ \mu_2 \\ \vdots \\ \mu_k \end{pmatrix} = \begin{pmatrix} \beta_0 \\ \beta_1 \\ \vdots \\ \beta_k \end{pmatrix}.$$

C is a $k \times k$ matrix. For instance, for the treatment coding example, C would be the 3×3 matrix `cMat`: the first row describes the intercept in terms of group means, and so on for subsequent rows. Multiplying both sides by the inverse of C, we see that C^{-1} describes group means in terms of the coefficients—and the first column of C^{-1} is all 1s (because the equation for every group mean includes β_0). That is

$$\begin{pmatrix} \beta_0 \\ \beta_1 \\ \vdots \\ \beta_k \end{pmatrix} = \underbrace{\begin{pmatrix} 1 & & \\ \vdots & M & \\ 1 & & \end{pmatrix}}_{C^{-1}} \begin{pmatrix} \mu_1 \\ \mu_2 \\ \vdots \\ \mu_k \end{pmatrix}.$$

The contrast matrix in R is M: the $k \times (k-1)$ matrix, where each column is one contrast.

Thus, to get C from the contrast matrix, we add a column of 1s and then invert. To get the contrast matrix corresponding to a given C, we invert and then remove the first column.

Example: Simple coding Sometimes it is useful in analyzing experimental data for the contrasts to have the "treatment" interpretation: level i versus level 1. This makes the contrasts nonorthogonal, but there is no reason we can't center them. This is called *simple coding*: contrasts with the same interpretation as in treatment coding, but where the intercept is the (unweighted) grand mean.

There is no base R function to make simple contrasts, but we can use the inverse trick. To determine the simple coding contrast matrix for a factor with three levels, first make the matrix corresponding to the interpretations of the intercept (row 1 = grand mean) and contrasts (row i = level $(i-1)$ minus level 1); then invert this matrix and remove the first column:

```
interpMat <- t(matrix(c(1 / 3, 1 / 3, 1 / 3, -1, 1, 0, -1, 0, 1),
  nrow = 3))
interpMat
##       [,1] [,2] [,3]
## [1,]  0.33 0.33 0.33
## [2,] -1.00 1.00 0.00
## [3,] -1.00 0.00 1.00
# Remove the intercept column
simpleCMat <- solve(interpMat)[, 2:3]
simpleCMat
##       [,1]  [,2]
## [1,] -0.33 -0.33
## [2,]  0.67 -0.33
## [3,] -0.33  0.67
```

Now, rerun the model with `place` coded with simple contrasts:

```
vot_michael$place_simp <- vot_michael$place
contrasts(vot_michael$place_simp) <- simpleCMat
update(vot_cc_mod_1, . ~ place_simp) %>% tidy()
##          term estimate std.error statistic  p.value
## 1 (Intercept)       45       1.3      33.8 5.8e-140
## 2 place_simp1       13       3.2       4.1  4.2e-05
## 3 place_simp2       24       3.2       7.7  4.9e-14
```

The two rows corresponding to the contrasts are the same as in the original model, where the contrasts were treatment-coded, but the intercept is different. Its estimate is the (unweighted) grand mean of `vot` (45, from `vot_means`).

7.2.7 Sum Coding

Next, we consider *sum coding* (or "deviation coding," "effect coding"). For a factor with k levels, the interpretation of the intercept and contrasts are

- Intercept: unweighted grand mean
- Contrast 1: LEVEL 1 – unweighted grand mean
- Contrast 2: LEVEL 2 – unweighted grand mean
- ...

Exercise 7.2 asks you to verify, using the inverse trick, that sum contrasts have this interpretation.

The `contr.sum()` function gives the contrast matrix in this case—for example, when $k = 3$:

```
contr.sum(3)
##   [,1] [,2]
```

```
## 1    1    0
## 2    0    1
## 3   -1   -1
```

(Note that the interpretation is not obvious from the contrast matrix.)

For the `vot_michael` example, let's define a version of `place` with this contrast matrix and refit the model:

```
vot_michael <- mutate(vot_michael, place_sum = place)
contrasts(vot_michael$place_sum) <- contr.sum(3)
update(vot_cc_mod_1, . ~ place_sum) %>% tidy()
##            term estimate std.error statistic  p.value
## 1 (Intercept)     44.85       1.3     33.77 5.8e-140
## 2  place_sum1    -12.52       1.8     -7.01  6.5e-12
## 3  place_sum2      0.66       1.9      0.34  7.3e-01
```

The intercept estimate is the same as for simple coding, where the intercept was also the (unweighted) grand mean. The interpretation of the first and second contrasts are "LABIAL – grand mean" and "ALVEOLAR – grand mean," which can be verified by calculating these quantities from the empirical means:

```
## Unweighted grand mean
gm <- vot_means["unweighted"]

## Contrast 1
vot_michael_place_summ["labial", "mean_vot"] - gm
##   unweighted
## 1        -13
## Contrast 2
vot_michael_place_summ["alveolar", "mean_vot"] - gm
##   unweighted
## 1       0.66
```

Note the high p-value for the second contrast, which indicates that ALVEOLAR stops aren't meaningfully different from the grand mean. This makes sense given the empirical data (figure 7.1, left).

Sum contrasts have the advantage of being centered, but they are not orthogonal. This can be verified using the contrast matrix (e.g., `contr.sum(4)`): each column sums to 0, but no two columns are uncorrelated.

7.2.8 Helmert Coding

We next consider *Helmert coding*, where each contrast corresponds to the difference between level k and the (unweighted) mean of the previous levels.[6]

6. This coding scheme is sometimes also called *reverse Helmert coding*, due to disagreement on which direction is "reverse," or *difference coding*; in this case each contrast is the difference between level k and the mean of subsequent levels.

As implemented in R (`contr.helmert()`), the interpretation of the intercept and contrasts are level k minus previous levels, but they are scaled by constants:

- Intercept: unweighted grand mean
- Contrast 1: $(\text{LEVEL 2} - \text{LEVEL 1}) \cdot \frac{1}{2}$
- Contrast 2: $(\text{LEVEL 3} - \frac{\text{LEVEL 1+LEVEL 2}}{2}) \cdot \frac{1}{3}$
- Contrast 3: $(\text{LEVEL 4} - \frac{\text{LEVEL 1+LEVEL 2+LEVEL 3}}{3}) \cdot \frac{1}{4}$

As an example, let's use Helmert coding to model how a word's lexical class (`lexical_class`) affects the probability it has been produced (`produces`) for the `french_cdi_24` data (the pattern in figure 7.1, right).

```
contrasts(french_cdi_24$lexical_class) <- contr.helmert(4)
cdi_cc_mod_1 <- glm(produces ~ lexical_class, data = french_cdi_24,
  family = "binomial")
tidy(cdi_cc_mod_1, conf.int = TRUE) %>%
  select(term, estimate, conf.low, conf.high)
##              term estimate conf.low conf.high
## 1     (Intercept)    -0.28   -0.489    -0.069
## 2 lexical_class1      0.46    0.153     0.784
## 3 lexical_class2      0.21    0.013     0.411
## 4 lexical_class3      0.21    0.125     0.302
```

The model's predictions for each `lexical_class` are shown in figure 7.2. To get the intuition behind Helmert contrasts, we just focus on the sign of each coefficient estimate and its 95% CI. The intercept is the unweighted grand mean, intuitively "the predicted log-odds an average word is produced," which is lower than chance.[7] The first contrast (row `lexical_class1`) means "VERBS are more likely to be produced than FUNCTION_WORDS." The second contrast means "ADJECTIVES are more likely to be produced than (the average of) VERBS and FUNCTION_WORDS." The third contrast means "NOUNS are more likely to be produced than non-NOUNS." We can be confident about the directions of these comparisons as no 95% CI includes 0.

Importantly, Helmert contrasts are both centered and orthogonal: knowing the value of one contrast doesn't tell you anything about the value of another contrast. For example, knowing that NOUNS > non-NOUNS for this data (positive contrast 3 coefficient) tells you nothing about how different non-nouns compare (contrasts 1, 2).

These properties mean *Helmert contrasts are a good default*, from a practical perspective (model interpretability, convergence). However they can be unintuitive to interpret, especially if you are used to thinking in terms of a "base value" (treatment contrasts); box 7.4 gives context. What is important is the general intuition ("level k compared to previous levels").

7. What the unweighted grand mean means here is actually less intuitive: the average of the log-odds that `produces = 1` in each group.

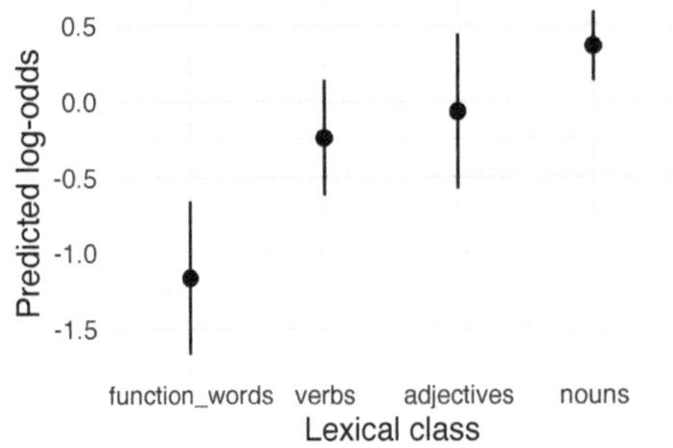

Figure 7.2
Model-predicted log-odds of word production as a function of lexical class, for the `french_cdi_24` data.

Box 7.4
Broader Context: Helmert Contrasts—Details

The interpretation for each contrast follows from the inverse trick (section 7.2.6):

```
solve(cbind(1, contrasts(french_cdi_24$lexical_class))) %>% fractions()
##       function_words verbs adjectives nouns
## [1,]   1/4              1/4    1/4        1/4
## [2,]  -1/2              1/2      0          0
## [3,]  -1/6             -1/6    1/3          0
## [4,] -1/12            -1/12 -1/12         1/4
```

After the intercept (row 1 = grand mean), the intuition for the contrasts follows from each row: the first contrast (row 2) is VERBS minus ADJECTIVES; the second contrast (row 3) column is adjectives minus VERBS/FUNCTION_WORDS; and so on.

The R implementation is confusing because each contrast is then multiplied by a scaling constant. For example, row 3 (contrast 2) is

$$\frac{1}{3} \cdot \left(\text{LEVEL } 3 - \frac{\text{LEVEL } 1 + \text{LEVEL } 2}{2} \right) = \frac{\text{ADJECTIVES}}{3} - \frac{\text{FUNCTION\_WORDS}}{6} - \frac{\text{VERBS}}{6}.$$

More sensible would be "level *k* minus average of previous levels," which is how Helmert contrasts are usually described. This is implemented in the `contr.helmert.unweighted()` function in the psycholing package (Fraundorf 2021), and the "weighted" version `contr.helmert.weighted()`. R's version is used in the text because it is widespread. (If you use the psycholing package, note that it must be installed from Github: see https://www.lrdc.pitt.edu/maplelab/statistics.html.)

Helmert contrasts are often used when there is an expected order of levels. A common mistake with this coding scheme is to interpret significant estimates all with the same sign as meaning that each level is (significantly) larger than the previous level: in our example, NOUNS > ADJECTIVES, ADJECTIVES > VERBS, and so on. These comparisons correspond to a different coding scheme: *successive difference contrasts* (`contr.sdif()` in the MASS package): the first contrast is LEVEL 2 – LEVEL 1, the second is LEVEL 3 – LEVEL 2, and so on. Successive difference contrasts are not orthogonal, so they are not a good default from a practical perspective, but sometimes they are a good choice if they capture research questions.

7.2.9 Weighted Helmert Contrasts

Recall the motivation for "standardizing" two-level factors, by converting to numeric and then subtracting the mean (section 5.5.4). This step makes the predictor average to 0 across the dataset, which improves interpretability of the intercept and main effects (in the presence of interactions). If the factor is balanced, this is the same as coding it as a single contrast with levels –0.5 and 0.5 (or –1 and 1), which is the same as in every centered contrast coding scheme we've considered so far. But for unbalanced data, it is not.

Factors with $k > 3$ levels have the same issue for the centered coding schemes considered so far (sum, simple, Helmert): each contrast will average to 0 for balanced data but will not do so for unbalanced data. The underlying issue is that the intercept in these coding schemes corresponds to the *unweighted* mean (e.g., $(\mu_1 + \mu_2 + \mu_3)/3$ for $k = 3$), which assumes balanced data, rather than the *weighted* mean, which does not.

Weighted contrasts deal with this issue by coding the intercept to correspond to the weighted mean. This is the equivalent of centering a continuous predictor, with similar advantages (model interpretability, fitting) and disadvantages (model results more tied to this particular dataset). Any contrast coding scheme can be made "weighted" (e.g., Cohen et al. 2002, section 8.4); here just *weighted Helmert contrasts* are shown as an example. In this coding scheme, the contrasts have the same interpretation as for (unweighted) Helmert contrasts (contrast i: difference between level $i + 1$ and unweighted mean of previous levels), but the intercept now corresponds to the weighted mean.

We can make our own weighted Helmert contrast matrix for a given dataset using the inverse trick: constructing the interpretation matrix and inverting.

```
## Weighted Helmert version of lexical_class
french_cdi_24$lexical_class_wh <- french_cdi_24$lexical_class

## 1. Make interpretation matrix: invert Helmert contrast matrix
interpMat <- ginv(cbind(1, contrasts(french_cdi_24$lexical_class)))

## 2. Change the intercept's interpretation (row 1)
## to weighted average
interpMat[1, ] <- count(french_cdi_24, lexical_class)$n /
  nrow(french_cdi_24)

# Invert and remove column 1 to get contrast matrix:
contrasts(french_cdi_24$lexical_class_wh) <- ginv(interpMat)[, 2:4]
```

Compare the weighted Helmert contrast matrix and the original (unweighted Helmert) version:

```
contrasts(french_cdi_24$lexical_class_wh)
##                   [,1]  [,2] [,3]
## function_words -1.045 -0.87 -2.2
## verbs           0.955 -0.87 -2.2
## adjectives     -0.045  2.13 -2.2
## nouns          -0.045  0.13  1.8
contrasts(french_cdi_24$lexical_class)
##                [,1] [,2] [,3]
```

```
## function_words    -1   -1   -1
## verbs              1   -1   -1
## adjectives         0    2   -1
## nouns              0    0    3
```

The contrast weights have been adjusted a bit in the weighted version, especially in cells related to NOUNS (in contrast 3), reflecting the fact that this is the most common level (approximately half the data).

We can also refit the regression with weighted Helmert contrasts and compare to the version with (unweighted) Helmert contrasts (`cdi_cc_mod_1`):

```
update(cdi_cc_mod_1, . ~ lexical_class_wh) %>%
  tidy(conf.int = TRUE) %>%
  select(term, estimate, conf.low, conf.high)
##                term estimate conf.low conf.high
## 1       (Intercept)    -0.03   -0.202      0.14
## 2 lexical_class_wh1     0.46    0.153      0.78
## 3 lexical_class_wh2     0.21    0.013      0.41
## 4 lexical_class_wh3     0.21    0.125      0.30
```

Only the intercept differs: it is now –0.03 (close to 50% probability), up from –0.28 (∼43% probability). These numbers correspond to the weighted and unweighted means (of log-odds that `produces` = 1), across the levels of `lexical_class`:

```
french_cdi_24 %>%
  group_by(lexical_class) %>%
  summarise(m = logit(mean(produces)), n = n()) %>%
  summarise(weighted = weighted.mean(m, n), unweighted = mean(m))
##   weighted unweighted
## 1    -0.03      -0.28
```

The weighted average seems more meaningful for this case: it is very similar to the overall proportion of words known in the data (50%), which is an intuitive measure (proportion of words on the questionnaire "known"), whereas the unweighted average doesn't correspond to anything intuitive.

More generally, weighted or unweighted contrasts could make more sense depending on the data/research questions (box 7.5). To simplify the presentation, this book will usually not use weighted contrasts, but they may make sense for your data.

Box 7.5
Broader Context: Unweighted/Weighted Contrasts or Means

Whether to use weighted versus unweighted contrasts is closely related to whether to use weighted or unweighted means when calculating model predictions (to marginalize over other predictors). In both cases, there are conceptual and practical reasons for either choice, and different packages use different defaults (e.g., `effects` and `emmeans` use weighted and unweighted means for model predictions, respectively).

> From a conceptual perspective, weighted contrasts make more sense when the dataset can be thought of as a sample from the "real" data distribution, so each contrast gives an estimate of what's expected (of some quantity) in new data; this is common for corpus data. Unweighted contrasts make more sense when different levels are equally valid, like treatment groups in laboratory experiments, and any imbalance is the result of incidental factors (e.g., cost of data collection); the contrast estimates what the quantity would be if the experiment were balanced.
>
> From a practical perspective, weighted Helmert contrasts are the best choice considered so far: they are orthogonal and centered, even for unbalanced data, and correspond to the standardization we are already doing by default (rescaling) for two-level factors. So arguably, weighted Helmert contrasts seem like a better default coding scheme than unweighted Helmert for factors, especially as imbalance is very common for linguistic data. The practical drawbacks are minor: weighted contrasts (Helmert or other) are not currently widely used for linguistic data, nor are they implemented in a standard R function. I am not sure how often the weighted/unweighted choice makes a real difference to model results; intuitively, we'd expect very little difference when analyzing data from more balanced data (e.g., laboratory experiments) and more for highly imbalanced data (e.g., corpus data).
>
> The `emmeans` "Basics of estimated marginal means" vignette and Cohen et al. (2002, sections 8.3.1, 8.4.1) give useful discussion.

7.2.10 Custom Contrasts

Recall that the primary motivation for choosing a contrast coding scheme is to match the model structure to the research questions (section 7.2.2). Often this can be done with a standard coding scheme; for example, in an experiment with a control condition and two test conditions, simple coding could capture two comparisons of interest (CONTROL vs. TEST1, CONTROL vs. TEST2).

But it is often useful to use a custom coding scheme. As an example, consider the `french_cdi_24` example (section 7.1.2), where the literature has focused on noun bias—meaning NOUNS versus predicates (VERBS/ADJECTIVES)—and function-word bias may be even stronger (meaning FUNCTION_WORDS vs. predicates). No difference between predicate types (ADJECTIVES, VERBS) is expected a priori. We can code these comparisons as three contrasts, letting the intercept be the unweighted grand mean (though this could easily be extended to a weighted version).

The rows of the interpretation matrix are

- Intercept: 0.25, 0.25, 0.25, 0.25 (unweighted mean)
- Contrast 1 (noun bias): 0, –0.5, –0.5, 1 (NOUNS vs. average of VERBS/ADJECTIVES)
- Contrast 2 (function-word bias): –1, 0.5, 0.5, 0 (predicates vs. FUNCTION_WORDS)
- Contrast 3 (within-predicates): 0, –1, 1, 0 (VERBS vs. ADJECTIVES)

Let's construct this matrix and then apply the inverse trick to get the corresponding contrast matrix:

```
## Interpretation matrix
customInterp <- t(matrix(c(
  0.25, 0.25, 0.25, 0.25,
  0, -0.5, -0.5, 1,
  -1, 0.5, 0.5, 0,
  0, -1, 1, 0
```

```
), ncol = 4))

## Corresponding contrast matrix: invert, then remove first column
customCont <- solve(customInterp)[, -1]

## New version of lexical_class with these contrasts
french_cdi_24 <- mutate(french_cdi_24,
  lexical_class_custom = lexical_class
)
contrasts(french_cdi_24$lexical_class_custom) <- customCont
```

We can then refit the `french_cdi_24` model with lexical class coded with these contrasts:

```
update(cdi_cc_mod_1, . ~ lexical_class_custom) %>%
  tidy(conf.int = TRUE) %>%
  select(term, estimate, conf.low, conf.high)
##                    term estimate conf.low conf.high
## 1           (Intercept)    -0.28    -0.49    -0.069
## 2 lexical_class_custom1     0.52     0.13     0.906
## 3 lexical_class_custom2     1.01     0.43     1.621
## 4 lexical_class_custom3     0.17    -0.46     0.807
```

Thus, there is significant noun bias and function-word bias in this data (contrasts 1, 2), but there is no definitive difference between types of predicates (compare to model predictions: figure 7.2). The advantage of using custom contrasts is that the answers to the research questions can just be read off from the model results.

7.2.11 Ordered Factors/Orthogonal Polynominal Contrasts

So far we have considered two types of variables as predictors in regression models: continuous and factors. Continuous variables are ordered, meaning that there are larger and smaller values of the variable. When a continuous x is used as a predictor in a regression, it is assumed that a unit change always has the same effect on the response y (the slope): increasing x from 1 to 2 has the same effect as increasing x from 3 to 4. Factors are discrete and unordered: no level of a factor is larger than other levels. When a factor is used as a predictor in a regression model, it is not assumed that changing from level 1 to level 2 has the same effect on the response as changing from level 2 to level 3 does.

Ordered factors lie in between these two types of variables—they are discrete, but ordered. It is assumed that level 1 is conceptually "less than" level 2 and level 2 "less than" level 3 (like a continuous variable), but it is not assumed that level 1→2 has the same effect as level 2→3 (like a factor). Ordered factors are often used for variables where the levels can be thought of as lying on a scale and take on few values (~3–7). Some examples are number of syllables in a word or level of second-language proficiency (NO KNOWLEDGE < BEGINNER...). For exposition, we assume here that `lexical_class` also makes sense to think of as ordered (subsequent levels are more noun-like).

Ordered factors are implemented as a separate data type in R (like `numeric` or `factor`). For example, `as.ordered()` converts `lexical_class` to an ordered factor:

```
french_cdi_24 <- mutate(french_cdi_24,
  lexical_class_ord = as.ordered(lexical_class)
)
french_cdi_24$lexical_class_ord %>% head()
## [1] nouns nouns nouns nouns nouns nouns
## Levels: function_words < verbs < adjectives < nouns
```

The R output shows the ordering of levels with <. Compare to the unordered version:

```
french_cdi_24$lexical_class %>% head()
## [1] nouns nouns nouns nouns nouns nouns
## attr(,"contrasts")
##                [,1] [,2] [,3]
## function_words   -1   -1   -1
## verbs             1   -1   -1
## adjectives        0    2   -1
## nouns             0    0    3
## Levels: function_words verbs adjectives nouns
```

Coding a factor as "ordered" just means using a particular contrast coding scheme: *orthogonal polynomial contrasts.*[8] In this scheme the intercept is the grand mean, and each contrast corresponds to a different kind of (polynomial) relationship that could hold between the factor and the response, if the factor were considered as a continuous variable (with values 1, 2, ...). The first contrast is the "linear trend" (how much does the relationship look like a line?), the second the "quadratic trend" (how much does the relationship look like a parabola?), and so on.

For example, for the `lexical_class` example there are three contrasts (linear, quadratic, cubic), which correspond to the trends shown in figure 7.3.

```
contrasts(french_cdi_24$lexical_class_ord)
##          .L   .Q    .C
## [1,] -0.67  0.5 -0.22
## [2,] -0.22 -0.5  0.67
## [3,]  0.22 -0.5 -0.67
## [4,]  0.67  0.5  0.22
```

Let's refit the model for the `french_cdi_24` data as a function of lexical class, with `lexical_class` coded as an ordered factor:

```
cdi_cc_mod_1 %>%
  update(. ~ lexical_class_ord) %>% tidy(conf.int = TRUE) %>%
  select(term, estimate, conf.low, conf.high)
##                  term estimate conf.low conf.high
```

8. The extra data type for "ordered factors" lets R functionality (e.g., for visualization) treat ordered and unordered factors differently in useful ways, but formally both are just factors. Note that `as.ordered()` assumes the factor's levels are already in the desired order, which they are for the `french_cdi_24` case (because we reordered the levels in section 7.1.2).

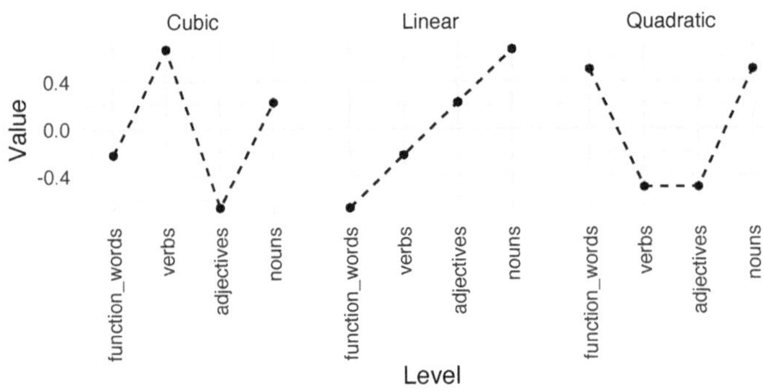

Figure 7.3
Orthogonal polynomial contrasts for `lexical_class` coded as an ordered factor.

```
## 1           (Intercept)    -0.28    -0.49    -0.069
## 2 lexical_class_ord.L     1.06     0.68     1.470
## 3 lexical_class_ord.Q    -0.25    -0.67     0.169
## 4 lexical_class_ord.C     0.23    -0.22     0.668
```

The linear trend has a CI well away from 0, but the quadratic and cubic trends are not meaningfully different from 0. This means the model predicts an essentially linear relationship between `lexical_class_ord` and `produces`. (This can be seen in figure 7.2, where the relationship is plausibly roughly linear, in log-odds space.)

Box 7.6
Practical Note: Ordered Factors, Dimensionality Reduction, and Accessing Contrasts

This example illustrates one motivation for ordered factors—dimensionality reduction—by reducing the number of contrasts needed to represent a factor's effect. If we wanted to simplify the model, it would be justified at this point to drop the quadratic and cubic trends and just use the linear contrast to represent the effect of `lexical_class_ord`:

```
french_cdi_24$lexical_class_lin <-
  model.matrix(~lexical_class_ord, data = french_cdi_24)[, 2]
update(cdi_cc_mod_1, . ~ lexical_class_lin) %>% tidy()
##                  term estimate std.error statistic p.value
## 1         (Intercept)    -0.26     0.099      -2.6 8.1e-03
## 2 lexical_class_lin      0.97     0.173       5.6 2.2e-08
```

Regardless of whether we actually remove the nonlinear contrasts, it is easier to interpret this model's result when `lexical_class` is coded as an ordered factor (one contrast matters) than as an unordered factor (all contrasts matter in model `cdi_cc_mod_1`) . When dealing with an ordered factor with many levels, or several ordered factors, dimensionality reduction can make model fitting and interpretation much easier.

Note that we extracted a single contrast from a factor, as a numeric predictor, by using `model.matrix()` (whose column 1 = the intercept, column 2 = contrast 1, and so on). This functionality will come up a lot in later chapters; it is often useful to decompose a factor into individual contrasts.

7.2.12 Discussion

We have covered a lot of ground on different contrast coding schemes. If you have a multilevel factor in your data, you will have to use contrast coding. The choice of a coding scheme is case-specific: it should be driven by research questions, practical properties of the coding scheme (whether orthogonal or centered), and ease of interpretability (depending on how complex your data are, norms in your area)—in that order. The remaining chapters will give some examples of choosing contrasts in particular cases.

Let's recap some general advice:

- If you are using contrasts to capture research questions, *choose a coding scheme that matches the research questions* and design custom contrasts if needed.
- Regardless of whether you will be interpreting contrasts with respect to research questions (for example, when choosing the coding scheme for control predictors): *don't use treatment coding without a good reason.* Although most common, this coding scheme is neither orthogonal nor centered.
- Instead, minimally *use a centered coding scheme* (e.g., simple, sum, Helmert, ordered). This pays dividends for model interpretability (and convergence, once we get to mixed-effects models).
- Preferably use an orthogonal coding scheme (e.g., Helmert, ordered), because this makes contrasts capture independent information, but this is less important in practice than centering.
- Weighted contrasts are an interesting option, especially for unbalanced data (which most linguistic data are), but weighted contrasts are nonstandard at this point.

Because treatment coding is the default in R, it is commonly recommended to change the default to a centered coding scheme (sum or Helmert), so you don't have to manually change every factor's contrasts. We'll use Helmert contrasts as our default going forward (and orthogonal polynomials for ordered factors):

```
options(contrasts = c("contr.helmert", "contr.poly"))
```

Thus, *our default for standardizing predictors* is now (building on section 5.5.4):

- Continuous predictors: center, divide by two SDs
- Binary factors: transform to 0 and 1, then center
- Factors with $k > 2$ levels: Helmert contrasts

If you are confused at this point, this is natural. Contrasts are tricky to use and interpret, and although it gets easier with practice, it is important to keep the big picture in mind: contrast coding schemes are just fancy ways of coding the relationship between a factor and the response; they all result in equivalent models that make the same predictions. There is no wrong contrast coding scheme, just better and worse ones for a particular case.

Box 7.7
Practical Note: Cell-Means Coding

It is useful to know about a final way of coding a categorical variable, which is to not use contrasts at all. In *cell-means coding* there is no intercept term; instead of coding as 1 intercept and $k-1$ contrasts, the variable is coded as k "indicator" predictors: the ith indicator x_i is 1 for level i observations and 0 otherwise.

The notation for cell-means coding in an R model formula is `-1 + X` (intuitively, the intercept has been "removed"); `0 + X` is equivalent notation.

For example, here we fit the `french_cdi_24` model with `lexical_class` coded in this way:

```
glm(produces ~ -1 + lexical_class,
  data = french_cdi_24, family = "binomial"
) %>% summary()
...
## Coefficients:
##                             Estimate Std. Error z value Pr(>|z|)
## lexical_classfunction_words  -1.1632     0.2562   -4.54  5.6e-06
## lexical_classverbs           -0.2397     0.1929   -1.24   0.2142
## lexical_classadjectives      -0.0667     0.2583   -0.26   0.7963
## lexical_classnouns            0.3622     0.1160    3.12   0.0018
...
# Equivalent ways to fit model:
# update(cdi_cc_mod_1, . ~ -1 + lexical_class)
# update(cdi_cc_mod_1, . ~ 0 + lexical_class)
```

The coefficients for the ith predictor are just the estimates for level i (μ_i), rather than a comparison *between* levels, as in contrast coding. This can also be thought of as a comparison between μ_i and 0, which is the hypothesis test reported for each coefficient (a z-value and p-value).

Cell-means coding is very intuitive, but it is not usually a good choice. The predictors are neither centered nor orthogonal (because any being 1 means all others must be 0) and coefficient interpretations get unintuitive quickly once other predictors are added.

Nonetheless it is good to understand that cell-means coding is just another way of coding a factor. Cell-means coding often comes up as an issue to work around due to how R implements factors. Cell-means coding is used in some subfields, such as traditional variationist sociolinguistics, where categorical predictors are coded this way by default in Varbrul/Goldvarb models.

7.3 Omnibus and Post Hoc Tests

We now turn to the alternative to contrast coding for examining how a factor x affects the response: asking "does x have any effect?" ("omnibus test"), followed by "which levels of x differ?" ("post hoc tests").

7.3.1 Omnibus Tests

For a continuous predictor, the question "does this predictor significantly affect the response?" is answered by the significance of the regression coefficient. For a factor, we would like to ask the same question (e.g., "does `lexical_class` affect the likelihood of a word being produced?"), but no one coefficient's p-value gives the answer, because the factor's effect on the response is jointly captured by all its contrasts. Instead, we assess whether the factor affects the response using model comparison (section 5.9).

For a factor with k levels, the contrasts correspond to $k-1$ predictors; there may be q other predictors in the model. We can compare the full model to the reduced model (containing just the q predictors) using the methods previously discussed for testing the contribution of $k-1$ predictors: an F test for linear regression or an LRT for logistic regression (section 6.7.1).

For example, this model comparison shows that `place` significantly ($p<0.001$) affects `vot` for the `vot_michael` data:

```
anova(vot_cc_mod_1, update(vot_cc_mod_1, . ~ 1))
...
## Model 1: vot ~ place
## Model 2: vot ~ 1
##   Res.Df    RSS Df Sum of Sq    F  Pr(>F)
## 1    590 604254
## 2    592 666243 -2    -61988 30.3 3.1e-13
```

In model `vot_cc_mod_1`, `place` was coded using treatment contrasts, but this does not matter. The result is the same if we use, for example, sum contrasts:

```
m1 <- update(vot_cc_mod_1, . ~ place_sum)
m2 <- update(vot_cc_mod_1, . ~ 1)
anova(m1, m2)
...
## Model 1: vot ~ place_sum
## Model 2: vot ~ 1
##   Res.Df    RSS Df Sum of Sq    F  Pr(>F)
## 1    590 604254
## 2    592 666243 -2    -61988 30.3 3.1e-13
```

Crucially, *the result of model comparison does not depend on the coding scheme used*, because model comparison addresses a question that does not depend on the coding scheme ("does this factor improve the model?"). Thus, the choice of a coding scheme and any associated concerns (e.g., centering, collinearity) don't apply if you are only interested in an omnibus effect.

Example: Omnibus tests versus contrast coding This example illustrates why it can matter whether contrast coding or the omnibus/post hoc method is used to assess the effect of a factor.

Let's return to the `diatones` data from chapter 6. An important predictor was `syll2_coda` (the number of consonants in the syllable coda), which we coded as numeric. This was in fact an odd choice from a theoretical perspective, because "number of coda consonants" rarely plays a role in phonology or sound change. Instead, what tends to matter is the "weight" of a syllable—whether it is "light," "heavy," or "superheavy": sometimes light and nonlight syllables behave differently, sometimes all three behave differently.

Here, these categories correspond to `syll2_coda` = 0, 1, and 2/3. It makes sense to recode `syll2_coda` as a factor with levels 0, C, and CC* (superheavy, heavy, light):

```
diatones <- diatones %>%
  mutate(
    syll2_coda_fact =
      fct_recode(syll2_coda_orig, `CC*` = "CC", `CC*` = "CCC")
  ) %>%
  mutate(syll2_coda_fact = fct_relevel(syll2_coda_fact, rev))
```

We have ordered the levels as CC*, C, O, so that the two Helmert contrasts (which are now assigned by default, following our `options(contrasts...)` command) correspond to theoretically motivated comparisons: heavy versus superheavy (contrast 1), and light versus heavy (contrast 2).

```
contrasts(diatones$syll2_coda_fact)
##      [,1] [,2]
## CC*    -1   -1
## C       1   -1
## O       0    2
```

Now, consider our first step in building up a model for the `diatones` data (section 6.7.2.1)—a model with main effects for all predictors:

```
mlogreg_mod_1_new <- glm(stress_shifted ~ syll2_coda_fact + syll2_td +
    frequency + syll1_coda, data = diatones, family = "binomial")
```

We could assess whether `syll2_coda_fact` "has an effect" in two ways, using the $\alpha = 0.05$ significance level. In the omnibus/post hoc test approach, we would first compare the models with and without this term:

```
drop1(mlogreg_mod_1_new, "syll2_coda_fact", test = "Chisq")
...
## Model:
## stress_shifted ~ syll2_coda_fact + syll2_td + frequency + syll1_coda
##                 Df Deviance AIC LRT Pr(>Chi)
## <none>                  105 117
## syll2_coda_fact  2      110 118 5.3    0.071
```

The LRT suggests that syllable 2 weight does *not* significantly affect a word's likelihood of shifting stress. We would conclude that `syll2_coda_fact` does not matter, and we could stop there.

In the contrast coding approach, we examine the rows of the regression table corresponding to this variable's contrasts:

```
tidy(mlogreg_mod_1_new) %>% filter(str_detect(term, "syll2_coda"))
##               term estimate std.error statistic p.value
## 1 syll2_coda_fact1     0.27      0.35      0.78   0.433
## 2 syll2_coda_fact2     0.65      0.28      2.36   0.018
```

The significant second contrast means that words with light second syllables (`syll2_coda_fact = 0`) are more likely to shift stress. In the same direction, heavy > superheavy (contrast 1), but the effect is not significant. We would conclude that `syll2_coda_fact` *does* matter.

This example shows how the omnibus/post hoc and contrast coding approaches can lead to different conclusions, when the effect of a factor is relatively weak: the right contrast(s) will detect the effect, whereas an omnibus test will not, because the latter has higher *df*.

This is the sense in which the contrast coding approach has higher power to detect effects of interest (Cohen et al. 2002, 340): if there is in reality a difference between light and nonlight syllables, failing to detect it using the omnibus test is a type II error. In a factor with enough levels *some* comparison will be significant by chance, even when in reality the factor has no effect, in which case choosing a coding scheme including that comparison would lead to a type I error. So it is important that the coding scheme chosen be theoretically motivated when possible, and as is often the case (section 3.3.5), there is a trade-off between type I and type II errors for different analysis methods, which comes into play for relatively weak effects.

Which conclusion we make about the effect of `syll2_coda_fact` at this point could lead to qualitatively different conclusions in the final analysis of the `diatones` data (exercises 7.3).

7.3.2 Post Hoc Tests

Post hoc tests test for differences between (combinations of) levels of a factor x after fitting the model. This is conceptually the same as contrast coding, except that instead of fixing in advance the $k-1$ differences to be tested, we test as many differences as we want but correct for multiple comparisons. The most common application is to test all pairwise comparisons of levels of x ("which levels differ?") after an omnibus test has shown an effect of x.

Box 7.8
Broader Context: Whither Multiple Comparisons?

When should we correct for multiple comparisons in regression analyses? It is customary to correct for multiple comparisons when carrying out post hoc tests on a fitted model—even if doing fewer than $k-1$ tests (for a factor with k levels)—but not when using contrast coding, or indeed for predictors in multiple regression models generally. One could imagine correcting the p-values in m rows of a regression table for m comparisons—or correcting all p-values in a paper (e.g., across several regression models)—but this is rarely done in practice.

I am not sure if there is a good reason for this distinction. It makes sense in the traditional "planned versus post hoc" comparison scenario, but this often doesn't apply in practice (box 7.1). As Dienes (2008, 77) notes, "the decision [to correct in some cases and not others] is basically arbitrary and tacit conventions determine practice." This is ultimately a choice between type I and type II errors, where neither extreme makes sense in all cases. In a regression including 10 predictors, all included for good theoretical reasons, correcting for multiple comparisons will lower the power to detect (presumably) real effects. At the same time, if you analyze the same data in several exploratory analyses including many predictors (most of which probably have no real effect), without correcting for multiple comparisons, you are likely to make a type I error. The tacit convention seems like a good compromise.

In practical terms, post hoc tests mean carrying out follow-up analyses on the fitted model rather than the analyses being part of the model itself (as contrasts). We use the emmeans package, which does post hoc tests in two steps.

First, model predictions are computed for each level of x (the "estimated marginal means," which are described in section 7.3.3). For the `french_cdi_24` example:

```
emm_cdi <- emmeans(cdi_cc_mod_1, ~lexical_class)
emm_cdi
##  lexical_class  emmean    SE  df asymp.LCL asymp.UCL
##  function_words  -1.16 0.256 Inf     -1.67     -0.66
##  verbs           -0.24 0.193 Inf     -0.62      0.14
##  adjectives      -0.07 0.258 Inf     -0.57      0.44
##  nouns            0.36 0.116 Inf      0.13      0.59
##
## Results are given on the logit (not the response) scale.
## Confidence level used: 0.95
```

Second, hypothesis tests comparing each pair of levels are carried out. To compute pairwise comparisons:

```
contrast(emm_cdi, "pairwise")
##  contrast                     estimate   SE  df z.ratio p.value
##  function_words - verbs          -0.92 0.32 Inf  -2.900  0.0200
##  function_words - adjectives     -1.10 0.36 Inf  -3.000  0.0100
##  function_words - nouns          -1.53 0.28 Inf  -5.400  <.0001
##  verbs - adjectives              -0.17 0.32 Inf  -0.500  0.9500
##  verbs - nouns                   -0.60 0.23 Inf  -2.700  0.0400
##  adjectives - nouns              -0.43 0.28 Inf  -1.500  0.4300
##
## Results are given on the log odds ratio (not the response) scale.
## P value adjustment: tukey method for comparing a family of 4 estimates
```

Each test corresponds to one contrast (comparison of levels)—so the table looks exactly like a regression table, except that p-values have been corrected for multiple comparisons.

Recall that there are different multiple comparison methods that trade off in type I/II errors (section 3.3.6). The `emmeans()` function selects a sensible default, in this case the Tukey method (which is designed for pairwise comparisons).

For this example, using the $\alpha = 0.05$ level, FUNCTION_WORDS are less likely to be acquired than VERBS, ADJECTIVES, or NOUNS, and VERBS are less likely to be acquired than NOUNS. These results are consistent with the model's predictions (figure 7.2, `emm_cdi`): levels whose 95% CIs do not overlap are "different."

Box 7.9
Practical Note: Reporting the Effect of a Multilevel Factor

If coded using custom contrasts, as in section 7.2.10, the effect of lexical class could be reported as follows (if using 95% CIs):

> [Coding of lexical class should have been described earlier.] A word's `lexical_class` affects its likelihood of being produced: FUNCTION_WORDS were less likely than predicates (VERBS and ADJECTIVES) to be produced (contrast 1: $\hat{\beta} = 0.46$, $z = 2.9$, $p = 0.004$, 95% CI [0.15, 0.78]), NOUNS were more likely than predicates (contrast 2: $\hat{\beta} = 0.21$, $z = 2.1$, $p = 0.037$, 95% CI [0.013, 0.41]), and VERBS and ADJECTIVES did not significantly differ ($\hat{\beta} = 0.21$, $z = 4.7$, $p < 0.001$, 95% CI [0.13, 0.3]).
>
> An omnibus/post hoc analysis could be reported as follows:
>
> Words differ significantly by `lexical_class` in their likelihood of being produced (likelihood ratio test: $\chi^2(3) = 36$, $p < 0.001$; odds ratio = 1.1). Pairwise post hoc tests were applied to determine which lexical classes significantly differ ($\alpha = 0.05$), with p-values corrected using the Tukey honestly significant difference method; results are shown in table XX (showing the output of `pairs(emm_cdi)`). FUNCTION_WORDS, VERBS, and NOUNS were progressively more likely to be produced (FUNCTION_WORDS < VERBS < NOUNS), while ADJECTIVES were more likely to be produced than FUNCTION_WORDS but don't significantly differ from VERBS or NOUNS.
>
> Note that we have included an appropriate effect size for the omnibus test (an odds ratio; see section 6.7.1); for all other tests the coefficient estimate is an effect size.
>
> The first example reports test results in the text and the second example in a table; either is fine. Text-only reports of effects of factors are hard to parse (and very common). For effects of interest it's a good idea to augment with a figure ("model predictions for each `lexical_class` are shown in figure 7.2") or summaries ("FUNCTION_WORDS < VERBS < NOUNS, FUNCTION_WORDS < ADJECTIVES").

For the VOT example, pairwise comparisons give a simpler result: LABIAL < ALVEOLAR < VELAR.

```
vot_emm <- emmeans(vot_cc_mod_1, ~place)
vot_emm
##  place    emmean   SE  df lower.CL upper.CL
##  labial       32 2.07 590       28       36
##  alveolar     46 2.43 590       41       50
##  velar        57 2.39 590       52       61
##
## Confidence level used: 0.95
```

The CIs of the estimate for each level don't overlap; equivalently, all levels significantly differ (`contrast(vot_emm, 'pairwise')`).

To summarize, for the French CDI example, words differ by lexical class in their likelihood of being produced, but the exact order of lexical classes is unclear (FUNCTION_WORDS < VERBS < NOUNS; ADJECTIVES unclear). For the VOT example, VOT differs by place of articulation in the order LABIAL < ALVEOLAR < VELAR.

7.3.3 More on Post Hoc Tests

Whereas the previous case is the most common application (pairwise comparisons after an omnibus test), the general two-step process used for post hoc tests has many useful applications. We explore some in this and the following section (section 7.4), using emmeans, one of several packages with this general functionality (multicomp is also popular).[9]

9. emmeans has broad functionality and excellent documentation, so this exploration only scratches the surface; I highly recommend the emmeans vignettes.

emmeans contains very general functionality for each of the two steps:

1. Make *model predictions* as predictors of interest vary, marginalizing over other predictors. emmeans calculates estimated marginal means (hence the name), which are very similar to marginal effects we have discussed previously in the context of model predictions (section 6.7.3). The main difference is that emmeans marginalizes over factors using unweighted averages, while tools used previously (sjPlot, margins) use weighted averages, by default (box 7.5).
2. *Make comparisons* of different predictions, which are often different levels of a factor.

For our purposes, these steps can always be thought of as plotting model predictions, then assessing whether particular patterns in the plot are real (via hypothesis tests).

Custom post hoc tests Rather than coding custom contrasts to capture research questions, as in section 7.2.10, one can just carry out post hoc tests on the fitted model corresponding to these contrasts. In emmeans this is done by specifying the contrasts in a list (rather than an interpretation matrix, but the idea is the same).

```
## Compare to customInterp rows 2-4
interpVecs <- list(
  nounBias = c(0, -0.5, -0.5, 1),
  fwBias = c(-1, 0.5, 0.5, 0),
  verbVAdj = c(0, -1, 1, 0)
)
```

We show the result of the post hoc tests with CIs rather than p-values, to compare to the regression table using custom contrasts (in section 7.2.10):

```
contrast(emm_cdi, method = interpVecs) %>% confint()
##  contrast estimate   SE  df asymp.LCL asymp.UCL
##  nounBias     0.52 0.20 Inf      0.13       0.9
##  fwBias       1.01 0.30 Inf      0.42       1.6
##  verbVAdj     0.17 0.32 Inf     -0.46       0.8
##
## Results are given on the log odds ratio (not the response) scale.
## Confidence level used: 0.95
```

The results are the same. This illustrates how contrast coding can be thought of as a special case of post hoc tests. This is useful because contrast coding can be difficult to implement or interpret. *It is always an option to just fit the model* with any coding scheme, *then use post hoc tests to assess the hypotheses of interest*—regardless of whether they were planned—and correcting for multiple comparisons if you examine more than $k-1$ tests.

7.4 Interpreting Interactions

Often the effects of interest in a regression model are interactions involving multilevel factors or including three or more predictors ("three-way" interactions). This section shows some examples of unpacking such interactions using the methods we've already discussed.

7.4.1 Post Hoc Trends

First, consider the case of a continuous x_1 interacting with a factor x_2. The regression coefficients for the interaction describe how the slope of x_1 changes across levels of the factor, corresponding to the contrasts, but it is often of interest to know the actual slope of x_1 at different levels of x_2. emmeans can calculate these "trends" post hoc.

For example, consider the final `diatones` model, `mlogreg_mod_3`, from section 6.7.2.2. There were significant interactions of `frequency` with two aspects of syllable structure (`syll1_coda`, `syll2_td`), pictured in figure 6.5, which describe how the frequency effect differs between groups of words. The general finding was that words with more "change-promoting structure" have a greater `frequency` slope. But in this case the actual value of the frequency effect for each group of words is of theoretical interest—a (significantly) positive or negative slope corresponds to different cognitive mechanisms.

To examine this, we first refit the model using the factor versions of the predictors `frequency` interacts with:

```
mlogreg_mod_4 <- update(mlogreg_mod_3, . ~ syll2_coda +
  frequency * syll1_coda_orig + syll2_td_orig * frequency)
```

(It is often good to refit the model with factors coded as factors rather than numeric before post hoc analyses; it makes the output cleaner.)

The `emtrends()` function determines the slope of a continuous predictor as other predictors of interest are varied (marginalizing over remaining predictors)—analogously to `emmeans()`'s estimates at each level of a categorical predictor. In this case:

```
emtrends(mlogreg_mod_4, ~ syll1_coda_orig + syll2_td_orig,
  var = "frequency")
## syll1_coda_orig syll2_td_orig frequency.trend   SE  df asymp.LCL
## no              no                       -2.4 0.77 Inf      -3.9
## yes             no                        1.4 1.28 Inf      -1.1
## no              yes                       0.1 0.86 Inf      -1.6
## yes             yes                       3.8 1.56 Inf       0.8
## asymp.UCL
##      -0.9
##       3.9
##       1.7
##       6.9
##
## Confidence level used: 0.95
```

From the 95% CIs: a *negative* slope for frequency is predicted for words with minimal change-promoting structure (no syllable 1 coda or t/d in syllable 2 coda), whereas a *positive* slope is predicted for words with maximal change-promoting structure (and for other words, the CI includes 0). So the post hoc trends let us conclude that the `frequency` effect actually does change sign across different groups of words—which is more theoretically meaningful than the conclusion just from the regression model, that the `frequency` effect changes as a function of syllable structure.

7.4.2 Example: Two-Way Interactions

Let's return to the final multiple regression model we fit for the vot_michael data in section 5.10.3.3 (model gh_mod_3), but now we'll let the variable place have three levels (previously we collapsed to place = LABIAL/non-LABIAL for simplicity).

We will focus on interactions with place in this model. The LABIAL < ALVEOLAR < VELAR ordering for VOT has been observed across many languages and is assumed to have a physiological basis; it is thus of theoretical interest whether this pattern *always* holds, for different kinds of words (e.g., voiced vs. voiceless stops). The ordering LABIAL < non-LABIAL is very consistent in previous work, but the relative order of ALVEOLAR and VELAR is less so. So it makes sense to code place using Helmert contrasts, whose interpretations are ALVEOLAR versus VELAR and LABIAL versus non-LABIAL.

```
## Reverse levels so Helmert coding works
vot_michael$place <-
  fct_relevel(vot_michael$place, c("velar", "alveolar", "labial"))

contrasts(vot_michael$place) <- contr.helmert(3)

## Give contrasts better names
colnames(contrasts(vot_michael$place)) <- c("TvK", "PvTK")
```

These contrast names abbreviate LABIAL/ALVEOLAR/VELAR as P/T/K.

Refitting the model with our current vot_michael dataframe, place will now be a three-level factor:

```
## Same model formula as gh_mod_3
gh_mod_4 <- lm(log_vot ~ voicing*(speaking_rate + cons_cluster +
    place) + foll_high_vowel*place + log_corpus_freq, data = vot_michael)
```

It is a good idea to look at the model summary (tidy(gh_mod_4)); we will highlight relevant parts in section 7.4.2.2.

To unpack the effect of place, as for any interaction effect, we examine model predictions together with the results of statistical tests.

7.4.2.1 Model predictions

We plot the marginal effects of the model as predictors of interest are varied (as in section 6.7.3). For example, the marginal effects as place and voicing are varied are:

```
emmeans(gh_mod_4, ~ place * voicing)
##  place    voicing   emmean    SE  df lower.CL upper.CL
##  velar    voiced       3.0 0.062 579      2.9      3.2
##  alveolar voiced       3.0 0.051 579      2.9      3.1
##  labial   voiced       2.3 0.043 579      2.2      2.4
##  velar    voiceless    4.3 0.044 579      4.2      4.3
##  alveolar voiceless    4.3 0.049 579      4.2      4.4
##  labial   voiceless    4.1 0.043 579      4.0      4.2
##
```

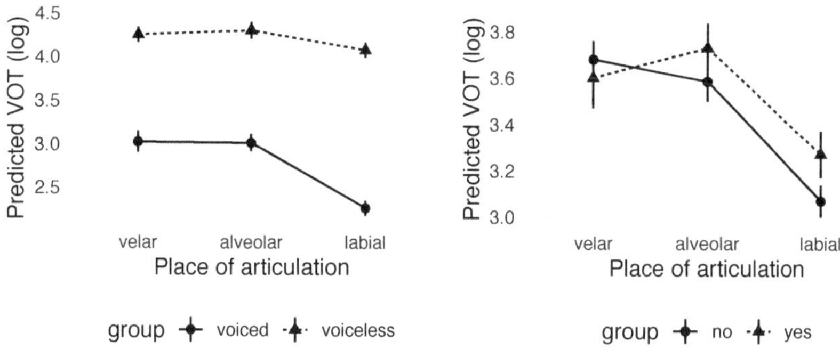

Figure 7.4
Interaction plots of predicted `vot` as a function of `place` and `voicing`, and of `place` and `foll_high_vowel`, marginalizing over other predictors.

```
## Results are averaged over the levels of: cons_cluster, foll_high_vowel
## Confidence level used: 0.95
```

Figure 7.4 (code not shown) plots these marginal effects, as well as those when `place` and `foll_high_vowel` are varied.

It looks like there is substantial variation in the `place` effect as a function of covariates. At this point we can proceed either via contrast coding or omnibus/post hoc tests to understand the interactions; in either case, the idea is to determine which patterns seen in the plots are "real."

7.4.2.2 Approach 1: Contrast coding

Consider the terms of the regression table involving `place`:

```
tidy(gh_mod_4) %>% filter(str_detect(term, "place"))
##                           term estimate std.error statistic p.value
## 1                     placeTvK   0.0086     0.026      0.33 7.4e-01
## 2                    placePvTK  -0.1592     0.013    -12.11 2.9e-30
## 3             voicing1:placeTvK   0.0152     0.024      0.63 5.3e-01
## 4            voicing1:placePvTK   0.0925     0.012      7.60 1.2e-13
## 5  placeTvK:foll_high_vowel1   0.0554     0.025      2.18 2.9e-02
## 6 placePvTK:foll_high_vowel1   0.0281     0.013      2.16 3.1e-02
```

`voicing` and `foll_high_vowel` have been "centered" automatically (via Helmert contrasts), so we can interpret the main effect of `place` as an "average" effect (marginalizing over voicing and vowel height). On average, LABIAL < non-LABIAL, but ALVEOLAR and VELAR do not reliably differ.

Turning to interactions: the `voicing:placeTvK` contrast has a small magnitude and high *p*-value, reflecting the very similar VELAR/ALVEOLAR difference for VOICED and VOICELESS stops. The significant `voicing:placePvTK` contrast captures the larger LABIAL/non-LABIAL difference for VOICED stops.

For interactions with vowel height, the ALVEOLAR/VELAR difference is larger when the following vowel is HIGH (contrast `foll_high_vowel:placeTvK`), and the LABIAL/non-LABIAL difference is smaller (contrast `foll_high_vowel:placePvTK`). (The sign of the latter is confusing just working from the regression table, but it's clear from the plot what effect is being captured.) Both contrasts are significantly different from 0.

```
c0 <- coefficients(gh_mod_4)[['(Intercept)']]
c1 <- coefficients(gh_mod_4)[['placePvTK']]
c2 <- coefficients(gh_mod_4)[['voicing1:placePvTK']]

minPvTK <- min(abs(c1-c2), abs(c1+c2))
maxPvTK <- max(abs(c1-c2), abs(c1+c2))

minPvTK_ms <- exp(c0 + minPvTK) - exp(c0)
maxPvTK_ms <- exp(c0 + maxPvTK) - exp(c0)
```

Qualitatively, the result is that the LABIAL > non-LABIAL pattern is maintained across different word types (contrast 2), but its magnitude differs, whereas the ALVEOLAR < VELAR pattern does not consistently hold (contrast 1).[10]

7.4.2.3 Approach 2: Omnibus/post hoc tests

An F-test comparing models with and without the `voicing:place` interaction shows that the place effect significantly differs between voiced and voiceless stops:

```
drop1(gh_mod_4, "voicing:place", test = "F")
...
##                Df Sum of Sq RSS   AIC F value Pr(>F)
## <none>                      103 -1010
## voicing:place  2       10.3 113  -957    28.9  1e-12
```

Similarly, the `place` effect differs between high and nonhigh vowels:

```
drop1(gh_mod_4, "place:foll_high_vowel", test = "F")
...
##                        Df Sum of Sq RSS   AIC F value Pr(>F)
## <none>                              103 -1010
## place:foll_high_vowel  2       1.58 105 -1005    4.43  0.012
```

Because both interactions are significant, we proceed to post hoc tests, examining which `place` levels differ, for each value of `voicing`:

```
emm <- emmeans(gh_mod_4, ~ place | voicing)
contrast(emm, "pairwise")
```

10. Specifically, we can calculate from the model coefficients that the magnitude of the LABIAL/non-LABIAL difference varies between 0.067 and 0.25 in `log_vot`, which corresponds to differences of 3.5 and 15 msec in `vot` (evaluated at the intercept). See code file for calculation.

```
## voicing = voiced:
##  contrast           estimate    SE  df t.ratio p.value
##  velar - alveolar       0.01 0.078 579   0.200  0.9800
##  velar - labial         0.76 0.073 579  10.500  <.0001
##  alveolar - labial      0.75 0.060 579  12.500  <.0001
##
## voicing = voiceless:
##  contrast           estimate    SE  df t.ratio p.value
##  velar - alveolar      -0.05 0.064 579  -0.700  0.7400
##  velar - labial         0.18 0.060 579   2.900  0.0100
##  alveolar - labial      0.22 0.064 579   3.500  <.0001
##
## Results are averaged over the levels of: cons_cluster, foll_high_vowel
## P value adjustment: tukey method for comparing a family of 3 estimates
```

Regardless of `voicing`, LABIAL stops have significantly lower VOT than VELAR or ALVEOLAR stops, but VELAR and ALVEOLAR stops do not reliably differ.

Post hoc tests for the `place:foll_high_vowel` interaction give a similar result:

```
contrast(emmeans(gh_mod_4, ~ place | foll_high_vowel), "pairwise")
## foll_high_vowel = no:
##  contrast           estimate    SE  df t.ratio p.value
##  velar - alveolar       0.09 0.057 579   1.600  0.2330
##  velar - labial         0.61 0.052 579  11.700  <.0001
##  alveolar - labial      0.52 0.054 579   9.600  <.0001
##
## foll_high_vowel = yes:
##  contrast           estimate    SE  df t.ratio p.value
##  velar - alveolar      -0.13 0.085 579  -1.500  0.2910
##  velar - labial         0.33 0.083 579   4.000  <.0001
##  alveolar - labial      0.46 0.072 579   6.400  <.0001
##
## Results are averaged over the levels of: voicing, cons_cluster
## P value adjustment: tukey method for comparing a family of 3 estimates
```

These post hoc tests can be summarized as "LABIAL < ALVEOLAR = VELAR—regardless of stop voicing or following vowel height." So VOT differences between places of articulation are modulated by stop voicing and following vowel height (omnibus test), but they are not modulated enough to actually change the qualitative `place` effect for any subset of words (post hoc tests).

7.4.3 Example: Three-Way Interaction

Let's now add a three-way interaction involving `place` to the VOT model continuing the Gelman and Hill method begun in section 5.10.3.3 (where we only considered two-way interactions) by considering interactions between predictors with large effects. We are interested in `place` and don't have a priori motivation for any three-way interaction, so we simply add a three-way interaction involving place that significantly improves the model (and builds on already significant interactions): `place:voicing:cons_cluster`.

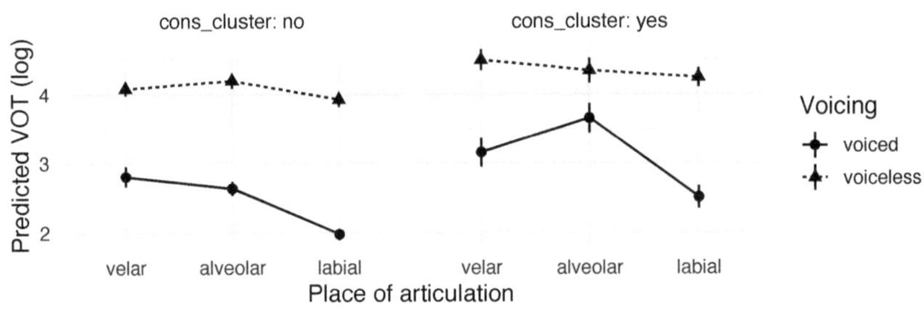

Figure 7.5
Interaction plot of `place`, `voicing`, and `cons_cluster` for model `gh_mod_5`, marginalizing over other predictors.

```
# F-test showing this term significantly improves the model
add1(gh_mod_4, "place:cons_cluster:voicing", test = "F")
...
##                            Df Sum of Sq   RSS   AIC F value
## <none>                                  103.0 -1010
## place:cons_cluster:voicing  4      3.55  99.5 -1023    5.14
##                             Pr(>F)
## <none>
## place:cons_cluster:voicing 0.00045
# New model
gh_mod_5 <- update(gh_mod_4, . ~ . +
  cons_cluster * place * voicing, data = vot_michael)
```

(We'll leave it until exercise 7.4 to verify there are no other three-way interactions that should be added.)

Figure 7.5 shows model predictions for this case. It looks like the `place` effect varies a lot as a function of consonant cluster and stop voicing, but it is hard to puzzle out from the regression table exactly what about this pattern the three-way interaction is capturing. Instead it is easier to do (pairwise) post hoc tests examining the `place` effect as the other two predictors are varied.

```
gh_mod_5_emm <- emmeans(gh_mod_5, ~ place | voicing + cons_cluster)
contrast(gh_mod_5_emm, "pairwise")
## voicing = voiced, cons_cluster = no:
##  contrast          estimate    SE  df t.ratio p.value
##  velar - alveolar      0.17 0.089 575   1.900  0.1300
##  velar - labial        0.83 0.085 575   9.900  <.0001
##  alveolar - labial     0.66 0.065 575  10.100  <.0001
##
## voicing = voiceless, cons_cluster = no:
##  contrast          estimate    SE  df t.ratio p.value
##  velar - alveolar     -0.11 0.073 575  -1.500  0.2700
##  velar - labial        0.16 0.070 575   2.300  0.0600
##  alveolar - labial     0.27 0.074 575   3.700  <.0001
##
```

```
## voicing = voiced, cons_cluster = yes:
##  contrast           estimate    SE  df t.ratio p.value
##  velar - alveolar      -0.50 0.150 575  -3.300  <.0001
##  velar - labial         0.65 0.134 575   4.800  <.0001
##  alveolar - labial      1.14 0.136 575   8.400  <.0001
##
## voicing = voiceless, cons_cluster = yes:
##  contrast           estimate    SE  df t.ratio p.value
##  velar - alveolar       0.15 0.118 575   1.300  0.4000
##  velar - labial         0.25 0.102 575   2.500  0.0300
##  alveolar - labial      0.10 0.114 575   0.900  0.6500
##
## Results are averaged over the levels of: foll_high_vowel
## P value adjustment: tukey method for comparing a family of 3 estimates
```

Abbreviating LABIAL/ALVEOLAR/VELAR as P/T/K, these post hoc tests can be summarized as

- `voiced`, no consonant cluster: P < T = K
- `voiceless`, no: P < T = K
- `voiced`, yes: P < K < T
- `voiceless`, yes: P < K

Across word types LABIAL < VELAR, but the placement of ALVEOLAR is inconsistent (but never lower than LABIAL: LABIAL $\leq$ ALVEOLAR). Interestingly, for one word type (voiced stops in consonant clusters) ALVEOLAR stops have the highest VOT, which is theoretically unexpected.

in

7.4.4 Two Interaction Fallacies

Interaction effects are often central to what a regression analysis says about research questions. It is therefore important to understand exactly what can be concluded from an interaction (whether using contrast coding or an omnibus test), versus post hoc tests for individual levels, to avoid two common fallacies.

7.4.4.1 A significant interaction does not imply qualitative differences between levels

Consider the significant interactions with `placeTvK` in the VOT model from `gh_mod_4`; the terms of the regression table involving `place` are in section 7.4.2.2 (or run `summary(gh_mod_4)`).

These interactions mean that the ALVEOLAR/VELAR VOT difference differs between groups of words (depending on stop voicing and following vowel height). This doesn't directly address the research question, which is what the relative order of levels are: here, ALVEOLAR and VELAR. Finding ALVEOLAR > VELAR for any group of words would be theoretically interesting and unexpected, but ALVEOLAR < VELAR is expected.

Given that the *main effect* is effectively 0 (`placeTvK`: $\hat{\beta} = 0.0086$, $p = 0.74$)—meaning the ALVEOLAR/VELAR difference, averaged across the data—it is tempting to infer that the interactions mean "ALVEOLAR < VELAR for some groups of words and ALVEOLAR > VELAR for others" (as voicing, foll_high_vowel are varied). But this is not the case: the post hoc tests (section 7.4.2.3) show that the ALVEOLAR/VELAR difference is not significant for any group of words. For this data there is enough evidence to conclude that the ALVEOLAR/VELAR difference varies between groups, but there is not enough to conclude that its sign changes. So the analysis is inconclusive with respect to the theoretical question.

7.4.4.2 Qualitative differences between levels do not imply a significant interaction

Showing that the effect of a predictor x_1 differs between groups A and B, indexed by predictor x_2, requires a (significant) $x_1 : x_2$ interaction. It is very common to instead test for an effect of x_1 separately in group A and in group B, find a significant effect in one group and not the other, then conclude that A and B differ in the effect of x_1. This is incorrect, because "the difference between significant and not significant is not itself necessarily significant" (Nieuwenhuis, Forstmann, and Wagenmakers 2011).[11]

For instance, consider the example from section 4.5.1: the voicing:prosodic_boundary interaction for the neutralization data (model mlr_mod_3). Of interest is whether the voicing pattern (VOICED > VOICELESS) differs depending whether there is a prosodic boundary. Suppose we are using $\alpha = 0.01$ to decide what effects are significant. Plots of this interaction (figure 4.4, left) suggest the expected voicing effect is present (= significantly different from 0) only when there is a prosodic boundary. We can confirm this using post hoc tests:

```
mlr_mod_3_emm <- emmeans(mlr_mod_3, ~ voicing | prosodic_boundary)
contrast(mlr_mod_3_emm, "pairwise")
## prosodic_boundary = no:
##  contrast            estimate  SE  df t.ratio p.value
##  voiced - voiceless       5.7 3.7 745   1.500  0.1220
##
## prosodic_boundary = yes:
##  contrast            estimate  SE  df t.ratio p.value
##  voiced - voiceless      18.7 5.4 745   3.500  0.0010
```

However, the actual voicing:prosodic_boundary interaction is not significant at the $\alpha = 0.01$ level (model table in section 4.5.1):

```
tidy(mlr_mod_3) %>%
  select(term, estimate, p.value) %>%
  # Extract row where 'term' contains ':' character
  filter(str_detect(term, "\\:"))
##                                    term estimate p.value
## 1 prosodic_boundaryyes:voicingvoiceless      -13   0.047
```

11. We consider the case where "separately" means "conduct post hoc tests." It is common to actually fit two different regressions, one testing the effect of x_1 in each group, in which case there are further issues (loss of power) but the basic problem is the same.

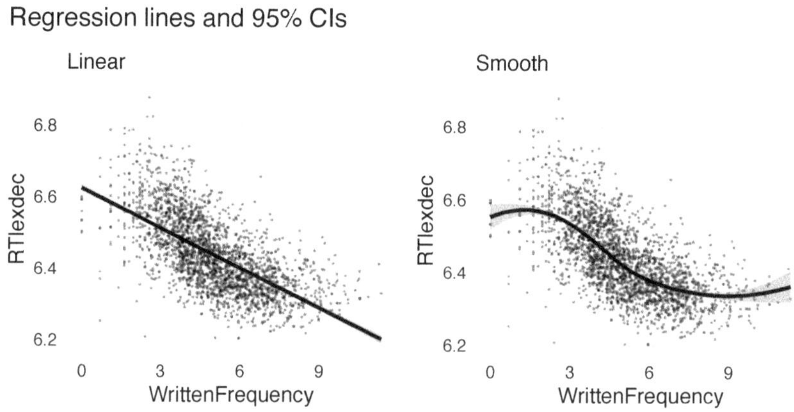

Figure 7.6
Least-squares regression line (*left*) and LOESS smooth (*right*) for `english_young`.

In other words, there is evidence that there is a `voicing` effect for stops before prosodic boundaries, but there's not enough to say anything about whether there is a `voicing` effect for other stops or whether a prosodic boundary modulates the `voicing` effect.

This general fallacy is extremely common in other fields (box 2.6) and probably in language sciences as well.

in

7.5 Nonlinear Effects

7.5.1 Motivation: Nonlinear Smoothers

So far we have assumed that continuous variables have a relationship with the response that is linear—well-described by a straight line. In reality this is often not the case.

Consider the plots in figure 7.6 of data from section 4.3.4, both of which are plotted with a "smoother" (with 95% CIs)—a function estimated by R to describe an x-y scatterplot (here, using `geom_smooth()` from ggplot2). The left plot shows the most straightforward smoother: the least-squares line of best fit, with its 95% CIs.

Box 7.10
Broader Context: More on Nonparametric Smoothers

LOESS-style smoothers are described widely, see, for example, Faraway (2016, section 14.3), and the tutorials herein introduce GAMs and the intuitions behind them. Both methods are "nonparametric" in the sense that the relationship between x and y can't be described by a set of parameters (like, e.g., a polynomial).

In addition to the linear versus nonlinear relationship, an important difference between parametric and nonparametric smoothers is in their CIs. For the linear smoother, the width of the CI increases for points further from the mean, because they are more influential for the fit (section 5.7.3), but for the nonlinear smoother, the width is determined by nearby points.

In the regression models presented in this book, a nonparametric smoother can only be used for visualization of the effect of x on y—capturing the effect in a model requires a parametric function. This approach gets complicated quickly, and typically we care about modeling the nonlinear relationship but not the exact parameters, so it would be nice to just include a nonparametric smoother in the model. GAMs are a generalization of regression models allowing this, which are increasingly popular across language sciences, especially for modeling time dependence (e.g., in pitch contours) or spatial dependence (e.g., dialect variation). Sóskuthy (2017), Wieling (2018), and Chuang et al. (2021) are recent tutorials using the mgcv package. GAMs can model non-linear relationships more easily, using "smoothing splines," but their results are harder to interpret (and they are trickier to fit for realistic cases, using current functionality). It is a bit harder to include nonlinear relationships in regression models as parametric terms (the option covered here), but the models are easier to interpret and fit.

Which option makes sense depends on context: is the additional complexity of GAMs worthwhile given the research questions? More important than the method used is that clearly nonlinear relationships in your data are *somehow* modeled as nonlinear.

The right plot shows a *nonparametric smoother*: a smooth curve interpolated based on nearby observations. There are different ways to do this; `geom_smooth` uses the locally estimated scatterplot smoothing (LOESS) method for small samples and generalized additive modeling (GAM) for large samples.[12] It is fine for our purposes to treat both as black boxes (box 7.10) The intuition in each case is similar: draw a smooth curve that fits the data in an x-y scatterplot "locally" (nearby points).

In any case, the right plot better characterizes the x-y relationship, which is *nonlinear* (not a line). To model a nonlinear relationship in a regression model, we need a way to represent it "parametrically"—by a set of k parameters (e.g., the coefficients of a quadratic function: $k=2$), which can be estimated as terms of a regression model. These parameters are multiplied by a series of "component" functions to describe the fitted x-y relationship; the set of functions is called the *basis*.

Two common bases used for nonlinear effects in regression models are polynomials and splines.

We can fit increasingly complex functions by increasing k (the *degree*), and so we need to choose a value that neither underfits (k is too small) or overfits (k is too large).

7.5.2 Polynomials

The simplest basis is regular polynomials. The components are just $x, x^2, \ldots, x^k$. (A linear relationship is the special case where $k=1$.) section 5.5.1 showed a cubic polynomial ($k=3$) fit to the `english_young` data. A polynomial effect of degree k can be fit by including either a series of terms like `I(WrittenFrequency2)` in the model formula (as in section 5.5.1) or a single `poly(x, k)` term.

Figure 7.7 shows increasingly complex polynomial functions fitted to the `english_young` data ($k=2, 4, 10$), fitted using these models:

```
## raw=TRUE: use regular polynomials
## The 'poly' term expands to WF + I(WF^2)
eng_poly2 <- lm(RTlexdec ~ poly(WrittenFrequency, 2, raw = TRUE),
  data = english_young)
eng_poly4 <- update(eng_poly2, . ~
    poly(WrittenFrequency, 4, raw = TRUE))
```

12. LOESS is better for visualization, but it is too slow for large samples (see `?geom_smooth`).

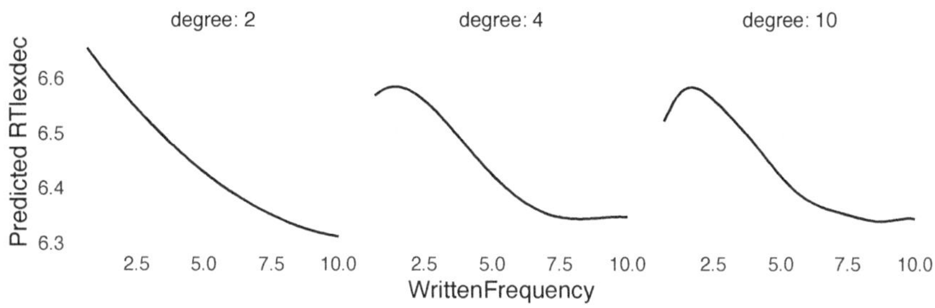

Figure 7.7
Polynomial effects of `WrittenFrequency` on `RTlexdec`, for the `english_young` data.

```
eng_poly10 <- update(eng_poly2, . ~
    poly(WrittenFrequency, 10, raw = TRUE))

alpha <- 0.01
upper99 <- quantile(english_young$WrittenFrequency, 1 - alpha / 2)
lower99 <- quantile(english_young$WrittenFrequency, alpha / 2)
```

Comparing to the plot of the empirical data (figure 7.6, right), it is visually clear that $k=2$ is not complex enough ("underfits") and $k=10$ is too complex ("overfits").

Choosing a basis for a nonlinear function means coding x as the components $f_1(x)=x$, $f_2(x)=x^2$, and so on, analogously to coding a categorical x using contrasts.

By the same logic as for contrast coding schemes (model interpretability, minimizing collinearity), it is desirable for components to be centered and orthogonal—here meaning that each component should average to 0 over the dataset, and any two components should be uncorrelated. Regular polynomials have neither of these properties, so they should typically be avoided.

Nonetheless, regular polynomials have the advantages of simplicity and coefficients that are easiest to interpret (x^2 always means the same thing, regardless of the dataset), and they are widely used.

7.5.3 Orthogonal Polynomials

Orthogonal polynomials are a basis where each component is constructed to be centered and orthogonal to other components on the dataset: $f_1(x)$ is a linear function (with mean 0), $f_2(x)$ is a quadratic function uncorrelated with $f_1(x)$ (and with mean 0), $f_3(x)$ a cubic function uncorrelated with $f_2(x)$ and $f_1(x)$ (and with mean 0), and so on. For example, figure 7.8 (left) shows components for the $k=4$ case for the `english_young` dataset.[13]

The following models fit increasingly complex nonlinear effects of `WrittenFrequency`, with degrees of 1–5, for the `english_young` case:

---

13. In the R implementation of orthogonal polynomials the components are chosen to each have norm of 1, meaning $f_i(x)^2$ sums to 1 over the dataset, but this is not necessary. Note that "orthogonal polynomial coding" for a factor with k levels is the same as converting to a numeric variable (1, 2, 3, ...) and then using orthogonal polynomials as predictors (e.g., figure 7.3).

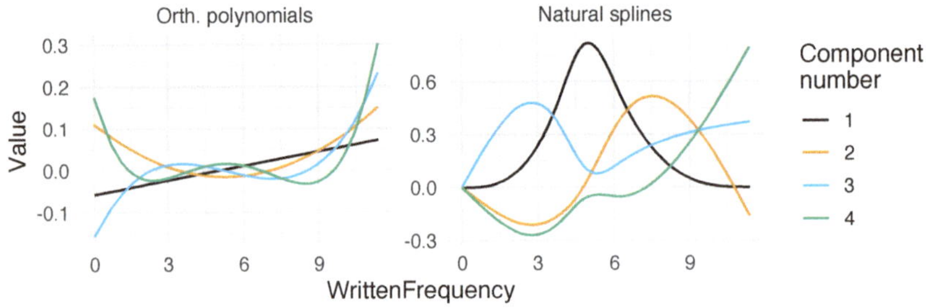

Figure 7.8
Orthogonal polynomial (*left*) and natural spline (*right*) components for the `english_young` data.

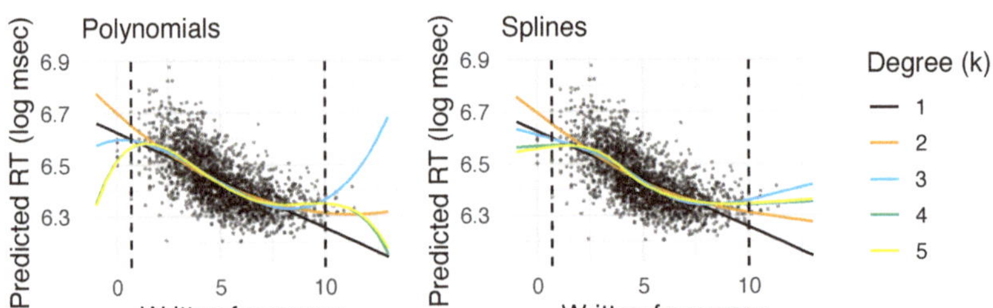

Figure 7.9
Predicted reaction time for YOUNG speakers as a function of frequency, in models of the `english` data with different nonlinear terms for `WrittenFrequency`: polynomial functions (*left*) and natural splines (*right*). Degree = 1 shows a linear fit (least-squares). Points are overlaid empirical data; dashed lines show 99% quantiles for `WrittenFrequency`.

```
eng_mod_lin <- lm(RTlexdec ~ poly(WrittenFrequency, 1),
  data = english_young)
eng_mod_poly2 <- update(eng_mod_lin, . ~ poly(WrittenFrequency, 2))
eng_mod_poly3 <- update(eng_mod_lin, . ~ poly(WrittenFrequency, 3))
eng_mod_poly4 <- update(eng_mod_lin, . ~ poly(WrittenFrequency, 4))
eng_mod_poly5 <- update(eng_mod_lin, . ~ poly(WrittenFrequency, 5))
```

Figure 7.9 (left) shows the predicted effect for each model. Over the values `WrittenFrequency` takes on in the dataset (between dashed lines), it looks like the quadratic, cubic, and quartic fits ($k = 2, 3, 4$) fit the data progressively better, while the $k = 5$ fit is similar to $k = 4$.

Deciding on k, including whether a nonlinear effect is justified at all, is a case of model comparison (section 5.9): we choose among candidate models using an appropriate metric, then sanity-check the result.

7.5.3.1 Nested model comparison

Given how the orthogonal polynomials are constructed, the components for $k = 3$ are a subset of those for $k = 4$, and so on. (The same is true for raw polynomials.) Thus, models with different k fitted to the same data are nested, so nested model comparison is appropriate for selecting among polynomial fits. For this example:

```
anova(eng_mod_lin, eng_mod_poly2, eng_mod_poly3,
  eng_mod_poly4, eng_mod_poly5)
## Analysis of Variance Table
##
## Model 1: RTlexdec ~ WrittenFrequency
## Model 2: RTlexdec ~ poly(WrittenFrequency, 2)
## Model 3: RTlexdec ~ poly(WrittenFrequency, 3)
## Model 4: RTlexdec ~ poly(WrittenFrequency, 4)
## Model 5: RTlexdec ~ poly(WrittenFrequency, 5)
##   Res.Df  RSS Df Sum of Sq     F  Pr(>F)
## 1   2282 15.1
## 2   2281 14.7  1     0.473 76.96 < 2e-16
## 3   2280 14.2  1     0.436 71.05 < 2e-16
## 4   2279 14.0  1     0.224 36.38 1.9e-09
## 5   2278 14.0  1     0.000  0.02     0.9
```

The significant difference between the linear model and the $k=2$ model means that a nonlinear effect is justified. Increasing k from 2 to 3 significantly improves the model (at $\alpha=0.05$), as does increasing from 3 to 4, but increasing from 4 to 5 does not. So we would choose the nonlinear relationship with $k=4$.

7.5.3.2 Nonnested model comparison

We can alternatively choose k using a measure of model quality, such as AIC, BIC, or adjusted R^2:

```
## Function from performance package for comparing several models
compare_performance(
  eng_mod_lin, eng_mod_poly2, eng_mod_poly3,
  eng_mod_poly4, eng_mod_poly5
) %>%
  select(Model, AIC, BIC, R2_adjusted)
## # Comparison of Model Performance Indices
##
## Model |       AIC |       BIC | R2 (adj.)
## ------------------------------------------
## lm    | -4971.705 | -4954.504 |     0.414
## lm    | -5042.230 | -5019.295 |     0.432
## lm    | -5109.294 | -5080.625 |     0.448
## lm    | -5143.485 | -5109.083 |     0.457
## lm    | -5141.502 | -5101.366 |     0.457
```

Using any of these metrics we would choose $k=4$.

7.5.3.3 Visual inspection

Whichever model selection method is used, the value of k should be sanity-checked by examining a nonlinear smoother of the relationship between x and y to eyeball whether the curve deviates from a line at all (in which case $k>1$) and if so to determine the number of inflection points (zero inflection points suggests $k=2$, one inflection point suggests $k=3$, and so on). For example, for the `english_young` plot (figure 7.6, right), it looks like there are one to two inflection points, so we'd choose $k=3$ or 4.

Visual inspection can also be used by itself, especially when it's not practical to do model selection, for example, when screening many predictors for nonlinear effects.

7.5.4 Splines

Intuitively, a spline is a curve made up of several polynomials glued together, which aims to (a) approximate the underlying relationship between x and y well and (b) be smooth, especially at the places where the polynomial pieces connect (the *knots*). You have probably encountered splines when drawing a curve or path in a computer program (e.g., PowerPoint, Photoshop) by specifying the endpoints of the curve and several points it must go through. There are many kinds of splines; the R default of *natural (cubic) splines* is a common choice: cubic splines (the polynomials are cubics) are constrained to (c) grow only linearly outside of the endpoints ("natural").

Splines have important advantages over polynomials (box 7.11). They are less prone to overfitting and approximate the underlying function better, especially at very high and low values of the variable being modeled (example below).

Box 7.11
Broader Context: More on Splines

Splines are less widely used than polynomials in language and behavioral sciences, but splines have important advantages for modeling nonlinearities common in linguistic data. Splines lie between a nonlinear smoother and polynomials and have the advantages of both. Like a nonlinear smoother, the fit is *local* and *smooth*: each section of the curve fits nearby points better and is not influenced by far-away points; smoothness conditions at the knots (typically equal first derivatives, zero second derivatives) mean the curve never changes too quickly, unless justified by the data. In a polynomial fit certain points (especially near the endpoints) are overly influential, and nothing forces the curve to be smooth: overfitted polynomials tend to look unreasonably wiggly (due to *Runge's phenomenon*; the internet has many good illustrations). Like a polynomial, splines are *parametric*, so they can be estimated in a regression model, and extrapolated beyond observed values of x—but they are less liable to "blowing up" when extrapolated than polynomials. These issues are exacerbated for smaller sample sizes and when observations are unevenly distributed among values of x, both of which are very common for linguistic data.

Fit quality near endpoints is an important issue in practice because these are often the most relevant data to the research questions. This is especially true for observational data, for example, the `vot_michael` example where the highest speaking rates are of interest, but also for controlled data, such as the lowest-frequency words in the `english_young` example.

There are many ways of defining splines, depending on the polynomials used, how knots are chosen, and so on, leading to a variety of names. R's `ns()` function ("natural splines") is the most basic and widely supported version (cubic polynomials, evenly spaced knots, linear outside of endpoints). "Restricted cubic splines" (`rcs()`, from the rms package; see Harrell [2015, section 2.4]) are widely used in linguistics, following Baayen (2008). They have the useful property that the first component is a line, so results can be reported as linear and nonlinear terms. We use `ns()` as our default because it is most compatible with other packages (for model predictions, etc.). "Smoothing splines" are a more complex, nonparametric method, such as in the `gam()` function used by `geom_smooth()`.

The downside of splines is that they are less intuitive than polynomials. Instead of x, x^2, and so on (or orthogonalized versions) the k spline components are cubic functions chosen *for this dataset* to satisfy conditions (b) and (c) given the distribution of x. It is not typically necessary or useful to examine spline components because they are not easily interpretable. Figure 7.8 (right) shows the components for `WrittenFrequency` when $k = 4$. Each component is a cubic function, which is linear at its endpoints, but higher-order components are more "bendy" (there are more inflection points).

k is the degree of the spline (i.e., its degrees of freedom), one greater than the number of knots; intuitively, this is the degree of the polynomial that would be needed to fit the same curve. So a spline with $k=2$ typically looks quadratic/has no inflection point, $k=3$ has one inflection point, and so on.

A natural spline term is included in a model using `ns()` (from the splines package). For example, this model uses a $k=4$ spline for the `english_young` data:

```
eng_mod_ns4 <- update(eng_mod_lin, . ~ ns(WrittenFrequency, 4))
```

Figure 7.9 (right) shows nonlinear effects fitted for this data, using natural splines with $k = 2$–5. Similarly to polynomials, visual inspection suggests k of at least 3 is justified (one or more inflection points). Note the difference in behavior between the spline and polynomial fits, which illustrate how polynomials tend to "blow up" when extrapolated beyond the range of x in the data used to fit the curve. All polynomial and spline fits in figure 7.9 look good for x between about 1 and 10, where 99% of the data lie, but the polynomial curves with $k=3, 4, 5$ wildly diverge for x extrapolated just outside of this range. In contrast, splines fitted to this data look reasonable when extrapolated outside of 1–10, because they are constrained to grow linearly.

In particular, note how the "good" fits of equivalent complexity (with $k \geq 4$) differ for words with frequency 0—these are real words, which were observed only once in the corpus (hence log-frequency of 0). In other words, their real frequencies are not 0 (if frequencies came from a larger corpus), and there are many other English words with frequencies in this range (most words of a language are low-frequency), so it is scientifically important what the predicted relationship is on the left edge of the graph. The spline fits predict a flat relationship: reaction time will be similar to observed zero-frequency words, which seems plausible. The polynomial fits are all problematic: $k=3$ predicts a flat relationship, but with `RTlexdec` different from observed zero-frequency words; $k>3$ predicts a sharply decreasing relationship, which is scientifically implausible. Avoiding such extrapolation errors is a major motivation for using splines (box 7.11).

Choosing model complexity To select the degree for a spline term, only non-nested model comparison should be used. This is because (natural) spline components are a function of the dataset and k—the components for $k=3$ are not a subset of those for $k=4$—so models fitted with spline terms of different degree k are not in a subset-superset relation.

Models with different k can be compared using AIC, BIC, and so on, exactly as for polynomial fits, and it is good practice to sanity-check with visual inspection. Exercise 7.5 asks you to fit and compare these models for the `english_young` case, showing that BIC, AIC, and adjusted R^2, select k of 4, 5, and 6. Visual inspection suggests that $k=4$ is a good choice.

7.5.5 Interpretation and Reporting

Consider the fitted models selected (using BIC) to describe the nonlinear relationship between `WrittenFrequency` and `RTlexdec`, using (orthogonal) polynomials and splines:

```
tidy(eng_mod_poly3)
##                         term estimate std.error statistic  p.value
## 1                (Intercept)     6.44    0.0017    3897.1  0.0e+00
## 2 poly(WrittenFrequency, 3)1    -3.27    0.0790     -41.4 7.5e-280
## 3 poly(WrittenFrequency, 3)2     0.69    0.0790       8.7  5.9e-18
## 4 poly(WrittenFrequency, 3)3     0.66    0.0790       8.4  1.0e-16
tidy(eng_mod_ns4)
##                       term estimate std.error statistic p.value
## 1              (Intercept)     6.57     0.016     416.0 0.0e+00
## 2 ns(WrittenFrequency, 4)1    -0.16     0.015     -10.5 2.0e-25
## 3 ns(WrittenFrequency, 4)2    -0.24     0.012     -19.6 2.5e-79
## 4 ns(WrittenFrequency, 4)3    -0.22     0.035      -6.2 5.3e-10
## 5 ns(WrittenFrequency, 4)4    -0.23     0.019     -12.0 5.3e-32
```

The intercept in each case is interpretable as the mean of `RTlexdec` (6.4), because both natural splines and orthogonal polynomials are centered. For all models we've fitted previously, without nonlinear effects, the regression coefficients have been interpretable as the slope of x or a difference between levels. What do the k coefficients describing each nonlinear effect mean?

The short answer is that these coefficients are hard to interpret, and it is usually fine to ignore them, instead interpreting the nonlinear effect by plotting model predictions (as in figure 7.9) and carrying out omnibus tests (to decide whether the nonlinear effect is significant).

For example, to compute an omnibus effect and associated effect size for the nonlinear effect of `WrittenFrequency` in model `eng_mod_ns4`:

```
# Intercept-only model
eng_mod_baseline <- update(eng_mod_ns4, . ~ 1)
anova(eng_mod_baseline, eng_mod_ns4)
## Analysis of Variance Table
##
## Model 1: RTlexdec ~ 1
## Model 2: RTlexdec ~ ns(WrittenFrequency, 4)
##   Res.Df  RSS Df Sum of Sq   F Pr(>F)
## 1   2283 25.8
## 2   2279 14.0  4      11.8 481 <2e-16
cohens_f2(eng_mod_baseline, eng_mod_ns4)
## [1] 0.84
```

The effect could be reported in a paper as follows:

> There was a nonlinear effect of frequency on reaction time, pictured in figure 7.6 (right), modeled using a natural spline of degree 4 (Cohen's $f = 0.84$; $F(2300, 4) = 480$, $p < 0.001$), with the spline's degree chosen using BIC and confirmed with visual inspection.

If you also want to report that a *nonlinear* effect in particular was justified, you could add a sentence:[14]

> A nonlinear relationship is clear from the empirical data (figure 7.6, right), and the nonlinear model improves on a linear effect of frequency (BIC = –3742, –5109 for linear, nonlinear models).

7.5.6 Examples

Let's consider two more realistic examples showing how nonlinear effects come up in multiple regression models.

7.5.6.1 Interactions with nonlinear effects

Consider our working model for the `vot_michael` data, gh_mod_4. Recall that an interaction between `speaking_rate` and `voicing` was included in the model, based on the empirical plot in figure 5.6 (bottom), to allow for the different speaking rate effect for VOICED and VOICELESS stops. The model captures the main pattern that there is a weaker effect for voiced stops. But the same plot suggests the speaking rate effect may be nonlinear.

To assess this, we compare the model with a linear `speaking_rate` effect (and its interaction with `voicing`) to models where this effect is parametrized as a natural spline with degree $k = 2$ or 3. (From visual inspection, $k = 3$ looks like the maximum justified degree.)

```
## Fit nonlinear models
vot_mod_ns2 <- update(gh_mod_4, . ~ . - voicing * speaking_rate +
  ns(speaking_rate, 2) * voicing)
vot_mod_ns3 <- update(gh_mod_4, . ~ . - voicing * speaking_rate +
  ns(speaking_rate, 3) * voicing)

## Compare linear and nonlinear models
compare_performance(gh_mod_4, vot_mod_ns2, vot_mod_ns3) %>%
  select(Model, AIC, BIC, R2_adjusted)
## # Comparison of Model Performance Indices
##
## Model |     AIC |     BIC | R2 (adj.)
## --------------------------------------
## lm    | 674.993 | 740.771 |     0.824
## lm    | 666.476 | 741.025 |     0.827
## lm    | 670.047 | 753.366 |     0.826
```

AIC selects the nonlinear model with degree 2, but BIC selects the linear model. Figure 7.10 shows the predicted speaking rate–voicing interaction for these two models. Both models capture the main empirical pattern that the `voicing` effect is smaller in faster speech. In the linear model this is the pattern across all speech rates, while in the nonlinear model it is only for fast speech that the `voicing` effect decreases, due to the direction of the rate effect reversing for VOICED stops. Both models seem reasonable given the empirical data, but the nonlinear model is odd on theoretical grounds: it would be surprising

14. The linear and nonlinear models are not nested in this case, so we can't easily assess the significance of the difference.

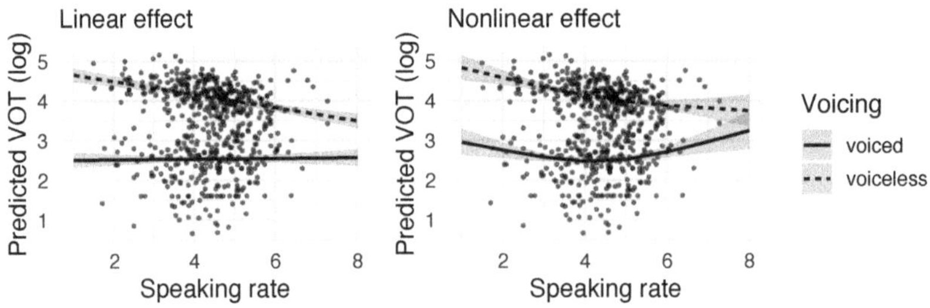

Figure 7.10
Interaction plots of `speaking_rate` and `voicing` for models of the `vot_michael` data with linear (*left*) and nonlinear (*right*) `speaking_rate` effects, with empirical data overplotted.

for any phonetic duration to be *longer* in faster speech. This suggests selecting the more conservative (linear) model.

7.5.6.2 Nonlinear effects in logistic regression

The examples so far have used linear regression, and although a nonlinear relationship was justified, the main qualitative conclusion about the x-y relationship was similar whether a linear or nonlinear effect was fitted. This example uses logistic regression in a case where nonlinearity matters for the research question.

We'll model verb regularity (`Regularity`) for the `regularity` dataset, as a function of predictors considered so far (`WrittenFrequency`, `Auxiliary`, `Ncountstem`; see sections 2.2.3 and 2.5.1.2), as well as an additional predictor: morphological family size (`FamilySize`: number of types in the word's morphological family). This is a simplified version of the model from Baayen (2008, section 6.3.1).

Of interest here is the `FamilySize` effect, for which figure 7.11 shows linear and nonlinear smooths to the empirical data. The linear and nonlinear smooths in figure 7.11 (left) both model the log-odds that $y=1$. The linear smooth is predictions from a simple logistic regression of $x=$ `Regularity` (as discussed in box 6.1); the nonlinear smooth is predictions from a GAM with a smoothing spline term for x.

One can imagine either a positive or negative family size effect, which could correspond to different hypotheses of how the mental lexicon influences language change.[15] The empirical data suggest a nonlinear effect is justified, which we can verify by fitting linear and nonlinear models (with splines, $k=2,3$):

```
reg_mod_lin <- glm(Regularity ~ WrittenFrequency + FamilySize +
  NcountStem + Auxiliary, data = regularity, family = "binomial")
reg_mod_nonlin2 <- update(reg_mod_lin, . ~ .
- FamilySize + ns(FamilySize, 2))
reg_mod_nonlin3 <- update(reg_mod_lin, . ~ .
- FamilySize + ns(FamilySize, 3))
compare_performance(reg_mod_lin, reg_mod_nonlin2, reg_mod_nonlin3) %>%
```

15. Words with larger morphological families may be processed more by analogy to the rest of the lexicon (promoting regularity) or may have their representation strengthened (promoting irregularity).

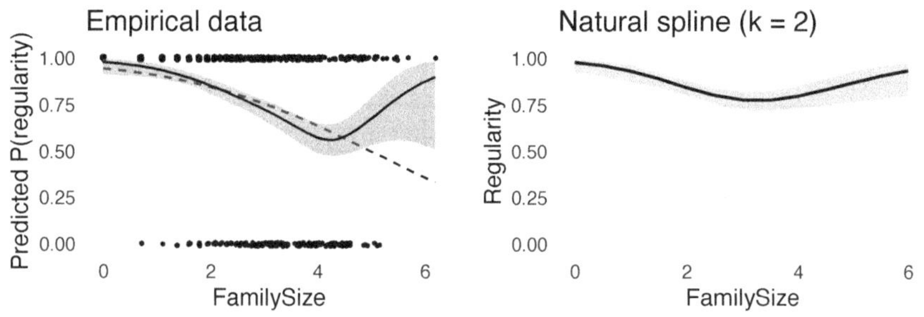

Figure 7.11
Left: Empirical plot of probability of verb `Regularity` as a function of `FamilySize`, using linear (dotted line) and nonlinear (GAM: solid line, with 95% CIs) smoothers, for the `regularity` data. *Right*: Partial effect of `FamilySize` from model `reg_mod_nonlin2`.

```
  select(Model, AIC, BIC)
## # Comparison of Model Performance Indices
##
## Model |     AIC |     BIC
## -------------------------
## glm   | 601.546 | 628.853
## glm   | 592.026 | 623.883
## glm   | 593.817 | 630.226
```

Both AIC and BIC select the nonlinear model with $k=2$. The predicted effect of `FamilySize` for this model is shown in figure 7.11 (right).

Note also that in the linear model `FamilySize` does not significantly contribute to the model, but in the nonlinear model it does (by LRTs):

```
reg_mod_none <- update(reg_mod_lin, . ~ . - FamilySize)
anova(reg_mod_none, reg_mod_lin, test = "Chisq")
...
##   Resid. Df Resid. Dev Df Deviance Pr(>Chi)
## 1       695        593
## 2       694        590  1     3.29     0.07
```

```
anova(reg_mod_none, reg_mod_nonlin2, test = "Chisq")
...
##   Resid. Df Resid. Dev Df Deviance Pr(>Chi)
## 1       695        593
## 2       693        578  2     14.8  0.00061
```

Thus, the linear and nonlinear models give qualitatively different results for an effect of interest. From the linear model we would conclude that morphological family size is not predictive of verb regularity, but in the nonlinear model we would conclude there is an interesting U-shaped effect of family size.

This illustrates one motivation for considering nonlinear effects: the qualitative conclusions about the predictor of interest may change.[16]

7.6 Collinearity Diagnostics Revisited

Section 5.6.3 introduced the VIF measure, which quantifies how much the variance of each predictor's estimate is "inflated" by the presence of others. This does not make sense when a subset of the predictors are related, as for multilevel factors or nonlinear effects, where a single variable corresponds to several predictors. Another way of stating this distinction is that these measures assume each variable has $df = 1$; multilevel factors or nonlinear effects have $df > 1$ (e.g., df = number of contrasts). Intuitively we would like a measure of how predictable each set of related predictors is from the others, for example, how predictable `place` ($df = 2$) is from other predictors for the `vot_michael` data.

The *generalized variance inflation factor* (GVIF) is a generalization of VIF to this case (Fox and Weisberg 2019, section 8.8), which is automatically computed by `vif` in the `car` package when applied to a model where some terms have $df > 1$.[17] $GVIF^{1/2 \cdot df}$ turns out to quantify how much the SE of the variable's effect is inflated due to collinearity; following Yanagida (2020), we call this the *SIF* (standard error inflation factor). The SIF is a generalization of $\sqrt{VIF}$ (when $df = 1$, $\sqrt{VIF} = SIF$).

For example, consider SIF values for the model of the `vot_michael` data with a nonlinear `speaking_rate` effect chosen by AIC (rightmost column):

```
car::vif(vot_mod_ns2)
##                                GVIF Df GVIF^(1/(2*Df))
## cons_cluster                    1.1  1             1.1
## place                           1.7  2             1.1
## foll_high_vowel                 1.1  1             1.0
## log_corpus_freq                 1.1  1             1.0
## ns(speaking_rate, 2)            1.1  2             1.0
## voicing                        28.7  1             5.4
## cons_cluster:voicing            1.4  1             1.2
## place:voicing                   1.2  2             1.1
## place:foll_high_vowel           1.6  2             1.1
## ns(speaking_rate, 2):voicing 29.7  2             2.3
```

In this model the multilevel factor `place` and the nonlinear effect of `speaking_rate`, each with $df = 2$, receive a single SIF value. The estimates of the `voicing` and `voicing:speaking_rate` effects are more strongly affected by collinearity than other predictors.

16. Another possibility is that the effects of *other* predictors change once a nonlinear effect is accounted for. However, in this and the VOT example, it can be verified from the model tables that the effects of other predictors are very similar in the linear and nonlinear models.

17. GVIF is the degree of inflation of the joint confidence region of the df regression coefficients in the presence of the other $k - df$ predictors.

If we used a cutoff of SIF = 3.2 (analogous to VIF = 10: $\sqrt{10} = 3.2$), we would conclude that there is potentially problematic collinearity in this model (maximum $SIF > 3.2$), in particular for estimating the `voicing:speaking_rate` effect of key theoretical interest. This would be relevant to report in a write-up when discussing the `voicing:speaking_rate` effect: although model selection suggests a linear (if using BIC) or slightly nonlinear (if using AIC) effect, the degree of collinearity suggests we should not put much confidence in the exact degree chosen. (In contrast, there is clear evidence for *some* `voicing:speaking_rate` interaction.)

Note that the other measure of collinearity we discussed in section 5.6.3, condition number (κ), generalizes to the $df > 1$ case because it applies to the *set* of predictors rather than to individual predictors.

7.7 Further Reading

Schad et al. (2020) discusses contrast coding in R, focusing on regression and linguistic data. More detailed treatments focus on the ANOVA case (e.g., Cohen et al. 2002, chapter 8; Baguley 2012, chapter 15) or R's implementation of contrast coding (e.g., Venables and Ripley 2002, section 6.2). UCLA Statistical Consulting Group (2011) is a useful online tutorial.

On marginal means and post hoc tests (and their use for interpreting interactions), the emmeans and margins vignettes are helpful (Lenth 2021; Leeper 2021). The marginaleffects package, which combines and extends emmeans/margins functionality, is worth checking out (Arel-Bundock 2021). Basic introductions to interpreting interactions are in most regression analysis books, including Winter (2019, chapter 8) and Gries (2021, section 5.2) for linguistic data in particular.

On nonlinear effects, Baayen (2008) goes into particular depth for linguistic data, using the tools from Harrell (2015) (e.g., section 2.4). More general introductions to the topics discussed (smoothers, polynomials, splines, etc.), from increasingly technical perspectives, include Zuur, Ieno, and Smith (2007, section 7.4); Wood (2017, chapter 4); Hastie, Tibshirani, and Friedman (2009).

Cohen et al. (2002) is a useful in-depth treatment of most of these topics for behavioral sciences generally, which exemplifies the vast methodological literature on these topics from related fields, to which we can't do justice. For example, Mize (2019) (sociology) and Preacher, Curran, and Bauer (2006) (behavioral sciences) discuss methods for interpreting interactions.

Exercises

7.1. Consider the intercept terms in the VOT models where `lexical_class` is coded using treatment contrasts (`vot_cc_mod_1`) and simple contrasts (section 7.2.6), which have different estimates corresponding to their different interpretations. Why is the intercept's SE higher in `vot_cc_mod_1`?

7.2. Starting from the contrast matrix for sum contrasts with four levels (`contr.sum(4)`), calculate the corresponding interpretation matrix and verify that the intercept's interpretation is the (unweighted) grand mean and the interpretation of contrast *i* is "level *i* minus the grand mean."

7.3. Section 7.3.1 showed how the first step of building up a model for the `diatones` data, a model including all main effects, can lead to two different conclusions about the effect of `syll2_coda_fact` depending on the analysis method used.

(a) Build up the model via the Gelman and Hill method for this case, as in section 6.7.2.2, assuming that `syll2_coda_fact` does have a significant main effect.

(b) Now do the same, assuming that `syll2_coda_fact` does not have a significant main effect.

(c) What is the qualitative difference between the two resulting "final" models? Does this difference affect what the model says about the research questions?

7.4. Verify, using the Gelman and Hill method applied to `gh_mod_4`, that the term added in section 7.4.3 (`cons_cluster:place:voicing`) is the only three-way interaction justified. (You can assume that there is no theoretical motivation for any three-way interaction.)

7.5. Fit models for the `english_young` case with natural splines, with degrees of $k = 1–7$. Show that BIC, AIC, and adjusted R^2 select k of 4, 5, and 6. Which k does visual inspection suggest is reasonable?

8 Mixed-Effects Models 1: Linear Regression

This chapter introduces regression models for clustered data, focusing on conceptual understanding, fitting, and interpretation of basic linear mixed-effects models. Chapter 9 extends this discussion to mixed-effects logistic regression. These chapters mostly assume the most common use case, where population-level effects ("fixed effects") are of primary interest. We assume that the set of terms to include in the models are known in advance and that there are no model fitting problems. Chapter 10 considers model selection, model convergence, and other "advanced" topics that come up frequently in practice.

8.1 Preliminaries

8.1.1 Packages

This chapter assumes that you have loaded several packages from previous chapters, as well as the lme4 package for fitting mixed-effects models (Bates, Mächler, et al. 2015) and various packages for working with lme4 models (broom.mixed, MuMIn, pbkrtest, RLRsim, lmerTest; see Bolker and Robinson [2021]; Barton [2020]; Halekoh and Højsgaard [2014]; Scheipl, Greven, and Kuechenhoff [2008]; Kuznetsova, Brockhoff, and Christensen [2017]):

```
library(tidyverse)
library(broom)
library(arm)
library(lme4)
library(car)
library(broom.mixed)
library(MuMIn)
library(pbkrtest)
library(RLRsim)
library(lmerTest)
## makes `lmer` be lme4, rather than lmerTest version
lmer <- lme4::lmer
```

The most important package here is lme4, for fitting mixed-effects models. This discussion uses version 1.1.27.1, but any version after 1.0, when major changes were made, should be fine.

8.1.2 Data

This discussion also assumes you have set the default contrasts for factors to Helmert contrasts (section 7.2.12), and loaded the `neutralization` and `vot` datasets (sections 3.1.2 and 5.1.2), followed by some processing done in previous chapters (sections 4.1.2, 4.5.3, 7.1.2):

```
options(contrasts = c("contr.helmert", "contr.poly"))

neutralization <- read.csv("data/neutralization_rmld.csv",
  stringsAsFactors = TRUE) %>%
  ## Factor version of voicing, with voiceless < voiced
  mutate(voicing_fact = fct_relevel(voicing, "voiceless")) %>%
  filter(!is.na(prosodic_boundary))

vot <- read.csv("data/vot_rmld.csv", stringsAsFactors = TRUE) %>%
  # Relevel place to be labial < alveolar < velar
  mutate(place = fct_relevel(place, "labial"))
```

Finally, for both datasets we need to standardize predictors we will use in models (by "rescaling"; see section 5.5.4):

```
## Two-level factors: change to 0/1 then center
## Continuous predictors: center and divide by 2 SD
neutralization <-
  mutate(neutralization,
    prosodic_boundary = rescale(prosodic_boundary),
    ## voiced > voiceless
    voicing = rescale(voicing_fact)
  )

vot <- vot %>% mutate(
  log_corpus_freq = rescale(log_corpus_freq),
  speaking_rate_dev = rescale(speaking_rate_dev),
  foll_high_vowel = rescale(foll_high_vowel),
  cons_cluster = rescale(cons_cluster),
  gender = rescale(gender)
)
```

8.2 Motivation: Grouped Data

Previously, in the context of *t*-tests and linear regression (sections 3.5.2 and 5.3.1), we argued that independence assumptions are the most important assumption underlying analysis of linguistic data. This is because most linguistic data have grouping structure(s)—by participants, items, words, and so on—that immediately implies nonindependence of observations. Grouping by one or more *grouping factors* is the norm in linguistic data—whether from a laboratory experiment, a speech or text corpus, or the output of a computational experiment. There are some exceptions (e.g., lexical data, as in the `regularity` and `diatones` datasets), but they prove the rule. Analyzing such data requires taking grouping structure into account, because observations from the same unit are typically not independent. For example, in a study of lexical decision RT as a function of participant age and word frequency, certain participants will be characteristically fast, compared to other participants of the same age (they may have slept more or be more motivated), and certain words will take longer to recognize, beyond the effect of word frequency (e.g., orthographically longer words).

One way to analyze such data is to average over observations to remove grouping structure (as in the `english` dataset); another is to analyze a subset where independence holds (as we did for the `vot_michael` dataset). Neither is a good option: at best these ways result in a loss of power.

Analyzing grouped data without taking grouping into account, or *pseudoreplication* (box 3.4), can easily result in type I or II errors, as illustrated in section 3.5.2 for the simplest case, *t*-tests (does *y* differ between two groups)? These errors resulted from the mismatch between the data's structure and assumptions of the analysis method, and analogous errors are a danger for *any analysis of grouped data* (linear regression, logistic regression, ANOVAs, etc.) not taking grouping structure into account. Pseudoreplication can also result in errors in the magnitude or sign of an effect (type M and S errors; see section 3.4), such as *Simpson's paradox*, where an effect (e.g., positive) observed across groups is reversed (e.g., negative) or disappears once grouping structure is taken into account.[1]

One method for analyzing grouped data is *repeated measures ANOVAs*, which were the norm in many subfields (e.g., psycholinguistics) until the early 2010s, and are still widely used. It is useful to know about them (box 8.3), and they are a fine analysis method for data from relatively simple experiments with balanced designs, but they have important limitations that do not make them a good default method.

Mixed-effects regression models, or *mixed models* (or multilevel models), are another way of analyzing grouped data, which have become the standard for analyzing grouped data in much of the language sciences, as well as other fields.

Mixed models have several advantages over other analysis methods for grouped data, which mirror the structure of linguistic data, making them a good default:

- *Multiple grouping factors* (e.g., participant, item) can be included in the same model.
- *Unbalanced data* are assumed (unequal number of observations per level).
- *Explicit modeling of variability* among levels of a grouping factor, such as by-participant differences, allowing the analyst to explore and test questions about the variability itself.

Practical notes Before fitting any mixed model, it is important to check that the variables to be used as grouping factors are indeed coded as factors. For example, participant and item IDs are often numeric in data from laboratory experiments such as `neutralization`:

```
select(neutralization, subject, item_pair) %>% head()
##   subject item_pair
## 1      15        24
```

1. Jaeger et al. (2011, figure 2) gives an elegant example from language typology.

```
## 2        15          22
...
```

```
# Recode subject and item_pair as factors
neutralization <- mutate(neutralization,
  item_pair = as.factor(item_pair),
  subject = as.factor(subject)
)
```

Coding these variables as factors ensures that functions for fitting and interpreting mixed-effects models will work properly. We also do this for the relevant variables (`speaker`, `word`) for the `vot` data for good measure. Depending on how the data are loaded or subsequently processed, these may be coded as characters rather than factors:[2]

```
vot <- mutate(vot,
  word = as.factor(word),
  speaker = as.factor(speaker)
)
```

We also explicitly set the contrasts for factors that will be predictors in our models:

```
contrasts(neutralization$voicing_fact) <- contr.helmert(2)
contrasts(neutralization$place) <- contr.helmert(3)
contrasts(neutralization$vowel) <- contr.helmert(5)

contrasts(vot$place) <- contr.helmert(3)
```

This is necessary for model predictions to work correctly in all cases, even though we have set unordered factors to default to Helmert contrasts. Rather than go into the underlying reason, which is very technical, it is easiest to just remember the rule, *always explicitly specify contrasts for predictors in your mixed-effects models*, to be safe.

8.3 Linear Mixed Models: Introduction

This discussion will introduce mixed models through a series of examples. We will start with the simplest case, where there is a single grouping factor and variability between groups just in the value of the intercept, to introduce fitting and interpreting a mixed model (sections 8.3.3 and 8.3.4), then turn to multiple grouping factors (section 8.3.5) and variability between groups in the effects of predictors (random slopes; see section 8.4).

Consider the `neutralization` data for a single participant:

```
neutralization_2 <- filter(neutralization, subject == 2)
```

2. For example, all tidyverse functions default to characters rather than factors.

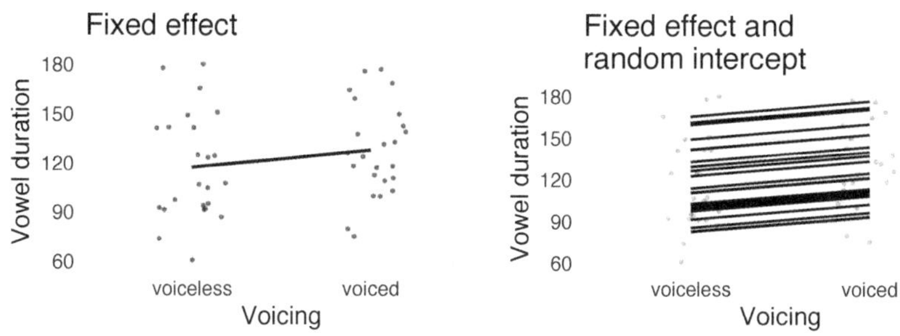

Figure 8.1
Predicted effect of `voicing` for two models of the `neutralization_2` data: a simple linear regression (*left*), and different items in a linear mixed model containing a by-item random intercept (*right*).

These data are grouped by item (grouping factor = `item_pair`): there are 24 pairs of words, each differing only in the final consonant's voicing (see section 3.1.2). We consider just this predictor (x = `voicing`, values –0.5 and 0.5, corresponding to VOICELESS/VOICED consonants) for simplicity; of interest is its effect on `vowel_dur` (y).

8.3.1 Simple Linear Regression

A simple linear regression model for this case, without accounting for by-item variability, would be (equation (4.3)):

$$y_i = \beta_0 + \beta_1 x_i + \varepsilon_i,$$
$$\varepsilon_i \sim N(0, \sigma), \quad \text{for } i = 1, \ldots, n. \tag{8.1}$$

This model assumes the same intercept (β_0) and slope (β_1) for all groups (i.e., all items), as schematized in figure 8.1 (left), which shows the predictions of the model fitted in R:

```
neut_2_mod_0 <- lm(vowel_dur ~ voicing, data = neutralization_2)
```

The same `vowel_dur` value is predicted for all observations with the same `voicing` value, regardless of which item an observation comes from. Because the data are grouped by item, this model violates the independent error assumption of linear regression. Items might differ in either:

1. Their mean `vowel_dur`
2. The effect of `voicing`

Our goal in fitting the model is to assess whether there is an effect of `voicing`, beyond by-item variability.

Specifically, based on this model we would conclude that the voiced/voiceless difference does not significantly differ from 0:

```
tidy(neut_2_mod_0)
##          term estimate std.error statistic p.value
## 1 (Intercept)      123       4.4      28.0 1.5e-30
```

```
## 2     voicing        11         8.8        1.2 2.2e-01
```

So failing to account for (1) or (2) could have resulted in a type II error (if there is in reality a voiced/voiceless difference).

Mixed models deal with nonindependence of errors by using two types of coefficients: *random effects* are coefficients that vary between groups, and *fixed effects* are coefficients that do not vary between groups.[3]

Fixed effects are also called *population-level* or *average effects*, because they can be thought of as the effect averaging across groups; these are often of primary interest. The *group-level* variability captured by random effects can be thought of either as "error" that must be controlled to correctly interpret fixed effects, or as phenomena of interest in their own right. "Random" refers to the assumption that levels of the grouping factor are randomly sampled from the population we want to generalize across. For example, in modeling the `neutralization_2` data, we are assuming that the items (= word pairs differing only in final consonant voicing) seen by participant 2 are randomly sampled from all such possible word pairs in German.

Fixed effects are essentially what we have referred to as "regression coefficients" in the linear and logistic regression models covered in chapters 4–7. Random effects can model by-group variability in the *intercept* or predictor *slopes* (as in (1), (2)); these two kinds are introduced in the next models we consider.

8.3.2 One Grouping Factor, Random Intercept

Let J be the number of groups indexed by the grouping factor, and $j[i]$ be the group of the ith observation. For example, for the `neutralization_2` data $J = 24$ (24 item pairs) and $j[1] = 10$ (observation 1 is from item 10).

The regression model for a simple linear regression where the intercept varies by group is

$$
\begin{aligned}
y_i &= \beta_0 + \alpha_{j[i]} + \beta_1 x_i + \varepsilon_i, \\
\varepsilon_i &\sim N(0, \sigma_E^2), \quad \text{for } i = 1, \ldots, n. \\
\alpha_j &\sim N(0, \sigma_I^2), \quad \text{for } j = 1, \ldots, J. \qquad (8.2)
\end{aligned}
$$

The first line looks like a simple linear regression (equation (8.1)), except for the *random intercept* $\alpha_{j[i]}$, which captures the offset of group $j[i]$ from the population-level intercept (β_0). The random intercept values for different items are assumed to be normally distributed, where σ_I captures the amount of between-item variation in the intercept value.

The errors ε_i are still assumed to be normally distributed, as in the simple linear regression. We now write the error variance as σ_E (instead of σ), to distinguish it from σ_I.

3. This discussion uses "fixed effect" and "random effect" because this is standard terminology across language sciences. Gelman and Hill (2007, section 11.4) discuss problems with these terms (inconsistent and ambiguous definitions across fields) and suggest more precise "constant"/"varying" terminology (e.g., "varying intercepts," "varying slopes"), which is nonstandard but used by some R packages. See box 9.3.

This model makes predictions shown in figure 8.1 (right): one line per item, differing in intercept but not in slope. (section 8.3.4 shows how these predictions are calculated.)

In other words, the regression describes the relationship between y and x at several levels. A different relationship is predicted for *each item*, for example, for item 10:

$$y = \beta_0 + \alpha_{10} + \beta_1 x,$$

where $(\beta_0 + \alpha_{10})$ is the value of the intercept for item 10.

The model for an *average item* is

$$y = \beta_0 + \beta_1 x.$$

Here, the intercept is β_0, the population-level effect (averaging across items). The model for an average item captures the relationship in figure 8.1 (left), identically to the simple linear regression without random effects (equation (8.1)).

So the mixed-effects model can be thought of as a generalization of simple linear regression, modeling both the average relationship and how it varies across groups.

8.3.3 Fitting and Interpreting the Model

Fitting a mixed-effects model requires estimating the fixed-effect coefficients (β_0, β_1) and the *variance components* describing the degree of different kinds of variability (σ_I^2, σ_E^2). (The random-effect coefficients α_j are not directly estimated by the model. This point is discussed in box 8.1.)

We use the lme4 package for fitting (linear) mixed effects models, which is the most widely used R package for such models. Functions in this package fit mixed models by one of two methods: ML, which tends to underestimate the extent of variability, giving biased variance component estimates, or *restricted maximum likelihood* (REML; see Pinheiro and Bates 2000, section 2.2.5), which gives unbiased variance component estimates and is typically the default. The difference between ML and REML only matters in practice when doing model comparison (section 8.5).

In either case, lme4 uses an optimization algorithm to find a model that maximizes the likelihood. Model-fitting problems—where the algorithm does not converge, or it gives a model with unreasonable parameter estimates ("singular"; see section 8.7.2)—are fairly common in practice. We will discuss these in more depth in section 10.3 and, until then, mostly abstract away from them by fitting convergent models and ignoring singular model warnings.

We fit the mixed model described using the `lmer()` function:

```
neut_2_mod_1 <- lmer(vowel_dur ~ voicing + (1 | item_pair),
  neutralization_2)
```

Here, `(1|item_pair)` is lme4 notation for a by-item random intercept (`1` means "intercept," `|item_pair` means "grouped by item").

This model's output is

```
summary(neut_2_mod_1)
## Linear mixed model fit by REML ['lmerMod']
## Formula: vowel_dur ~ voicing + (1 | item_pair)
##    Data: neutralization_2
##
## REML criterion at convergence: 428
##
## Scaled residuals:
##     Min      1Q  Median      3Q     Max
## -2.2079 -0.5324  0.0582  0.4183  1.7190
##
## Random effects:
##  Groups    Name        Variance Std.Dev.
##  item_pair (Intercept) 732      27.1
##  Residual              193      13.9
## Number of obs: 48, groups:  item_pair, 24
##
## Fixed effects:
##             Estimate Std. Error t value
## (Intercept)   122.92       5.88   20.92
## voicing        10.80       4.01    2.69
##
## Correlation of Fixed Effects:
##         (Intr)
## voicing 0.000
```

The model has estimated the fixed-effect coefficients (under `Fixed effects`) and the variance components (under `Random effects`). The estimated intercept is $\hat{\beta}_0 = 123$; the slope of `voicing` is $\hat{\beta}_1 = 11$; the estimated degree of by-participant variability in the intercept is $\hat{\sigma}_i^2 = 732$; and the estimated residual error is $\hat{\sigma}_e^2 = 193$.

Typically the information outside of these two blocks is not of interest, and we suppress it in the output. `Correlation of Fixed Effects` are the estimated correlations between fixed-effect estimates—analogous to the correlations between coefficient estimates for a linear regression (section 5.6.2.2).

The fixed-effects coefficients mean that the model predicts the relationship for an "average item" to be

$$\texttt{vowel\_dur} = 123 + 11 \cdot \texttt{voicing}$$

There are no p-values, for reasons discussed in section 8.5, but $|t| > 2$ can be used as a rough rule of thumb for which fixed effects are significant at the $\alpha = 0.05$ level (section 8.5.1.1). Alternatively we can compute 95% CIs for the model parameters:

```
## 'parm' argument gives CIs only for fixed effects
confint(neut_2_mod_1, parm = "beta_")
##             2.5 % 97.5 %
## (Intercept) 111.2    135
## voicing       2.8     19
```

By either method there is a “real” `voicing` effect in this model but not in the simple linear regression, suggesting that we made a type II error using the latter.

Note that the estimated intercept and slope (of `voicing`) are almost exactly the same in the simple linear regression:

```
tidy(neut_2_mod_0) %>% select(estimate, std.error)
##   estimate std.error
## 1      123       4.4
## 2       11       8.8
```

This makes sense, as both models are predicting the average effect. However the SEs differ. The slope estimate is more certain in the mixed-effects model because the effect is clearer once by-item variability in (average) `vowel_dur` has been taken into account. This is very similar to what happens when a paired two-sample *t*-test is used instead of an unpaired test, as in section 3.5.2.2 for the same data.

Note that mixed models can also be summarized and processed using the same tidy functions as used previously (now from the broom.mixed package): `tidy()` for cleaner model output, `glance()`, `augment()`, and so on:

```
tidy(neut_2_mod_1)
##     effect     group             term estimate std.error statistic
## 1    fixed      <NA>      (Intercept)      123       5.9      20.9
## 2    fixed      <NA>          voicing       11       4.0       2.7
## 3 ran_pars item_pair sd__(Intercept)       27        NA        NA
## 4 ran_pars  Residual sd__Observation       14        NA        NA
```

To get a sense of what $\hat{\sigma}_I = 27$ means, note that because the random intercepts (the α_j) are normally distributed, we expect roughly 95% of items to have intercepts within $2\sigma_I$ of the population-level intercept, β_0. Thus, the model predicts that 95% of items in the *population* have intercepts between 69 and 177 msec:

$$\text{Lower bound:} \quad \hat{\beta}_0 - 2 \cdot \hat{\sigma}_I = 123 - 2 \cdot 27 = 69,$$

$$\text{Upper bound:} \quad \hat{\beta}_0 + 2 \cdot \hat{\sigma}_I = 177.$$

8.3.4 Predicted Random Effects

It is often of interest to extract the *predicted random effects*, for example, to plot model predictions for individual items. In the current example this would be $\hat{\alpha}_1, \ldots, \hat{\alpha}_{24}$, the predicted offset of each item's intercept from the population value. These values, often called best linear unbiased predictors (or BLUPs; see box 8.1), can be extracted using the `ranef()` function:

```
ranef(neut_2_mod_1)$item_pair
##    (Intercept)
## 1          9.2
## 2        -31.7
## 3         15.6
```

```
...
```

Thus $\hat{\alpha}_1 = 9.2$ and so on. It can be useful to instead use the predicted coefficient values for each level (which here would be $\hat{\beta}_0 + \hat{\alpha}_1$ for item 1 and so on) using, for example, the `coefficients()` function:

```
coefficients(neut_2_mod_1)$item_pair
##     (Intercept) voicing
## 1           132      11
## 2            91      11
## 3           139      11
...
```

Here we see explicitly that the intercept varies by item (e.g., item 1's predicted intercept is 132 msec), while the `voicing` slope does not.

Box 8.1
Broader Context: Predicted Random Effects

It is not immediately clear what "predicted random effects" means, because the random effects (e.g., α_j) are not parameters estimated by the model (as opposed to the variances of the random effects, which are).

Because these are so often of interest for model interpretation, it is standard to define the *predicted random effects* to be the most likely values of the random effects when the fixed-effect coefficients are held at their estimated values (here, $\hat{\beta}_0$ and $\hat{\beta}_1$); these are called the BLUPs or "conditional modes" (e.g., Bates, Mächler, et al. 2015, section 5.1). These are part of fitted lme4 models, along with approximate SEs, which can be extracted with `ranef()`, or `tidy()` (from broom.mixed):

```
tidy(neut_2_mod_1, effects = "ran_vals")
##       effect     group level        term estimate std.error
## 1   ran_vals item_pair     1 (Intercept)      9.2       9.2
## 2   ran_vals item_pair     2 (Intercept)    -31.7       9.2
## 3   ran_vals item_pair     3 (Intercept)     15.6       9.2
...
```

The predicted random effects are used to make model predictions (e.g., using equation (8.3)) by other functions. For example, `fortify.merMod()` (from lme4) adds the model's prediction to the original dataframe (as column `.fitted`):

```
fortify.merMod(neut_2_mod_1)
```

And here `predict()` is used to get predicted values of `vowel_dur` for each item for both `voicing` values, shown in figure 8.1 (right):

```
## Define the prediction dataframe
pred_df <- neutralization %>% expand(voicing, item_pair)
## tidyr::expand: make dataframe with all combinations of variables
```

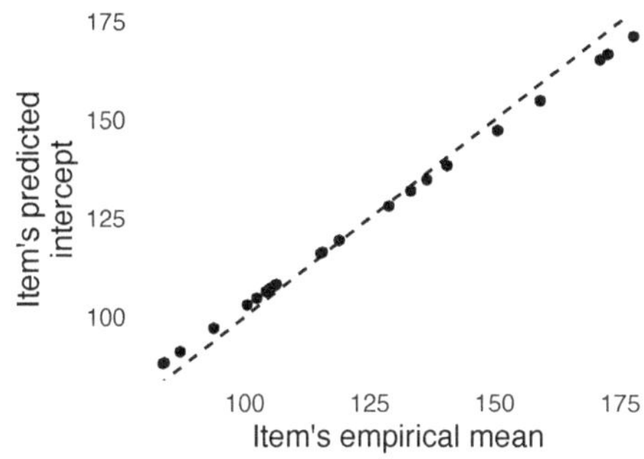

Figure 8.2
Example of shrinkage: empirical means of `vowel_dur` versus model predictions (from model `neut_2_mod_1`), for each item in the `neutralization_2` data.

```
# Get predicted values
pred_df$pred <- predict(neut_2_mod_1, newdata = pred_df)
```

Box 8.2
Broader Context: Shrinkage and Partial Pooling

To get a sense of what the random-intercept model is doing, we can compare the predicted intercept for each item ($\hat{\beta}_0 + \hat{\alpha}_j$ for item j) with the empirical mean for each item (by-item average `vowel_dur`), shown in figure 8.2.

Each item's predicted intercept is closer to the overall mean (of `vowel_dur`) than it is to the empirical values, a phenomenon called *shrinkage*, because the (absolute values of the) random intercepts "shrank" from the empirical means toward the population-level estimate (the fixed effect).

Shrinkage is an important property of all random effects, which improves estimates of the degree of variability (here, $\hat{\sigma}_I$) and thus generalization of the model to new data (here, items). Surprisingly, shrinkage also makes the BLUPs a better estimate of each item's (true) mean value—better than the empirical means, on average (Snijders and Bosker 2011, section 4.8).

The intuition behind shrinkage is that random effects are estimated by "partial pooling," which trades off between two extremes of how by-item variability could be accounted for in a vanilla regression model: "no pooling," where the model estimates the mean for each item separately (e.g., a 24-level factor for `item_pair`), and "full pooling," where grouping by item is ignored and only one average mean is estimated.

Each random effect can be thought of as a weighted average of the empirical effect for one group level and the population-level effect, with more weight given to the empirical effect for levels with more observations. (For example, in the `vot` data the by-word intercept for the word *and* will effectively be its empirical average.) This relates to two important features of mixed-effects models: imbalanced data are not a problem, because there is no assumption that levels have the same number of observations, and the model can predict for unseen levels (e.g., a new item), which is simply the extreme case of zero observations.

These concepts are discussed further in Gelman and Hill (2007, chapter 12).

8.3.5 Multiply Grouped Data

The preceding mixed model considered one kind of variability, by-item variability in the average of `vowel_dur`. Typically in linguistic data there is more than one grouping factor—some common cases are laboratory data where both participants and items are sampled from

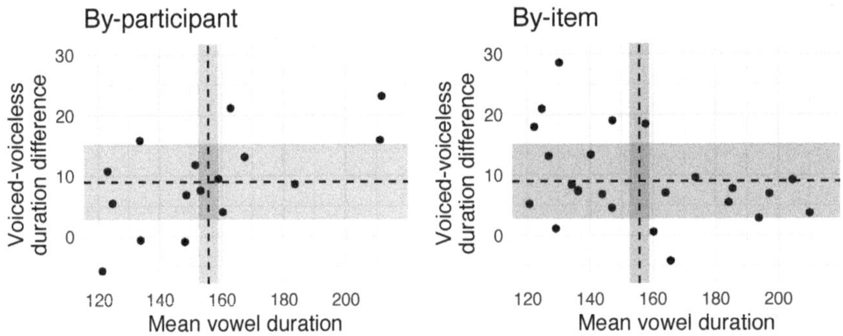

Figure 8.3
Points are the empirical estimates of the intercept (mean `vowel_dur`) and the slope (mean VOICED/VOICELESS difference in `vowel_dur`), for each item and particiant in the `neutralization` data, calculated from voiced/voieless pairs for each item and participant. To help assess the degree of variability, dotted lines and shading show the estimated intercept and `voicing` slope, with 95% CIs, from a simple linear regression (i.e., not taking item/participant variability into account).

a larger population and corpus data where both authors/speakers and the linguistic unit of interest (e.g., words) are sampled from larger populations (e.g., the `vot` data).

Consider the full `neutralization` dataset, a typical laboratory dataset grouped by both participant and item, where the `voicing` effect is of interest. Figure 8.3 shows the empirical estimates of the intercept (mean `vowel_dur`) and the slope (VOICED/VOICELESS difference in `vowel_dur`), for each item and each participant. For example, the lowest-left point in the left panel corresponds to a participant with average vowel duration of ~120 msec and with a VOICED/VOICELESS difference of ~–10 msec (in the opposite direction to other participants). Participants and items show substantial variability in intercept and slope: the `voicing` effect of interest varies from ~0 to 30 msec, which turns out to be a large effect for this case (section 3.2.3.2).

Most linguistic data show such clear variability by using two or more grouping factors. In mixed-effects models it is possible to account for both kinds of variability simultaneously, by including by-participant and by-item random effects. This is a case of *crossed* random effects: multiple grouping factors that vary independently. (The alternative is *nested* random effects, discussed in section 10.2.) Crossed random-effect structure is necessary to model even simple linguistic experiments, so the facility of fitting crossed random effects in lme4 makes it well-suited for modeling linguistic data (box 8.3).

Box 8.3
Broader Context: Repeated Measures Data in Language Sciences and Beyond

The importance of accounting for both kinds of variability goes back to at least Clark (1973), who famously termed not accounting for by-item variability the "language-as-fixed-effect fallacy." Following Clark, the standard in some subfields (e.g., psycholinguistics, phonetics) has been repeated measures ANOVA analyses where separate by-participant and by-item models are fit and then interpreted together (e.g., Meteyard and Davies 2020, 3–7; Baayen 2008, section 7.2; Johnson 2008, chapter 4; Gries 2013, chapter 5). Although this method is not wrong, it has issues including inflexibility to more complex designs and lowered statistical power.

> Mixed-effects models were introduced for this case for psycholinguistics using lme4 in Baayen, Davidson, and Bates (2008), which was an extremely influential paper across language sciences. (Earlier papers, such as Quené and Van den Bergh [2004] and Brysbaert [2007] had introduced mixed-effects models without lme4.) Mixed models had been long used before this in other fields with grouped data, such as education and ecology, and as McElreath (2020, 400) notes, "multilevel regression deserves to be the default approach [across fields]. There are certainly contexts in which it would be better to use an old-fashioned single-level model. But the contexts in which multilevel models are superior are much more numerous."

8.3.6 Two Grouping Factors, Random Intercepts

We will begin with a model with by-item and by-participant random intercepts, for the `neutralization` case. These data have $J = 16$ participants and $K = 24$ items; the data are fully crossed (every participant sees both `voicing` values from every item), with $n = 749$.[4] Let $j[i]$ and $k[i]$ now denote the participant and item for the ith observation. The simplest regression model for this case, with a single fixed-effect term (x: `voicing`), would be

$$
\begin{aligned}
y_i &= \beta_0 + \alpha_{P,j[i]} + \alpha_{I,k[i]} + \beta_1 x_i + \varepsilon_i, \\
\varepsilon_i &\sim N(0, \sigma_E^2), \quad \text{for } i = 1, \ldots, 749, \\
\alpha_{P,j} &\sim N(0, \sigma_{P,0}^2), \quad \text{for } j = 1, \ldots, 16, \\
\alpha_{I,k} &\sim N(0, \sigma_{I,0}^2), \quad \text{for } k = 1, \ldots, 24.
\end{aligned} \tag{8.3}
$$

The first line looks similar to the previous model (equation (8.2)), except there are now two random-effect terms: $\alpha_{I,k[i]}$ is the offset of item $k[i]$ from the population-level intercept (β_0), whereas $\alpha_{P,j[i]}$ is the offset of participant $j[i]$. Participants' offsets (or "by-participant random intercepts") and items' offsets are each assumed to be normally distributed, with $\sigma_{P,0}$ and $\sigma_{I,0}$ capturing the degree of variability in participants' and items' intercepts, respectively. (The 0 subscript is to distinguish these from random slopes; see section 8.4.)

This model predicts a different line for each participant/item pair, with differing intercepts but the same slope, as in figure 8.4. For example, the relationship predicted for participant 5 for item 10 would be $y = \beta_0 + \alpha_{P,5} + \alpha_{I,10} + \beta_1 x$. The population-level relationship is still $y = \beta_0 + \beta_1 x$, but now β_0 is the intercept for an "average participant and item."

To fit and summarize this model, using `lmer()`:

```
neut_mod_0 <- lmer(vowel_dur ~ voicing + (1 | subject) +
    (1 | item_pair), data = neutralization)
summary(neut_mod_0)
...
## Random effects:
##  Groups    Name        Variance Std.Dev.
##  item_pair (Intercept) 773      27.8
##  subject   (Intercept) 765      27.7
```

4. Note that this is a bit less than $2 \cdot 16 \cdot 24 = 768$ due to some missing data.

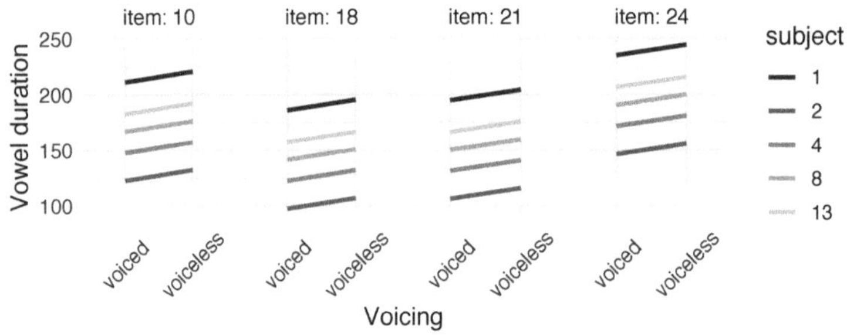

Figure 8.4
Predicted `voicing` effects for a subset of participants and items from model `neut_mod_0`, where intercepts vary by participant and item.

```
## Residual                  412       20.3
## Number of obs: 749, groups:  item_pair, 24; subject, 16
##
## Fixed effects:
##              Estimate Std. Error t value
## (Intercept)   155.74       8.97   17.36
## voicing         9.27       1.48    6.25
##
...
```

This model suggests that vowels before VOICED stops are significantly longer than before VOICELESS ($|t| > 2$).

Recall that the effect of stop voicing on vowel duration is expected to be small—the research questions are whether an effect exists at all, and if so, its exact size. The answer to either could be affected by not accounting for other factors affecting vowel duration.

To get closer to a realistic model for these data, let's refit the model with some control predictors as fixed effects: `prosodic_boundary`, `place`, and `vowel` (factors with two, three, and five levels)—and include these controls in subsequent models. This model, which is similar to the original model of these data reported by Roettger et al. (2014, experiment 1), corresponds to adding additional fixed-effect terms to the model in equation (8.3) (e.g., $\beta_2 x_{2,i}$ for `prosodic_boundary` for observation i).[5]

This model is fitted using `lmer()` as follows:

```
neut_mod_1 <- lmer(vowel_dur ~ voicing + prosodic_boundary + place +
    vowel + (1 | subject) + (1 | item_pair), data = neutralization)
summary(neut_mod_1)
...
## Random effects:
##  Groups    Name        Variance Std.Dev.
##  item_pair (Intercept) 71.9      8.48
```

5. These controls account for the presence of a prosodic boundary and the identity of the vowel and its following stop, all of which strongly affect the vowel's duration.

```
## subject (Intercept) 669.3 25.87
## Residual 402.5 20.06
## Number of obs: 749, groups: item_pair, 24; subject, 16
##
## Fixed effects:
## Estimate Std. Error t value
## (Intercept) 157.175 6.741 23.32
## voicing 9.599 1.469 6.53
## prosodic_boundary 12.392 2.612 4.74
## place1 -9.373 2.428 -3.86
## place2 1.604 1.321 1.21
## vowel1 -0.288 3.125 -0.09
## vowel2 -18.239 1.722 -10.59
## vowel3 -3.724 1.160 -3.21
## vowel4 -6.757 0.897 -7.53
##
...
```

The `voicing` effect remains significantly positive ($|t|>2$) and is in fact "clearer" once the control predictors are included: the fixed-effect coefficient estimate is further from 0 and has a higher t-value. The control predictors also all have significant effects, but these are not of interest here (exercise 8.1).

Turning to the random effects, note that the estimated degree of by-item variability ($\hat{\sigma}_I = 8.5$) is lower than the degree of by-participant variability ($\hat{\sigma}_P = 26$). This "participants vary more than items" pattern is common for data from laboratory experiments, because items are chosen from a constrained set by the researcher (here, words of a particular phonotactic shape) and participants are chosen from a less constrained set (e.g., all undergraduates).

8.3.7 Types of Predictors

Mixed models are also called "hierarchical" or "multilevel" models because they predict relationships at different *levels*: individual observations, individual items, or individual participants—that is, any value of a grouping factor.

A predictor that describes a property of participants (e.g., gender) is *participant-level*; a predictor that describes a property of items (e.g., word frequency) is *item-level*; and so on. Participant-level predictors are sometimes called "between-participant," because they do not vary within participants, whereas a predictor that varies within participants is "within-participant" (and similarly for "between-item," etc.). A predictor that can vary within each grouping factor—for example, varies within both items and participants—is called *observation-level*.

For the `neutralization` data, every item (`item_pair`) corresponds to a pair of pseudowords ending in *VCe* and *VC*$_2$*e* (vowel-consonant-"e"), which are identical except for the voicing of the consonant. For example, item 24 is 'stauge' and 'stauke' (V = /au/, C = /g/ or /k/). Thus, `vowel` (the identity of V) and `place` (the place of articulation of C) are item-level: they are properties of the item. All other predictors of interest are observation-level: `voicing` varies within item and within participant, but `prosodic_boundary` relates to how this particular observation was produced. (Speaker `gender` is a participant-level predictor in the dataset but it is not included in our analysis.) As this example illustrates, working out the levels of predictors requires a good understanding of the structure of the data.

8.3.7.1 Example: Changes in variance components

It is often useful for model fitting and interpretation to think of predictors in terms of their level. This helps us to understand what random-effect terms mean and why variance components differ between models of the same data.

First, consider what a random-effect term like “by-item random intercept” means. For example, earlier we described e.g., α_j as the offset for item j from the mean, but this is not quite right—it is the offset *after* controlling for item-level predictors. For example, in model `neut_mod_1`, which contains item-level predictors `place` and `vowel`, the by-item random intercept for the ‘stauCe’ item is how much its `vowel_dur` differs from what is expected given that `place=VELAR` and `vowel=AU`.

This explains the main difference between models `neut_mod_0` and `neut_mod_1` that resulted from adding the control predictors: the estimated degree of by-item variability ($\hat{\sigma}_I$) drops from 28 to 8.5, whereas the estimated by-participant variability and the residual error change little. This is because two item-level predictors were added to the model, so the meaning of $\hat{\sigma}_I$ changed from “degree of variability among word pairs” to “degree of variability among word pairs with the same `place` and `vowel` values.” In other words, some variance that looked like “random” variation among items turned out to be attributable to properties of the items.

Finally, consider the residual error, which quantifies “how much variance is left over.” The residual error is similar for the two models fitted to the full `neutralization` dataset (`neut_mod_0`: 20, `neut_mod_1`: 20). This is because the models differ in the addition of control predictors, which did not change how much variance in the data was “explained,” but what source it is ascribed to: all random versus some random and some item-level predictors.

In contrast, consider the two models fitted for participant 2’s data from section 8.3. The value for the mixed model (`neut_2_mod_1`), $\hat{\sigma}_E = 14$, is much lower than the value for the simple linear regression:

```
glance(neut_2_mod_0)$sigma
## [1] 30
```

The residual error is much smaller in the random-intercept model because the “leftover” variance of the simple linear regression model has been partitioned into item-level (the α_j) and observation-level (σ_E) variance. Intuitively, some error has been given to the participant level.

8.4 Random Slopes

Models discussed thus far allowed the intercept to differ between groups—each participant or item’s “baseline” is different. In addition, the slope of an effect could differ by group, which is captured in mixed models by *random slope* terms. For the `neutralization` case, the empirical data suggest the effect of `voicing` might vary between both participants and items (figure 8.3).

8.4.1 Introduction: One Random Slope

To introduce random slopes, let's start with a simple example: adding a single random slope term to the random-intercepts model without controls (`neut_mod_0`) to account for by-participant variation: a "by-participant random slope of `voicing`." The regression model in this case is

$$\begin{aligned} y_i &= \beta_0 + \alpha_{P,j[i]} + \alpha_{I,k[i]} + (\beta_1 + \gamma_{P,j[i]})x_i + \varepsilon_i, \\ \varepsilon_i &\sim N(0, \sigma_E^2), \quad \text{for } i = 1, \ldots, 749, \\ \alpha_{P,j} &\sim N(0, \sigma_{P,0}^2), \quad \text{for } j = 1, \ldots, 16, \\ \gamma_{P,j} &\sim N(0, \sigma_{P,1}^2), \\ \alpha_{I,k} &\sim N(0, \sigma_{I,0}^2), \quad \text{for } k = 1, \ldots, 24. \end{aligned} \tag{8.4}$$

This is similar to equation (8.3), but with an extra random slope term: $\gamma_{P,j}$ captures how much each participant's slope (for `voicing`) differs from the average slope across participants (β_1), and these by-participant offsets are normally distributed (variance $\sigma_{P,1}^2$). All other terms have the same interpretations as before (equation (8.3)).

The model predicts a different line for each participant (averaging across items), differing in both intercept and slope. For example, for participant 5:

$$y = \beta_0 + \alpha_{P,5} + (\beta_1 + \gamma_{P,5})x$$

.

These by-participant predictions (code in section 8.8.1.2) are shown in figure 8.5, which are from the fitted model:

```
neut_mod_2 <- lmer(vowel_dur ~ voicing + (1 + voicing || subject) +
  (1 | item_pair), data = neutralization)
```

In this formula, `(1 + voicing || subject)`, read as "uncorrelated by-subject random intercept and by-subject random slope," is lme4 notation for the following:

- Random intercept (`1` means "intercept")
- Random slope of `voicing` (random effects go to the left of the pipe symbol, `|`)
- Grouped by participant (grouping factor goes to the right of the pipe)
- The random intercept and slope are *uncorrelated*—indicated by the double pipe, `||`

We'll discuss the last point further when correlated random effects are introduced (section 8.7); until then, we'll assume uncorrelated random effects in all models.

The model's parameter estimates are

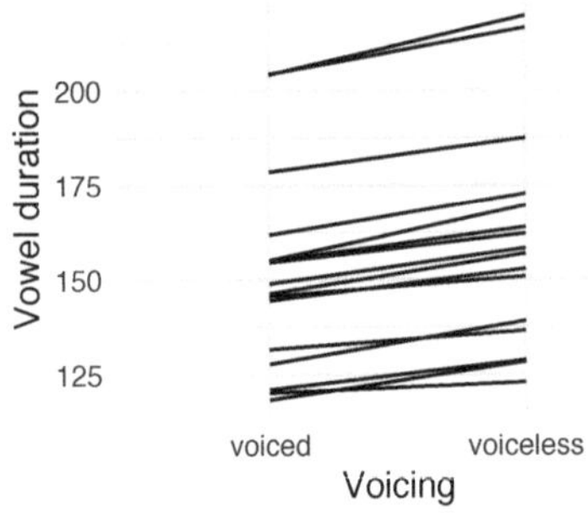

Figure 8.5
Predicted `voicing` effect for each participant from model `neut_mod_2` (averaging across items).

```
summary(neut_mod_2)
...
## Random effects:
##  Groups     Name        Variance Std.Dev.
##  item_pair (Intercept) 773      27.8
##  subject    voicing      26       5.1
##  subject.1 (Intercept) 766      27.7
##  Residual               406      20.1
## Number of obs: 749, groups:  item_pair, 24; subject, 16
##
## Fixed effects:
##             Estimate Std. Error t value
## (Intercept)   155.74       8.98   17.35
## voicing         9.28       1.95    4.76
...
```

Most of the rows here have the same interpretation as in the random intercepts-only model (`neut_mod_1`). What is new is the random slope variance component, which is estimated to be $\hat{\sigma}_{P,1} = 5.1$. This is the degree of by-participant variability in the `voicing` effect.

The degree of variability in the slope of x (the random slope term) is only meaningful relative to the size of the population-level slope of x (the fixed-effect term): if $\sigma_{\hat{P},1} = 0.2 \cdot \hat{\beta}_1$, then participants are predicted to have slopes that differ in magnitude but not sign (because 95% of participants lie in $\hat{\beta}_1 \pm 1.96 \cdot \hat{\sigma}_{P,1}$), whereas if $\hat{\sigma}_{P,1} = 1.5 \cdot \hat{\beta}_1$, participant slopes are predicted to differ in both magnitude and sign, and so on.

For this reason, if reporting a random-effect variance it is useful to summarize the distribution with (approximately) a 95% or 99% interval, like "95% of participants are predicted to have slopes between –0.72 and 19" ($= \hat{\beta}_1 \pm 2 \cdot \hat{\sigma}_{P,1}$)—which is the prediction for `neut_mod_2`.

At the end of section 8.4, after we build a more realistic model for these data, we'll return to discussing how to interpret "the `voicing` effect" in a mixed model (i.e., the presence of both a fixed effect and random slopes).

Notation: Uncorrelated random effects In lme4, uncorrelated random effects can be written in several ways in the model formula:

1. `(1 + speaking_rate_dev || speaker)`
2. `(1 | speaker) + ( 0 + speaking_rate_dev | speaker)`
3. `(1 | speaker) + (-1 + speaking_rate_dev | speaker)`

Expression (1) is useful shorthand when writing model formulas, but under the hood lme4 formulas represent uncorrelated random slopes using (2). This is why you will see this model formula in the full `summary(neut_mod2)` output:

```
formula(neut_mod_2)
## vowel_dur ~ voicing + ((1 | subject) + (0 + voicing | subject)) +
##     (1 | item_pair)
```

Expressions (3) and (2) are equivalent ways to exclude the intercept from R model formulas (box 7.7). When referring to just an uncorrelated random slope, for example, in a hypothesis test, you use (2) or (3).

8.4.2 Multiple Random Slopes

We now build a more realistic model, accounting for both by-participant and by-item variation in the effect of `voicing` (figure 8.3), by adding by-participant and by-item random slopes to the random-intercepts model *with* control predictors (`neut_mod_1`). The first line of the regression model in this case looks similar to equation (8.4):

$$y_i = \underbrace{\beta_0 + \alpha_{P,j[i]} + \alpha_{I,k[i]}}_{\text{intercept}} + \underbrace{(\beta_1 + \gamma_{P,j[i]} + \gamma_{I,k[i]})}_{\texttt{voicing}\text{ slope}} x_{1,i} + \text{(control predictors)} + \varepsilon_i. \tag{8.5}$$

Here $x_{1,i}$ is the value of `voicing` for observation i, and we have omitted the control predictors x_2–x_8, which would correspond to β_2–β_8, for clarity. In this model, both the intercept and the `voicing` slope now differ by participant and by item, and each kind of offset is normally distributed (variances $\hat{\sigma}^2_{P,0}$, $\hat{\sigma}^2_{I,0}$, $\hat{\sigma}^2_{P,1}$, $\hat{\sigma}^2_{I,1}$). To fit this model in R:

```
neut_mod_3 <- lmer(vowel_dur ~ voicing + prosodic_boundary + place +
  vowel + (1 + voicing || subject) + (1 + voicing || item_pair),
data = neutralization)

summary(neut_mod_3)
...
## Random effects:
##  Groups      Name        Variance Std.Dev.
##  item_pair   voicing       2.13    1.46
##  item_pair.1 (Intercept)  72.15    8.49
##  subject     voicing      19.69    4.44
##  subject.1   (Intercept) 671.48   25.91
##  Residual                397.04   19.93
## Number of obs: 749, groups:  item_pair, 24; subject, 16
##
## Fixed effects:
##                    Estimate Std. Error t value
```

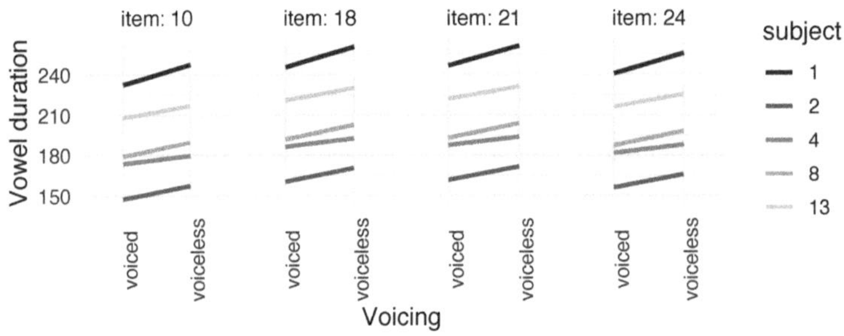

Figure 8.6
Predicted `voicing` effects for a subset of participants and items from model `neut_mod_3`, where both intercepts and slopes vary by participant and items.

```
## (Intercept)          157.176      6.751   23.28
## voicing                9.606      1.857    5.17
## prosodic_boundary     12.203      2.605    4.68
## place1                -9.352      2.429   -3.85
## place2                 1.608      1.322    1.22
## vowel1                -0.294      3.127   -0.09
...
```

As shown in figure 8.6 (prediction/plot code hidden), the model predicts a different relationship (intercept and slope) for each participant/item pair.

Focusing on the `voicing` slope (of primary interest) in particular, the model makes predictions at several levels. At the population level, vowel duration is greater for VOICED than VOICELESS stops, for an average participant and item (`voicing` fixed effect: $\hat{\beta}_1 = 9.6$, $|t| > 2$).

At the level of individual participants and items, the model predicts the `voicing` slope (= VOICED-VOICELESS difference) for participant j (averaging over items) or item k (averaging over participants).[6] For example, the `voicing` slopes for the first two participants and items are

```
tidy(neut_mod_3, effects = "ran_coefs") %>%
  filter(level %in% c("1", "2"), term == "voicing")
##       effect     group level    term estimate
## 1 ran_coefs item_pair     1 voicing      9.6
## 2 ran_coefs item_pair     2 voicing      9.8
## 3 ran_coefs   subject     1 voicing     15.0
## 4 ran_coefs   subject     2 voicing     10.1
```

The parameters $\hat{\sigma}_{P,1} = 26$ and $\hat{\sigma}_{I,1} = 8.5$ describe the degree of variability in these `voicing` slopes across participants and items.

6. These would be $\hat{\beta}_1 + \gamma_{P,j}$ for participant j and $\hat{\beta}_1 + (\text{control predictors}) + \gamma_{I,k}$—where the control predictors would account for the `place` and `vowel` of item k (both item-level predictors). In practice you can just make predictions with, e.g., `predict` without worrying about how they are calculated.

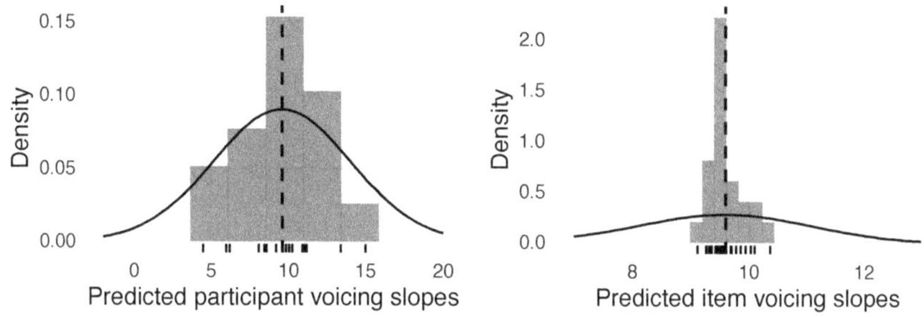

Figure 8.7
By-participant and by-item variation in the `voicing` effect for `neut_mod_3`. Distribution of model-predicted `voicing` coefficient estimates (histograms, rug plots) for each participant (*left*) and each item (*right*) with overlaid estimates for `voicing` population-level effect (dotted lines) and predicted distribution of `voicing` effects across participants/items (solid lines).

To help get an intuition for these parameters, figure 8.7 shows the predicted distributions of `voicing` slopes predicted by the model across participants and items in the population (solid lines): normal distributions centered at $\hat{\beta}_1 = 9.6$, with widths $\hat{\sigma}_{P,1} = 4.4$ and $\hat{\sigma}_{I,1} = 1.5$. These are superimposed on the predicted slopes for observed participants/items (histograms). Every row in a mixed model's "random effects" table (i.e., a variance component estimate) can be thought of as the width of such a normal distribution, describing how much some predictor's effect (or the intercept) is predicted to vary in the population.

In this case, the `voicing` slope is predicted to vary substantially in magnitude across participants, but not direction, and to vary only slightly across items (exercise 8.2).

The overall picture is of a robust population-level `voicing` effect of magnitude of ~ 9 msec (95% CI = [6, 13]). Individual participants, and to a smaller extent items, vary in the magnitude but not direction of the effect.

8.4.3 What Does Adding a Random Slope Do?

Random slopes are crucial to using mixed models in practice, because they affect estimates of fixed effects (which are often of most interest), and determining which slopes to include is typically the hardest part of model selection. We'll discuss the first issue here and will return to model selection in section 10.5.

Consider models `neut_mod_0` and `neut_mod_2` fitted for the `neutralization` data. The models each have a single fixed effect (for `voicing`) and differ only in the by-participant random effects:

- Model `neut_mod_0`: by-participant random intercept (only)
- Model `neut_mod_2`: by-participant random intercept, random slope of `voicing`

`neut_mod_0` models `vowel_dur` as a linear function of `voicing`, with the *same slope* for every participant. Thus, every participant's data are contributing to estimating one number (related to `voicing`): β_1, the population-level effect.

`neut_mod_1` models `vowel_dur` as a linear function of `voicing`, with a *different slope* for every participant. Thus, every participant's data are contributing to estimating two numbers: β_1 and the offset of that participant's slope from β_1. Intuitively, this results in the model being less certain about the population-level effect (β_1).

We can see the result by comparing the `voicing` fixed-effect rows in the two models (code hidden):

```
##           model estimate std.error   t
## 1     Intercept      9.6       1.5 6.5
## 2 Int. + slope       9.3       1.9 4.8
```

Both models show similar estimated population-level effects ($\hat{\beta}_1$) for `voicing`. But the intercept + slope model has a larger SE for this effect, resulting in a smaller t value. It makes sense that the SE of the `voicing` fixed effect goes up when a random slope is added: the model detects significant variability among participants in the effect of `voicing`, which makes it less sure of the average effect of `voicing`. Typically *adding a random slope increases the SE* of the corresponding fixed effect.

Adding a random slope could also change the estimated population-level effect, especially if the data are highly imbalanced by level of the grouping factor (as in corpus data) or some levels have outlier effects, neither of which is the case for the `neutralization` data.

8.4.3.1 Example: Speaking rate for voiced stops for `vot` data

To see the practical relevance of all this, let's turn to an example using the `vot` data where whether the random slope is included changes the qualitative conclusion from the model. In addition this example uses corpus data, which is an important use case for mixed-effects models. Different issues tend to come up when fitting mixed-effects models to data from corpora or laboratory experiments, so we will use examples of both going forward (here, `vot` and `neutralization`).

Sonderegger, Bane, and Graff (2017) model `vot` in this dataset for all speakers and words, as a function of many predictors, separately for voiceless and voiced stops. To keep things relatively simple we will simplify the data a bit, to a subset `vot_voiced_core`: voiced stops, restricted to just 11 *speakers* with the most data and to *words* that are one syllable long (so we don't have to consider variables like syllable `stress`):

```
core_speakers <- c(
  "dale", "darnell", "lisa", "luke", "michael",
  "mohamed", "rachel", "rebecca", "rex", "sara", "stuart"
)

vot_voiced_core <- vot %>%
  filter(
    speaker %in% core_speakers & syll_length == 1 & voicing == "voiced"
  ) %>%
  droplevels()
```

Our research question will be: Is there a (population-level) speaking rate effect (predictor `speaking_rate_dev`) for voiced stops? This is of interest for theories of the phonetics-phonology interface, and the "real" answer is actually unclear: the true effect size may be either 0 or negative (but small). Consider the following two models of VOT: both contain a reasonable set of predictors for this case (fixed effects) and differ in random effects—intercepts only versus intercepts plus a random slope of the predictor of interest (`speaking_rate_dev`):[7]

```
vot_mod_1 <- lmer(log_vot ~ speaking_rate_dev + foll_high_vowel +
  cons_cluster + log_corpus_freq + place + gender +
    (1 | word) + (1 | speaker), data = vot_voiced_core)

vot_mod_2 <- lmer(log_vot ~ speaking_rate_dev + foll_high_vowel +
    cons_cluster + log_corpus_freq + place + gender +
    (1 | word) + (1 + speaking_rate_dev || speaker),
  data = vot_voiced_core)
```

The output for the model with the random slope is

```
summary(vot_mod_2)
...
## Random effects:
##  Groups    Name              Variance Std.Dev.
##  word      (Intercept)       0.01761  0.1327
##  speaker   (Intercept)       0.04630  0.2152
##  speaker.1 speaking_rate_dev 0.00304  0.0552
##  Residual                    0.32090  0.5665
## Number of obs: 7175, groups:  word, 272; speaker, 11
##
## Fixed effects:
##                   Estimate Std. Error t value
## (Intercept)         2.6199     0.0684   38.29
## speaking_rate_dev  -0.0402     0.0216   -1.86
## foll_high_vowel     0.0873     0.0372    2.34
## cons_cluster        0.4646     0.0416   11.16
## log_corpus_freq     0.0421     0.0396    1.06
## place1              0.3237     0.0198   16.38
## place2              0.1282     0.0123   10.39
## gender             -0.3316     0.1358   -2.44
##
...
```

Compare the speaking rate effect in this model (row `speaking_rate_dev`) and in the intercepts-only model:

7. In this chapter "speaking rate" usually means column `speaking_rate_dev`, which is speaking rate relative to the speaker's mean rate, rather than column `speaking_rate`, which is confounded with the speaker's mean rate (see section 8.10.2.3). To keep things simple, the fixed effects in models `vot_mod_1` and `vot_mod_2` are a subset of those in the full model of these data reported in Sonderegger, Bane, and Graff (2017, table A1).

```
filter(tidy(vot_mod_1), term == "speaking_rate_dev") %>%
  select(-group)
##    effect                term estimate std.error statistic
## 1  fixed speaking_rate_dev   -0.039     0.013      -2.9
```

Again, the fixed-effect estimate is similar in each model (~ -0.04), but its SE is higher when a random slope is added, leading to a lower $|t|$ value (–1.9 vs. –2.9). By the $|t| > 2$ heuristic we would conclude there is a significant speaking rate effect for the intercepts-only model but not for the intercepts + slope model. It is not clear what the real `speaking_rate_dev` effect is for this case; the two possibilities illustrate the two possible outcomes of adding a random slope.

8.4.3.2 Type I/II error trade-off

If the true effect is null, then the effect detected in `vot_mod_1` is spurious; this is a *type I error*. From the perspective of avoiding type I errors, including random slopes is a good thing. In general, participants (and items) will differ in the effect of a given predictor x. Thus, if a by-participant (or item) random slope for x is not included, we are underestimating the uncertainty in the fixed effect for x and, therefore, can easily falsely conclude there is a significant effect—for example, based on a subset of participants who show large effects. In general, not including a possible random slope for x is *anticonservative* (for evaluating whether the fixed-effect coefficient of x is 0).

On the other hand, if the true effect is not null, then `vot_mod_2` fails to detect it; this is a *type II error*. In other words, adding a random slope is *conservative*: it tends to lower power to detect a true (fixed) effect of x. This is especially true if there is little by-participant (or item, etc.) variability in the effect of X.

8.4.4 Random-Effects Model Selection: Basics

So adding a random slope term can be good (fewer type I errors) or bad (lower power). There are two corresponding philosophies for deciding on random-effect structure, which in practice usually means what random slope terms (and correlations; see section 8.7) to include in a model:

- "Maximal": add as many random slopes as possible (Barr et al. 2013).
- "Data-driven": add random slopes that significantly improve the model (e.g., LRT with $\alpha = 0.2$; see Matuschek et al. [2017]).[8]

We will return to this issue in section 10.5 when we'll discuss model selection, where choosing random-effect structure is the biggest issue in practice.

Until then, we'll fit models assuming the random-effect structure is known. There are three important guidelines on random-effect structure that are necessary to fit models that make sense at all (which is a lower bar than choosing the "best" random-effect structure, in model selection):

8. Matuschek et al. (2017) discuss why α should be higher than 0.05, essentially to err toward more complex random effect structure and give some motivation for $\alpha = 0.2$ (though the exact value is arbitrary).

1. If the data are grouped by z (e.g., participants, items), a by-z random intercept term should be included.
2. All possible random slope terms should be considered for effects of theoretical interest—either by including them in the model (maximal perspective) or testing whether they should be added (data-driven perspective). Considering random slopes for predictors not directly related to research questions (controls) is less important.
3. If a random slope for an interaction is included, the corresponding random slopes for all subsets of the interaction should be included—for the same reason that you must include all subsets of an interaction as fixed effects (box 4.7). (For example, a by-participant `x1:x2` random slope would mean that by-participant `x1` and `x2` random slopes must be included.)

This requires clarifying two issues: what a possible random slope is and how to test a random-effect term's contribution to a model, which is addressed in section 8.5.2.

8.4.5 Possible and Impossible Random Slopes

Which random slope terms are possible for a predictor x depends on its level (section 8.3.7). A by-speaker random slope of x models speaker variability in the effect of x. Thus, this term only makes sense if x *can* vary within speakers—it is not speaker-level. More generally, by-z random slopes are only possible for predictors that are not z-level. (See Brauer and Curtin 2018, 13–15, and Barr et al. 2013 for more details.)

For example, consider the speaker `gender` predictor in model `vot_mod_2`, which estimates the difference between male and female speakers. The effect of this predictor cannot vary within speakers: each speaker has only one value of gender, so it doesn't make sense to estimate "difference between MALE and FEMALE for speaker 3." Thus, there can be no by-speaker random slope term for `gender`. More generally, there can be no by-speaker random slopes for participant-level predictors. By similar logic, there can be no by-item random slopes for item-level predictors, and so on.

In model `vot_mod_2`, `speaking_rate_dev` is observation-level, `gender` is speaker-level, and all other predictors are word-level. Thus the possible random slopes are

- By-speaker: all predictors except `gender`
- By-word: `speaking_rate_dev`, `gender`

Thus, the maximal model in this case, assuming uncorrelated random effects, would have the following random-effect structure:

```
(1 + speaking_rate_dev + foll_high_vowel + cons_cluster +
    log_corpus_freq + place || speaker) +
  (1 + speaking_rate_dev + gender || word)
```

We instead consider a much simpler model, following guidelines (1)–(3). By the first guideline, we should consider by-speaker and by-word random slopes of `speaking_rate_dev`, the predictor of theoretical interest. The by-word random slope turns out to not significantly contribute to `vot_mod_2` (exercise 8.5). So we follow the data-driven approach and do not include this term, proceeding instead with vot_mod_2, which contains only the single (by-speaker, voicing) random slope term.

8.4.6 Random Slopes for Multiple Predictors

In this chapter we mostly consider models with just one random slope, for one predictor, to keep things simple. But models of real data typically contain multiple random slopes, for multiple predictors, so it is worth having an example in mind to think through how methods we cover apply to this case.

Consider the same model as vot_mod_2, but with all possible random slopes added (uncorrelated) for terms whose fixed effects are near the $|t| = 2$ cutoff:

```
vot_mod_2_more_slopes <-
  lmer(log_vot ~ speaking_rate_dev + foll_high_vowel + cons_cluster +
      log_corpus_freq + place + gender +
      (1 + gender + speaking_rate_dev || word) +
      (1 + speaking_rate_dev + foll_high_vowel || speaker),
    data = vot_voiced_core)
```

This model's structure is somewhat arbitrary, to work with tools we have so far, but it could make sense as a candidate model if a research question was "which factors significantly affect VOT?"

Note the notation used for multiple random slope terms: (1 + A + B || speaker) means "by-speaker random intercept, random slope of A, and random slope of B, all uncorrelated."

This model assumes six kinds of variability, three each for speakers and words:

- Speakers differ in overall VOT and in the effects of speaking_rate_dev and foll_high_vowel.
- Words differ in overall VOT and in the effects of speaking_rate_dev and speaker gender.

The new model's "random effects" block contains a row for each:

```
summary(vot_mod_2_more_slopes)
...
##
## Random effects:
##  Groups    Name              Variance Std.Dev.
##  word      (Intercept)       0.01354  0.1164
##  word.1    gender            0.07426  0.2725
##  word.2    speaking_rate_dev 0.00116  0.0341
##  speaker   (Intercept)       0.04745  0.2178
##  speaker.1 speaking_rate_dev 0.00270  0.0520
##  speaker.2 foll_high_vowel   0.00352  0.0593
##  Residual                    0.30724  0.5543
...
```

Comparing the output of this model to vot_mod_2 (not shown—exercise 8.3):

```
tidy(vot_mod_2)
tidy(vot_mod_2_more_slopes)
```

We see that the model with multiple random slopes predicts substantial by-speaker variability in the effect of a following high vowel on VOT and by-word variability in the effect of speaker gender. Accordingly, the SEs of both fixed-effect terms (`foll_high_vowel`, `gender`) increase.

8.5 Hypothesis Tests

It is of interest for mixed models to obtain *p*-values corresponding to hypothesis tests that each fixed-effect and random-effect term is different from 0. However these are not shown in the standard output of an `lmer()` model (see any `summary()` or the previous `tidy()` output). Under `Random effects` there are no test statistics or *p*-values, and under `Fixed effects` there are test statistics but no *p*-values.

This is because how to calculate *p*-values for the contribution of a fixed- or random-effect term to a model is somewhat contentious—especially for fixed-effect terms.[9]

The upshot is that for fixed or random effects, there are several methods to calculate *p*-values that trade off between accuracy and computation time. It is good to be aware of these options and the issues in calculating *p*-values, because different methods come up in practice and computing them illustrates more generally useful tools. If you would like quick advice without going into details, you could skip to section 8.5.1.3 and just read about the "approximate *df*" methods, which are a reasonable default (following Luke 2017; Meteyard and Davies 2020).

It is important practical background for hypothesis testing that models with different fixed effects must be fitted using ML for most model comparison methods to make sense (e.g., LRT, AIC, BIC). In all other cases, REML fits are better (box 8.4).

Box 8.4
Practical Note: REML versus ML Fits

When using various functions to fit or compare (generalized) linear mixed models ([G]LMMs) in R you may see messages related to refitting using ML. This is because typical model comparison methods (LRTs, AIC, BIC) do not make theoretical sense for comparison of two LMMs *with different fixed effects* that were fitted using REML, because their likelihood functions are not comparable (e.g., Faraway 2016, section 10.2; Wood 2017, section 2.4.5). Because this is the default, the models must be first refit using ML (option `REML=FALSE`). This is done automatically when you compare two `lmer()` models with an LRT (`anova()`). Otherwise, for example when comparing non-nested models using AIC, you must refit the models yourself.

Note that **any** model comparison of LMMs using `anova()` leads to the models being refit using ML by default, even in cases where REML fits would make more sense, like models differing only in random-effect structure. (I think this is just because differing fixed effects is the most common case.) In such cases you force a REML fit using `refit=FALSE` (see section 8.5.2).

9. Essentially, if mixed-effects models are thought of from a frequentist perspective it's unclear what *df* should be for *t*/*F* tests and from a Bayesian perspective *p*-values are not a meaningful concept. For details see lme4 documentation (`?pvalues`), Bolker (2021, "Testing hypotheses"), and elsewhere online (google "lme4 p-values").

> Although comparing `REML` fits when `ML` fits should be used doesn't make theoretical sense, in practice this usually just gives anticonservative p-values (more so for smaller sample sizes). Similarly, using `ML` fits when `REML` fits should be used gives conservative p-values or CIs. So this technical-sounding issue is really just another grain of salt added to any p-values/CIs you report.

8.5.1 Hypothesis Tests: Fixed Effects

Options for calculating p-values for fixed-effect terms include the following, listed in increasing order of accuracy and computation time:

1. Wald z-test
2. LRT
3. t/F tests with approximate df
4. Parametric bootstrapping

Options (1)–(4) are discussed in more detail, with a simulation study, by Luke (2017).

8.5.1.1 Wald z-test

The mixed-effects model table shows a t-statistic for each coefficient—its estimated value divided by its SE. It is unclear what the df should be for each t, which can be thought of as the number of independent observations that go into calculating the mean in the t-statistic. df could be thought of as k, because the mean is over k groups (each of which has their own group mean), or $n-1$ (the answer in linear regression), the number of independent observations used to calculate the mean. In reality df is somewhere between k and $n-1$ and can be approximated using different methods.

The first is to just assume that df is large, so t follows a normal (z) distribution, and calculate a p-value using a two-sided Wald z-test. For example, for the fixed effects in model `vot_mod_2`:

```
tidy(vot_mod_2) %>%
  filter(effect == "fixed") %>%
  mutate(p = 2 * (1 - pnorm(abs(statistic))))
##   effect group              term estimate std.error statistic
## 1  fixed  <NA>       (Intercept)    2.620     0.068      38.3
## 2  fixed  <NA> speaking_rate_dev   -0.040     0.022      -1.9
## 3  fixed  <NA>   foll_high_vowel    0.087     0.037       2.3
## 4  fixed  <NA>      cons_cluster    0.465     0.042      11.2
## 5  fixed  <NA>   log_corpus_freq    0.042     0.040       1.1
## 6  fixed  <NA>            place1    0.324     0.020      16.4
## 7  fixed  <NA>            place2    0.128     0.012      10.4
## 8  fixed  <NA>            gender   -0.332     0.136      -2.4
##       p
## 1 0.000
## 2 0.062
## 3 0.019
## 4 0.000
## 5 0.288
## 6 0.000
```

```
## 7 0.000
## 8 0.015
```

For example, $p = 0.062$ for the `speaking_rate_dev` coefficient and $p = 0.015$ for the `gender` coefficient.

This is not a great way to calculate a p-value, because it is very approximate and in general the p-value will be anticonservative (because df is too high). However, it has a useful corollary: observing that 95% of a t-distribution with high df has $|t| < 2$ gives the rule of thumb that fixed effects with $|t| > 2$ are (roughly!) significant at the $p < 0.05$ level.

Thus, if $|t|$ is much larger than 2 (say $|t| > 4$), the effect "is significant" and you don't need to bother using a more exact method to get a p-value. Similarly if $|t| < 1$ (say) the effect is unlikely to "be significant" no matter how p-values are computed.

8.5.1.2 Likelihood-ratio tests

This works like testing the significance of one or more terms for a logistic regression (section 6.5.2.2): fit models with and without k terms, and compare them using an LRT (via `anova()` or `drop1()`). For large enough datasets, the difference in deviance between the two models follows a $\chi^2(k)$ distribution under the null hypothesis that the fixed-effect coefficients (for the omitted predictors) are 0 (Pinheiro and Bates 2000, section 2.4.2). Note that the k terms could be fixed or random effects, so this test is very generally applicable. It is common in papers to use "likelihood-ratio test" to mean this large-sample chi-squared test; this discussion will often use this shorthand.[10]

LRT-based p-values are anticonservative, but typically less so than Wald z-test p-values (e.g., Luke 2017). They take longer to compute because the two models must be refitted using ML (as their fixed effects differ) instead of REML.

Single terms For example, to test the contribution of the speaking rate and vowel height terms to `vot_mod_2`, separately:

```
## Models are automatically refitted with ML
drop1(vot_mod_2, "speaking_rate_dev", test = "Chisq")
...
##                    npar   AIC  LRT Pr(Chi)
## <none>                  12395
## speaking_rate_dev     1 12397 3.32   0.068
```

```
drop1(vot_mod_2, "gender", test = "Chisq")
...
##          npar   AIC  LRT Pr(Chi)
## <none>        12395
## gender      1 12399 5.56   0.018
```

10. Technically "likelihood-ratio test" could also mean a hypothesis test using the likelihood carried out using a more exact method, such as parametric bootstrapping.

Note that the p-values are slightly higher than those from Wald z-tests, as expected.

Multiple terms An LRT can also be used to test the contribution of multiple fixed-effect terms, for example, for a factor with two or more levels. For example to test the contribution of `place` for our neutralization model:

```
## Refit model without the fixed effect of place
neut_mod_3_noplace <- update(neut_mod_3, . ~ . - place)
## Likelihood-ratio test
anova(neut_mod_3, neut_mod_3_noplace)
...
##                    npar  AIC  BIC logLik deviance Chisq Df
## neut_mod_3_noplace   12 6759 6815  -3368     6735
## neut_mod_3           14 6748 6812  -3360     6720  15.7  2
##                    Pr(>Chisq)
## neut_mod_3_noplace
## neut_mod_3            0.00039
```

This would be summarized as "`place` significantly contributes to the model, by an LRT ($\chi^2(2) = 16, p < 0.001$)."

8.5.1.3 *t*/*F*-tests with approximate *df*

Another option to compute p-values uses the same logic as the Wald z approach, but approximating df rather than assuming $df = \infty$. For single fixed-effect terms, the *Satterthwaite approximation* gives df for two-sided t-tests; for multiple terms, it gives approximate dfs for an F test (as for linear regression; see section 5.9.1).

These methods have become popular in part due to their handy implementation in the lmerTest package, which redefines the `lmer()` command to calculate (Satterthwaite) df and p-values and show them in the model summary for each fixed effect, as well as in the output of `anova()` applied to a model.

The Satterthwaite approximation requires calculating additional information about the fitted model.[11] Hence these p-values take a bit longer to compute than those from Wald z-tests or LRTs (see `?lmerTest`), but are more accurate (less anticonservative). For example, to add p-values for each coefficient of the `vot_mod_2` model:

```
vot_mod_2_lmerTest <- lmerTest::lmer(log_vot ~ speaking_rate_dev +
    foll_high_vowel + cons_cluster + log_corpus_freq + place + gender +
    (1 | word) + (1 + speaking_rate_dev || speaker),
  data = vot_voiced_core)
summary(vot_mod_2_lmerTest)
...
##
```

11. Intuitively, the Satterthwaite approximation estimates how large *df* should be given the different sources of information that go into the estimate of a fixed effect, based on combining the estimated variances (= uncertainty) of each source. For example for a participant-level predictor, *df* will be similar to the number of participants, all else being equal. These estimates require evaluating complex partial derivatives on the fitted model (Kuznetsova, Brockhoff, and Christensen 2017).

```
## Fixed effects:
##                    Estimate Std. Error       df t value Pr(>|t|)
## (Intercept)          2.6199     0.0684  10.8354   38.29  6.5e-13
## speaking_rate_dev   -0.0402     0.0216  10.4114   -1.86    0.091
## foll_high_vowel      0.0873     0.0372  60.8091    2.34    0.022
## cons_cluster         0.4646     0.0416 140.5613   11.16  < 2e-16
## log_corpus_freq      0.0421     0.0396  70.3528    1.06    0.291
## place1               0.3237     0.0198  67.2073   16.38  < 2e-16
## place2               0.1282     0.0123  48.2057   10.39  6.8e-14
## gender              -0.3316     0.1358   9.0503   -2.44    0.037
##
...
```

(Note that because lmerTest redefines the `lmer()` command, we manually set `lmer()` to mean the lme4 version at the beginning of the chapter, and we now have to use `lmerTest::lmer()` here to get the lmerTest version.)

To test the contribution of `place` to the `neut_mod_3` model:

```
neut_mod_3_lmerTest <- lmerTest::lmer(vowel_dur ~ voicing +
    prosodic_boundary + place + vowel + (1 + voicing || subject) +
    (1 + voicing || item_pair), data = neutralization)
## Uses an F test when applied to an lmerTest model
drop1(neut_mod_3_lmerTest, "place")
...
##         Sum Sq Mean Sq NumDF DenDF F value Pr(>F)
## place     6649    3324     2  16.8    8.37  0.003
```

These p-values are in general higher than those from Wald z-tests or LRTs because they are more conservative (and more accurate, hence type I errors are less likely). For example, compare the p-values in this section for `speaking_rate_dev`, `gender` (in model table) and `place` (from F-test) to the p-values from LRTs in the previous section.

Satterthwaite-based p-values are a good default for calculating p-values for fixed-effect terms in linear mixed-effects models, as an easy and reasonably accurate option (Luke 2017; Meteyard and Davies 2020).

A few practical aspects of using lmerTest are

- After loading lmerTest, your `lmer()` models will take about twice as long to run. This is only an issue if you are fitting complicated models or analyzing large datasets.
- Loading lmerTest (after loading lme4) redefines the `lmer()` command, so the fitted models are now of type `lmerModLmerTest` rather than `lmerMod`. This usually doesn't matter, but sometimes the small differences between these model types cause errors in a helper function that should work on an `lmer()` model.
- If you want to use the lme4 version of `lmer()` after loading lmerTest (e.g., to debug such an error or to run a model that doesn't take twice as long), you use `lme4::lmer()`.

Box 8.5
Practical Note: Testing the Contribution of Multiple Fixed (and Random) Effects

Testing the contribution of multiple fixed-effect terms is conceptually the same whether the terms are part of a factor (e.g., `place` in the previous example) or are not. But lmerTest appears to only have functionality for the former case.

To test the contribution of multiple terms more generally (whether fixed-effect or fixed- and random-effect terms), an F-test can be carried out using a different approximation to *df* (the *Kenward-Rogers approximation*) by model comparison using the `KRmodcomp()` function, from the pbkrtest package (see Halekoh and Højsgaard 2014 for mathematical details).

For example, to assess the contribution of all fixed-effect terms to model `neut_mod_3` (analogous to the F-test reported at the bottom of an `lm()` model), we compare to a model with the fixed-effect terms removed:

```
neut_mod_3_noFixef <- update(neut_mod_3, . ~ . -
    voicing - prosodic_boundary - place - vowel)
KRmodcomp(neut_mod_3, neut_mod_3_noFixef)
...
##         stat  ndf   ddf F.scaling p.value
## Ftest 29.0  8.0 25.2     0.974 7.6e-11
```

8.5.1.4 Parametric bootstrap

Parametric bootstrapping (PB) is a very accurate and very slow method for calculating p-values. One useful implementation of PB is in the pbkrtest package. For more general PB computations, the `bootMer()` function of lme4 can be used.[12]

Roughly, what PB is doing is simulating many new datasets from your model, assuming a given fixed-effect coefficient is set to 0, but all other coefficients are kept at their fitted values (section 5). It then fits the original model to each of the new datasets, resulting in a distribution of values for the fixed-effect coefficient of interest. The proportion of these refitted models where the t-statistic of the coefficient (which should be 0) is as large or larger than the original estimate of t for this coefficient—the first time you ran the model, on the real data—is the p-value. This is a direct implementation of the meaning of type I error ("how often would I wrongly conclude the coefficient was at least this large, if I redid the experiment many times?").

Thus, PB effectively needs to rerun your model `nsim` times, and the higher `nsim` is, the more accurate the p-values are. If your machine has multiple cores it is a good idea to use them here to speed up computation.[13]

For example, to calculate p-values for the `speaking_rate_dev` and `gender` terms for model `vot_mod_2`, with `nsim = 1000`:

12. Refitting the model with `mixed()` from the afex package is a useful shortcut to obtain PB p-values for all terms.

13. The code example uses all cores on my machine, which may not be a good idea if you have other programs running. A reviewer suggests always leaving two cores free (`nc <- max(1,detectCores()-2)`).

```
## library(parallel)
## nc <- detectCores()
## cl <- makeCluster(rep("localhost", nc))
##
## vot_mod_2_no_rate_fixef <- update(vot_mod_2, ~ . - speaking_rate_dev)
## vot_mod_2_noFhvFixef <- update(vot_mod_2, ~ . - gender)
## ## Parametric bootstraps
## pb_vot_1 <- PBmodcomp(vot_mod_2, vot_mod_2_no_rate_fixef, cl = cl,
##   nsim = 1000)
## pb_vot_2 <- PBmodcomp(vot_mod_2, vot_mod_2_noGender, cl = cl,
##   nsim = 1000)
## ## Full output of PBmodcomp
pb_vot_1$test
##        stat df p.value
## LRT     3.3  1   0.071
## PBtest  3.3 NA   0.107
## Just the p-value for the PB test
pb_vot_2$test["PBtest", "p.value"]
## [1] 0.038
```

The PBmodcomp calls take about five minutes, compared to a few seconds for previous methods.

The same PB idea can be used to calculate a p-value for multiple fixed-effect terms (new datasets are now simulated assuming all these terms' coefficients are 0), for example, for place in model neut_mod_3:

```
## pb_neut_1 <- PBmodcomp(neut_mod_3, neut_mod_3_noplace,
##   cl = cl, nsim = 1000)
pb_neut_1$test["PBtest", "p.value"]
## [1] 0.006
```

Box 8.6
Broader Context: p-Values for Mixed Models

We have discussed several ways to calculate p-values that are used in practice and recommended "approximate *df*" methods as a default. But does it actually matter which method you use, especially for fixed effects? Yes and no.

Simulation studies (e.g., Luke 2017) and theory (Pinheiro and Bates 2000) suggest that slower methods give p-values that are more accurate and less anticonservative, especially for small enough *df*. For mixed-effects models *df* is smallest for participant/item level effects, where it is primarily determined by the number of groups (and secondarily by observations per group). My intuition is *df* is "small enough" for many linguistic datasets (e.g., less than 24 participants or items, in Luke's simulations; <40 levels is "small" in Snijders and Bosker 2011, section 6.2). Because we typically care more about type I error, this seems like an important issue.

However, p-values from different methods rarely are hugely different. For example, in the simulations by Luke (2017), which use the four methods covered here, "good" and "bad" methods give type I errors of 0.045 to 0.08 at worst (compared to a "correct" value of $\alpha = 0.05$). You can also verify that the p-values we obtained by four methods for fixed-effect terms for the VOT model are similar—all within one decimal place of each other (for the same term). This is not nothing, but small enough that if it really matters what method you are using, your finding may be shaky anyway.

My practical advice is to take p-values with a grain of salt, as always, and try to use a "more accurate" method in the final write-up of your results.

8.5.2 Hypothesis Tests: Random Effects

As for fixed effects there are several options that trade off accuracy and computation time: LRT, exact LRT, and PB. Let's use as an example testing the contribution of the by-speaker random slope of `speaking_rate_dev` to `vot_mod_2`—intuitively, "does the speaking rate effect significantly differ by speaker?"

8.5.2.1 Likelihood-ratio test

By the logic from section 8.5.1.2, models differing in k random-effect terms can be compared using an LRT, which gives an approximate p-value (more accurate for larger samples).

For example, for the `speaking_rate_dev` random slope:

```
anova(vot_mod_1, vot_mod_2, refit = FALSE)
...
##            npar   AIC   BIC logLik deviance Chisq Df Pr(>Chisq)
## vot_mod_1    11 12439 12515  -6208    12417
## vot_mod_2    12 12434 12517  -6205    12410  6.65  1     0.0099
```

This could be reported as "adding a by-participant random slope for speaking rate significantly improved the model, by an LRT ($\chi^2(1)=6.6$, $p=0.0099$)."

The `refit = FALSE` flag makes the model comparison use REML fits, which have better estimates of the variance components (e.g., Zuur 2009, section 5.7).

This method doesn't work for testing the significance of a random effect in a model with just one random effect, such as model `neut_2_mod_1` (section 8.3.3), because the models being compared are not of the same type (one is a mixed model and the other isn't, so the likelihoods are not directly comparable). In this case, the exact restricted LRT must be used.

8.5.2.2 Exact LRT

The LRT turns out to give *conservative* p-values (p-values too high) when comparing models differing in random-effect terms.[14]

More accurate p-values can be calculated for excluding k random-effect terms using an "exact" LRT, by simulation, implemented in the RLRsim package. This test (in `exactRLRT`) requires three models: the models with and without the k terms (as for a normal LRT) and the model with only these terms. For example, for the `speaking_rate_dev` slope:

14. This is related to the null hypothesis (e.g., $\sigma_{P,1}=0$) being on the boundary of the hypothesis space (Pinheiro and Bates 2000, section 2.4.1). The idea is that because a variance component can only be positive, a p-value computed using a two-sided test is too high.

```
## Remove random effects except the by-speaker speaking_rate_dev slope
vot_mod_2_justRateSlope <- update(vot_mod_2, . ~ . -
    (1 | word) - (1 | speaker))
exactRLRT(vot_mod_2_justRateSlope, vot_mod_2, vot_mod_1)
##
##  simulated finite sample distribution of RLRT.
##
##  (p-value based on 10000 simulated values)
##
## data:
## RLRT = 7, p-value = 0.003
```

This method takes a bit longer than the LRT (fitting an extra model, simulation) but is much more accurate.

Box 8.7
Practical Advice: Testing Random Effects

Note how more accurate methods for the `speaking_rate_dev` slope example lead to lower p-values (LRT (ML) > LRT (REML) > exact LRT), as expected—they are less conservative. A parametric bootstrap could also be used and would presumably give an even more accurate p-value.

Which method should be used in practice? Even when random effects aren't of direct interest, this matters because of the role such tests play in model selection (section 10.5). Using the data-driven approach, more conservative tests will give sparser random-effect structure, which will (typically) give more type I errors for *fixed*-effect terms.

I advise using LRTs with REML by default (which is also the fastest method) and exact LRTs or PB any time the results matter (e.g., when the research question asks "do speakers differ?") or in model selection when p is near the α cutoff.

8.5.3 Fixed and Random Effects

A final case is assessing the contribution of fixed- and random-effect terms in a model, which requires getting into an issue we'll return to in section 8.5.3: what is the best way to assess "does predictor x significantly improve the model?"

When a model contains both a fixed effect and random slope(s) for x, the answer is not clear. Testing the contribution of the fixed effect only is the most common method in language sciences/psychology and makes sense as this is the aspect of how x affects the response ("overall") that we care about, but this does not take into account variability in the effect among participants or items. It is possible for a predictor to show no overall effect (across participants and items), but to show significant variability among participants or items; indeed, this may be why there is no significant overall effect (Rights and Sterba 2019, 12).

An alternative method is to assess the contribution of x by comparing the full model with a model where all terms involving x—fixed and random effects—are excluded. There are several options for obtaining a p-value for this comparison that will (again) trade off in accuracy and speed: an LRT, an F-test using approximate *df* (using Kenward-Roger; box 8.5), or a parametric bootstrap.

For example, for our VOT case, one way to address the research question "does speaking rate matter for voiced VOT?" is to test whether the fixed effect and by-speaker random slope for `speaking_rate_dev` contribute to model `vot_mod_2`.

We can obtain a p-value for this example using a LRT comparing `vot_mod_2` to the model with the two `speaking_rate_dev` terms excluded:

```
## speaking_rate_dev fixed effect and by-speaker random slope excluded
vot_mod_2_noRate <- update(vot_mod_2, . ~ . - speaking_rate_dev -
    (0 + speaking_rate_dev | speaker))
anova(vot_mod_2, vot_mod_2_noRate)
...
##                    npar   AIC   BIC logLik deviance Chisq Df
## vot_mod_2_noRate     10 12405 12474  -6193    12385
## vot_mod_2            12 12395 12478  -6186    12371  14.3  2
##                    Pr(>Chisq)
## vot_mod_2_noRate
## vot_mod_2               8e-04
```

(Note that using ML fits is appropriate here because the models differ in fixed-effect structure.)

Thus, adding information about speaking rate *does* significantly improve the model ($\chi^2(2)=14$, $p<0.001$) even though there is not a significant overall effect (`speaking_rate_dev` fixed effect $p>0.05$). This reflects substantial speaker variability (as shown in section 8.5.2). In this case, which model comparison we perform changes how we would answer the research question ("does speaking rate matter for voiced stops?").

A more accurate p-value could be computed using a slower method—an F-test using Kenward-Roger *df* or a parametric bootstrap—but p is low enough that the qualitative conclusion will not change (exercise 8.7).

8.5.4 Confidence Intervals

CIs are easier to compute for mixed-model parameters (fixed and random effects), using the `confint()` function.

The options again trade off in speed and accuracy:

- *Wald z-based* CIs ($\hat{\beta} \pm 1.96 \cdot SE$)—for fixed-effect coefficients only
- *Profile* CIs based on the likelihood profile (as for logistic regression; see section 6.5.2.2)
- *Parametric bootstrap* CIs: computed from the distribution of values for the coefficient(s) of interest, across bootstrap samples

The default is profile CIs, which take some time to compute (seconds to minutes) but are reasonably accurate.

For example, consider the `voicing` effect in `neut_mod_3`, which is of primary theoretical interest, calculated in these three ways:

```
confint(neut_mod_3, parm = "voicing", method = "Wald")
##          2.5 % 97.5 %
## voicing      6     13
confint(neut_mod_3, parm = "voicing", method = "profile")
```

```
##          2.5 % 97.5 %
## voicing   5.9     13
```

```
## use multiple cores
## takes ~2 min on my 8-core laptop
confint(neut_mod_3,
  parm = "voicing", method = "boot",
  nsim = 5000, parallel = "multicore", ncpus = nc
)
```

```
##          2.5 % 97.5 %
## voicing   5.9     13
```

The two more accurate methods give (very slightly) wider CIs than the Wald method, because they correspond to less anticonservative p-values.

CIs can also be computed for random effects (box 8.8).

Box 8.8
Broader Context: Confidence Intervals for Variance Components

Reporting CIs for random effects is currently not common in practice (in language sciences), but seeing an example illustrates some useful points. 95% profile CIs for the variance components of model `vot_mod_2` are

```
## `parm', `oldNames' arguments gives CIs only for variance components
## and give them intuitive names.
confint(vot_mod_2, method = "profile", parm = "theta_", oldNames = F)
##                                2.5 % 97.5 %
## sd_(Intercept)|word            0.083   0.17
## sd_(Intercept)|speaker         0.134   0.32
## sd_speaking_rate_dev|speaker 0.017   0.10
## sigma                          0.557   0.58
```

All three CIs don't include 0. This should be functionally equivalent to significant ($p<0.05$) LRTs of the contribution of each random-effect term to the model, whereas a 95% CI including 0 would correspond to a nonsignificant test.

Note how large the CIs are: for example, the degree of speaker variability in the `speaking_rate_dev` slope could be small enough that speakers effectively all have the same directionality for this effect, or large enough that they differ roughly 50/50 in the effect's direction. In general, accurately estimating the degree of variability in effects (random effects) requires much more data than estimating population-level effects (fixed effects) does.

A corollary is that we should be cautious about drawing inferences from the random-effect estimates reported in `lmer()` output if these relate to research questions. For example, suppose it is of interest whether speakers or words differ more in VOT (after controlling for speaker- and word-level predictors). Because the estimated degree of by-speaker intercept variability (0.22) is larger than the by-word value (0.13), we might be temped to conclude that speakers differ more than words, but the 95% CIs for the estimates overlap, meaning we can't be certain.

8.6 Model Summaries

As for non-mixed-effects models it is useful to define quantitative summaries of the model—goodness of fit (like R^2) or measures of model likelihood/"information" (like AIC)—which are used in variable selection or reporting the model.

8.6.1 Goodness of Fit

We would like a measure of goodness-of-fit for LMMs that intuitively relates to "proportion of variation explained" (like R^2 for linear regression). There is no such single measure for LMMs, because they contain more than one kind of variation that can be explained: residual variation (error in each observation), variation in the intercept (e.g., by speaker, item: random intercepts), variation in predictor slopes (random slopes), and so on. Thus many R^2 measures can be defined, and in general capturing goodness of fit requires more than one measure. This is an active research area (box 8.9), but two measures are most commonly used in practice.

These are *marginal* R^2 (R^2_m), intuitively the proportion of variance explained by fixed effects, and *conditional* R^2 (R^2_c), the proportion of variance explained by fixed and random effects (Nakagawa and Schielzeth 2013; Johnson 2014); both lie between 0 and 1 (no/all variance explained). A reasonable default is to report these two measures, which convey different information. If only fixed effects are of interest or space is limited, it could make sense to report just marginal R^2.

Box 8.9
Broader Context: R^2 Measures for LMMs

Rights and Sterba (2019) summarize the tangled literature on R^2-like measures for mixed-effects models, discuss why the measure used matters (see also Snijders and Bosker 2011), and present a general framework that encompasses many slightly different R^2 measures proposed across different fields. Rights and Sterba (2020) extend this discussion to assessing the effect size of one or several predictors.

The central issue is whether only variance attributable to fixed effects is treated as "explained" or if random effects count as well (and if so how to calculate their contribution). This is conceptually similar to the issue we saw in section 8.5.3 in defining whether a predictor x "has an effect": Should we assess just its fixed effect or by-x random slopes as well?

Our "marginal R^2" and "conditional R^2" are columns 3 and 7 of Rights and Sterba (2019, table 4). R^2_m is the proportion of variance in y explained by just using the fixed-effect coefficient estimates, which can be thought of as "marginalizing" across random effects. R^2_C is the proportion of variance explained when intercepts and slopes are adjusted using the estimated random effects (the BLUPs), that is, model predictions are "conditioned" on random-effect values.

One option for R^2 that should probably *not* be used is the squared correlation of an LMM's predictions ($\hat{y}_i$) with the observations (y_i). This recipe is widely used (e.g., Bolker 2021, "Simple/crude solutions"), and was recommended in an earlier version of this book (Sonderegger, Wagner, and Torreira 2018, section 7.5.4), but it does not properly account for the contribution of random effects. (Intuitively, this measure uses BLUPs, which are not fitted model parameters, rather than the variance components, which are.)

These can be calculated using `r.squaredGLMM` from the MuMIn package. For example, for the simplest model of the `neutralization` data (`neut_mod_0`; see section 8.3.6):

```
r.squaredGLMM(neut_mod_0)
##          R2m  R2c
## [1,] 0.011 0.79
```

Here R_m^2 is low (near 0) because the single fixed effect (`voicing`) explains little variance, whereas R_c^2 is much higher because by-item and by-participant variability explain much of the variance in `vowel_dur` in the data. In other words, most variance is explained by the participant and item, but we don't know *why*: what properties of participants and items matter or whether this is random variation. R_c^2 is often rather high for LMMs, simply because much variability in the data comes down to (e.g.,) by-participant and by-item variability—even if the model would not actually have much predictive power on new data (unseen participants or items).

To get a better sense of what the two R^2 measures mean, we compare their values for all four `neutralization` models fitted in the preceding discussion (code hidden):

```
##          model          fixed                      random   R2m  R2c
## 1 neut_mod_0        voicing                  intercepts 0.011 0.79
## 2 neut_mod_1 v. + controls                  intercepts 0.388 0.78
## 3 neut_mod_2        voicing I + by-participant slope 0.011 0.79
## 4 neut_mod_3 v. + controls   I + by-par/item slopes 0.387 0.79
```

Comparing models 0 and 1: R_m^2 increases because the control predictors have been added (to `voicing`) as fixed effects. R_c^2 is similar because the total variance explained (by fixed + random effects) hasn't changed: the control predictors are mostly item-level, so some variance is now just explained by fixed instead of random effects.

Comparing models 1 and 3: R_m^2 is the same because the fixed effects are the same. Adding a random-effect term (the by-participant random slope) leads to a small increase in R_c^2.

8.6.2 Information Criteria

As for nonmixed models (section 5.9.2), various information criteria can be defined that quantify the trade-off between model likelihood (L) and model complexity (k, number of parameters); see Vaida and Blanchard (2005) and Müller, Scealy, and Welsh (2013) for details. Let's consider AIC, AICc, and BIC, which are defined using the same equations as for nonmixed models (equations (5.5)–(5.7)). For example, AIC is still:

$$AIC = -2 \cdot L + 2 \cdot k. \tag{8.6}$$

How to define L and k depends on the "level of focus" at which predictions are being made (Vaida and Blanchard 2005; Saefken et al. 2018), similarly to defining marginal versus conditional R^2.

If interest is in the fixed effects (prediction for an average participant/item), you use *marginal AIC*, where L is just the likelihood (ML or REML) of the fitted model and k is the number of fixed-effect coefficients and variance components. If interest is in the random effects (predictions for individual participants/items), you use *conditional AIC*, whose definition is more complex; a correct implementation is in the cAIC4 package (see Saefken et al. 2018 for discussion of this case). These measures use the "marginal likelihood" and "conditional likelihood"; see Bolker (2015, section 13.3.2) or Bolker (2021) for an intuitive explanation. Corrected (marginal) AIC (AICc), used for smaller samples, can be calculated with `AICc()` from MuMIn. Here we'll only consider marginal AIC, which is what is calculated by default for lme4 models.

Similarly, (marginal) BIC is defined by substituting $\log(n)$ for 2 in equation (8.6) and is the default for lme4 models. Many other *IC options exist (e.g., DIC, a Bayesian interpretation of AIC; see Gelman and Hill [2007, section 24.3]).

As for nonmixed models, when used for model selection, AIC/AICc/BIC tend to select progressively smaller (mixed-effects) models.

For example, to calculate the three criteria for the neutralization models `neut_mod_0` and `neut_mod_2`, which differ in whether the by-speaker random slope of voicing is included:

```
AIC(neut_mod_0, neut_mod_2)
##            df  AIC
## neut_mod_0  5 6800
## neut_mod_2  6 6800
MuMIn::AICc(neut_mod_0, neut_mod_2)
##            df AICc
## neut_mod_0  5 6801
## neut_mod_2  6 6800
BIC(neut_mod_0, neut_mod_2)
##            df  BIC
## neut_mod_0  5 6824
## neut_mod_2  6 6827
```

We would keep this term if using AIC or AICc, but not if using BIC.

8.7 Random-Effect Correlations

In examples so far we have always used "uncorrelated" random effects, which assume there is no relationship between different random-effect terms for the same grouping factor. For example, in model `vot_mod_2` it is assumed there is no relationship between (1) a speaker's offset from the overall intercept (random intercept) and (2) their offset from the overall slope of `speaking_rate_dev`.

Box 8.10
Broader Context: Uncorrelated Random Effects and Centering Predictors

There is an important caveat to models with uncorrelated random effects: predictors for which random slopes are included should typically be centered, meaning continuous variables have mean 0 and all factors use a "centered" (and ideally orthogonal) coding scheme. There are two reasons for this. First, this helps minimize collinearity between the predictors and between the predictors and the intercept, which makes zero correlations between random effects a more plausible assumption. A more subtle reason is that assuming uncorrelated random effects makes the fitted models *not* invariant to linear shifts in the predictors (e.g., adding 1 to `speaking_rate_dev`), which means the model predictions become tied to the particular scale each predictor was measured on (Snijders and Bosker 2011, section 5.1.2; Bates, Mächler, et al. 2015). As a result, Snijders and Bosker recommend never forcing correlations to be 0, and Bates, Mächler, et al. recommend only using uncorrelated random effects for predictors that are "measured on a ratio scale ... i.e.,

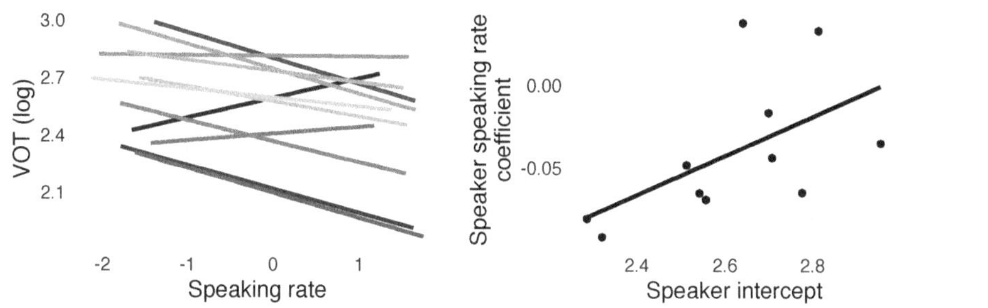

Figure 8.8
Empirical effect of speaking rate on log VOT (linear smooth) for each speaker (*left*) in the `vot_voiced_core` data, and predicted intercept and speaking rate coefficient for each speaker from model `vot_mod_2` (*right*, with linear smoothing).

> the zero point on the scale is meaningful" (8). The simplest way to do this is just to center all predictors. This is another argument for centering your predictors by default.
>
> Sometimes a subset of predictors are intrinsically related, and uncorrelated random effects make less sense. An example is a nonlinear effect of x coded as several polynomial or spline terms. The contrasts in a factor could also be thought of as intrinsically related, even when an orthogonal coding scheme is used (Snijders and Bosker 2011, section 6.4). Using uncorrelated random effects in these cases—or indeed, when predictors have not been centered—is not wrong per se, as the model can still learn random effects that happen to be correlated if the data support this, but special hacks are needed to code the model (sections 8.7.4, 10.7). Alternatively, it could make sense for your simplest model to have correlated random effects for random slopes for such intrinsically related predictors (and the intercept), and uncorrelated random slopes for other predictors.

If there were a relationship between (1) and (2), it would take the form of a positive or negative correlation ("correlated random effects"), for example, "speakers with higher average VOT have higher `speaking_rate_dev` effects" (positive correlation).

To determine whether the uncorrelated random effect assumption is realistic in this case, consider the by-speaker plot of `vot` versus `speaking_rate_dev` (figure 8.8, left). A nonzero correlation here would correspond to a relationship between the height of a line (= speaker's intercept) and its slope (= speaker's slope). A relationship doesn't jump out, but we can check this more carefully by plotting the relationship between speaker's random intercepts and their `speaking_rate_dev` random slopes (figure 8.8, right).

There may be a weak positive correlation here, where speakers with higher average VOT may also have larger speaking rate effects:

```
coeffs <- coefficients(vot_mod_2)$speaker
cor.test(coeffs[["(Intercept)"]], coeffs[["speaking_rate_dev"]]) %>%
  tidy(conf.int=TRUE) %>%
  ## Show correlation, p-value, 95% CI
  select(estimate, p.value, conf.low, conf.high)
##   estimate p.value conf.low conf.high
## 1     0.56   0.072   -0.058      0.87
```

Alternatively, this may be a spurious correlation, not significantly different from no correlation. We can test whether there's evidence for a "real" correlation between participants' slopes and intercepts by including this term in the mixed model.

There is a single new term in this model, compared to model `vot_mod_2`: ρ, the correlation between participants' intercepts and slopes of `speaking_rate_dev`. The only difference is that `vot_mod_2` assumes that $\rho = 0$, whereas `vot_mod_3` fits $\rho \in (-1, 1)$.

Box 8.11
Broader Context: Random-Effect Correlations

To see how random-effect correlations change a model's definition, it is easier to start from a simpler model: `neut_mod_2`, the model for the `neutralization` data with a single fixed effect (`voicing`) and a by-participant intercept and random slope (as well as a by-item random intercept), which are assumed to be uncorrelated. This model's formula is in equation (8.4). Removing the assumption of no correlation, the lines of this formula involving by-participant random intercepts ($\alpha_{P,j}$) and `voicing` slopes ($\gamma_{P,j}$) become

$$\begin{pmatrix} \alpha_{P,j} \\ \gamma_{P,j} \end{pmatrix} \sim N\left(\begin{pmatrix} 0 \\ 0 \end{pmatrix}, \begin{pmatrix} \sigma_{P,0} & \sigma_{P,01} \\ \sigma_{P,01} & \sigma_{P,1} \end{pmatrix}\right), \quad j = 1, \ldots, 16.$$

These random effects now follow a multivariate normal distribution. The new random-effect parameter, $\sigma_{P,01}$, is their covariance. That is, a covariance matrix is being fitted for the by-participant random effects:

$$\Sigma_P = \begin{pmatrix} \sigma_{P,0} & \sigma_{P,01} \\ \sigma_{P,01} & \sigma_{P,1} \end{pmatrix}.$$

Any covariance matrix for k variables is equivalent to a *set of variances* (one for each variable) *and correlations* (between distinct variables). These are the parameters `lmer()` reports under "random effects" (here $\rho = \sigma_{P,01}/(\sigma_{P,0}\sigma_{P,1})$ is the single correlation term). For each additional grouping factor with correlated random effects (e.g., by-item), another covariance matrix is fitted.

When $k > 1$, Σ will correspond to $k(k+1)/2$ terms: the random intercept variance, $k-1$ random-slope variances, and $k(k-1)/2$ correlation terms. For example, for the model `vot_mod2_more_slopes`, which contains correlated by-item and by-participant random slopes, each with $k = 2$, there are two random intercepts, four random slope variances, and three correlation terms. The number of correlation terms grows quickly as k increases (e.g., 15 correlations for $k = 5$ random slopes). This greatly affects the speed of model fitting, and—at least for typically sized linguistic datasets—whether the model converges at all, when correlated random effects are used.

In R, the "single pipe" notation (`|`) is used for correlated random effects. To fit `vot_mod_3`:

```
## Change from (1+speaking_rate_dev||speaker)
## to (1+speaking_rate_dev|speaker)  (uncorr. to correlated)
vot_mod_3 <- update(vot_mod_2, . ~ . - (1 | speaker) -
  (0 + speaking_rate_dev | speaker) +
    (1 + speaking_rate_dev | speaker))
```

Note the syntax for correlated random effects: `(1 + A | speaker)` (or `(1 + A + B | speaker)`, etc.). Compare the formulas for the two models:

```
formula(vot_mod_2)
## log_vot ~ speaking_rate_dev + foll_high_vowel + cons_cluster +
##     log_corpus_freq + place + gender + (1 | word) + ((1 | speaker) +
```

```
##     (0 + speaking_rate_dev | speaker))
formula(vot_mod_3)
## log_vot ~ speaking_rate_dev + foll_high_vowel + cons_cluster +
##     log_corpus_freq + place + gender + (1 | word) +
##     (1 + speaking_rate_dev | speaker)
```

The new model's output is

```
summary(vot_mod_3)
...
## Random effects:
##  Groups   Name              Variance Std.Dev. Corr
##  word     (Intercept)       0.01767  0.1329
##  speaker  (Intercept)       0.04515  0.2125
##           speaking_rate_dev 0.00288  0.0537   0.68
##  Residual                   0.32090  0.5665
## Number of obs: 7175, groups:  word, 272; speaker, 11
##
## Fixed effects:
##                    Estimate Std. Error t value
## (Intercept)          2.6198     0.0676   38.73
## speaking_rate_dev   -0.0397     0.0212   -1.88
## foll_high_vowel      0.0871     0.0372    2.34
## cons_cluster         0.4641     0.0416   11.15
## log_corpus_freq      0.0424     0.0396    1.07
## place1               0.3238     0.0198   16.38
## place2               0.1284     0.0124   10.40
## gender              -0.3486     0.1144   -3.05
##
...
```

(Compare to `vot_mod_2` output in section 8.4.3.1.)

We see a positive correlation term under `Random effects/Corr` ($\rho = 0.68$)—the direction guessed from the scatterplot of random effects for `vot_mod_2` (figure 8.8, right).

We can repeat that plot using the random effects from `vot_mod_3`, to see the relationship predicted by this model between participant offsets for intercept and `speaking_rate_dev` slope (figure 8.9).

It seems that there is a stronger relationship that is detected by including a correlation term. To evaluate whether this correlation is significantly different from 0, we can do an LRT of the models:

```
anova(vot_mod_2, vot_mod_3, refit = FALSE)
...
##           npar   AIC   BIC logLik deviance Chisq Df Pr(>Chisq)
## vot_mod_2   12 12434 12517  -6205    12410
## vot_mod_3   13 12433 12522  -6204    12407  3.29  1       0.07
```

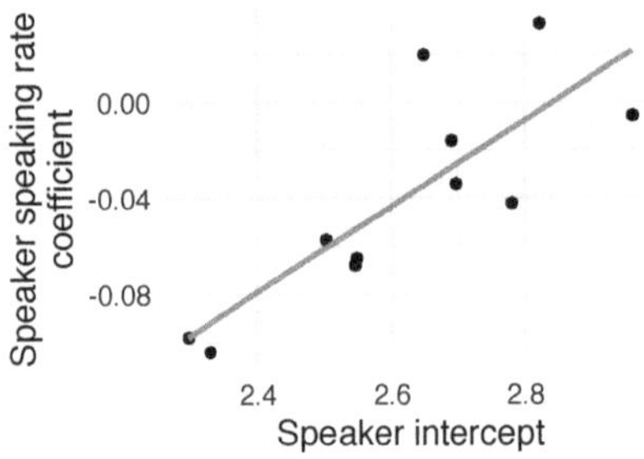

Figure 8.9
Predicted intercept and speaking rate coefficient for each speaker from model `vot_mod_3`, with linear smoothing.

(Here the `refit = FALSE` flag makes REML fits be used, because the models differ only in random-effect terms; see section 8.5.2.1.) This test suggests that speakers' rate effects and intercepts are not significantly correlated.

Despite the correlation term, note that the fixed effects for the two models (with and without this term) are very similar, in terms of coefficients and SEs, including the intercept and `speaking_rate_dev` terms:

```
## Model without correlation
tidy(vot_mod_2)[1:2, ] %>% select(-group)
##    effect                term estimate std.error statistic
## 1  fixed          (Intercept)     2.62     0.068      38.3
## 2  fixed speaking_rate_dev      -0.04     0.022      -1.9
## Model with correlation
tidy(vot_mod_3)[1:2, ] %>% select(-group)
##    effect                term estimate std.error statistic
## 1  fixed          (Intercept)     2.62     0.068      38.7
## 2  fixed speaking_rate_dev      -0.04     0.021      -1.9
```

The random-effect terms don't differ much, beyond the obvious addition of a correlation $\rho > 0$ for model `vot_mod_3`.

8.7.1 Multiple Random Slopes

Correlated random effects extend straightforwardly to the case of multiple random slopes. For example, we can refit model `vot_mod2_more_slopes` (section 8.4.6) with correlated random effects:

```
vot_mod_2_more_slopes_corr <-
  lmer(log_vot ~ speaking_rate_dev + foll_high_vowel + cons_cluster +
    log_corpus_freq + place + gender +
    (1 + gender + speaking_rate_dev | word) +
    (1 + speaking_rate_dev + foll_high_vowel | speaker),
  data = vot_voiced_core)
summary(vot_mod_2_more_slopes_corr)
...
## Groups   Name                 Variance Std.Dev. Corr
```

```
##  word     (Intercept)        0.01501  0.1225
##           gender             0.07036  0.2653    0.67
##           speaking_rate_dev  0.00109  0.0331   -0.03 -0.11
##  speaker  (Intercept)        0.04591  0.2143
##           speaking_rate_dev  0.00251  0.0501   0.72
##           foll_high_vowel    0.00334  0.0578   0.57  0.60
##  Residual                    0.30715  0.5542
## Number of obs: 7175, groups:  word, 272; speaker, 11
##
## Fixed effects:
##                   Estimate Std. Error t value
## (Intercept)         2.6334     0.0673   39.14
## speaking_rate_dev  -0.0401     0.0212   -1.90
## foll_high_vowel     0.0818     0.0377    2.17
## cons_cluster        0.4519     0.0409   11.04
## log_corpus_freq     0.0545     0.0359    1.52
## place1              0.3745     0.0176   21.25
## place2              0.1149     0.0109   10.53
## gender             -0.3210     0.1136   -2.83
##
...
```

This model contains six correlation terms: one for each possible relationship between by-participant random intercepts/slopes (=3) and similarly for by-item random effects (=3). For example, the correlation between the by-speaker random intercept and `speaking_rate_dev` random slope is 0.72; the correlation between the by-word random slopes of `gender` and `speaking_rate_dev` is –0.11.

These terms have significantly improved the model, by an LRT:

```
anova(vot_mod_2_more_slopes_corr, vot_mod_2, refit = FALSE)
...
##                            npar   AIC   BIC logLik deviance Chisq
## vot_mod_2                    12 12434 12517  -6205    12410
## vot_mod_2_more_slopes_corr   21 12234 12378  -6096    12192   218
##                            Df Pr(>Chisq)
## vot_mod_2
## vot_mod_2_more_slopes_corr  9     <2e-16
```

Compare to the output of `vot_mod2_more_slopes` (shown in section 8.4.6):

```
summary(vot_mod_2_more_slopes)
```

The estimated random effects are again similar (except for the addition of correlation terms). However the fixed-effect estimates are a bit different: most have slightly higher magnitudes and lower SEs in the model with correlations (and thus slightly lower *p*-values, by any method). In other words, adding correlations has led to clearer effects.

8.7.2 Singular Models

The preceding examples illustrate how adding correlation terms to random-effect structure tends to change a model's estimates, especially fixed effects: very little or toward "clearer" estimates.[15] In addition, using uncorrelated random effects technically only makes sense provided all predictors are conceptually "centered" (box 8.10), which takes some extra effort. So it makes sense that correlated random effects are the default in lme4—the "true" model likely does have some nonzero correlations.

However in practice it is common for models with correlated random effects to not converge or to have a related issue. For example, consider adding correlations to model `neut_mod_3`, which contains random slopes only for the predictor of primary interest (`voicing`):

```
neut_mod_4 <- lmer(vowel_dur ~ voicing + prosodic_boundary +
  place + vowel + (1 + voicing | subject) +
  (1 + voicing | item_pair), data = neutralization)

## boundary (singular) fit: see ?isSingular

summary(neut_mod_4, correlation = FALSE)

...
## Random effects:
##  Groups    Name        Variance Std.Dev. Corr
##  item_pair (Intercept)  76.57    8.75
##            voicing       8.35    2.89    -1.00
##  subject   (Intercept) 674.27   25.97
##            voicing      18.86    4.34     1.00
##  Residual              395.35   19.88
## Number of obs: 749, groups:  item_pair, 24; subject, 16
##
## Fixed effects:
##                   Estimate Std. Error t value
## (Intercept)        157.085      6.777   23.18
## voicing              9.604      1.910    5.03
## prosodic_boundary   11.768      2.579    4.56
## place1             -10.592      2.336   -4.53
## place2               2.120      1.269    1.67
## vowel1              -0.400      3.013   -0.13
## vowel2             -17.542      1.659  -10.58
## vowel3              -3.185      1.115   -2.86
## vowel4              -6.207      0.868   -7.15
## optimizer (nloptwrap) convergence code: 0 (OK)
## boundary (singular) fit: see ?isSingular
```

This model is *singular*, meaning that a "dimension" of the random effects has been estimated as 0. Typically this means variances that are (near) 0, or correlation parameters near 1 or –1. We'll discuss singularity further in section 10.4. For now, the crucial points are that

15. This statement is purely empirical: provided predictors are "centered" (box 8.10), I have rarely seen a case where adding random-effect correlations to a model changed a qualitative conclusion. (Except about the degree of correlations between random effects, of course.)

singular fits are (1) suspect, especially if due to perfect correlations, and (2) common when modeling linguistic data, especially using correlated random effects—typically because the random-effect structure is too complex to be estimated given the data.

8.7.3 Correlated versus Uncorrelated Random Effects

As we saw in the preceding examples, adding random-effect correlations rarely changes qualitative conclusions, but doing so often seriously affects model fitting (because the number of correlation terms grows quadratically with the number of random-effect terms).

In contrast, which random *slope* terms are included is very important, as described in section 8.4.3—in particular, excluding a random slope term is anticonservative and could change qualitative conclusions from a model.

To deal with these issues, it is often useful to *first try models with uncorrelated random effects*: including random slopes but no correlations.[16] This makes it less likely the random-effect structure will be too complex for the data, while avoiding dropping random slopes.[17] You can then add in correlations as needed (e.g., significantly increases model likelihood) or fit a final "full" model with correlated random effects if your data support it.

Section 10.5.4 will elaborate on this "uncorrelated first" strategy. Note that this strategy is my personal opinion (shared by Seedorff, Oleson, and McMurray 2019). Other sources (e.g., Brauer and Curtin 2018; Barr et al. 2013) recommend fitting models with correlated random effects by default, then removing correlations as needed; this is also a reasonable default. Model-fitting issues due to complex random-effect structure is one of the most frequent issues that researchers have in practice using mixed-effects models, and they are the topic of ongoing debate. We'll return to these issues when we discuss model fitting and model selection in chapter 10. Until then we will usually fit models with uncorrelated random-effect structure and assume this is OK.

8.7.4 Random Slopes for Factors

Technical issues, related to R's implementation of factors, come up when factors are used as predictors in models with uncorrelated random effects (Singmann and Kellen 2019). This is an important issue in practice if you're building models with uncorrelated random effects (which we are, by default, but if you are not this section is skippable).

8.7.4.1 Two-level factors

First, consider two-level factors. Suppose we modeled the `neutralization` data using voicing coded as a two-level factor (`voicing_fact`) rather than as a numeric (`voicing`) factor. We first refit the simplest model (`neut_mod_0`):

```
neut_mod_0_fact <- lmer(vowel_dur ~ voicing_fact + (1 | subject) +
    (1 | item_pair), data = neutralization)
```

16. But remember, to do this your predictors should be centered.

17. Seedorff, Oleson, and McMurray (2019) reach a similar conclusion, in a simulation study. Barr et al. (2013, online appendix, 37) also conclude from a simulation study that "removing correlations will not harm the overall performance of the model"; but they do not recommend *starting* with a model without correlations.

The output of this model is identical to neut_mod_0 (summary(neut_mod_0_fact)) as expected.[18] We now try to add a by-participant random slope to the model, with uncorrelated random effects, as in neut_mod_2:

```
## Change from (1|item_pair) + (1|subject) to
## (1+voicing_fact||subject) + (1|item_pair)
neut_mod_2_fact <- update(neut_mod_0_fact, . ~ . +
    (0 + voicing_fact || subject))
summary(neut_mod_2_fact)
...
##  Groups    Name                   Variance Std.Dev. Corr
##  item_pair (Intercept)            773.4    27.81
##  subject   voicing_factvoiceless  616.9    24.84
##            voicing_factvoiced     894.0    29.90    1.00
##  subject.1 (Intercept)             17.1     4.14
##  Residual                         405.5    20.14
## Number of obs: 749, groups:  item_pair, 24; subject, 16
##
## Fixed effects:
##                Estimate Std. Error t value
## (Intercept)     155.758      8.981   17.34
## voicing_fact1     4.646      0.971    4.78
##
...
```

The random effects in this model do not make sense. (Sometimes you'll get an opaque warning, like `Model is nearly unidentifiable...`.) Rather than the random-effect table showing one row for the intercept and one row for the single `voicing_fact` contrast, it contains one row per *level* of `voicing_fact` (in addition to the intercept). This happens because R is interpreting the random-effect formula as referring to all levels of the factor, rather than its contrast.[19]

This issue will arise whenever a factor X is used in an uncorrelated random-effect formula (whether using `||`, `(-1 + X | Y)`, or `(0 + X | Y)` notation; see section 8.4.1).

To include factors in a model with uncorrelated random effects, we must first turn them into numeric variables corresponding to the contrasts. (This is what R usually does under the hood for factors, but it does not do so in this case.) One way to do this is to use the model matrix of the random-intercepts-only model, which contains these contrasts:

```
mm <- model.matrix(neut_mod_0_fact) %>% data.frame()
mm
##     X.Intercept. voicing_fact1
## 1              1             1
## 2              1            -1
## 3              1            -1
```

18. Modulo the `voicing` coefficient being divided by 2 due to how `voicing_fact` is coded.

19. This follows from the way uncorrelated random effects are treated in `lmer()` and how R deals with factors in models without intercept terms. A `(1 + X+ ... | Z)` term becomes a `(1 | Z) + (0 + X | Z) + ...` term in the model formula. Factors `0 + X` indicate cell-means coding (box 7.7).

```
## 4              1             1
...
```

We then extract the numeric variable for each contrast, add it to the original dataframe, then fit the model with random slopes using these variables in the random-effect formula rather than the actual factor:

```
neutralization <- add_column(neutralization,
  voicing_fact1 = mm$voicing_fact1)

neut_mod_2_fact <- lmer(vowel_dur ~ voicing_fact +
  (1 + voicing_fact1 || subject) + (1 | item_pair),
data = neutralization)
summary(neut_mod_2_fact)
...
##  Groups    Name          Variance Std.Dev.
##  item_pair (Intercept)   773.07   27.80
##  subject   voicing_fact1   6.51    2.55
##  subject.1 (Intercept)   765.93   27.68
##  Residual                405.51   20.14
## Number of obs: 749, groups:  item_pair, 24; subject, 16
##
## Fixed effects:
##               Estimate Std. Error t value
## (Intercept)    155.747      8.979   17.35
## voicing_fact1    4.639      0.974    4.76
##
...
```

The random effects for this model look fine. This is essentially the same model as neut_mod_2, as we'd expect.[20]

8.7.4.2 Multilevel factors

The same issue arises for factors with three or more levels. Suppose that we wanted to add by-speaker random slopes of place to model vot_mod_2, keeping uncorrelated random effects. Doing this with the usual notation for an uncorrelated random slope again gives a model with nonsensical random effects:

```
# change from (1|item_pair) + (1|speaker) to
# (1|item_pair) + (1|place||speaker)
vot_mod_2_place_slope <- update(vot_mod_2, . ~ . +
    (0 + place | speaker))
summary(vot_mod_2_place_slope)
...
```

20. An even easier route to fit models with factors and uncorrelated random effects is the lmer_alt function from afex, which automates the process of extracting numeric predictors for factors and fitting the model. This function doesn't extend to GLMMs or to models where you want some random effects correlated and others uncorrelated, so it's still good to know the "by-hand" method.

```
##
## Random effects:
##  Groups    Name              Variance Std.Dev. Corr
##  word      (Intercept)       0.01463  0.1209
##  speaker   (Intercept)       0.02500  0.1581
##  speaker.1 speaking_rate_dev 0.00331  0.0576
##  speaker.2 placelabial       0.04039  0.2010
##            placealveolar     0.08861  0.2977   0.13
##            placevelar        0.00814  0.0902   0.98 0.34
...
```

There is one row per level of `place` rather than one row for each of its (two) contrasts. We can again extract these contrasts manually, then use them in the random-effect structure to refit the model. (Shown here is a second way to extract the contrasts, which doesn't require having an intercepts-only model fitted.)

```
mm <- model.matrix(~place, vot_voiced_core)
# Contrasts 1 and 2  (column 1 of `mm' is the intercept)
vot_voiced_core$place1 <- mm[, 2] ## contrast 1
vot_voiced_core$place2 <- mm[, 3] ## contrast 2
...
##  Groups    Name              Variance Std.Dev.
##  word      (Intercept)       0.01446  0.1202
##  speaker   (Intercept)       0.04401  0.2098
##  speaker.1 speaking_rate_dev 0.00333  0.0577
##  speaker.2 place1            0.02784  0.1668
##  speaker.3 place2            0.00190  0.0436
##  Residual                    0.30431  0.5516
## Number of obs: 7175, groups:  word, 272; speaker, 11
...
```

The random-effect estimates for this model look fine. Note that it was OK to use uncorrelated random effects here because `place` is coded with Helmert contrasts (a "centered" coding scheme), so the `place1` and `place2` predictors are centered.

Box 8.12
Practical Note: Random Slopes for Factors with Correlations

To include random slopes for factors in a model with *correlated* random effects, the usual formula can be used (`(1 + X + ... | Y)`) and no tricks are needed. For example, to fit a correlated version of the first of the preceding models:

```
neut_mod_2_fact_corr <- lmer(vowel_dur ~ voicing_fact +
  (1 + voicing_fact | subject) + (1 | item_pair),
data = neutralization)
```

However the possible issues in models with multiple random slopes (and correlated random effects) discussed apply: (near)-singular models (see `summary(neut_mod2_fact_corr)`) or very slow fits.

8.8 Model Predictions

As for non-mixed-effects models, it is very useful to generate model predictions for LMMs and use these to visualize the model's results. We'll consider cases here that are most common in practice:

- *Fixed effects only*: partial-effect plots or interaction plots, showing the model's predictions as one or more fixed-effect predictors are varied.
- *Fixed and random effects*: predictions for different levels of a grouping factor (e.g., different participants)—in the partial effect of a given predictor ("slope"), the average effect ("intercept"), or something more complex.

We'll first cover getting predicted values, which is straightforward and similar to non-mixed-effects-models, then turn to adding uncertainties, where different options exist.

8.8.1 Predicted Values (Only)

8.8.1.1 Fixed effects

For partial-effect and interaction plots, we can use functionality we've seen already to extract model predictions as specified fixed-effect predictors are varied, either holding other fixed-effect predictors constant or marginalizing over them. This can be done using existing packages (e.g., `ggpredict()` or `ggemmeans()` from ggeffects/effects/emmeans) or by hand (using e.g., `predict()`; box 8.13). Typically we want to marginalize over random effects as well, so the prediction is for an "average participant," and so on.

For example, we could use `ggpredict()` to compute the partial effect of speaking rate for the VOT model (`speaking_rate_dev`, model `vot_mod_2`):

```
## `type' argument: ``fe'' = fixed-effects
vot_pred_1 <- ggpredict(vot_mod_2, terms = c("speaking_rate_dev"),
  type = "fe")
vot_pred_1
## # Predicted values of log_vot
##
## speaking_rate_dev | Predicted |       95% CI
## ---------------------------------------------
##                -3 |      2.25 | [2.06, 2.44]
##                -2 |      2.21 | [2.04, 2.38]
##                -1 |      2.17 | [2.02, 2.32]
##                 0 |      2.13 | [1.99, 2.27]
##                 1 |      2.09 | [1.94, 2.24]
##                 2 |      2.05 | [1.88, 2.22]
##
## Adjusted for:
## * foll_high_vowel =  -0.03
## *    cons_cluster =  -0.05
## * log_corpus_freq =   0.13
## *           place = labial
## *          gender =   0.05
## *            word = 0 (population-level)
## *         speaker = 0 (population-level)
```

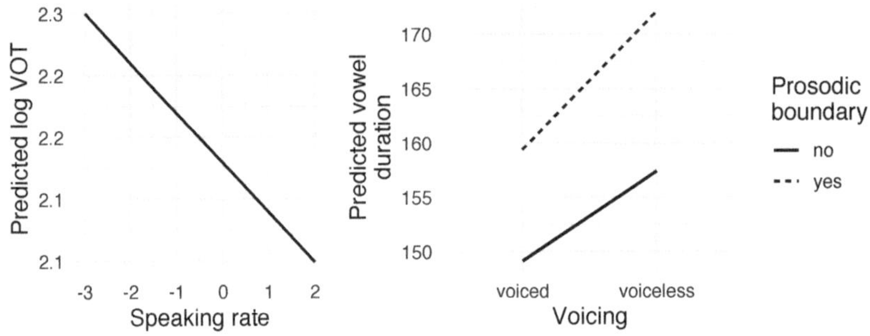

Figure 8.10
Partial effect of `speaking_rate_dev` in model `vot_mod_2` (holding other predictors constant); interaction plot of `voicing` and `prosodic_boundary` for model `neut_mod3_ixn` (marginalizing over other predictors).

This is the speaking rate effect for an average speaker and word, with other predictors held constant (figure 8.10, left).

To exemplify an interaction plot, we first fit a new `neutralization` model by adding a `voicing:prosodic_boundary` interaction to `neut_mod_3` (as in section 4.5.1):

```
neut_mod_3_ixn <- update(neut_mod_3, . ~ . +
    voicing:prosodic_boundary)
```

We could then use ggeffects functions (`ggemmeans()` or `ggpredict()`) to compute the model's predictions because `voicing` and `prosodic_boundary` are varied, for an average participant and item, marginalizing over other predictors (place, vowel). Figure 8.10 (right) shows the corresponding interaction plot.

```
neut_pred_1 <- ggemmeans(neut_mod_3_ixn,
  terms = c("voicing", "prosodic_boundary"))
```

Note that you can make quick partial-effect plots by just plotting the output of `ggpredict()`/`ggemmeans()`, for example, `plot(vot_pred_1)`, `plot(neut_pred_1)`. Figure 8.10 is just these plots customized to look nicer.[21]

8.8.1.2 Fixed and random effects

The same functionality can be used to compute group-level predictions, for example, for each participant, by just treating the grouping factor (e.g., `subject`) as another predictor being varied. For example, figure 8.5 showed the predicted voicing effect for each participant for model `neut_mod_2`, averaging across items; these predictions were made using `ggpredict()`:

21. As always you can see the plot code in the source file for this chapter. For the interaction plot, where we are varying two categorical predictors coded as numeric, it would be better to first refit the model with the `voicing` and `prosodic_boundary` fixed effects coded as factors (as in section 6.7.3).

```
## `type' argumnt: make predictions for each level of `subject'
ggpredict(neut_mod_2, terms = c("voicing", "subject"), type = "re")
## # Predicted values of vowel_dur
##
## # subject = 1
##
## voicing | Predicted |           95% CI
## --------------------------------------
##   -0.50 |    204.18 | [160.92, 247.43]
##    0.50 |    219.73 | [176.48, 262.99]
##
## # subject = 2
...
```

Box 8.13
Practical Note: Model Predictions by Hand

The text mostly uses pre-existing packages to make model predictions. It is also useful to know how to make model predictions yourself, as pre-existing packages make various assumptions, don't always give exactly what you want, and sometimes are buggy. Making predictions by hand is also the best way to understand what such packages are doing, which is crucial to figuring out how to use future packages. (By the time you read this, the ggeffects code may no longer work!)

The basic steps are the same as for nonmixed models: (1) make a new dataframe of values you want to predict at; (2) get the model's predictions for these values (using, e.g., `predict`); (2′) calculate and add CIs (if using); (3) make plots using the predictions.

Some examples are in Sonderegger, Wagner, and Torreira (2018, sections 9.4, 9.6), Bolker (2021, "Predictions and/or confidence intervals ..."), or many other places online. An example using simulation from the fitted model for (2)/(2′) rather than `predict()` is shown in section 8.8.2.

8.8.2 Predicted Values and Uncertainties

Generally we want to show model predictions with a measure of uncertainty, such as the 95% CIs computed for predictions from linear regressions (section 4.5.2).

Obtaining uncertainties is less straightforward than obtaining just model predictions because (analogously to our discussion of R^2 measures) mixed models capture several kinds of uncertainty that could be accounted for. Should we take into account just uncertainty in fixed effects (similarly to CIs for non-mixed-effects models)? Or should we account for random effects as well? If the latter, how? The issue is complex enough that lme4 simply does not provide errors on predictions (see `?predict.merMod`), and best practices are still evolving (details in box 8.14).

That said, there are two basic options, versions of which are straightforward to compute with existing packages: confidence and prediction intervals.

Confidence and prediction intervals The simplest option is to incorporate just uncertainty in the fixed-effect coefficients. The resulting CIs use the error in the population-level prediction, for example, "for an average participant/item" and are analogous to CIs for a nonmixed model. One implementation that calculates these automatically is `ggpredict()`

(or `ggemmeans()`) with the `type = 'fe'` option. These are the CIs shown above in `vot_pred_1` and `neut_pred_1` (section 8.8.1.1).

The other option is to calculate uncertainty for *predictions*, meaning new observations, by incorporating the uncertainty in the fixed-effect coefficients, as well as in individual observations. There are different ways to define this uncertainty: either just using the residual variance, or the random-effect variances (the variance components), or both. The resulting *prediction interval* (PI), however defined, captures the error in a new observation as opposed to an "average" observation and is thus always larger than the CI.

One useful method, which works very generally, is to *simulate from the fitted model* using the `simulate()` method of lme4 models. This function uses the model's definition in terms of probability distributions (e.g., equation (8.2)), which gives a probability distribution over possible y for an observation given its values for the predictors and coefficient values for the model. Suppose we have set up a "prediction dataframe," which specifies the predictor values for observations for which y is to be predicted (as in section 4.4.1 for linear regression). New model coefficients are simulated N times, then used to get N predictions of y for row i of the dataframe; the 95% quantiles are the PI. (See `?simulate.merMod` for more.)

It is possible to generate simulation-based PIs using functions from other packages that are "wrappers" for `simulate.merMod()`, such as `ggpredict()` (from ggeffects) using the `type = 'sim'` option. However, given the very different ways prediction error can be defined, it is safer to use `simulate()` directly so you know what is being computed, as shown here.[22]

For example, for the partial effect of `speaking_rate_dev` in `vot_mod_2`, we would first set up a dataframe varying this predictor holding others fixed:

```
vot_pi_df <- data.frame(
  speaking_rate_dev = seq(-3, 2, by = 0.1),
  gender = 0,
  foll_high_vowel = 0,
  cons_cluster = 0,
  log_corpus_freq = 0,
  place = "labial",
  word = "XXX",
  speaker = "XXX"
)
```

Here we have made the grouping factors `word` and `speaker` have new (unobserved) levels because this makes `simulate.merMod()` return population-level predictions (i.e., "values of 0 will be used for the random effects").

We then simulate many ($N = 10000$) predictions for each row and use the 95% quantiles as the PIs (and the median as the predictions):

22. This advice differs from the usual practice in this book of using functionality from existing packages when available, because I have found that existing functions for automatically computing PIs are not yet stable enough to recommend.

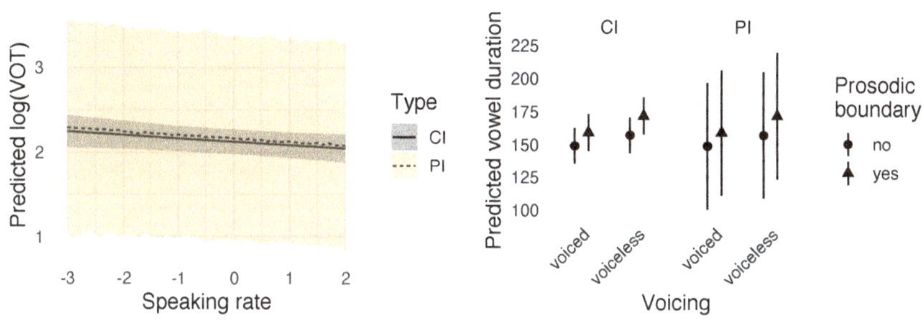

Figure 8.11
Partial-effect and interaction plots from figure 8.10, with confidence intervals and prediction intervals added.

```
## allow.new.levels=TRUE : together with speaker=XXX and word=XXX,
## makes the predictions population-level ("average speaker/word").
sims <- simulate(vot_mod_2, newdata = vot_pi_df, nsim = 10000,
  allow.new.levels = TRUE)

vot_pi_df$conf.low <- apply(sims, 1, function(x) {
  quantile(x, 0.025)
})
vot_pi_df$conf.high <- apply(sims, 1, function(x) {
  quantile(x, 0.975)
})
vot_pi_df$predicted <- apply(sims, 1, mean)

vot_pi_df %>% head(n=2)
##   speaking_rate_dev gender foll_high_vowel cons_cluster
## 1              -3.0      0               0            0
## 2              -2.9      0               0            0
##   log_corpus_freq  place word speaker conf.low conf.high predicted
## 1               0 labial  XXX     XXX     1.03       3.5       2.3
## 2               0 labial  XXX     XXX     0.99       3.6       2.3
```

These PIs and the previously computed CIs, for the `speaking_rate_dev` effect in `vot_mod_2`, are shown in figure 8.11 (left). We see that the PIs are much larger. The same is true for the `voicing:prosodic` boundary effect (right), for which PIs are generated similarly (code not shown).[23]

Box 8.14
Broader Context: Confidence and Prediction Intervals

CIs (as calculated by `ggpredict()` or `ggemmeans()`, here) involve computing an SE SE_i for each prediction $\hat{y}_i$; an approximate 95% interval is then $[\hat{y}_i - 1.96 \cdot SE_i, \hat{y}_i + 1.96 \cdot SE_i]$.

23. The CIs and PIs in the VOT example have slightly different "predictions" (compare dotted/dashed lines in figure 8.11, left); this is because the PIs do not marginalize over multilevel factors, here `place`. We could do this by defining `vot_pi_df` to contain all levels of `place` for each `speaking_rate_dev` value, then average the predictions across `place` levels. This doesn't matter for our focus here which is the relative width of CIs and PIs.

If you know some linear algebra, it can be useful to understand how these SEs are defined. The simplest way, which requires no computation beyond the fitted model, is done similarly to a linear regression. Let $\hat{\beta}$ be the estimated fixed-effect coefficients; $\hat{\Sigma}$ be the variance-covariance matrix of the fixed-effect coefficient estimates; and X the design matrix describing new data for which predictions will be made (each row = one observation, each column = one fixed effect). Then $\hat{y} = X\hat{\beta}$ are the predictions and $X\hat{\Sigma}X^t$ is their variance-covariance matrix. The diagonal elements of this matrix are approximately the (squared) error of each prediction, based just on error in the fixed-effect coefficients. In other words,

$$
\begin{aligned}
SE_i &= \text{diag}\left(\text{var}\left(\hat{y}\right)\right)_i = \text{diag}\left(\text{var}\left(X\hat{\beta}\right)\right)_i \\
&= \text{diag}(X\Sigma X^t)_i.
\end{aligned}
$$

This is the source of formulas like the following, which you will see in online recipes (e.g., Bolker 2021) or the CIs returned by functions in the ggeffects package (see "Short technical note" in the "ggeffects" vignette).

```
predvar <- diag(DesignMat \%*\% vcov(model) \%*\% t(Designmat))
```

This equation is useful to know for constructing (approximate) CIs for most statistical models.

PIs incorporate an estimate of observation-level error, which can be defined in different ways. The simplest option just adds the residual variance to each SE_i (as in Bolker 2021) and can be done by hand. The gold standard is PB (using `bootMer()`) to calculate a 95% PI for each observation, which takes all sources of error into account but is very computationally intensive. Section 10.6.4 shows an example. (If only error from fixed effects is taken into account, these are 95% CIs, which are more precise than the approximation described.) The "Prediction Intervals from merMod Objects" vignette in the merTools package (Knowles and Frederick 2020) gives useful discussion (and yet another PI implementation). Simulation from the fitted model, as described in section 8.8.2, is a middle way.

Box 8.15
Practical Note: Which Uncertainty to Show?

Which of confidence or prediction intervals should you use in practice to convey uncertainty in model predictions depends on the case. CIs are more commonly used, more intuitive, and are OK as long as we care mostly about fixed-effect estimates, that is, average values. But as soon as we care about anything besides average values, (some version of) PIs will be more appropriate. For example, if we were calculating the degree of overlap between `voicing` = VOICED and VOICELESS stops predicted by a model of VOT, it would make sense to use PIs incorporating all sources of uncertainty, which would capture the model's predicted VOT range for new observations.

It is good to be aware of the basic distinction between confidence and prediction intervals because different packages for making model predictions (e.g., ggeffects, emmeans, merTools) do different things by default, and what a model's effects "look like" when plotted has a big impact on the conclusions you draw from the model.

We've now seen how to add uncertainties to predictions for the most common case (partial-effect and interaction plots). This is also possible for by-participant predictions (e.g., figure 8.5), but this is a less common case, so we'll leave it until section 10.6.

8.9 Reporting the Fitted Model

The following discussion assumes that you have arrived at a final model to be reported. For broader discussion of what to report about the entire analysis process when using mixed-effects models—including the software used, your model selection process, and so on—I recommend Meteyard and Davies (2020, around table 5), on which the following discussion is mostly based.

In general the minimal report of a fitted model should include

1. A *definition of the model*, including specifying its fixed- and random-effect terms and any additional information needed (e.g., whether random-effect correlations are included). A simple way to do this is providing the model formula in lme4 format. Alternatively the definition could be written using equations (e.g., equation (8.2)), especially for readers not familiar with lme4 syntax. Verbal description of the model is useful, but it is not sufficient to unambiguously define the model.
2. *Fixed effects*, which can be provided in a table with one row per coefficient, containing its estimate, SE, and/or CI, the associated test statistic, and p-value.
3. *Random effects*, which can be provided in a table with the SDs and/or variances for each variance component and the values of any correlations. Reporting the residual term (SD and/or variance) is optional. This information is important to report even when the random effects are not of direct interest, as is often the case.[24]
4. *Sample size*, which is the number of observations, as well as the number of levels for each grouping factor.
5. One or more *quantitative summaries* of the model, including goodness of fit (e.g., marginal and conditional R^2). It is common to include AIC, BIC, log-likelihood, and/or deviance as measure(s) of model fit; however, these are only informative relative to other models of the same data.

Although everything needed for this minimum report is shown by various functions for summarizing the model we have discussed (`summary()`, `tidy()`, `glance()`, `r.squaredGLMM()`), none of these functions actually outputs *all* of this information, and none of them outputs well-formatted tables.[25]

As an example, tables 8.1 and 8.2 show all this information for `vot_mod_3` as "fixed-effect" and "random-effect" tables.

Reporting mixed-effects models is more complex than reporting non-mixed-effects models is, and the exact form of your report will depend a lot on your data and research questions. In addition to the five minimal pieces just discussed, additional information may be appropriate in your report: more detail (especially visualizations) of effects relevant for your research questions; effect-size measures or CIs for the most important effects; descriptive statistics of the whole dataset or individual groups (e.g., by-participant means for the effect of interest); and so on.

---

24. It is fine to report such information in an appendix or online supplement.

25. Various packages, such as stargazer and texreg, exist for formatting model output as tables for input into LaTeX or Word documents.

Table 8.1
Example report of fixed effects and model fit, for `vot_mod_3`. The model formula and R^2 values could alternatively be given in your text.

Fixed effects

Coefficient	$\hat{\beta}$	$SE(\hat{\beta})$	t	df	p
Intercept	2.62	0.07	38.73	11.40	<0.001
Speaking rate deviation	–0.04	0.02	–1.88	10.25	0.089
Following high vowel	0.09	0.04	2.34	60.94	0.023
Consonant cluster	0.46	0.04	11.15	140.48	<0.001
Corpus frequency (log)	0.04	0.04	1.07	70.42	0.29
Place1 (alveolar – labial)	0.32	0.02	16.38	67.35	<0.001
Place2 (velar – non-velar)	0.13	0.01	10.40	48.30	<0.001
Gender	–0.35	0.11	–3.05	9.38	0.013

Number of observations: 7175, groups: word (272), speaker (272). p-values/df calculated using the Satterthwaite approximation.
Model formula: `log_vot ~ speaking_rate_dev + foll_high_vowel + cons_cluster + log_corpus_freq + place + gender + (1 | word) + (1 + speaking_rate_dev | speaker).`
Marginal $R^2 = 0.28$, adjusted $R^2 = 0.4$.

Table 8.2
Example report of random effects, for `vot_mod_3`

Random effects

Group	Term	Variance	SD	Correlation
Word	Intercept	0.018	0.13	
Speaker	Intercept	0.045	0.21	
	Speaking rate deviation	0.0029	0.054	0.68
Residual		0.32	0.57	

8.10 Model Validation

LMMs model y at different levels—individual *observations* or among *groups* (e.g., by-participant variability)—and potential problems in the data and model may exist at each level. This means there are more possible ways to validate a mixed-effects model, but the goal of model validation remains the same: to "check whether the model is not grossly wrong" (section 5.2).

8.10.1 Observation Level

Possible problems with LMMs at the observation level, and the diagnostic tools used to address them, are mostly similar to (non-mixed-effects) linear regression (sections 5.2–5.7): Q-Q plots to evaluate normality of residuals, examining Cook's distance (CD) for influential observations, and so on.

Here we show some examples using our final model of the VOT data (`vot_mod_3`). Note that this is not a full model validation, because the concepts are similar to the non-mixed-effects case.

It will be useful to refer to the model equation for observation i:

$$y_i = \beta_0 + \alpha_{S,j[i]} + \alpha_{W,k[i]} + \underbrace{(\beta_1 + \gamma_{S,j[i]})x_{1,i}}_{\text{speaking rate}} +$$

$$+ \underbrace{(\beta_2 x_{2,i} + \cdots + \beta_7 x_{7,i})}_{\text{other predictors}} + \varepsilon_i, \tag{8.7}$$

where the residuals ε_i are normally distributed, as are the by-word random intercepts $\alpha_{W,k}$ for word k; the by-speaker random intercepts and slopes $\alpha_{S,j}$, $\gamma_{S,j}$ for speaker j follow a bivariate normal distribution. β_1 is the fixed-effect coefficient for `speaking_rate_dev`, the predictor of primary interest; $x_{1,i}$ is the value of this predictor for observation i. "Other predictors" refers to the other fixed effects (`foll_high_vowel`, etc.).

We add information useful for model validation to the dataframe: fitted values and residuals for each observation (columns `.fitted`, `.resid`, from `fortify.merMod()`), followed by CD and leverage:

```
vot_mod_3_df <- fortify.merMod(vot_mod_3)
vot_mod_3_df$.cooksd <- cooks.distance(vot_mod_3)
vot_mod_3_df$.hat <- hatvalues(vot_mod_3)
```

Box 8.16
Broader Context: More on LMM Diagnostics

Note that, like everything for mixed-effects models, residuals, leverage, and influence measures (e.g., CD) can be defined in more than one way. The values used here are the lme4 defaults, which are the simplest option (i.e., residuals are just $y_i - \hat{y}_i$) and evaluate quickly; see Fox and Weisberg (2019) for details. Loy and Hofmann (2014, and references therein) discuss other possibilities. Their paper describes the excellent HLMdiag package, which contains the most extensive functionality for mixed-effects model diagnostics, but (as of this writing) does not fully work for models with crossed random effects.

8.10.1.1 Residual plots

The augmented dataframe can then be used to make residual diagnostic plots for the observation-level residuals (ε_i in equation (8.7)) such as those in figure 8.12. Also possible would be fixed-effect-predictor versus residual plots, which are not shown.

The Q-Q plot (top right) suggests the residuals are not normally distributed. Inspecting a histogram of the residuals (top left) suggests the distribution is slightly right-skewed, but it is not unreasonably far from normal (exercise 8.8).

A fitted value-residual plot (bottom left) looks OK—it is reasonably close to a horizontal line. A scale-location plot (bottom right) suggests a small deviation from model assumptions; the residuals may have lower variance for very high `log_vot`.

So there are some small issues to look into, particularly for very high or very low y. These may not signal any problem because we are looking for large deviations, and residual plots for (G)LMMs tend to show some deviation from the ideal even for correctly specified models (Loy, Hofmann, and Cook 2017). Alternatively the issues we observe might be

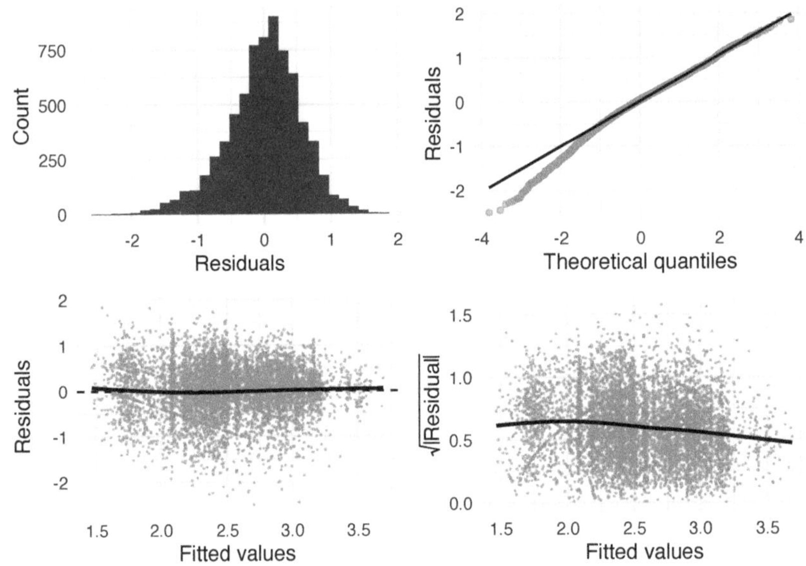

Figure 8.12
Histogram of residuals (*top left*) and three residual diagnostic plots for `vot_mod_3`: Q-Q plot of residuals (*top right*), fitted value-residual plot (*bottom left*), scale-location plot (*bottom right*).

related to other predictors/interactions predictive of log(VOT), which are used in the model reported in Sonderegger, Bane, and Graff (2017, table A1) but omitted here for simplicity.

8.10.1.2 Checking collinearity, influential observations

The metrics introduced for non-mixed-effects models, condition number (κ) and the SIF (sections 5.6.3.1, 7.6), generalize to mixed-effects models for assessing the effect of collinearity on *fixed-effect* estimates (Fox and Weisberg 2019, section 8.8).

For example, for `vot_mod_3`, we can calculate the condition number for the fixed-effect predictors ($x_1, \ldots, x_7$ in equation (8.7)):

```
## Matrix of fixed-effect predictor values, minus the intercept
my_data <- data.frame(model.matrix(vot_mod_3))[, -1]
collin.fnc(my_data)$cnumber
## [1] 2.1
```

Or we can calculate an SIF value for each of their coefficients ($\beta_1, \ldots, \beta_7$):

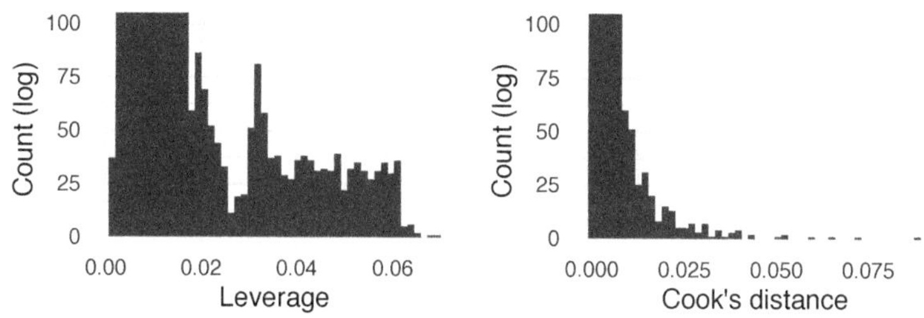

Figure 8.13
Distribution of observation-level leverage and influence measures for model `vot_mod_3`.

```
## SIF = rightmost column
car::vif(vot_mod_3)
##                    GVIF Df GVIF^(1/(2*Df))
## speaking_rate_dev   1.0  1             1.0
## foll_high_vowel     1.0  1             1.0
## cons_cluster        1.2  1             1.1
## log_corpus_freq     1.2  1             1.1
## place               1.1  2             1.0
## gender              1.0  1             1.0
```

Both measures indicate low collinearity ($\kappa < 5$, highest SIF<3.2).

To assess unusual observations, figure 8.13 shows histograms of leverage (`.hat`) and influence (`.cooksd`) of each observation, with the y-axis cut off for legibility given the large number of observations (most points have `.hat`/`.cooksd` near 0). Neither plot suggests a subset of observations that are much more influential than others.

8.10.2 Group Level

The new aspect of model validation for LMMs concerns group-level variability. There are three kinds of assumptions to be checked here (Loy and Hofmann 2014; Snijders and Bosker 2011, sections 4.6, 10.2, 10.6–10.7; Antonakis, Bastardoz, and Rönkkö 2021):

1. Random-effect terms are normally distributed (e.g., the by-word random intercepts: $\gamma_{W,j}$ in equation (8.7)).
2. No group level is overly influential (e.g., observations from one word).
3. Random-effect terms are uncorrelated with the predictors (e.g., `speaking_rate_dev` is uncorrelated with by-speaker random intercepts, across observations: $cov(x_{1,i}, \gamma_{S,J[i]}) = 0$ in equation (8.7)).

8.10.2.1 Random-effect distributions

LMMs always assume that random effects are drawn from a normal distribution, with mean 0 and some variance. We can check this assumption by examining the distribution of the random-effect estimates (the BLUPs) for a model using histograms or Q-Q plots. This check is imperfect, because this distribution often looks nonnormal even when the model is correct (e.g., simulated data), due to confounding between different random-effect terms (Loy and

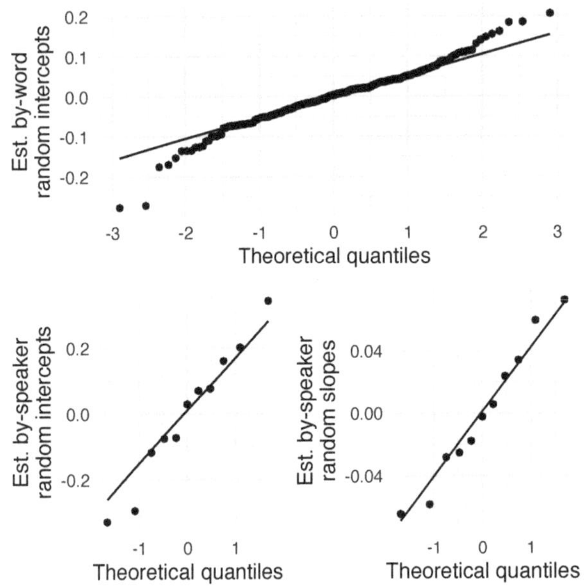

Figure 8.14
Q-Q plots for estimated random effects (BLUPs) from `vot_mod_3`: by-word (*top*) and speaker (*bottom left*) random intercepts and by-speaker `speaking_rate_dev` random slopes (*bottom right*).

Hofmann 2015). In addition, LMMs are surprisingly robust to nonnormal random-effect distributions (e.g., skewed, bimodal): simulation studies suggest that fixed-effect coefficient estimates/SEs at least are typically not affected (box 8.17). So if we are mostly interested in fixed effects, some nonnormality of random effects is OK.

Nonetheless, like checking residual plots, checking normality of random effects has benefits besides making sure your coefficient estimates are correct—it also often shows issues with the data or the model that should be investigated further.

For example, this code extracts BLUPs for the by-word random intercepts for model `vot_mod_3`:

```
## broom.mixed method to extract BLUPs in tidy format
vot_mod_3_ranef <- tidy(vot_mod_3, effects = "ran_vals")

vot_mod_3_ranef
##        effect  group    level              term estimate std.error
## 1  ran_vals   word        b       (Intercept) -0.02340     0.077
## 2  ran_vals   word     back       (Intercept) -0.06023     0.055
## 3  ran_vals   word    backs       (Intercept) -0.00597     0.120
...
# alternatively can use e.g., ranef(vot_mod_3)$word
# to get by-word random effect BLUPs
```

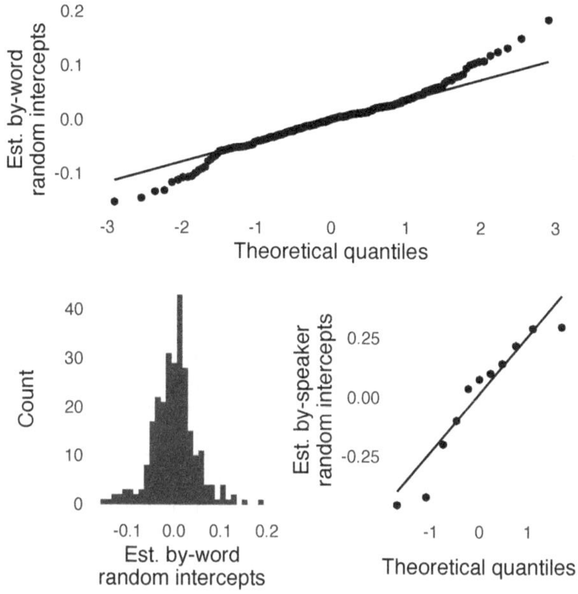

Figure 8.15
Q-Q plot (*top*) and histogram (*bottom left*) of estimated by-word random effects for `vot_mod_3` after excluding observations with `word` = DUDE or GLAD; Q-Q plot (*bottom right*) of estimated by-speaker random intercepts for the model refit without the `gender` fixed effect.

The Q-Q plots for all three types of random effects in `vot_mod_3` are shown in figure 8.14. The distributions of the by-speaker random effects (bottom panels) look basically normal. The distribution of the by-word random intercepts (top panel) looks further from normal, due especially to the two words with the lowest values:

```
filter(vot_mod_3_ranef, group == "word") %>% arrange(estimate)
##        effect group   level        term estimate std.error
## 1   ran_vals  word    dude (Intercept) -0.27521     0.104
## 2   ran_vals  word    glad (Intercept) -0.27049     0.097
...
```

These words should be flagged for further consideration. Excluding observations corresponding to these words from the model does not change any qualitative results (exercise 8.9), but still gives nonnormally distributed by-word random intercepts; the histogram and Q-Q plot are shown in figure 8.15 (top, bottom left). It looks like there are “tails” on both the left and right, suggesting there may be word-level variability that we are not accounting for in these data (i.e., missing word-level predictors).

Such deviations from normality can be useful for finding problematic data, for example, a participant who didn’t do the task correctly (in experimental data). They may also signal an important omitted speaker-level (or item-level, etc.) variable. For example, figure 8.15 (bottom right) shows the (nonnormal) distribution of by-speaker random intercepts when a model without a `gender` term is fitted to these data. This plot suggests a possible missing speaker-level variable. Supposing we didn’t know which one was missing, we could make different plots of candidate variables versus the random intercepts, and we'd eventually find that the random intercepts divide almost perfectly by speaker `gender`, suggesting this is an important omitted variable.

Box 8.17
Broader Context: The Effect of Nonnormal Random Effects

Simulation studies have addressed how robust LMMs, and to a lesser extent GLMMs, are to nonnormal random-effect distributions (and to violations of distributional assumptions generally; see Schielzeth et al. [2020] and Maas and Hox [2004], and references therein), such as skewed or bimodal distributions. In general, fixed-effect coefficient estimates are little-affected (at least for LMMs), which makes intuitive sense—the population-level effects, which average across group levels, should be roughly independent of the exact structure of group-level variability. Random-effect variance estimates are unbiased but less reliable (higher SEs), and estimated random effects (BLUPs for individual levels) can be greatly affected. This is all based on simulations that may not generalize to your data; for example, most simulation studies assume that 30 levels for a grouping factor is "low" whereas linguistic data frequently have <30 levels.

But the general points seem likely to hold, so it is a reasonable working assumption that nonnormality of random effects doesn't matter much if you are just interested in fixed effects. This assumption does not absolve you of checking random-effect distributions, but it should offer reassurance that your model's results are robust to an assumption that is difficult to check and often doesn't hold in practice.

As a concrete example, consider model `vot_mod_3` (exercise 8.9): when data from just the words DUDE and GLAD are excluded (out of 272 words), the estimated variance of the by-word random intercepts decreases by 1/3, and the CI narrows (lower SE). This makes sense: because the model assumes random effects are normally distributed, the only way to account for a few outlier words is to increase the random-effect variance. However, the fixed effects (estimates, SEs) barely change. Sonderegger, Wagner, and Torreira (2018, section 7.9) shows another example.

8.10.2.2 Influence

For nonmixed models (section 5.7.3) we used CD as a measure of influence, roughly defined as "how much do the model's coefficients ($\hat{\beta}$s) change when observation i is removed?" The related DFBETA measure quantifies influence with respect to one predictor (x_1): "How much does the coefficient $\hat{\beta}_1$ change when observation i is removed?"

CD and DFBETAs can be similarly defined for mixed-effects models, at each level of the model:

- How much do the fixed-effect coefficients (or a particular coefficient) change when observation i is removed?
- Or: when participant j is removed?
- Or: when item k is removed?

That is, there can be influential participants, items (or level of any grouping factor: words, etc.), or observations. Let's assume here that fixed-effect coefficients are of interest, but similar measures can be computed to quantify influence on the variance components. These can all be computed for lme4 models using functions from the car package.

We already considered influential observations; typically it is of more interest to check for influential groups (e.g., speakers). Unlike for nonmixed models, there is no formula to compute CD (or DFBETA) for this case; the model must be refitted without a given participant/item/and so on. to check how much the model changes. The `influence()` function from the car package (see `?influence.mixed.models`) automates this process by using some optimizations that make it faster than simply refitting the model J times for J speakers.

For model `vot_mod_3`, we first compute objects that contain all the information to compute by-speaker and by-word influence diagnostics:

```
alt_est_spk <- influence(model = vot_mod_3, group = "speaker")
alt_est_word <- influence(model = vot_mod_3, group = "word")
```

The following code extracts by-speaker CD and by-word DFBETA for the `speaking_rate_dev` fixed-effect coefficient (which is of primary interest) and puts them into a legible dataframe:

```
## Extract dfbetas as a numeric vector
dbS <- data.frame(dfbeta(alt_est_spk))[, "speaking_rate_dev"]
spk_infl_df <- data.frame(
  dfbeta = dbS,
  cd = cooks.distance(alt_est_spk),
  speaker = alt_est_spk$deleted
)
```

Similarly for by-word influence measures:

```
dbS <- data.frame(dfbeta(alt_est_word))[, "speaking_rate_dev"]
word_infl_df <- data.frame(
  dfbeta = dbS,
  cd = cooks.distance(alt_est_word),
  word = alt_est_word$deleted
)
```

For example, the speakers with the highest DFBETA values and the words with the highest CD values are

```
arrange(spk_infl_df, -abs(dfbeta))
##       dfbeta   cd speaker
## 1 -0.01250 0.46    dale
## 2  0.00935 1.20 darnell
## 3 -0.00906 0.89 michael
## 4  0.00612 0.51 rebecca
...
```

```
arrange(word_infl_df, -abs(cd))
##        dfbeta      cd    word
## 1   1.0e-04 2.0e-01   drunk
## 2  -1.2e-02 1.8e-01     but
## 3   6.4e-04 1.7e-01    glad
## 4   8.2e-04 1.7e-01    dude
...
```

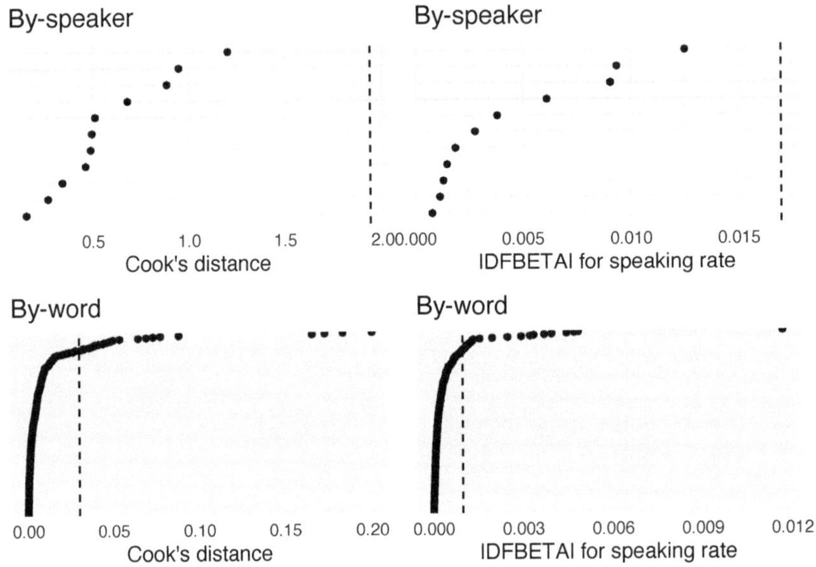

Figure 8.16
Dotplots of measures of influence of each speaker (*top row*) and each word (*bottom row*) on the fixed effects of model `vot_mod_3`: CD (*left column*) and DFBETA for the `speaking_rate_dev` coefficient (*right column*). Vertical lines are $Q_3 + 3 \cdot IQR$ cutoffs.

Figure 8.16 shows the values of CD and |DFBETA| (for `speaking_rate_dev`) for speakers and words, using dotplots, which are a useful way to visualize influence measures (Loy and Hofmann 2014).[26]

As for nonmixed models: we can use either visual inspection or a cutoff value to decide which speakers/words to flag as overly influential. Some possible cutoffs are (Belsley, Kuh, and Welsch 2004; Loy and Hofmann 2014)

- CD: $4/k$, where k is the number of levels of the grouping factor
- |DFBETA|: $2/\sqrt{k}$
- Either CD or |DFBETA|: $Q_3 + 3 \cdot IQR$ (e.g., 75th percentile + 3 · interquartile range of the CD values)

Figure 8.16 shows $Q_3 + 3 \cdot IQR$ cutoffs. It is visually clear that the cutoffs aren't very helpful here: they flag no speakers and too many words to practically investigate. Using visual inspection instead and looking for gaps in the distributions, we could flag for further consideration the speakers with the highest CD and |DFBETA| values and the words with very high CD and/or |DFBETA| values:

```
filter(word_infl_df, cd > 0.12 | abs(dfbeta) > 0.009) %>%
  select(word)
```

26. CD values are always positive, whereas DFBETA values can be negative or positive ($\hat{\beta}_1$ decreases/increases when word i is removed). It is usually more helpful to plot the absolute value of DFBETA values.

```
##     word
## 1    but
## 2 drunk
## 3  glad
## 4  dude
filter(spk_infl_df, cd > 1.0 | abs(dfbeta) > 0.01) %>%
  select(speaker)
##   speaker
## 1 darnell
## 2    dale
```

(One could alternatively use a more conservative visual cutoff and select more words/speakers.)

Note that these include the words we flagged previously based on nonnormal residuals.

At this point we have flagged data from four levels of `word` (DUDE, GLAD, BUT, DRUNK) and two levels of `speaker` (DALE, DARNELL) as potentially problematic (very influential or nonnormal residuals) for model `vot_mod_3`. Exactly what to do next is harder; as for non-mixed-effects models (section 5.7.4), there are different philosophies, and what makes sense depends on the particular data/research questions, but there are some minimum steps for good model hygiene.

For our VOT example, because the `speaking_rate_dev` fixed-effect coefficient is of primary interest, it makes sense to minimally check how the flagged speakers/words affect this term.

For example, excluding data from either flagged speaker dramatically changes the `speaking_rate_dev` estimate and its $|t|$ value:

```
## Original model
tidy(vot_mod_3) %>%
  filter(effect == "fixed" & term == "speaking_rate_dev")
##   effect group              term estimate std.error statistic
## 1  fixed  <NA> speaking_rate_dev    -0.04     0.021      -1.9


## Exclude speaker='dale'
update(vot_mod_3, subset = speaker != "dale") %>%
  tidy() %>%
  filter(effect == "fixed" & term == "speaking_rate_dev")
##   effect group              term estimate std.error statistic
## 1  fixed  <NA> speaking_rate_dev   -0.052     0.019      -2.7


## Exclude speaker='darnell'
update(vot_mod_3, subset = speaker != "darnell") %>%
  tidy() %>%
  filter(effect == "fixed" & term == "speaking_rate_dev")
##   effect group              term estimate std.error statistic
## 1  fixed  <NA> speaking_rate_dev    -0.03     0.021      -1.4
```

The estimate $\hat{\beta}_1$ varies by about 25% and $|t|$ goes from "significant" to "not significant" (by the $|t|<2$ rule), depending on which speaker is included. Excluding words also has some effect (exercise 8.9). Thus, the qualitative conclusion we would draw from this model about our research question ("is there a speaking rate effect?") is not robust to excluding influential speakers. This should be reported in any write-up of these data.

8.10.2.3 Uncorrelated random effects/predictors

Mixed-effects models assume that the random effects are uncorrelated with the predictors. Intuitively, any kind of by-group variability modeled in the random effects (e.g., by-participant intercept) should not be correlated with a predictor (e.g., speaking rate). Such correlations most commonly occur when the predictor differs systematically by group (e.g., speakers differ in mean speaking rate).

For example, consider a simpler version of `vot_mod_3`, where we model the effect of x_1 = `speaking_rate` on y = VOT, using a fixed effect of x_1 and a by-speaker random intercept.[27] If speakers differ in `speaking_rate`, as seems likely, the model can't tell whether the effect of x_1 on y is due to change in x_1 itself (the fixed effect) or speaker differences (the random effect). Technically this problem is called "endogeneity bias," and it is potentially serious for the fixed-effect coefficients: we lose any possibility of a causal interpretation of β_1, and its estimate can be inaccurate (Antonakis, Bastardoz, and Rönkkö 2021). Let's call this the *nonendogeneity assumption*.

Fortunately, there is a simple solution: adding by-group averages of y as a predictor. This changes the interpretation of the random effect, in a similar way to other group-level predictors (section 8.3.7.1), so they can no longer be correlated with x_1. For the `vot_mod_3` example we would add the speaker-level predictor `speaking_rate_mean`, which is the speaker's average for `speaking_rate` across the dataset:[28]

```
vot_mod_4 <- update(vot_mod_3, . ~ . + speaking_rate_mean)
```

The nonendogeneity assumption can be tested by a model comparison:

```
anova(vot_mod_4, vot_mod_3)
...
##            npar   AIC   BIC logLik deviance Chisq Df Pr(>Chisq)
## vot_mod_3    13 12393 12483  -6184    12367
## vot_mod_4    14 12394 12490  -6183    12366  1.53  1       0.22
```

Because `speaking_rate_mean` does not significantly improve the model, we can leave it out and consider the nonendogeneity assumption satisfied in this case.

The same problem extends to random slopes of x_1, with a similar solution. For our example, by-speaker variability in the effect of `speaking_rate` could be due to its effect differing depending on a speaker's mean speaking rate ($\bar{x}_1$). To check whether this is a

27. Note that `speaking_rate` is different from `speaking_rate_dev`; see note 7.

28. The interpretation of each speaker's random intercept is now "offset in `log_vot` from what is expected from their `speaking_rate_mean` and `gender`."

problem, we can test whether adding an $x_1 : \bar{x}_1$ interaction improves the model. For model vot_mod_3:

```
vot_mod_5 <- update(vot_mod_3, . ~ . +
    speaking_rate_mean * speaking_rate_dev)
anova(vot_mod_3, vot_mod_5)
...
##             npar   AIC   BIC logLik deviance Chisq Df Pr(>Chisq)
## vot_mod_3    13 12393 12483  -6184    12367
## vot_mod_5    15 12395 12498  -6183    12365  1.94  2       0.38
```

The high p-value suggests that endogeneity bias is not a problem for interpreting the speaking_rate_dev term of interest, so we can proceed with vot_mod_3.

Box 8.18
Broader Context: Random Effect-Predictor Correlations, Group-Level Averages

Endogeneity bias is an issue that receives a lot of attention in data analysis in some other fields—such as economics (where it is called the "random effects assumption")—on par with multicollinearity. Endogeneity bias is rarely checked for linguistic data. Determining whether endogeneity bias is important for research questions typically asked of linguistic data seems like a good direction for future work.

Because there are many possible sources of endogeneity in models of realistic data (multiple predictors, for each random-effect term) it seems unrealistic to check for all of them. In practice it seems most important to check endogeneity bias for predictors of interest for the research questions if they can reasonably vary by-group. For example, for vot_mod_3 it makes sense to consider by-speaker means of speaking_rate, but it makes less sense to consider by-word means (there is no reason words should vary in speaking rate).

Incorporating group-level averages as predictors is not only a way to check endogeneity assumptions: when adding these predictors is justified, one's findings can change in interesting ways. Snijders and Bosker (2011, sections 3.6, 4.6) discuss how the within-group and between-group predictors for the same variable (e.g., speaking_rate_dev, speaking_rate_mean) can capture qualitatively different aspects of the same phenomenon (e.g., "how does speech rate affect VOT?"). A nice example from recent work explores speaking rate effects on phonetic reduction (Cohen Priva and Gleason 2018): both "faster speakers" and "faster speech" matter in explaining degree of reduction (e.g., deleted segments) in spoken English; the authors use this to argue that faster speakers may help initiate sound changes.

8.11 Further Reading

Mixed-effects models are widely used in behavioral and social sciences, and there are many full-length treatments and tutorials on specific aspects. Some I have found particularly useful for this chapter are Snijders and Bosker (2011), Zuur (2009), McElreath (2020, chapter 13), and especially Gelman and Hill (2007, parts 2, 3), and the tutorials in the next paragraph. Model validation in particular is discussed in Snijders and Bosker (2011, chapter 10), Fox and Weisberg (2019, chapter 8), and Loy and Hofmann (2014).

Meteyard and Davies (2020) give a comprehensive best-practices tutorial for using LMMs in psychological sciences in 2020, including a long list of books and tutorials useful for language scientists. Brauer and Curtin (2018) is another excellent tutorial from this perspective. (More tutorials in this space appear frequently, such as DeBruine and Barr [2021] and Singmann and Kellen [2019].) Both focus on practical issues and summarize important primary literature. Tutorials by Bolker et al. (2009) and Bolker (2015) on (G)LMMs, though written for ecologists, are excellent. Many textbooks on statistics for linguists in particular cover mixed models, at different levels of detail (e.g., Baayen 2008, chapter 7, Gries 2021, chapter 6, Johnson 2008, chapter 7); Vasishth et al. (2021) is especially comprehensive, and Winter (2019, chapters 14, 15) is especially good for beginning/intermediate readers. Speelman, Heylen, and Geeraerts (2018) is an edited volume of case studies that apply mixed-effects models to linguistic data.

The lme4 documentation is extensive and contains useful practical advice. Pinheiro and Bates (2000) is an expansive introduction to mixed models in R using the earlier `lme` package, which typically carries over to lme4. Bates, Mächler, et al. (2015) describes the mathematical details and technical implementation of lme4. Many questions you may have about lme4 have been addressed on the `r-sig-mixed-models` mailing list. A very useful page for practical help with lme4 (and other mixed models in R) is Ben Bolker's "GLMM FAQ" webpage (Bolker 2021), which mostly summarizes answers to questions from this list.

Exercises

8.1. Calculate effect sizes for all predictors in model `neut_mod_1`, using Cohen's F (e.g., `cohens_f` in the effectsize package). Does `voicing` have a small effect, as expected (given the goal of the study)? If you know a bit about phonetics: Which predictor has the largest effect, and why does this make sense?

8.2. For model `neut_mod_3`, what range of values is the slope of `voicing` predicted to take across 95% of participants? Across 95% of items?

8.3. Comparing the output of models `vot_mod_2` and `vot_mod2_more_slopes`, verify this statement from section 8.4.6: "the model with multiple random slopes predicts substantial by-speaker variability in the effect of a following high vowel on VOT, and by-word variability in the effect of speaker gender. Accordingly, the SEs of both fixed-effect terms (`foll_high_vowel`, `gender`) increase."

8.4. Test whether speakers significantly differ in the slope of `speaking_rate_dev` in the VOT dataset, using model `vot_mod_1`. Try all methods from section 8.5.2. Does it matter which method is used if $\alpha = 0.01$? If $\alpha = 0.05$? Does it seem more likely to you that speakers do or do not differ in the speaking rate effect?

8.5. Check whether adding a by-word random slope of `voicing` to `vot_mod_2` significantly improves the model (using an LRT on REML fits, with $\alpha = 0.2$).

8.6. In exercise 8.4 we used $\alpha = 0.01$ or $\alpha = 0.05$, to test a hypothesis of speakers differing in an effect of a predictor. But in exercise 8.5, we use a higher α value ($\alpha = 0.2$) for testing whether a random-slope term should be added to a model. Both tests just involve adding a single random slope term to a model. Why are these two hypothesis tests conceptually different, and why does it make sense to use a higher α for the second one?

8.7. Redo the model comparison in section 8.5.3, assessing whether speaking rate matters for model `vot_mod_2` by using an F-test (with Kenward-Roger approximated-*df*) and a parametric bootstrap. How do the p-values from these methods compare to the value from the text, where a less exact method (LRT) is used?

8.8. This exercise addresses the Q-Q plot for model `vot_mod_3` (section 8.10.1.1).

(a) Examine a histogram of `log_vot` for these data. You should find a spike in this histogram, around a value of `vot` = k, where the distribution looks different to the left and right of k. What is k? (Hint: it should be a multiple of 5 msec.)

(b) It turns out that k was an important value in the measurement procedure for these data, which involved VOT being set to k by default, after which annotators could adjust upward or downward. In practice annotators mostly adjusted upward, so it is unclear what VOT really is for observations with `vot` = k or lower. It is unclear how to deal with this issue, but suppose we took a brute-force approach, excluding all data with `vot` $\leq k$. Refit the model for this subset of the data.

(c) Does the Q-Q plot look better for this new model? Do any qualitative conclusions, from the fixed-effect coefficients, differ between the two models?

8.9. Follow up on sections 8.10.2.1 and 8.10.2.2 by checking what effect excluding data from influential words has on `vot_mod_3`. (This includes words whose random intercepts were flagged as outliers.)

(a) As one case, refit the model without observations from the words DUDE and GLAD—these were the two words flagged in a Q-Q plot of by-word random intercepts.

(b) Verify this statement from box 8.17: "the estimated variance of the by-word random intercepts decreases ... and the CI narrows ... however, the fixed effects (estimates, SEs) barely change." Can you explain why?

9 Mixed-Effects Models 2: Logistic Regression

This chapter introduces mixed-effects logistic regression, first covering the same aspects as LMMs (sections 9.2–9.6): fitting and interpreting models, hypothesis testing, and so on. Then it discusses some practical topics—factors, nonlinear effects, variable importance—for mixed-effects models generally (sections 9.7–9.8).

9.1 Preliminaries

9.1.1 Packages

This chapter assumes that you have loaded various packages from previous chapters, as well as the afex, DHARMa, partR2, and marginaleffects packages (Singmann et al. 2021; Hartig 2021; Stoffel, Nakagawa, and Schielzeth 2021; Arel-Bundock 2021):

```
library(tidyverse)
library(broom)
library(arm)
library(lme4)
library(car)
library(broom.mixed)
library(MuMIn)
library(pbkrtest)
library(emmeans)
library(ggeffects)
library(margins)
library(ModelMetrics)
library(afex)
library(DHARMa)
library(partR2)
library(marginaleffects)

## ensures that `rescale' is not the version from the `scales' package
rescale <- arm::rescale
```

9.1.2 Data

This discussion also assumes you have set the default contrasts for factors to Helmert contrasts (section 7.2.12); loaded the `diatones` dataset from chapter 6, processed to end up with the same standardized versions of the predictors; and loaded the `french_cdi_24` data from chapter 7:

```
options(contrasts = c("contr.helmert", "contr.poly"))

diatones <- read.csv("data/diatones_rmld.csv", stringsAsFactors = TRUE)

diatones <- diatones %>% mutate(
  syll1_coda_orig = syll1_coda,
  syll2_coda_orig = syll2_coda,
  syll2_td_orig = syll2_td,
  ## turns no/yes -> 0/1, then center
  syll1_coda = rescale(syll1_coda_orig),
  ## code '0'/'C'/'CC'/'CCC' as ordered factor-> 0/1/2/3,
  ## then standardize
  syll2_coda = rescale(str_count(syll2_coda_orig, "C")),
  syll2_td = rescale(syll2_td_orig),
  frequency = rescale(frequency)
)

french_cdi_24 <- read.csv("data/french_cdi_24_rmld.csv",
  stringsAsFactors = TRUE)
```

The `givenness` data

Load the `givenness` dataset:

```
givenness <- read.csv("data/givenness_rmld.csv", stringsAsFactors = TRUE)
```

`givenness` contains data from a speech production experiment reported in Wagner (2012), examining how information structure affects which words in a sentence are pronounced with more emphasis ("prominence"). Sonderegger, Wagner, and Torreira (2018, section 10.6) describes this data and the scientific context in detail.

The data have 382 rows, each corresponding to one sentence read by a participant. The response variable is `stressshift`, which describes whether prominence on the target NP (noun phrase: intuitively, this is basically a noun) has been shifted (levels NOSHIFT, SHIFT); the numeric version is `shifted` (values 0 and 1). Of primary interest is whether the probability of shifting prominence is affected by experimental condition `conditionLabel` (levels CONTRAST, WILLIAMS); a higher probability in WILLIAMS than CONTRAST is called the *Williams effect*. Controls are the type of NP (levels FULL, PRONOUN), the `voice` of the sentence (levels ACTIVE, PASSIVE), and the sentence's `order` (how far the participant is in the experiment).

The data are grouped by `item` (16 levels), where `conditionLabel` and `npType` vary within item (each item is a 2×2 set), and each item has a fixed `voice` value. The data are also grouped by `participant` (27 levels); each participant saw none or one sentence from each `item` set:

```
xtabs(~ item + participant, data = givenness)
##     participant
## item 24 297 432 524 529 530 540 541 542 544 547 548 549 550 552
##    1  1   1   1   1   1   1   1   1   1   1   1   1   1   1   1
##    2  1   1   1   0   1   0   1   1   1   0   1   1   1   1   1
##    3  1   1   1   0   1   1   0   1   1   0   1   1   1   1   1
...
```

`conditionLabel`, `npType`, and `order` are thus observation-level (they vary within participants and items), while `voice` is item-level.

We'll mostly use versions of predictions that are standardized by "rescaling" (section 5.5.4): `clabel.williams`, `npType.pronoun`, `voice.passive`, `order.std`:

```
givenness <- mutate(givenness,
  clabel.williams = rescale(conditionLabel),
  npType.pronoun = rescale(npType),
  voice.passive = rescale(voice),
  order.std = rescale(order),
  shifted = (as.numeric(stressshift) - 1),
  ## make sure that grouping factors are factors
  item = as.factor(item),
  participant = as.factor(participant)
)
```

Figures 9.1 and 9.2 show the empirical effect of each predictor, as well as an interaction that will be included in our model.

9.2 Introduction

Mixed-effects logistic regression (MELR) is to logistic regression as linear mixed-effects models are to linear regression. MELRs combine pieces we have seen previously in chapters 6 and 8 on logistic regression and linear mixed-effects models. The response, y, is binary—it can take on values 1 or 0—and the data are grouped by one or more grouping factors. We model the log-odds that $y = 1$ as a function of fixed effects ("predictors" from logistic regression), which capture the average effects of predictors (across groups), and random effects, which capture by-group variability.

9.2.1 A First Model: `diatones` with a Random Intercept

To introduce notation and concepts, let's consider first an example with a single random-effect term using the `diatones` data.

Recall our final model for this data from section 6.7.2.2, `mlogreg_mod_3`, which we refit here as `diatones_fixef_mod`:

```
diatones_fixef_mod <- glm(formula = stress_shifted ~ syll2_coda +
  syll2_td + frequency + syll1_coda + syll2_td:frequency +
  frequency:syll1_coda, family = "binomial", data = diatones)
```

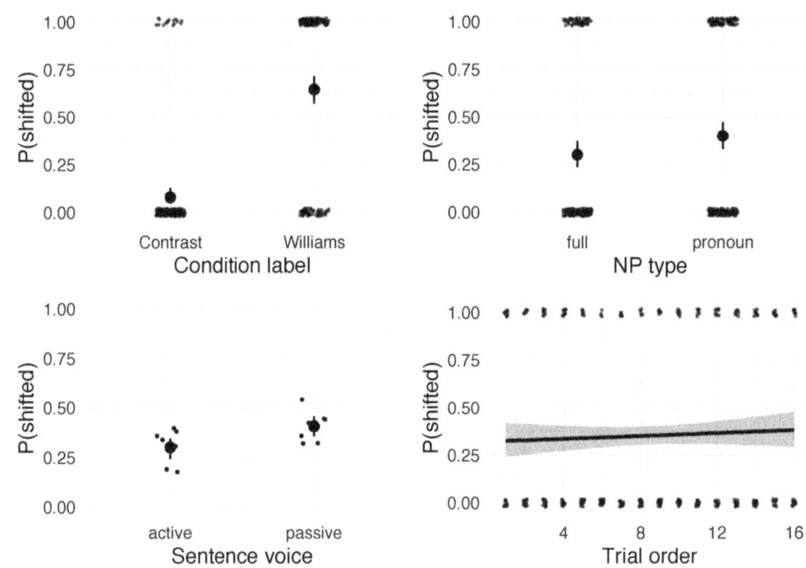

Figure 9.1
Empirical effect of each predictor, on the proportion of sentences with shifted prominence, for the `givenness` dataset. For observation-level predictors `conditionLabel`, `npType`, and `order`, points are individual observations (at $y = 0, 1$); dots/errorbars are means and exact 95% CIs; and the line/shading are predictions and 95% CIs from simple logistic regression. For item-level predictor `voice`, points and dots/errorbars are by-item proportions and corresponding means/bootstrapped 95% CIs.

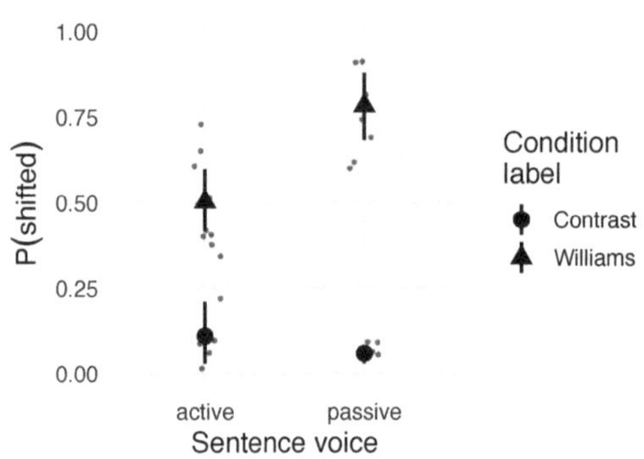

Figure 9.2
Empirical interaction plot of `voice` and `conditionLabel` for the `givenness` dataset. Points and dots/errorbars are by-item proportions and corresponding means/bootstrapped 95% CIs.

A diagnostic plot for this model (figure 6.8) suggested that words with the same prefix (predictor `prefix`) might not be independent.

We can address this nonindependence by fitting a mixed-effects logistic regression, with `prefix` as a grouping factor, and a by-`prefix` random intercept term, using familiar notation:

- As for logistic regression: $n = 130$ is the number of observations; y_i is the value of the response variable for observation i (0 or 1); x_{ij} is the value of the jth predictor.
- As for an LMM: the k predictors are fixed effects; their coefficients are $\beta_1, \ldots, \beta_k$ (in addition to the intercept, β_0).
- There are J levels of the grouping factor; $\alpha_{j[i]}$ is the random intercept for the group that observation i belongs to.

The model for the `diatones` example is then ($k = 6$, $J = 13$, $n = 130$):

$$\begin{aligned} \text{logit}\,[P(y_i = 1)] &= \beta_0 + \alpha_{j[i]} + \beta_1 x_{i1} + \cdots + \beta_6 x_{i6} && \text{for } i = 1, \ldots, 130, \\ \alpha_j &\sim N(0, \sigma_P^2), && \text{for } j = 1, \ldots, 13. \end{aligned} \quad (9.1)$$

This is the same model as `diatones_fixef_mod`, except it is now assumed that prefixes differ in their log-odds of shifting stress.

Mixed-effects logistic regressions are one kind of GLMM—analogously to logistic regression being one kind of GLM. To fit a MELR model in the lme4 package, you use the `glmer()` function (generalized linear mixed-effects regression), with a `family="binomial"` argument, similarly to fitting a logistic regression using the `glm()` function.

To fit the preceding model:

```
diatones_melr <- glmer(formula = stress_shifted ~ syll2_coda +
  syll2_td + frequency + syll1_coda + syll2_td:frequency +
  frequency:syll1_coda + (1 | prefix), family = "binomial",
  data = diatones)

summary(diatones_melr)
## Generalized linear mixed model fit by maximum likelihood
##   (Laplace Approximation) [glmerMod]
##  Family: binomial  ( logit )
## Formula:
## stress_shifted ~ syll2_coda + syll2_td + frequency + syll1_coda +
##     syll2_td:frequency + frequency:syll1_coda + (1 | prefix)
##    Data: diatones
##
##      AIC      BIC   logLik deviance df.resid
##      109      132      -47       93      122
##
## Scaled residuals:
##    Min     1Q Median     3Q    Max
## -1.055 -0.425 -0.216 -0.093  5.585
##
## Random effects:
##  Groups Name        Variance Std.Dev.
##  prefix (Intercept) 1.95     1.4
## Number of obs: 130, groups:  prefix, 13
##
## Fixed effects:
```

```
##                        Estimate Std. Error z value Pr(>|z|)
## (Intercept)             -2.252      0.662   -3.40  0.00067
## syll2_coda              -1.993      0.875   -2.28  0.02280
## syll2_td                 1.933      0.863    2.24  0.02507
## frequency               -0.857      0.642   -1.34  0.18176
## syll1_coda               2.398      1.292    1.86  0.06350
## syll2_td:frequency       2.965      1.374    2.16  0.03095
## frequency:syll1_coda     4.384      1.939    2.26  0.02379
##
## Correlation of Fixed Effects:
##             (Intr) syll2_c syll2_t frqncy syll1_ syl2_:
## syll2_coda   0.381
## syll2_td    -0.367 -0.567
## frequency    0.231  0.185  -0.302
## syll1_coda  -0.508 -0.424   0.468  -0.400
## syll2_td:fr -0.399 -0.333   0.398  -0.206  0.484
## frqncy:sy1_ -0.343 -0.300   0.495   0.000  0.336  0.336
```

As for LMMs/`lmer()`, the output of `glmer()` is verbose, and this discussion typically only shows the fixed-effect and random-effect tables. The fixed-effect and random-effect terms are interpreted similarly to an LMM, with the difference that the model predicts changes in log-odds.

9.2.2 Interpreting the Model

Fixed effects To get an intuitive sense of what adding the random effect does to the fixed effects of primary interest, compare the (fixed-effect) results for the model with this term (`Fixed effects` block) and the model without this term:

```
## fixed-effects-only model
tidy(diatones_fixef_mod)
##                    term estimate std.error statistic p.value
## 1           (Intercept)    -2.17      0.35      -6.3 3.8e-10
## 2            syll2_coda    -1.76      0.74      -2.4 1.7e-02
## 3              syll2_td     1.66      0.70       2.4 1.7e-02
## 4             frequency    -0.59      0.52      -1.1 2.5e-01
## 5            syll1_coda     1.77      0.68       2.6 8.7e-03
## 6    syll2_td:frequency     2.43      1.13       2.1 3.2e-02
## 7 frequency:syll1_coda     3.79      1.46       2.6 9.6e-03
```

To help compare, figure 9.3 shows the estimate and 95% CI for each coefficient. The results are mostly qualitatively similar, in the sense that no sign or significance of a fixed effect changes greatly: for most coefficients the 95% CIs either overlap 0 (e.g., `frequency`) or don't (e.g., `syll2_td`) in both models. (Overlapping 0 corresponds to the $|z| > 2$ threshold.) The one exception is `syll1_coda`, whose CI overlaps 0 in the mixed-effects model but not in the fixed-effects model. This makes sense: adding a by-prefix random intercept increases uncertainty in the estimate for the sole prefix-level predictor (`syll1_coda`).

The SEs are all higher in the mixed-effects model (= wider CIs), reflecting increased uncertainty about the estimates, which also raises *p*-values. This effectively means that the

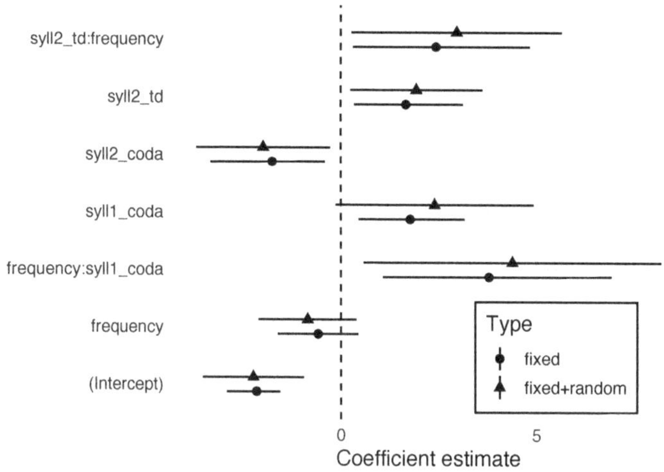

Figure 9.3
Coefficient estimates and 95% CIs, from models of the `diatones` data with and without a by-prefix random intercept (`diatones_melr`, `diatones_fixef_mod`: triangles, dots).

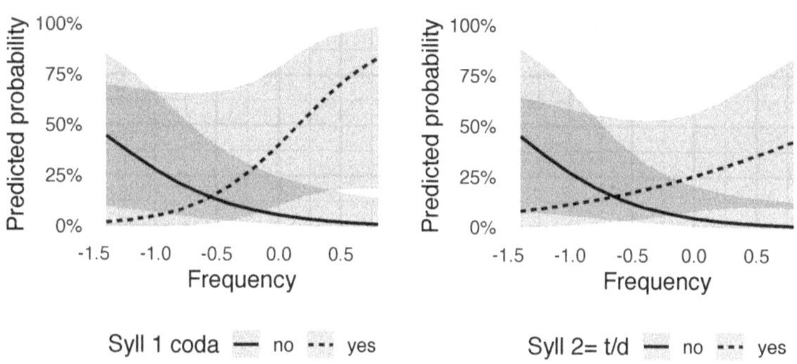

Figure 9.4
Interaction plots for each interaction in model `diatones_melr`, marginalizing over other predictors.

model's predictions look similar, but have wider CIs. For example, compare the interaction effects in figure 6.5 to figure 9.4.

Box 9.1
Practical Note: Model Prediction Tips

As for non-mixed-effects models, it is often useful to refit the model with two-level factors coded as factors (instead of their recoded numeric versions) to make prediction plots. For figure 9.4, this looks like the following:

```
diatones_melr_fact <- glmer(formula = stress_shifted ~ syll2_coda +
    syll2_td_orig + frequency + syll1_coda_orig +
    syll2_td_orig:frequency + frequency:syll1_coda_orig +
    (1 | prefix), family = "binomial", data = diatones)
```

Model predictions can then be made and plotted using similar functionality to regular logistic regression (section 6.7.3), as we will discuss more in section 9.7. Code for plots in figure 9.4 is in the code file.

Note that refitting the model with factors won't work if you have random slopes as factors and are using uncorrelated random effects (our default case). In this case you can just make model predictions as normal (e.g., `ggeffect(diatones_melr, terms=c('syll1_coda'))`), then process the resulting dataframe to turn the columns for, for example, `syll1_coda` to factors.

Random effects The random-intercept variance is new. Without worrying about exactly what the estimate in log-odds means, we can see that prefixes are predicted to differ a lot in "baseline" probability of stress shift by examining these values for each prefix ($\hat{\beta}_0 + \hat{\alpha}_j$, transformed from log-odds to probability):[1]

```
## Extracts by-word random intercept + fixed-effect intercept,
## in log-odds.
coefficients(diatones_melr)$prefix %>%
  rownames_to_column("prefix") %>%
  ## transform to probabilities
  mutate(p = invlogit(`(Intercept)`)) %>%
  select(prefix, p)
##    prefix     p
## 1      aC 0.087
## 2    comn 0.042
## 3      de 0.053
## 4     dis 0.068
## 5      eC 0.032
## 6      ex 0.056
## 7      iN 0.278
## 8     mis 0.071
## 9     pre 0.091
## 10    pro 0.449
## 11     re 0.102
## 12    sup 0.069
## 13    sur 0.393
```

Prefixes vary from roughly 3.2% probability of stress shift to 45%. The prefixes with the highest intercepts (most susceptible to stress shift) are iN, PRO, and SUR—the same ones flagged in the diagnostic plot (figure 6.8).

1. These predictions don't account for `syll1_coda`, which is a prefix-level predictor.

9.3 Two Grouping Factors, Random Intercepts and Slopes

To continue our introduction to MELRs, let's consider a more realistic example, involving crossed random effects (intercepts and slopes) and multiple predictors: modeling prominence shift (`stressshift`) in the `givenness` data (section 10.1.2), which are grouped by `participant` and `item`.

Let's fit a model, with four fixed-effect predictors: `clabel.williams` (of primary interest), `npType.pronoun`, `voice.passive` (controls), and the interaction `clabel.williams:voice.passive`, which may modulate the effect of primary interest. We'll motivate this model structure in sections 10.4.2 and 10.5.5, where we'll discuss it in the context of a model selection example. Here we'll take it as given.

The random effects contain by-participant and by-item intercepts, and all possible random slopes ("maximal"), uncorrelated. Because `clabel.williams` and `npType.pronoun` are observation-level (they vary across participants and items) whereas `voice.passive` is item-level (it varies across participants, but not items), the possible random slopes are

- Participant: `clabel.williams`, `npType.pronoun`, `voice.passive`, `clabel.williams:voice.passive`
- Item: `clabel.williams`, `npType.pronoun`

Box 9.2
Practical Note: Random Slopes for Interaction Terms

Note the interaction term in the participant list. An interaction term in the fixed effects of a mixed-effects model is itself a predictor, for which random slope terms may be possible. For example, the effect of `clabel.williams:voice.passive` can vary between participants, because both `clabel.williams` and `voice.passive` can vary, so a by-participant random slope is possible. It cannot vary between items, because `voice.passive` cannot vary, so a by-item random slope is not possible. More generally, if a predictor x is z-level (e.g., z = item), a by-z random slope is not possible for any interaction involving x.

This is a crucial point because interaction terms are often most important for the research questions. It is a common error to omit random slopes for an interaction term ($x_1{:}x_2$, say), perhaps thinking that the random slopes for subset terms (x_1, x_2) are sufficient. This raises the risk of a type I error (p-value too low), as for any predictor.

It may be helpful for understanding to explicitly write out the model equations. To do so, we define:[2]

- β_0, β_1, β_2, β_3, β_4: fixed-effect coefficients for the intercept, `clabel.williams`, `npType.pronoun`, `voice.passive`, and `clabel.williams:voice.passive`
- $x_{1,i}$,: value of `clabel.williams` for observation i ($x_{2,i}$, $x_{3,i}$, $x_{4,i}$ defined similarly)
- $\alpha_{P,j}$: random intercept for participant j
- $\alpha_{I,k}$: random intercept for item k
- $\gamma_{1,j}$: random slope of `clabel.williams` for participant j
- $\delta_{1,k}$: random slope of `clabel.williams` for item k

2. Our notation here is similar to section 8.4 for LMMs, but generalized to more complex model structure.

- Other random slopes defined similarly: $\gamma_{2,j}$, $\gamma_{3,j}$, $\gamma_{4,j}$, $\delta_{2,k}$
- $\sigma^2_{P,0}$, $\sigma^2_{I,0}$: variances of by-participant and by-item random intercepts
- $\sigma^2_{P,1}$, $\sigma^2_{P,2}$, and so on: variances of by-participant random slopes of `clabel.williams`, `npType.pronoun`, and so on.
- $\sigma^2_{I,1}$, $\sigma^2_{I,2}$: variances of by-item random slopes of `clabel.williams`, `npType.pronoun`

The model for observation i is then

$$\begin{aligned}\text{logit}\,(P(y_i = 1)) = &\left(\beta_0 + \alpha_{P,j[i]} + \alpha_{I,k[i]}\right) + \left(\beta_1 + \gamma_{1,j[i]} + \delta_{1,k[i]}\right) x_{1,i} \\ &+ \left(\beta_2 + \gamma_{2,j[i]} + \delta_{2,k[i]}\right) x_{2,i} + \left(\beta_3 + \gamma_{3,j[i]}\right) x_{3,i} \\ &+ \left(\beta_4 + \gamma_{4,j[i]}\right) x_{4,i}, \qquad \text{for } i = 1, \ldots, 382. \end{aligned} \tag{9.2}$$

The random effects all follow normal distributions:

$$\begin{aligned} &\alpha_{P,j} \sim N(0, \sigma_{P,0}), \quad \gamma_{1,j} \sim N(0, \sigma_{P,1}), \quad \gamma_{2,j} \sim N(0, \sigma_{P,2}), \\ &\gamma_{3,j} \sim N(0, \sigma_{P,3}) \quad \gamma_{4,j} \sim N(0, \sigma_{P,4}), \quad \text{for } j = 1, \ldots, 27. \end{aligned} \tag{9.3}$$

$$\begin{aligned} &\alpha_{I,k} \sim N(0, \sigma_{I,0}), \quad \gamma_{1,k} \sim N(0, \sigma_{I,1}), \\ &\gamma_{2,k} \sim N(0, \sigma_{I,2}), \quad \text{for } k = 1, \ldots, 16. \end{aligned} \tag{9.4}$$

To unpack this formula, note that

- $\beta_0 + \alpha_{P,j[i]} + \alpha_{I,k[i]}$ is the intercept for participant $j[i]$ and item $k[i]$.
- $\beta_1 + \gamma_{1,j[i]} + \delta_{1,k[i]}$ is the slope for `clabel.williams` for participant $j[i]$, item $k[i]$.
- $\beta_2 + \gamma_{2,j[i]} + \delta_{2,k[i]}$ is the slope for `npType.pronoun` for participant $j[i]$, item $k[i]$.
- $\beta_3 + \alpha_{3,j[i]}$ and $\beta_4 + \alpha_{4,j[i]}$ are the slopes for participant $j[i]$ for `voice.passive` and `clabel.williams:voice.passive`.

(Recall that observation i is from participant $j[i]$ and item $k[i]$.)

This suggests an interesting way we can think of a MELR, as a logistic regression with coefficients that vary between groups (box 9.3).

Box 9.3
Broader Context: Varying Coefficients/Hierarchical Bayesian Models

So far we have almost always written mixed-effects regression equations in one of two ways: using `lmer()`/`glmer()` formulas, or as one giant equation for y_i plus the distributions of random effects (which is essentially what the formulas are shorthand for).

It can be very useful for understanding to instead write mixed-effects regressions in different ways, as a set of equations that are being solved together. Gelman and Hill (2007, sections 12.5, 13.1) show various ways to write the same mixed-effects model, including as a classic regression (i.e., $y = \beta_0 + \beta_1 x$) whose coefficients vary across groups.

The MELR model in equation (9.2) can be rewritten as follows:

$$\text{logit}\,(P(y_i=1)) = \beta_0^{j[i],k[i]} + \beta_1^{j[i],k[i]} x_{1,i} + \beta_2^{j[i],k[i]} x_{2,i} + \beta_3^{j[i]} x_{3,i} + \beta_4^{j[i]} x_{4,i}$$

$$i = 1, \ldots, 382, \tag{9.5}$$

$$\beta_0^{j,k} = \beta_0 + \alpha_{P,j} + \alpha_{I,k}, \tag{9.6}$$

$$\beta_1^{j,k} = \beta_1 + \gamma_{1,j} + \delta_{1,k}, \tag{9.7}$$

$$\beta_2^{j,k} = \beta_2 + \gamma_{2,j} + \delta_{2,k}, \tag{9.8}$$

$$\beta_3^{j} = \beta_3 + \gamma_{3,j}, \tag{9.9}$$

$$\beta_4^{j} = \beta_4 + \gamma_{4,j}, \quad j = 1, \ldots 27, \quad k = 1, \ldots, 16. \tag{9.10}$$

(Equations (9.3) and (9.4) describing the random-effect distributions still hold.)

That is, the MELR can be thought of as a group of models—a logistic regression (equation (9.5)) with an intercept and three slopes—and those models (equations 9.6–9.10) describe how the intercept and slope coefficients vary, as a function of participant (β_0–β_4) and item (β_0–β_2). Gelman and Hill (2007, section 12.5) call this a "varying-intercept, varying-slope" model; a model with just a random intercept would be a "varying-intercept model."

Thinking of mixed-effects regressions in this way is useful for understanding exactly what they are doing; what they cannot do—for example, variance components differing by group, or autocorrelated errors; and as a stepping stone to more advanced regression models. Gelman and Hill (2007, section 12.5) describe other ways of writing mixed-effects regressions, as several equations, where it is easier to see how the equations should be changed to accommodate such extensions—but an lme4 formula cannot accommodate them.

Such extensions are possible in a more general "hierarchical Bayesian model" framework, which contains mixed-effects models as a subset. These models are increasingly used to model linguistic data and are a natural next step from mixed-effects models (see chapter 10's Further Reading section).

The terms fitted in this model—and hence shown in the regression model output—are the four fixed-effects coefficients (β_0–β_4), and the seven variance components (2 random-intercepts and 6 random-slopes: $\sigma^2_{P,0}$, $\sigma^2_{I,2}$, etc.).

Note that unlike linear mixed-effects models, there is no residual error variance term, for the same reasons as were given for logistic regression (see box 6.6).

9.3.1 Fitting the Model

To fit this model in R:

```
givenness_m1 <- glmer(stressshift ~ clabel.williams * voice.passive +
  npType.pronoun + (1 + clabel.williams + npType.pronoun || item) +
  (1 + clabel.williams * voice.passive + npType.pronoun || participant),
data = givenness, family = "binomial")

## Warning in checkConv(attr(opt, "derivs"), opt$par,
ctrl = control$checkConv, : Model failed to converge with
max|grad| = 0.209594 (tol = 0.002, component 1)
```

We get a convergence error. For MELRs, this happens a lot with the `glmer()` default settings and is commonly resolved by just changing the optimizer `glmer()` uses from the default to `bobyqa` and/or the maximum number of iterations the optimizer uses (box 9.4).

This chapter we use these solutions when necessary to fit models; in the next chapter (section 10.3), we'll discuss convergence issues in more detail.

For the current model:

```
givenness_m1 <- glmer(stressshift ~ clabel.williams * voice.passive +
    npType.pronoun + (1 + clabel.williams + npType.pronoun || item) +
    (1 + clabel.williams * voice.passive + npType.pronoun || participant),
  data = givenness, family = "binomial",
  control = glmerControl(optimizer = "bobyqa")
)

## boundary (singular) fit: see ?isSingular
```

The singular model warnings, due to some zero random-effect terms, are likely because there are not enough data relative to the number of random-effect terms (section 8.7.2). We will just ignore singular model warnings in this chapter, for simplicity, and return to these data in chapter 10. Models refitted without the offending random-effect terms are nearly identical.

Box 9.4
Broader Context: Details of GLMM and `glmer()`

We're only considering logistic regression, but as for GLMs, there are other types of GLMM (e.g., Poisson, multinomial) to which `glmer()` straightforwardly applies. See references in listed in section 9.10; these also cover mathematical details of GLMMs. In language sciences, there are few treatments of nonbinomial GLMMs (e.g. Winter and Bürkner 2021; Coupé 2018; Lo and Andrews 2015), but they are used in an increasing number of papers. I have chosen to spend a chapter just on MELR models because they are the most commonly used mixed-effects model for linguistic data.

For details of the `glmer()` implementation, the lme4 documentation contains some information, as does Bolker (2021), but there is currently no good write-up, analogous to Bates, Mächler, et al. (2015) for lme4. Still, it is useful to know a bit about the implementation in order to debug fitting problems and for perspective on why GLMMs are "harder" than LMMs.

Fitting a GLMM entails finding parameters that maximize a likelihood function, as for LMMs. As the top of `summary(diatones_melr)` says, "fit by maximum likelihood (Laplace approximation)." Note that GLMMs are fitted only by maximum likelihood, not REML (as for LMMs), because it is harder to define or compute REML for GLMMs (Bolker 2015, section 13.3.2). This means that the variance component estimates for GLMMs are a bit biased, although the bias is small except for very small sample size relative to the number of parameters (e.g., Snijders and Bosker 2011, section 4.7).

For LMMs it is possible to actually write out the likelihood. For GLMMs, it is not, and an approximation algorithm must be used to estimate the likelihood. There are several standard algorithms that trade off in speed and accuracy (Bolker et al. 2009); lme4 uses the compromise Laplace approximation by default.

This (approximate) likelihood must then be maximized, as for LMMs, using an optimization algorithm. lme4 has many possible options with many parameters; see `?glmerControl` for details. As of this writing (lme4 1.1.27.1), the most important default settings are `bobyqa` and `Nelder_Mead` for stages 1 and 2 of optimization, with `maxfun` = 10000, the maximum number of steps the algorithm tries before giving up.

This is a complex optimization problem, harder than for LMMs, and it is more common for GLMMs that the optimizer does not find a solution using the default settings ("the model doesn't converge")—even when there is nothing "wrong," the optimization parameters just need to be tweaked. Chapter 10 will go into convergence in more detail, but when fitting logistic regressions with `glmer()`, convergence problems are often resolved by setting the optimizer for both stages to `bobyqa` (as we do for `givenness_m1`) and/or increasing `maxfun`.

The model output is

```
summary(givenness_m1)
...
##
## Random effects:
##  Groups        Name                          Variance Std.Dev.
##  participant   clabel.williams:voice.passive 4.672    2.162
##  participant.1 npType.pronoun                0.847    0.920
##  participant.2 voice.passive                 2.103    1.450
##  participant.3 clabel.williams               0.000    0.000
##  participant.4 (Intercept)                   0.173    0.416
##  item          npType.pronoun                0.546    0.739
##  item.1        clabel.williams               0.656    0.810
##  item.2        (Intercept)                   0.000    0.000
## Number of obs: 382, groups:  participant, 27; item, 16
##
## Fixed effects:
##                               Estimate Std. Error z value Pr(>|z|)
## (Intercept)                     -1.184      0.234   -5.07  4.0e-07
## clabel.williams                  4.026      0.594    6.77  1.3e-11
## voice.passive                    0.407      0.456    0.89    0.373
## npType.pronoun                   0.870      0.426    2.04    0.041
## clabel.williams:voice.passive    2.328      0.941    2.47    0.013
...
```

9.3.2 Interpreting the Model

9.3.2.1 Fixed effects

These are interpreted similarly to non-mixed-effects logistic regression coefficients (section 6.4.2) in log-odds, odds, or probabilities.

For example, the estimated intercept is $\hat{\beta}_0 = -1.2$: this is the log-odds of stress shifting for an average participant and average item, with all predictors held at mean values, corresponding to a probability of

```
beta_0 <- fixef(givenness_m1)[["(Intercept)"]]
invlogit(beta_0)
## [1] 0.23
```

The slope for `clabel.williams` is $\hat{\beta}_1 = 4$, meaning that for an average participant and item, the log-odds of shifting stress are this much higher in the WILLIAMS condition than in the CONTRAST condition (averaging across `voice.passive`). (Alternatively, the odds of shifting stress are 56 times higher [$= e^4$].)

To interpret the coefficients in probability space (section 9.8.3), we could calculate their average marginal effects:

```
margins(givenness_m1)
##  clabel.williams voice.passive npType.pronoun
##           0.4096        0.1017        0.09966
```

The `clabel.williams` effect (of primary interest) corresponds to a change in probability of about 0.41, which is large.

9.3.2.2 Random effects

First, for random intercepts, the estimated by-participant random intercept SD is $\hat{\sigma}_{P,0} = 0.42$. This corresponds to 95% of participants having baseline log-odds of shifting stress in $\hat{\beta}_0 \pm 1.96 \cdot \hat{\sigma}_{P,0}$, which is

```
## Extract random-effect estimate from the model table printed above:
sig_P0 <- attr(VarCorr(givenness_m1)$participant.4, "stddev")
sig_P0 <- as.numeric(sig_P0)

## Lower and upper bounds on 95% of participants:
c(beta_0 - 1.96 * sig_P0, beta_0 + 1.96 * sig_P0)
## [1] -2.00 -0.37
```

This corresponds to 95% of participants having baseline probability of shifting stress in

```
c(invlogit(beta_0 - 1.96 * sig_P0), invlogit(beta_0 + 1.96 * sig_P0))
## [1] 0.12 0.41
```

In contrast, there is no (detectable) variability among items in the baseline log-odds of shifting stress ($\hat{\sigma}_{I,0} \approx 0$).

For random slopes, consider the `conditionLabel.williams` effect of primary interest. The degrees of variability among participants and items in the `clabel.williams` slope are $\hat{\sigma}_{P,1} = 0$ and $\hat{\sigma}_{I,1} = 0.81$ (in log-odds).

The predicted `clabel.williams` slopes for different participants and items are shown, with the population-level effect, in figure 9.5. These predictions are extracted similarly to LMMs (section 8.3.4):

```
beta <- fixef(givenness_m1)[["clabel.williams"]]

particSlopes <- beta + ranef(givenness_m1)$participant$clabel.williams
## Alternatively:
## coefficients(givenness_m1)$participant$clabel.williams

itemSlopes <- beta + ranef(givenness_m1)$item$clabel.williams
## exp(particSlopes) are the slopes in odds-ratios, which are
## plotted in histograms.
```

The model predicts that the size of the Williams effect varies greatly among items—but not participants—and is robust, in the sense of having the same direction across all participants and items.

Thus, participants (but not items) differ in the intercept, while items (but not participants) differ in the Williams effect (the `clabel.williams` slope).

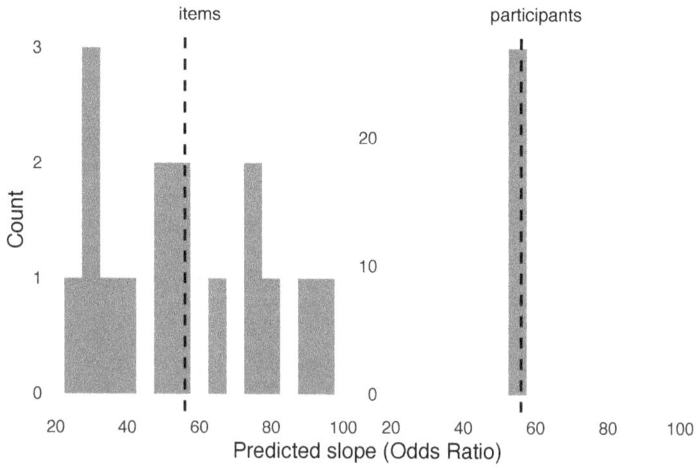

Figure 9.5
By-participant and by-item variation in the model-predicted `clabel.williams` effect for `givenness_mod1` (as odds ratios), with overlaid estimate for `clabel.williams` population-level effect (dotted lines).

> **Box 9.5**
> **Practical Note: Model Predictions by Group**
>
> There is more than one way to define a mixed-effects model's by-group predictions, such as each item's intercept or `clabel.williams` slope for the `givenness_m1` model. The most common way, which we've mostly used in this chapter and which is used by most helper functions (e.g., `coefficients()` or `ggpredict()` applied to a `(g)lmer` model), is to simply add together the relevant fixed-effect and random-effect term. This gives each item's predicted intercept and slope *without* accounting for any item-level predictors, which is unintuitive, but easier to compute. For our example, items also differ in slope because of the `clabel.williams:voicing.passive` interaction term and in their intercept due to the `voice.passive` term (recall that `voice.passive` is item-level). That is, items differ both because of their sentence voice and "random" differences. The more intuitive definition of each item's predicted intercept and slope would add in the contribution of the `clabel.williams:voicing.passive` and `voice.passive` fixed effects. "Predicted intercept for item 1" then actually means "the log-odds predicted for this item, if we obtained more data." An example is shown in section 9.7.6.
>
> For the current example, a more accurate report of the result would be "participants differ in the intercept, while items do not differ beyond what is expected from their voicing; items differ in the Williams effect, both due to their voicing and beyond this, while participants do not differ in the Williams effect."

9.4 Hypothesis Tests

As for LMMs (section 8.5), there are several options for calculating uncertainties for MELRs—SEs, *p*-values, and CIs—in different cases of interest (fixed effects, random effects, both), that trade off between accuracy and computation time:

- Wald *z*-test
- Likelihood-ratio tests
- Parametric bootstrap

Box 9.6
Broader Context: p-Values for GLMMs

The picture here is simpler than for LMMs: we don't need to worry about REML versus ML fits, and there is no method using approximate *df* (like the Satterthwaite method, which was recommended for LMMs). This is not because GLMM p-values are better-founded, but because the relevant statistical issues are less settled for GLMMs, for example, how to define a restricted likelihood or approximate *df*s (Bolker et al. 2009; Bolker 2015). In particular, it is not clear that there is a good reason why `glmer()` shows p-values and `lmer()` does not. There is not an equivalent study to Luke (2017) rigorously evaluating different p-value options for GLMMs (or MELRs). The upshot is that one should treat p-values for GLMMs and LMMs with the same caveats, in both directions (box 8.6): be skeptical of exact values and understand that the exact method used is unlikely to matter tremendously. In examples throughout section 9.4, as well as for the `diatones_melr` model (not shown), p-values computed using the three methods are within a factor of 2–3.

9.4.1 Fixed Effects

Recall that for LMMs, no p-value is shown in the `lmer()` output, reflecting different possible methods to calculate p-values for fixed-effect coefficients and whether doing so is a good idea. The picture for GLMMs is similar (box 9.6), though the methods differ slightly.

9.4.1.1 Wald tests

The roughest method (section 8.5.1.1) was to assume that the estimated coefficient value divided by its SE ($\hat{\beta}/SE(\hat{\beta})$) was normally distributed and use this to calculate a p-value (using a two-sided Wald test).

These are the p-values (and SEs, z-values) shown by default in the `glmer()` output. For example, for the `npType.pronoun` fixed-effect coefficient from `givenness_m1`:

```
tidy(givenness_m1) %>%
  filter(effect == "fixed" & term == "npType.pronoun") %>%
  select(-effect, -group)
##                term estimate std.error statistic p.value
## 1 npType.pronoun     0.87      0.43          2   0.041
```

Wald-based p-values are slightly more kosher than those for LMMs, but they are still approximations: they are only correct in the limit of a large number of groups and observations, and they are particularly bad for binomial data that show complete or quasi-complete separation (Bolker 2015, section 13.6.2).

9.4.1.2 Likelihood-ratio tests

Similarly to LMMs (section 8.5.1.2), p-values from likelihood-ratio tests (LRTs) can be calculated by refitting the model excluding k terms (fixed- or random-effect), then comparing to the full model. These are somewhat more accurate than Wald tests, but slower, because they require refitting the model.

The LRT relies on the same large-sample assumption as usual—the difference in deviance between the full and reduced model when k terms are removed follows a χ^2_k

distribution—and for smaller sample sizes, the p-values can be less accurate/anticonservative.[3] Still, they are better than Wald tests and are a reasonable default when writing up results for GLMMs.

Single terms For example, to calculate an LRT p-value for the `npType.pronoun` fixed-effect term:

```
giv_m1_noNpFixef <- update(givenness_m1, . ~ . - npType.pronoun)
anova(givenness_m1, giv_m1_noNpFixef)
...
##                  npar AIC BIC logLik deviance Chisq Df Pr(>Chisq)
## giv_m1_noNpFixef   12 347 394   -162      323
## givenness_m1       13 345 397   -160      319  3.77  1      0.052
```

This p-value is slightly higher than for the Wald test used previously.

To calculate LR-based p-values for all fixed effects in a model, the `mixed()` function from afex is useful (it automates fitting subset models and doing model comparisons):

```
givenness_m1_lr <- mixed(stressshift ~
    clabel.williams * voice.passive + npType.pronoun +
    (1 + clabel.williams + npType.pronoun || item) +
    (1 + clabel.williams * voice.passive + npType.pronoun || participant),
  data = givenness,
  family = "binomial", method = "LRT",
control = glmerControl(optimizer = "bobyqa")
)
## Fitting 5 (g)lmer() models:
## [.....]
```

```
givenness_m1_lr
...
##                                Effect df     Chisq p.value
## 1                     clabel.williams  1 33.25 ***   <.001
## 2                       voice.passive  1      0.79    .375
## 3                      npType.pronoun  1    3.77 +    .052
## 4 clabel.williams:voice.passive  1    5.86 *    .015
...
```

Multiple terms An LRT can also be used to test the contribution of $k > 1$ fixed-effect terms. For example, to test whether word frequency information contributes to the `diatones_melr` model, we would exclude the $k = 3$ terms involving `frequency`:

```
diatones_melr_noFreqFixef <- update(diatones_melr, . ~ . -
    frequency - syll2_td:frequency - frequency:syll1_coda)
```

3. A typical rule of thumb is that LRTs are only recommended when the models have 40+ levels for each grouping factor (Bolker 2015), which is often not the case for linguistic data.

```
anova(diatones_melr, diatones_melr_noFreqFixef)
...
##                          npar AIC BIC logLik deviance Chisq Df
## diatones_melr_noFreqFixef   5 117 131  -53.5    106.9
## diatones_melr               8 109 132  -46.7     93.3  13.6  3
##                          Pr(>Chisq)
## diatones_melr_noFreqFixef
## diatones_melr               0.0036
```

Thus, even though no single fixed effect involving `frequency` has a particularly low p-value (minimum $p = 0.024$), we can be confident that word frequency does help predict stress shift ($\chi^2(3) = 14$, $p = 0.0036$), after controlling for word prefix. This is a "credit assignment problem" (section 5.6.2.2).

The same method would work to test the contribution of a factor with three or more levels (as in section 9.7.3).

9.4.1.3 Parametric bootstraps

The most accurate p-values come from PB, which takes far longer. These can be calculated the same way as for LMMs (section 8.5.1.4).

Single terms For example for the `npType.pronoun` term in `givenness_m1`:

```
## ## code for running PB commented out because it takes a
## ## long time (15 minutes on my laptop, on 8 cores).
## library(parallel)
## nc <- detectCores()
## cl <- makeCluster(rep("localhost", nc))
##
## pb_giv_1 <- PBmodcomp(givenness_m1, givenness_m1_noNpFixef)
##
## ##  output the p-value for the PB test:
pb_giv_1$test["PBtest", "p.value"]
## [1] 0.082
```

This is higher than the values from the Wald or LR methods.

Multiple terms We could obtain a PB-based p-value for the example of excluding `frequency` and its interactions:

```
## ## this takes 5-10 min on my laptop
## diatones_melr_pb_1 <- PBmodcomp(diatones_melr,
##   diatones_melr_noFreqFixef, cl = cl, nsim = 1000)
diatones_melr_pb_1$test["PBtest", "p.value"]
## [1] 0.0052
```

This value is higher than the LR-based p-value.

9.4.1.4 Summary

LR-based p-values are a reasonable default when writing up results for GLMMs, but PB-based p-values are better when feasible—especially if p is near the significance threshold.

9.4.2 Random Effects

For obtaining p-values for random-effect terms in GLMMs, two of the methods used for LMMs work (section 8.5.2): LRTs and PB, with LR-based p-values conservative by the same logic as for LMMs (section 8.5.2.2).[4]

For example, we can assess whether items significantly differ in the `clabel.williams` effect in model `givenness_m1`, using an LRT to compare the models with and without the by-item `clabel.williams` random slope:

```
giv_m1_noClabelRanef <- glmer(stressshift ~
    clabel.williams * voice.passive + npType.pronoun +
    (1 + npType.pronoun || item) +
    (1 + clabel.williams * voice.passive + npType.pronoun || participant),
  data = givenness, family = "binomial",
  control = glmerControl(optimizer = "bobyqa")
)

anova(givenness_m1, giv_m1_noClabelRanef)
...
##                      npar AIC BIC logLik deviance Chisq Df
## giv_m1_noClabelRanef   12 344 392   -160      320
## givenness_m1           13 345 397   -160      319  1.08  1
##                      Pr(>Chisq)
## giv_m1_noClabelRanef
## givenness_m1                0.3
```

We could alternatively compare the models using PB:

```
## ## takes ~15 minutes on my laptop (on 8 cores)
## pb_giv_2 <- PBmodcomp(givenness_m1, giv_m1_noClabelRanef,
##    cl = cl, nsim = 1000)
##
## ##  output the p-value for the PB test:
pb_giv_2$test["PBtest", "p.value"]
## [1] 0.052
```

Note that the LR-based p-value is higher, as expected (it is conservative). By either method (with $\alpha = 0.05$), we would conclude that items do not significantly differ in the Williams effect.

4. At the time of writing, there is no stable package for testing variance components in GLMMs using "exact" tests, intermediate between LRTs and PB (like RLRsim for LMMs).

9.4.3 Other

This discussion does not cover other aspects of hypothesis testing for GLMMs, as they are very similar to LMMs.

The contribution of *fixed and random effects* could also be tested, using LRTs or PB, as for LMMs (section 8.5.3). Examples are presented in section 9.8.2.

As for LMMs (section 8.5.4), calculating *confidence intervals* is straightforward using `confint()`, with the progressively more accurate/slower Wald, profile likelihood (the default), and PB methods.

For fixed effects, we would expect the more accurate methods to give progressively wider CIs, corresponding to less anticonservative p-values.[5] For random-effect terms (variance components), we would expect profile likelihood to give wider CIs than PB, corresponding to more conservative p-values.

9.5 Model Summaries

As for LMMs (section 8.6), we can define quantitative summaries of GLMMs, such as goodness-of-fit measures or information criteria. An important difference from LMMs is that that none of these metrics are useful in isolation (as opposed to, e.g., R^2 measures for LMMs), because they do not have a straightforward interpretation. They can only be used to compare models of the same data—either a full model versus a baseline model or different candidate models.

9.5.1 Goodness of Fit

As for logistic regression (section 6.6), we consider pseudo-R^2 and classification-based measures of goodness of fit.

9.5.1.1 Pseudo-R^2 measures

Defining R^2-like measures of goodness of fit for mixed-effects logistic regression models inherits the issues with defining such measures for logistic regression and LMMs: (1) defining "variance explained," given there are no residuals for logistic regressions; (2) there are different possible ways to define "variance explained" for a mixed-effects model.

Recall that (2) was addressed for LMMs using the marginal and conditional measures R^2_m and R^2_c (section 8.6.1; Nakagawa and Schielzeth 2013; Johnson 2014; Nakagawa, Johnson, and Schielzeth 2017), which use the fitted variance components to quantify "variance explained by fixed effects" and "variance explained by fixed and random effects" in one particular way. Their definition extends to GLMMs, by defining a "residual variance" for each kind of GLM (which addresses (1)). An important difference from R^2 for LMMs is that R^2 values for GLMMs, including MELRs, are $0 \leq R^2 < 1$: they can never reach 1, and they tend to be lower (closer to 0) than for comparable LMMs.[6]

5. Note that profile likelihood-based CIs are closely related to LR-based p-values.

6. For logistic regressions, the "residual variance" can be defined in two different ways (Nakagawa, Johnson, and Schielzeth 2017, section 7), which correspond to the theoretical and delta values outputted by `r.squaredGLMM()`. This variance goes in the denominator of R^2_m and R^2_c, as extra variance that can never be explained (Nakagawa and Schielzeth 2013, 138), hence R^2_m and R^2_c are always < 1.

We can again use the `r.squaredGLMM()` function from MuMIn to compute R^2_m and R^2_c. For exposition, let's consider three MELR models for the `givenness` data:

- M1: A baseline model without fixed effects (random intercepts only: `givenness_m0`)
- M2: A model without random slopes (`givenness_m2`)
- M3: The model with random slopes (`givenness_m1`)

We first fit models M1 and M2:

```
givenness_m0 <- update(givenness_m1, . ~ (1 | participant) +
    (1 | item))
givenness_m2 <- update(givenness_m1, . ~
    clabel.williams * voice.passive + npType.pronoun + (1 | item) +
    (1 | participant))
```

To calculate R^2_m and R^2_c for the three models:

```
# baseline model
r.squaredGLMM(givenness_m0)
##             R2m R2c
## theoretical   0   0
## delta         0   0
## model without random slopes
r.squaredGLMM(givenness_m2, null = givenness_m0)
##              R2m  R2c
## theoretical 0.47 0.48
## delta       0.40 0.41
## model with random slopes
r.squaredGLMM(givenness_m1, null = givenness_m0)
##              R2m  R2c
## theoretical 0.49 0.65
## delta       0.44 0.58
```

The "theoretical" and "delta" rows use different methods to address (1) (see footnote 6). Here we'll just consider the theoretical values.

Comparing the R^2_m and R^2_c measures for these models, we see

- R^2_c is much higher in M2 and M3 than in M1: adding information about the predictors, whether as fixed or random effects, improves greatly over the baseline.
- $R^2_c - R^2_m$ is much smaller in M2 than in M3: most of the explanatory power of random effects for these data comes from random slopes; little comes from random intercepts.
- R^2_m is similar for M2 and M3, whereas R^2_c is higher for M3: adding in random slopes doesn't change the variance explained by fixed effects, but does increase the overall variance explained.

On point (1), note that $R^2_m = 0$ always for a baseline model, because it contains no fixed effects, so only R^2_c is meaningful to compare with the baseline. It is a fluke for these data that $R^2_c = 0$ for the baseline model. Usually it will be larger than 0. Exercise 9.2 gives an example, using the French CDI data.

9.5.1.2 Classification-based measures

As for non-mixed-effects logistic regression, we can alternatively measure goodness of fit by evaluating the MELR model's performance as a classifier. Two measures we ended up considering were accuracy (more intuitive, but sensitive to class sizes) and AUC (less intuitive, better-founded).

I have not found a good discussion of classification-based metrics for GLMMs (including MELRs), but I assume that we run again into issue (2): different accuracy/AUC/and so on values can be defined, depending on what predictions from the model are used. We could define "marginal" and "conditional" values, based on predictions from just the fixed effects or the fixed and random effects (held at their most likely values).

For example, `auc()` from the ModelMetrics package computes the latter. For the `diatones_melr` model and its corresponding baseline (`diatones_melr_0`), (conditional) AUC is

```
diatones_melr_0 <- update(diatones_melr, . ~ (1 | prefix),
  control = glmerControl(optimizer = "bobyqa"))
auc(diatones_melr_0)
## [1] 0.72
auc(diatones_melr)
## [1] 0.88
```

We see that the full model has high predictive power (on the original dataset), but some of this comes from simple by-prefix variability (the baseline model's AUC is well above 0.5).

We could also calculate "conditional accuracy" (in the sense defined in section 6.6.1) and "marginal accuracy" by writing a simple function:[7]

```
melr_acc <- function(mod, type = "conditional") {
  ## conditional or marginal R2?
  stopifnot(type %in% c("conditional", "marginal"))
  ## type='resp':  predictions from fixed + ran ef;
  ## type='response': preds from just fixed eff

  ## predicted responses: 0/1
  resps <- mod@resp$y # 0/1
  ## predicted probabilties
  if (type == "marginal") {
    probs <- predict(mod, type = "response", re.form = NA)
  } else {
    probs <- predict(mod, type = "response")
  }
  ## 'expected accuracy': see ?performance_pcp
  mean(resps * probs + (1 - resps) * (1 - probs))
}
```

7. This code adapts `performance_pcp()` from the performance package, which does not work for `glmer()` models at present.

```
melr_acc(diatones_melr)
## [1] 0.81
melr_acc(diatones_melr_0)
## [1] 0.76
```

We see that the full model has conditional accuracy around 0.81, which improves on the baseline model.

It would also be possible to define marginal versions of AUC and accuracy by hand (box 9.7).

Box 9.7
Practical Note: More on GLMM Classification Measures

We can compute marginal accuracy by using predicted probabilities based on fixed effects only (= marginalizing over the random effects). The `melr_acc` function does this

```
melr_acc(diatones_melr, type = "marginal")
## [1] 0.78
melr_acc(diatones_melr_0, type = "marginal")
## [1] 0.71
```

We can calculate AUC for any set of n predicted probabilities ($\hat{p}_i$), corresponding to n observations of 0 or 1 (y_i), using the `auc()` function. For marginal AUC, we use the predicted probabilities based on fixed effects only:

```
## predicted responses: 0/1
resps <- diatones_melr@resp$y

## predictions without random effects:
## full and baseline models
probs <- predict(diatones_melr, type = "response", re.form = NA)
probs_0 <- predict(diatones_melr_0, type = "response", re.form = NA)

## AUC for full model:
auc(resps, probs)
## [1] 0.78
## AUC for baseline model:
auc(resps, probs_0)
## [1] 0.5
```

Both AUC and accuracy for `diatones_melr` are lower than the conditional values, and they are better able to indicate what we could expect to happen once we apply the model to unseen data (a new word).

Reporting these "marginal" AUC/accuracy values in addition to the "conditional" values makes sense to me, by the same logic as for R^2 for LMMs (section 8.6.1), given that they convey different information and we are often most interested in "average" (fixed) effects. But I have not yet found discussion of this issue, and it seems cumbersome to report four values (conditional/marginal $\times$ full model/baseline). Reporting just the marginal or conditional values, depending on which makes more sense given your research questions, seems OK if there isn't space.

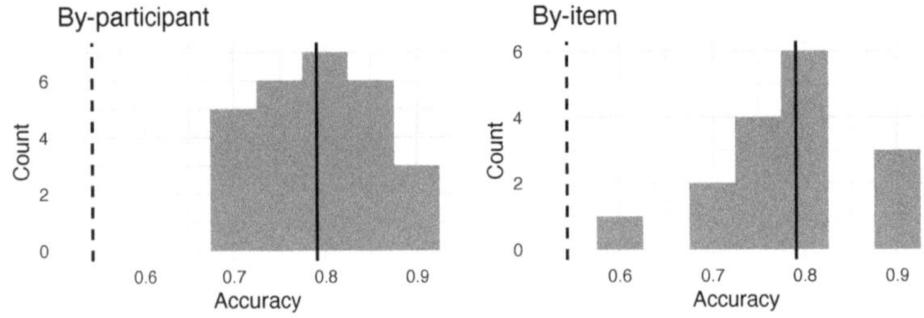

Figure 9.6
Histogram of (conditional) accuracy of model `givenness_m1` by-participant (*left*) and by-item (*right*). Dotted and solid lines are baseline and full model accuracies.

9.5.1.3 By-group goodness-of-fit

All the measures just discussed (accuracy, etc.) are observation-level: they evaluate the model's performance on classifying observations. These are easiest to report, but often we are interested in generalizing to new levels of the grouping factor (e.g., a new participant for the `givenness` data)—not observations, per se. It is possible to examine any of these measures by participant or item, to get a sense of whether the model fits some levels much better than others, and how the model might predict for a new participant/item.[8]

For example, this code calculates (conditional) accuracy for model `givenness_m1` on data from each participant and item:

```
## original dataframe with predictions added
giv_df <- fortify.merMod(givenness_m1) %>%
  mutate(prob = plogis(.fitted))

# 0/1 responses
giv_df$resp <- givenness_m1@resp$y

# by-participant accuracy
giv_by_par_df <- giv_df %>%
  group_by(participant) %>%
  summarise(acc = mean(resp * prob + (1 - resp) * (1 - prob)))

# by-item accuracy
giv_by_item_df <- giv_df %>%
  group_by(item) %>%
  summarise(acc = mean(resp * prob + (1 - resp) * (1 - prob)))
```

Their distributions are plotted in figure 9.6. There is considerable variability in accuracy for different participants and items, relative to the overall accuracy (solid lines: 0.79), but all are above baseline accuracy (dotted lines: 0.54).

8. It would be better to examine these questions by simulating new participants/items, but this is more involved; see, e.g., `simulate.merMod()`, functionality in the merTools package.

9.5.2 Information Criteria

The information criteria defined for LMMs in section 8.6.2 (marginal and conditional AIC, AICc, BIC, etc.) are defined similarly for GLMMs—with similar caveats on how a model's degrees of freedom are counted, which is more uncertain for GLMMs than for LMMs (e.g., Bolker 2021, "Model selection"). These criteria are nonetheless widely used for GLMMs, particularly the marginal versions of AIC/AICc/BIC. These are computed the same way for GLMMs as for LMMs (`AIC()`, `BIC()` default for `glmer()` models; `AICc()` from the MuMIn package) and can be used for model comparison (i.e., as an alternative to LRTs).

9.6 Model Validation

The issues motivating model validation for MELRs are similar to (non-mixed-effects) logistic regressions and LMMs (sections 6.8 and 8.10), and the methods covered there broadly apply: residual plots using deviance/binned residuals, checking for influential participants/items, and so on. Therefore, this discussion does not go into much detail here, but a couple examples are given where there is something new to say.

9.6.1 Scaled Residuals

Recall that residuals must be defined for a logistic regression model to perform model validation. The most common option is (binned) deviance residuals (section 6.8.1.1); these can also be defined for a MELR model (e.g., the `.resid` column of `fortify.merMod(myModel)`) and used to make fitted-residual plots, and so on. Unlike LMMs, R functionality to automate this process is limited for GLMMs, but this is an active development area and more should become available in the coming years.[9]

An intriguing alternative method for defining residuals for (G)LMMs is *scaled (quantile) residuals*, which have an excellent implementation in the DHARMa package (Hartig 2021), including a thorough vignette, from which I am paraphrasing.

Scaled residuals are defined by simulating from the fitted model, as in section 8.8.2. For each observation i, n new values are simulated of the fitted value ($\hat{y}_i$) to obtain a distribution. The "residual" ε_i is how far out on this distribution the observed value y_i lies, as a quantile ($0 \leq \varepsilon_i \leq 1$): $\varepsilon_i = 0.1$ means 10% of the distribution is below y_i, 0.5 means 50%, and so on.[10]

The expected distribution of the scaled residuals ε_i, if the model is correct, is uniform on [0, 1], regardless of the model structure (existence of random effects, GLM vs. LM, etc.): we expect 1% of them to lie between 0.60 and 0.61, and so on. These residuals can thus be used for all the same diagnostics as linear regression, using "uniform" as the reference distribution rather than normal.

For example, let's generate scaled residuals for model `givenness_m1` (with $n = 250$, by default):

```
sim <- simulateResiduals(givenness_m1, plot = F)
```

9. Some functionality is available in the performance package, e.g., `binned_residuals()`.

10. "Simulating from the model" can mean several things. By default DHARMa resimulates all random effects, i.e., every participant gets a new random intercept value in each of the n simulations.

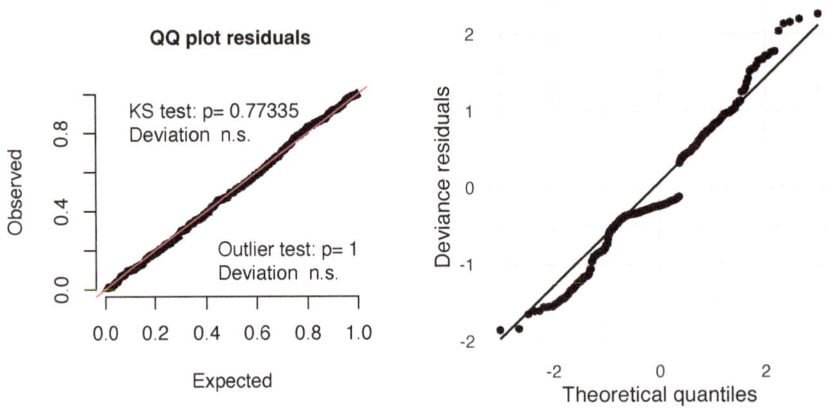

Figure 9.7
Q-Q plots for model `givenness_m1`, using scaled (*left*) and deviance (*right*) residuals.

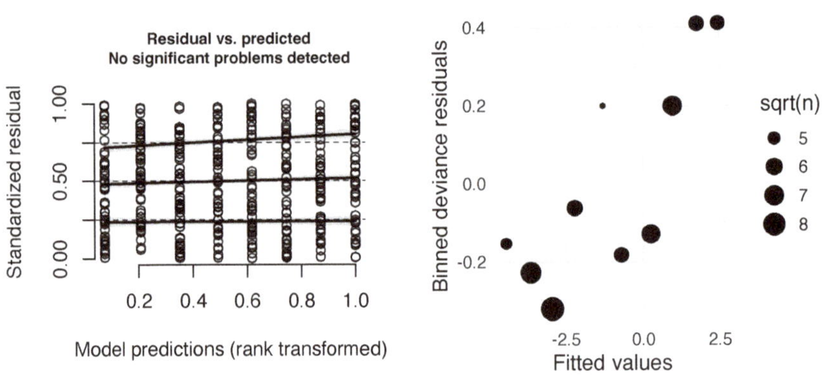

Figure 9.8
Fitted-residual plots for model `givenness_m1` using scaled residuals (*left*) and binned deviance residuals (*right*).

Figure 9.7 shows Q-Q plots using these residuals (left: using the `plotQQunif()` function) and deviance residuals (right), judged against a uniform (left) or normal (right) distribution.

The scaled residual plot is more useful because it does not indicate significant deviation from uniformity (a line), which is confirmed by a quantitative measure (a Kolmogorov-Smirnov test, used to check deviation from a reference distribution). Also shown is a summary of a test for outliers (`testOutliers()`), which we'd want to check for anyway. The deviance residual Q-Q plot is less useful because it always shows two curves, corresponding to 0 and 1 predictions, even for a correct model.

As another example, figure 9.8 shows fitted versus residual plots using scaled residuals (left: using `plotResiduals()`) and binned deviance residuals (right). In the scaled residual plot, we expect a roughly equal distribution on the y-axis from 0 to 1, regardless of the fitted value (x-axis); the plotted lines are 0.25/0.5/0.75 quantiles and should all be flat. The plot suggests no major issues. This is easy to judge visually with one point per observation.

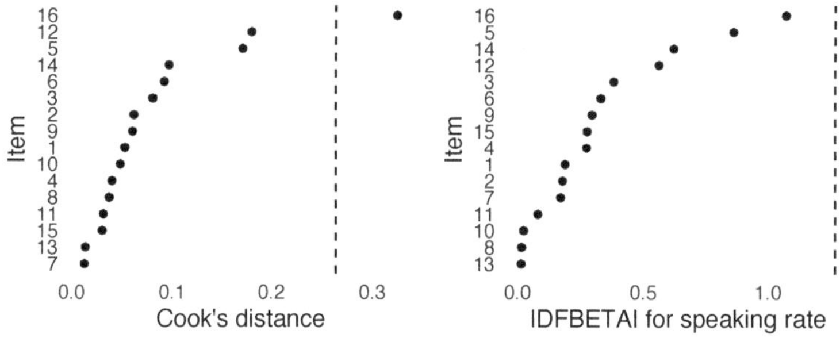

Figure 9.9
Dotplots of measures of influence of each item on the fixed effects of model `givenness_m1`: Cook's distance (*left*) and DFBETA for the `clabel.williams` coefficient (*right*). Vertical lines are $Q_3 + 3 \cdot IQR$ cutoffs.

The deviance residual plot is harder to interpret. We expect the residuals to show roughly constant mean and variance for a correct model (as in section 6.8.1.1), which is probably not the case here. But unfortunately deviations are also common for correctly specified models, and in any case with so few points (= bins), it is hard to judge. This kind of plot can be more useful for larger datasets where more bins are used (exercise 9.3).

The DHARMa vignette shows other useful plots/tests, such as predictor-residual plots.

9.6.2 Random Effects: Influence

As for LMMs, an important model validation step for the random effects is checking for influential groups (e.g., participants, items).

For example, for `givenness_m1`, we saw that items vary a lot in the key `clabel.williams` effect of interest (figure 9.5), so it makes sense to check by-item influence. We can calculate influence diagnostics as for LMMs, then extract two of interest: CD and DFBETA for the `clabel.williams` effect.

```
giv_inf_item <- influence(givenness_m1, "item")
cw_betas <- data.frame(dfbetas(giv_inf_item))[, "clabel.williams"]

item_inf_df <- data.frame(
  item = giv_inf_item$deleted,
  clabel.williams = cw_betas,
  CD = cooks.distance(giv_inf_item)
)
item_inf_df
##     item clabel.williams    CD
## 1      1          -0.191 0.053
## 2     10           0.025 0.049
## 3     11          -0.081 0.032
...
```

Figure 9.9 plots these, as for LMMs in figure 8.16 (dotplots with $Q_3 + 3 \cdot IQR$ cutoffs). There is one item (16) with high CD. The two items with high $|DFBETA|$ are 5 and 16; these are visually separate, though neither is above the cutoff.

To see what effect these items have on the model, consider the fixed-effect coefficients before and after excluding item 16 (using `compareCoefs()` from the car package):

```
compareCoefs(givenness_m1, update(givenness_m1,
  subset = item != "16"), zvals = TRUE)
...
##
##                              Model 1 Model 2
## (Intercept)                   -1.184  -1.466
## SE                             0.234   0.319
## z                              -5.07   -4.59
##
## clabel.williams                4.026   4.895
## SE                             0.594   0.803
## z                               6.77    6.09
##
...
```

The `clabel.williams` effect is stronger when this item is excluded: it has a larger effect size and similar z (and thus p). It turns out that other influential items and participants have the same effect on the model (exercise 9.4). So we can be confident that the effect of interest is robust. When writing up the model, this should be reported (briefly), for example: "To assess robustness of the key `clabel.williams` effect, we checked for influential participants and items (visual inspection of Cook's distance and DFBETA) and found that refitting the model without these groups gave a stronger effect (higher $\hat{\beta}$, similar z)."

This differs from our LMM influence example in section 8.10.2.2, where we found the effect was driven by a couple speakers.

9.7 Nonlinear and Factor Effects

In this and the next section, we'll turn to "practical" topics, which we already discussed for non-mixed models (mostly in chapter 7) but have avoided so far for LMMs and GLMMs: multilevel factors, post hoc tests, nonlinear effects, and variable importance.

Much of this functionality extends straightforwardly to the mixed-model context, but examples may be useful.

9.7.1 CDI Data

Let's return to the `french_cdi_24` data, but now for all children (aged 24 months), rather than restricting to a single child. This lets us consider child-level variability. First, process the data similarly to section 7.1.2:

```
## exclude lexical_class = other
french_cdi_24 <- french_cdi_24 %>%
  filter(lexical_class != "other") %>%
  mutate(child = as.factor(data_id)) %>%
  droplevels() %>%
  tibble()
```

```
## set lexical_class to intuitive order
french_cdi_24 <- french_cdi_24 %>%
  mutate(lexical_class = fct_relevel(
    lexical_class,
    "function_words", "verbs", "adjectives", "nouns"
  ))
```

The research question remains how a word's `lexical_class` (factor with four levels) affects its probability of being produced (`produced`: values 0, 1), in particular noun bias and function word bias. As control predictors, we include child `sex`, as well as three properties of words—(log) frequency (`frequency`), length in phones (`num_phons`), and mean utterance length (`MLU`)—which we make standardized versions of:

```
french_cdi_24 <- french_cdi_24 %>%
  mutate(
    ## frequency: already log-transformed
    freq.std = rescale(frequency),
    ## log-transform num_phons due to right-skewed distribution
    log_nphones.std = rescale(log(num_phons)),
    MLU.std = rescale(MLU)
  )
```

Intuitively, words that are more frequent, shorter, or tend to appear in shorter utterances are expected to be acquired earlier.

These predictors are only defined for about half of the words, to which we'll restrict the data for convenience:

```
# Restrict to words where frequency, num_phones not NA
french_cdi_24 <- filter(french_cdi_24,
  !(is.na(freq.std) | is.na(log_nphones.std) | is.na(MLU.std)))
```

The resulting dataset is $n = 16146$ observations, from $J = 54$ children (`child`) and $K = 299$ words (`definition`); these are the grouping factors.

We also explicitly set the contrasts for two factors that will be predictors in our models (discussed in section 8.2):

```
contrasts(french_cdi_24$lexical_class) <- contr.helmert(4)
contrasts(french_cdi_24$sex) <- contr.helmert(2)
```

9.7.2 A Nonlinear Effect

Figure 9.10 shows the empirical effect of each predictor. Here we sketch a rationale for the model we'll fit for these data; in reality you would build up the model more rigorously, as in section 10.5.

For fixed effects, the model must include `sex` and `lexical_class`. There are three continuous predictors. For each one nonlinearity can be assessed similarly to non-mixed models, using empirical plots and comparing models with different degree nonlinear terms.

This would be done with intercepts-only random-effect structure, assuming you are following a fixed-effect-first model selection strategy (our default, as discussed in section 10.5). In this case, linear effects of word length and MLU are reasonable from the empirical plot. For word frequency, a nonlinear effect is possible; we choose a spline with $k=2$ as having lowest BIC and best reflecting the empirical trend (exercise 9.7).

For random effects, we need at least by-child and by-word random intercepts. We also add by-child random slopes for `lexical_class`, as the predictor of primary interest, by our minimal guidelines for random-effect structure (section 8.4.4). For a "real" model further random slopes should be considered, but we'll keep this one simpler for feasibility.

To fit this model:

```
cdi_mod <- glmer(produces ~ lexical_class + sex + ns(freq.std, 2) +
  log_nphones.std + (1 + lexical_class | child) + (1 | definition),
data = french_cdi_24, family = "binomial",
control = glmerControl(optimizer = "bobyqa")
)
```

It will take a while to fit this model (~15 min to fit on my laptop), and you should try not to do it more than once (box 9.8).

Box 9.8
Practical Note: Saving/Loading Models

Whenever you fit a model in R that takes a long time (minutes to hours)—which is very common once you are fitting realistic mixed models—it is good practice to only fit it once. The easiest way to do this is to save your fitted model as an .rds object:

```
## objects/ is where large saved objects for this
## book go.
saveRDS(cdi_mod, file = "objects/cdi_m2.rds")
```

You can then load it next time you need it, which is what we do in the source file for this chapter:

```
cdi_mod <- readRDS("objects/cdi_m2.rds")
```

It is very common to just fit a bunch of models in one script, which are refit every time you rerun the script. Don't do this! The time you spend writing code to save and load computed objects will massively pay off. Researchers sometimes think that something is wrong if a model does not fit instantaneously, but running time is not a good diagnostic. Fitting a GLMM is a complex optimization problem, and a reasonable model can take minutes to fit for even medium dataset size/model complexity.

The same "don't recompute" advice holds for any nontrivial computation. For example, the PB calls in section 8.5.1.4 and the simulations to create figure 2.1 (in chapter 2) take a while, so the source code for these chapters checks to see if the objects exist, loads them if they do, and only runs the computations if they don't.

The fitted model is

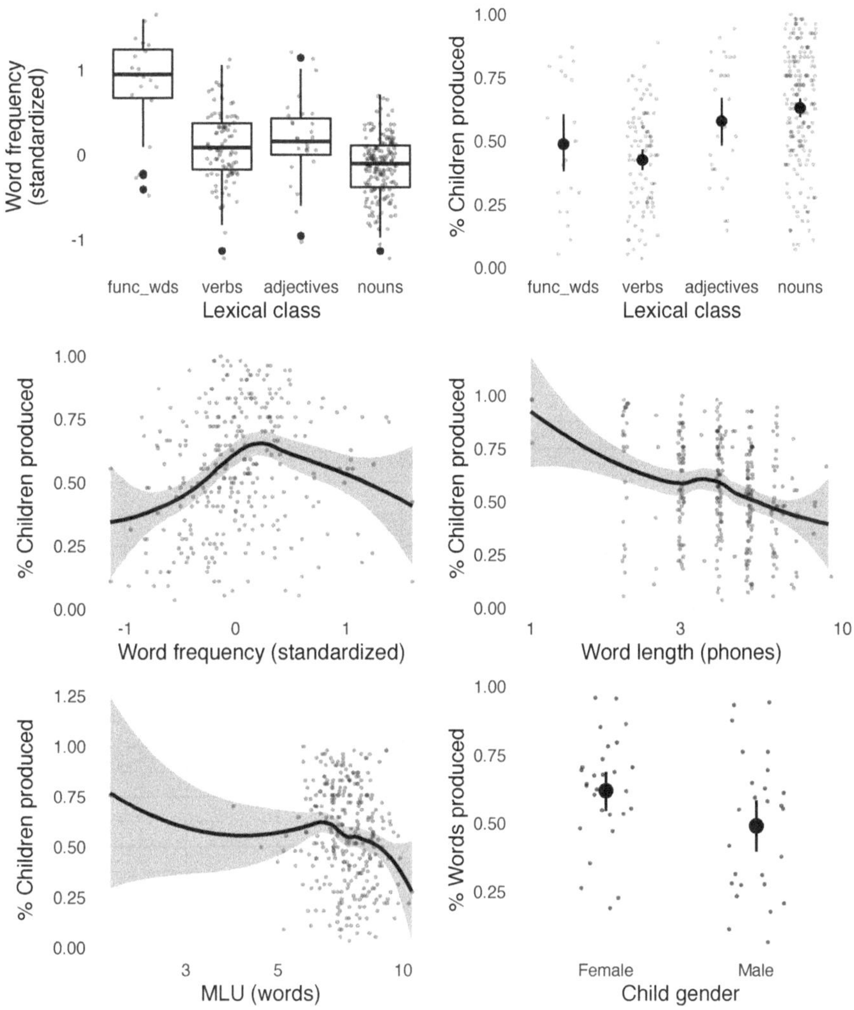

Figure 9.10
Top left: Empirical relationship between word frequency and lexical class (one point per word). *Other plots*: Empirical effect of five predictors on the proportion of words produced, for the `french_cdi_24` dataset. Points are by-child proportions for child-level predictor `sex` and by-word proportions for word-level predictors (all other predictors); dots/errorbars are means and bootstrapped 95% CIs over points, and lines/shading are LOESS smooths with 95% CIs.

```
summary(cdi_mod)
...
## Random effects:
##  Groups     Name           Variance Std.Dev. Corr
##  definition (Intercept)    2.4612   1.569
##  child      (Intercept)    3.0858   1.757
##             lexical_class1 0.3350   0.579     0.70
##             lexical_class2 0.0288   0.170     0.17 -0.37
```

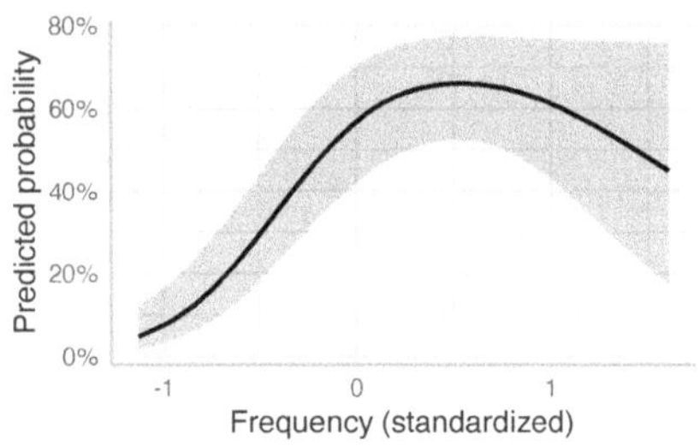

Figure 9.11
Predicted effect of `freq.std` for model `cdi_mod`, with 95% CIs, marginalizing over other predictors.

```
##              lexical_class3 0.0480    0.219    -0.01  0.17 -0.10
## Number of obs: 16146, groups:  definition, 299; child, 54
##
## Fixed effects:
##                  Estimate Std. Error z value Pr(>|z|)
## (Intercept)       -3.0320     0.5253   -5.77  7.8e-09
## lexical_class1     0.2071     0.2733    0.76  0.44869
## lexical_class2     0.3249     0.1314    2.47  0.01343
## lexical_class3     0.3963     0.0715    5.54  3.0e-08
## sex1              -0.1696     0.2040   -0.83  0.40577
## ns(freq.std, 2)1   6.4625     0.9497    6.80  1.0e-11
## ns(freq.std, 2)2   0.9170     0.7572    1.21  0.22590
## log_nphones.std   -0.9041     0.2364   -3.82  0.00013
## MLU.std           -0.7540     0.1907   -3.95  7.7e-05
##
...
```

For example, the `log_nphones.std` and `MLU.std` effects are negative (and significant), as expected. We interpret other terms throughout this discussion.

Recall that the individual rows of the model table for a nonlinear term are not very interpretable. Figure 9.11 shows the predicted effect of word frequency (using `ggemmeans()`), marginalizing over other predictors. (An example of making predictions from scratch for this model is in section 9.7.6.)

The model predicts that the `freq.std` effect is definitely positive for low-mid frequency words, then becomes less certain, possibly reversing for high-frequency words. This probably reflects function words, which on average have higher frequency but are less likely to be produced than other lexical classes (figure 9.10, top left). There are also fewer high-frequency words generally (by Zipf's law), so there are fewer data from which to infer the frequency effect.

9.7.3 Factors, Marginal Means, and Post Hoc Tests

Most functionality for factors, including different contrast coding schemes, carries over from non-mixed-effects models (sections 7.2–7.4) to fixed effects for mixed models. By default we are coding fixed effects using Helmert contrasts (as in previous chapters, especially section 7.2.12), but other schemes are possible.

For example, the sex effect in the model means that male children may be slightly less likely than female children to know an average word, but the effect is not significantly different from 0.

We can confirm that lexical_class significantly contributes to the model ("omnibus test") using an LRT:

```
## fit model without any lexical_class terms
cdi_mod_noLexClass <- update(cdi_mod, . ~ . -
    (1 + lexical_class | child) + (1 | child))
anova(cdi_mod, cdi_mod_noLexClass)
...
##                    npar   AIC   BIC logLik deviance Chisq Df
## cdi_mod_noLexClass   11 13675 13760  -6827    13653
## cdi_mod              20 13375 13528  -6667    13335   318  9
##                    Pr(>Chisq)
## cdi_mod_noLexClass
## cdi_mod                <2e-16
```

It would be possible to interpret the individual contrasts for lexical_class (e.g., lexical_class3: an average child is more likely to know NOUNS than words from other lexical classes); instead let's turn to the easier option.

9.7.4 Omnibus and Post Hoc Tests

Multilevel factors are more easily interpreted by plotting model predictions and comparing marginal means ("post hoc tests"). The emmeans package (section 7.3.2) contains functionality for mixed-effects models as well (see "Sophisticated models in emmeans" vignette).[11]

For example, we can get the "marginal means" for lexical_class—the predicted log-odds for each level, for an average child, marginalizing over other predictors (plotted in figure 9.12, left):

```
emm_cdi_1 <- emmeans(cdi_mod, ~lexical_class)
emm_cdi_1
##  lexical_class  emmean   SE  df asymp.LCL asymp.UCL
##  function_words  -0.62 0.55 Inf     -1.69      0.45
##  verbs           -0.21 0.36 Inf     -0.91      0.49
##  adjectives       0.56 0.42 Inf     -0.25      1.38
##  nouns            1.50 0.29 Inf      0.92      2.07
##
## Results are averaged over the levels of: sex
## Results are given on the logit (not the response) scale.
## Confidence level used: 0.95
```

We could then do pairwise comparisons:

11. Note that by default emmeans uses fast but less accurate p-value estimates—Wald z for GLMMs, t-as-z for LMMs unless the dataset is small—so bear in mind the p-values are a little low.

```
contrast(emm_cdi_1, method = "pairwise")
##  contrast                      estimate   SE  df z.ratio p.value
##  function_words - verbs           -0.41 0.55 Inf  -0.800  0.8700
##  function_words - adjectives      -1.18 0.56 Inf  -2.100  0.1400
##  function_words - nouns           -2.12 0.53 Inf  -4.000  <.0001
##  verbs - adjectives               -0.77 0.39 Inf  -2.000  0.2000
##  verbs - nouns                    -1.70 0.27 Inf  -6.200  <.0001
##  adjectives - nouns               -0.94 0.37 Inf  -2.500  0.0600
##
## Results are averaged over the levels of: sex
## Results are given on the log odds ratio (not the response) scale.
## P value adjustment: tukey method for comparing a family of 4 estimates
```

For a logistic regression, it may be useful to get model predictions and do post hoc tests on the probability scale—that is, as probabilities and odds ratios—rather than log-odds. This requires doing all calculations in log-odds (model predictions, CIs, etc.), then transforming, which emmeans does for you:

```
## probability space: type = 'response'
emm_cdi_2 <- emmeans(cdi_mod, ~lexical_class, type = "response")

## odds ratio, rather than difference in log-odds
## results shown using tidy() method to clean up output
contrast(emm_cdi_2, method = "pairwise") %>%
  tidy() %>%
  select(-term, -null.value, -df)
##                      contrast odds.ratio std.error null statistic
## 1      function_words / verbs       0.66     0.361    1     -0.76
## 2 function_words / adjectives       0.31     0.170    1     -2.13
## 3      function_words / nouns       0.12     0.064    1     -3.97
## 4          verbs / adjectives       0.46     0.181    1     -1.97
## 5               verbs / nouns       0.18     0.050    1     -6.24
## 6          adjectives / nouns       0.39     0.147    1     -2.50
##   adj.p.value
## 1     8.7e-01
## 2     1.4e-01
## 3     4.2e-04
## 4     2.0e-01
## 5     2.6e-09
## 6     5.9e-02
```

Either way, we would conclude that FUNCTION WORDS, VERBS < NOUNS (and nothing about ADJECTIVES). This is reflected in the model predictions, shown in log-odds and probability in figure 9.12.

While the predicted probabilities are in the expected order (FUNCTION_WORDS, VERBS, ADJECTIVES, NOUNS), there is not enough evidence to conclude that all levels significantly differ for an average child.

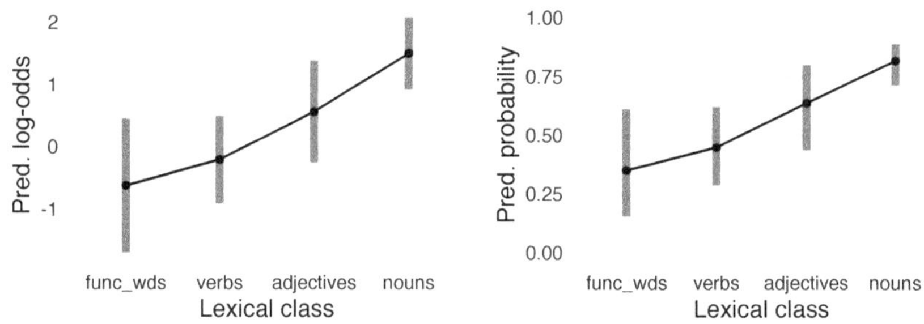

Figure 9.12
Predicted effect of `lexical_class` for model `cdi_mod` in log-odds (*left*) and probability (*right*), with 95% CIs, marginalizing over other predictors.

9.7.5 Custom Contrasts

To assess the (fixed) effect of `lexical_class`, we could alternatively set the same "custom contrasts" as for the smaller version of this dataset (section 7.3.3), to test targeted hypotheses: whether, on average, children show a noun bias (NOUNS – ADJECTIVES/VERBS) or function word bias (ADJECTIVES/VERBS – FUNCTION_WORDS):

```
interpVecs <- list(
  nounBias = c(0, -0.5, -0.5, 1),
  fwBias = c(-1, 0.5, 0.5, 0),
  verbVAdj = c(0, -1, 1, 0)
)
## verbVAdj is just a third orthogonal contrast

contrast(emm_cdi_1, method = interpVecs)
##  contrast estimate   SE  df z.ratio p.value
##  nounBias     1.32 0.26 Inf   5.000  <.0001
##  fwBias       0.80 0.52 Inf   1.500  0.1220
##  verbVAdj     0.77 0.39 Inf   2.000  0.0490
##
## Results are averaged over the levels of: sex
## Results are given on the log odds ratio (not the response) scale.
```

The post hoc test using these contrasts suggests that, (1) on average, children show noun bias and verb bias in the expected direction, and (2) only noun bias is significantly different from 0.

9.7.6 By-Group Effects

What is new with mixed-effects models is by-group variability in the effect of a factor. In this example, children vary in the effect of `lexical_class`, and we could calculate the model's prediction for each lexical class for each child. Let's predict this by hand for an "average word":

```
# set up dataframe
newdata <- french_cdi_24 %>%
```

```
  tidyr::expand(child, lexical_class) %>%
  mutate(lexical_class = fct_relevel(
    lexical_class, "function_words",
    "verbs", "adjectives", "nouns"
  )) %>%
  ## `average word': all word-level preds = 0
  mutate(log_nphones.std = 0, freq.std = 0, MLU.std = 0)

## get child-level predictors (here, just sex)
## and merge into the prediction dataframe
newdata <- french_cdi_24 %>%
  distinct(child, .keep_all = TRUE) %>%
  select(child, sex) %>%
  right_join(newdata)
contrasts(newdata$lexical_class) <- contr.helmert(4)
contrasts(newdata$sex) <- contr.helmert(2)

cdi_child_preds <- newdata %>%
  ## random-effects: by-child only
  ## no by-word REs = for an `average noun', verb, etc.
  mutate(pred_logit = predict(cdi_mod,
    newdata = newdata,
    re.form = ~ (1 + lexical_class | child)
  ))
```

These are plotted in figure 9.13 (left). Note that these predictions account for child-level variables (here, just `sex`), so the predictions really are by-child. (Functions for making model predictions, e.g., from `ggeffects()` or `effects()`, don't usually account for group-level predictors; see box 9.5.) It is visually clear that children vary massively in the intercept—meaning how many words they have acquired, total—reflecting the large by-child intercept estimate. They also differ to some extent in the `lexical_class` effect (exercise 9.8).

To relate more explicitly to the research question, we can calculate noun bias and function word bias values for each child, in log-odds:

```
## make there be one row per child, with `noun',
## `verb', etc. columns for predicted logit
cdi_child_preds_2 <- cdi_child_preds %>%
  pivot_wider(names_from = lexical_class, values_from = pred_logit) %>%
  ## calculate by-child noun/function word bias
  mutate(
    noun_bias = nouns - 0.5 * verbs - 0.5 * adjectives,
    fw_bias = 0.5 * verbs + 0.5 * adjectives - function_words
  )

cdi_child_preds_2 %>%
  select(child, noun_bias, fw_bias)
##      child noun_bias fw_bias
## 1   140272     0.974   0.517
## 2   140273     1.346   0.648
```

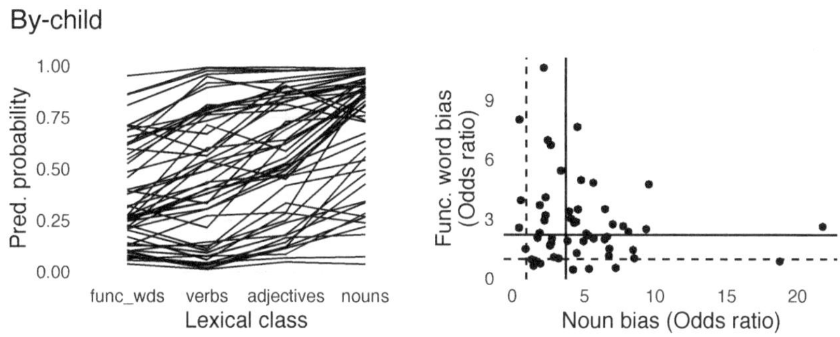

Figure 9.13
Predicted by-child probabilities of word production as a function of `lexical_class` (*left*), and noun bias and function word bias values (*right*), marginalizing over other predictors. Dotted/solid lines indicate null/population-level effects.

```
## 3  140274      1.512   2.037
...
```

These values are plotted in figure 9.13 (right). Most children show the expected direction for noun bias (positive *x*-axis values); a majority show the expected function word bias direction (positive *y*-axis values). What is clear is that children vary greatly in the effect size of their noun and verb biases.

This reflects a lot of by-child variability in the `lexical_class` effect. We can check by model comparison that children do significantly differ:

```
cdi_mod_noRanSlope <- update(cdi_mod, . ~ . -
    (1 + lexical_class | child) + (1 | child))
anova(cdi_mod, cdi_mod_noRanSlope)
...
##                    npar   AIC   BIC logLik deviance Chisq Df
## cdi_mod_noRanSlope   11 13675 13760  -6827    13653
## cdi_mod              20 13375 13528  -6667    13335   318  9
##                    Pr(>Chisq)
## cdi_mod_noRanSlope
## cdi_mod               <2e-16
```

9.7.7 Post Hoc Trends

We can also compute post hoc trends (section 7.4.1) for mixed-effects models to help interpret interactions, such as the modulation of the Williams effect by sentence voice (figure 9.2). To calculate the `clabel.williams` effect for each value of `voice.passive`:

```
giv_emt <- emtrends(givenness_m1, ~voice.passive, var = "clabel.williams")
giv_emt
##  voice.passive clabel.williams.trend   SE  df asymp.LCL asymp.UCL
##          -0.51                   2.8 0.66 Inf       1.5       4.1
```

```
##            0.49                       5.2 0.84 Inf        3.5        6.8
##
## Results are averaged over the levels of: clabel.williams, npType.pronoun
## Confidence level used: 0.95
```

The `clabel.williams` effect is positive for ACTIVE and PASSIVE sentence voice (corresponding to negative and positive `voice.passive`), averaging over other predictors—the 95% CIs don't overlap 0. We can conclude that the Williams effect exists regardless of sentence voice; only its magnitude changes.

9.8 Variable Importance

As for non-mixed-effects models (section 6.9.2), there are different possible ways to quantify "how important are different predictors" for mixed-effects models. This is very important for interpreting and reporting models with less focus on *p*-values and more on predicted effects/sizes.

9.8.1 Coefficient Sizes

For simple cases—where we are only interested in fixed effects, all predictors are standardized, and there are no multilevel factors (more than two levels) or interactions—the $|\hat{\beta}|$ values (absolute values of fixed-effect coefficient estimates) can be used, as for non-mixed-effects models. None of the mixed-effects models we've considered are this simple, so no example is provided here.

For anything more complex, there are different options to define "variable importance" for mixed-effects models. This is a difficult problem for the general case (see, e.g., Rights and Sterba 2020; Gelman and Hill 2007, section 21.4; functionality in the `merTools` package)—due partially to the ambiguity of "importance" for mixed-effects models, as we've seen a few times now, and partially to a lack of implementations that work across a wide class of models (e.g., **GLMMs**; factor or nonlinear effects; random slopes, crossed, or uncorrelated random effects).

This discussion will sketch out some options I know of, and hopefully more functionality will exist by the time you read this.

9.8.2 Model Comparison

As in section 6.9.2.1, the contribution of each predictor could be assessed by comparing the full model with a model where all its terms are dropped—for a mixed-effects model this means dropping fixed effects for the predictor, any interaction terms, and corresponding random slopes. The "significance" of each comparison can be assessed by an appropriate hypothesis test. For example, for the `givenness_m1` model we would do three model comparisons using LRTs (or PB, if we wanted to be more accurate), one for each predictor:

```
## original model formula :
## stressshift ~ clabel.williams + npType.pronoun + voice.passive +
## (1 + clabel.williams + npType.pronoun || item) +
## (1 + clabel.williams +  voice.passive + npType.pronoun || participant)
```

```
## exclude all fixed and random-effect terms
## involving clabel.williams
giv_m1_noClabel <- update(givenness_m1, . ~ npType.pronoun + voice.passive +
    (1 + npType.pronoun || item) +
    (1 + voice.passive + npType.pronoun || participant))

## same for voice.passive
giv_m1_noVoice <- update(givenness_m1, . ~ clabel.williams +
    npType.pronoun + (1 + clabel.williams + npType.pronoun || item) +
    (1 + clabel.williams + npType.pronoun || participant))

## same for npType.pronoun
giv_m1_noNpType <- update(givenness_m1, . ~
    clabel.williams * voice.passive +
    (1 + clabel.williams || item) +
    (1 + clabel.williams * voice.passive || participant))
```

```
## p-values for the LR test of each model
## vs. the full model
anova(givenness_m1, giv_m1_noClabel)$`Pr(>Chisq)`[2]
## [1] 1.6e-34
anova(givenness_m1, giv_m1_noVoice)$`Pr(>Chisq)`[2]
## [1] 0.0033
anova(givenness_m1, giv_m1_noNpType)$`Pr(>Chisq)`[2]
## [1] 0.033
```

The p-values of the three tests measure the significance of each variable's contributions and give the order of predictor importance:

$$\texttt{clabel.williams} > \texttt{voice.passive} > \texttt{npType.pronoun}$$

Alternatively, an appropriate effect size can be calculated for each model comparison, as a change in the value of any model quality metric that can be interpreted as an effect size (e.g., R^2_m, R^2_c; for logistic regression, AUC, accuracy). For example, the changes in (conditional) AUC for each preceding model comparison are

```
auc(givenness_m1) - auc(giv_m1_noClabel)
## [1] 0.32
auc(givenness_m1) - auc(giv_m1_noNpType)
## [1] 0.025
auc(givenness_m1) - auc(giv_m1_noVoice)
## [1] 0.025
```

By this ΔAUC metric, the order of predictor importance is

$$\texttt{clabel.williams} > \texttt{npType.pronoun} \geq \texttt{voice.passive}$$

(The $\geq$ sign is because the AUCs are almost equal.) This is similar to the conclusion we'd reach from the average marginal effects for predictors in this model (shown in section 9.3.2.1).

Partial R^2 A closely related notion is defining "partial" measures, which attempt to decompose, for example, R_m^2 or R_c^2 for the full model into the amount attributable to each predictor. Such standardized effect sizes are tricky to define and interpret (see, e.g., Stoffel, Nakagawa, and Schielzeth 2021; Rights and Sterba 2020; Bolker 2021, "Variable importance" for partial ΔR^2). Nonetheless, they seem useful as rough measures.

For R^2 measures, one could just calculate each variable's importance as ΔR^2 between the full and subset models, using `r.squaredGLMM()`. The partR2 package (Stoffel, Nakagawa, and Schielzeth 2021) does this properly, to give marginal or conditional ΔR^2, with CIs possible. Unfortunately this package doesn't currently accept models with random slopes, so let's do an example using the `diatones_melr` model.

We would like R^2 corresponding to dropping four sets of coefficients—`frequency` + interactions, and so on—which we specify using the `partbatch` argument.

```
## set up parallel processing as required by partR2
library(furrr)
plan(multisession, workers = 8)
batches <- list(
  c("syll1_coda", "frequency:syll1_coda"),
  c("syll2_coda"), c("syll2_td", "syll2_td:frequency"),
  c("frequency", "syll2_td:frequency", "frequency:syll1_coda")
)
res <- partR2(diatones_melr, partbatch = batches, nboot = 15,
  parallel = TRUE)

## marginal R2
res$R2[1:5, ]

res2 <- partR2(diatones_melr, R2_type = "conditional",
  partbatch = batches, nboot = 15, parallel = TRUE)

## conditional R2
res2$R2[1:5, ]
```

```
##                                                      term estimate
## 1                                                    Full    0.210
## 2                         syll1_coda+frequency:syll1_coda    0.146
## 3                                              syll2_coda    0.095
## 4                             syll2_td+syll2_td:frequency    0.130
## 5 frequency+syll2_td:frequency+frequency:syll1_coda    0.168
##   CI_lower CI_upper ndf
## 1    0.177     0.47   7
## 2    0.108     0.41   5
## 3    0.052     0.37   6
## 4    0.091     0.40   5
```

```
## 5    0.132     0.43   4
##                                                        term estimate
## 1                                                      Full     0.37
## 2                        syll1_coda+frequency:syll1_coda     0.31
## 3                                                syll2_coda     0.26
## 4                             syll2_td+syll2_td:frequency     0.30
## 5 frequency+syll2_td:frequency+frequency:syll1_coda     0.33
##   CI_lower CI_upper ndf
## 1    0.221     0.92   7
## 2    0.129     0.91   5
## 3    0.055     0.90   6
## 4    0.106     0.91   5
## 5    0.160     0.91   4
```

The importance of syll1_coda is $\Delta R^2_m = 0.146$, corresponding to the row for syll1_coda+frequency:syll1_coda, and so on. Using either marginal or conditional R^2, the order of predictor importance is

frequency > syll1_coda > syll2_td > syll2_coda

(However, the overlapping CIs suggest that we shouldn't put too much stock in this ordering.)

This is a similar order to what we found for the non-mixed-effects model (section 6.9.2.1) using an analogous method (odds ratio, which would be similar to calculating Δ accuracy), except that frequency is now the most important predictor instead of syll1_coda. It makes sense that adding the by-prefix random intercept to this model (which gives diatones_melr) decreases the importance of syll1_coda, because this predictor is prefix-level.

9.8.3 Other Options

Other approaches use the fitted model to quantify a predictor x's importance in terms of how much changing x affects model predictions. These methods are particularly useful when working with a model that takes a while to fit—such as cdi_mod, which takes 15+ min—because they don't require refitting subset models.

Average marginal effects (AMEs), introduced in section 9.8.3, can be calculated for mixed-effects models using the margins package. The AME is how much a small change in x affects the model's prediction, averaging across the dataset.[12]

At present the function (margins() from package margins) we previously used for AMEs doesn't work for models with nonlinear terms. Instead this example demonstrates functionality from the closely related marginaleffects package, which calculates AMEs similarly to margins and outputs them in tidy format.

12. Note that this means what "average" in AMEs means is different from usual for mixed-effects models: average (change in predictions) across all participants/items, not "for an average participant/item." The latter could be obtained using nondefault settings for margins.

```
## Get marginal effects for each predictor, for each observation
marginaleffects(cdi_mod) %>%
  ## summarize for each predictor to give AMEs
  tidy() %>%
  ## show just estimate and 95% CIs
  select(term, estimate, conf.low, conf.high)
##              term estimate conf.low conf.high
## 1        freq.std    0.185    0.126     0.244
## 2   lexical_class    0.161    0.017     0.305
## 3   lexical_class    0.283    0.147     0.418
## 4   lexical_class    0.066   -0.073     0.206
## 5 log_nphones.std   -0.112   -0.169    -0.054
## 6         MLU.std   -0.093   -0.139    -0.048
## 7             sex   -0.042   -0.142     0.057
```

Each AME is in probability space, for example, a male child has about 0.04 lower probability of knowing a word than a female child does. We can use the absolute values of the AMEs as a measure of variable importance, taking the largest value for `lexical_class` (0.28) as its importance: this corresponds to an increase of probability by 0.28 between FUNCTION WORDS and NOUNS.[13]

By this metric, the variable importances are

`lexical_class` > frequency > word length > MLU > `sex`

This could be important for the research question: a word's lexical class not only has a significant effect on when it is learned, after accounting for other word-level properties, but its effect size is also larger than those of the other word-level properties.

9.9 Reporting a Mixed-Effects Logistic Regression

Reporting a MELR mostly follows the same principles as for reporting LMMs (section 8.9). Part of your report, as for an LMM, should generally be one or more quantitative summaries of the model (section 9.5). Remember that whether classification-based (AUC, accuracy, etc.) or likelihood-based (AIC, R^2, etc.), these summaries are uninterpretable in isolation for a MELR model, so they should somehow be contextualized (e.g., reporting the same summaries for a baseline model).

9.10 Further Reading

Many of the resources discussed for LMMs (section 8.11) cover GLMMs as well, typically in less detail. For MELR in particular, Gelman and Hill (2007, chapter 14) and Snijders and Bosker (2011, chapter 17) are good general references. Tutorials and books on mixed-effects models for ecologists (Bolker et al. 2009; Bolker 2015; Zuur 2009) are useful for

13. By default, for factors `marginaleffects()` (or `margins()`, in the margins package) calculates an AME for the difference between each level and the base level, i.e., treatment coding. I think the largest AME can be thought of as a "range" measure of the factor's effect size, provided your base level is intuitively the smallest (here: FUNCTION_WORDS).

focusing on the GLMM case. Shorter introductions to MELR or GLMMs for language scientists in particular, focusing on different subfields, include Jaeger (2008), Johnson 2008, section 7.4, Levshina (2015, chapter 12), Gries (2015), and Johnson (2009). Kimball et al. (2019) discusses separation and quasi-separation for MELRs.

Exercises

9.1. To test whether the random-effect term is justified in a model containing a *single* random effect, such as `diatones_melr`, we cannot just compare the models with and without the term (e.g., using an LRT or AIC), as their likelihoods are not comparable. It would be ideal to use an exact LRT, as we did for LMMs for this case, but a stable R implementation does not exist for GLMMs. Instead, calculate a 95% CI for the by-prefix random intercept variance for this model, using a decent method (profile likelihood or parametric bootstrap), and check if it includes 0. What does the CI suggest in this case?

9.2. Calculate marginal and conditional R^2 for three models of the French CDI data: (1) the full model (`cdi_mod`), (2) that model without the by-child random slope of `lexical_class`, and (3) the baseline model, containing just by-child and by-word random intercepts (no fixed effects). What conclusions can you reach about "variance explained" by comparing the R^2 values for (1)–(3)?

9.3. Fitted-binned-residual plots can be useful for MELR models of sufficiently large datasets, where it is possible to derive CIs. Use `binned_residuals()` from performance (or `binned.residuals()` from arm) to make a plot for model `cdi_mod`. If the model is correct, ~95% of the (binned) residuals should fall inside the CIs.

(a) What are the residuals here? (Hint: they are not deviance residuals). This relates to why the CI is circle-shaped.

(b) Does the plot suggest any (big) problem with the model? If so, can you figure out which observations should be investigated further?

9.4. Verify this statement about model `givenness_m1`, from section 9.6.2: "It turns out that other influential items and participants have the same effect [larger effect size, similar or larger z] on the model."

9.5. Examine the full output of the `compareCoefs` call from section 9.6.2. You should find that excluding item 16 affects our conclusion about the `clabel.williams:voice.passive` effect. Overall, what can we conclude about the robustness of the Williams effect, including whether it varies depending on sentence structure?

9.6. Check for influential observations (= words) for the `diatones_melr` model. You should find there are some words that greatly affect the model. In what way(s) is the model affected? Does this change what qualitative conclusions can be drawn from the model?

9.7. Determine what nonlinear effect of `freq.std` to use in the French CDI model (section 9.7.2).

(a) Fit several candidate models of the `french_cdi_24` data, with different `freq.std` effects: linear, natural splines with $k=2$, $k=3$, and $k=4$. The model structure should be as described in section 9.7.2: fixed-effect terms for `lexical_class`, and so on, and by-child and by-word random intercepts (no random slopes).

(b) Compare the models using AIC and BIC. Which model would you select, using each criterion? Call these k_{AIC} and k_{BIC}.

(c) Decide which k makes most sense, by plotting the predicted `freq.std` effects for the models with k_{AIC} and k_{BIC}, comparing the model tables, and/or comparing to the empirical `freq.std` effect (in figure 9.10). In the text we assume $k=2$, but you don't need to reach the same conclusion.

9.8. For `cdi_mod`: quantify how much the model predicts that children vary in the intercept—a proxy for their overall vocabulary size—in an interpretable way, that refers to probabilities or odds ratios (i.e., not "95% of children have log-odds between X and Y"), from the by-child random intercept variance. For example, you could determine the model-predicted intercept value for each child, transformed to probability.

(a) It is visually clear from figure 9.13 that children vary in the `lexical_class` effect as well. What model parameters, that you see in `summary(cdi_mod)`, capture this variation? Why is it not straightforward to read off from these an interpretable summary of "how much do children vary"?

(b) Figure out such an interpretable summary using visualization and/or numbers.

10 Mixed-Effects Models 3: Practical and Advanced Topics

This chapter covers practical aspects of using mixed-effects models that were abstracted away from in chapters 8 and 9: dealing with fitting problems (convergence/singularity; see sections 10.3–10.4) and choosing the terms to include in a model (model selection; see section 10.5). This chapter also covers some topics that go beyond the standard use case for mixed-effects models in linguistics, where fixed effects are of primary interest: additional types/uses of random effects (section 10.2) and obtaining model predictions/uncertainties for individual levels (i.e., by speaker) via simulation or bootstrapping (section 10.6), including for nonlinear effects (section 10.7).

10.1 Preliminaries

10.1.1 Packages

This chapter assumes that you have loaded various packages from previous chapters.

```
library(tidyverse)
library(broom)
library(arm)
library(lme4)
library(car)
library(broom.mixed)
library(languageR)
library(splines)
rescale <- arm::rescale
```

10.1.2 Data

It also assumes you have set the default contrasts for factors to Helmert contrasts (section 7.2.12) and loaded various datasets used in chapters 8 and 9, with associated processing (sections 8.1, 9.1, 9.7.1):

```
options(contrasts = c("contr.helmert", "contr.poly"))

vot <- read.csv("data/vot_rmld.csv", stringsAsFactors = TRUE) %>%
  mutate(
    place = fct_relevel(place, "labial"),
```

```
    log_corpus_freq = rescale(log_corpus_freq),
    speaking_rate_dev = rescale(speaking_rate_dev),
    foll_high_vowel = rescale(foll_high_vowel),
    cons_cluster = rescale(cons_cluster),
    gender = rescale(gender),
    stress = rescale(stress)
  )

contrasts(vot$place) <- contr.helmert(3)

diatones <- read.csv("data/diatones_rmld.csv",
  stringsAsFactors = TRUE
) %>%
  mutate(
    syll1_coda_orig = syll1_coda,
    syll2_coda_orig = syll2_coda,
    syll2_td_orig = syll2_td,
    syll1_coda = rescale(syll1_coda_orig),
    syll2_coda = rescale(str_count(syll2_coda_orig, "C")),
    syll2_td = rescale(syll2_td_orig),
    frequency = rescale(frequency)
  )

neutralization <- read.csv("data/neutralization_rmld.csv",
  stringsAsFactors = TRUE
) %>%
  mutate(voicing_fact = fct_relevel(voicing, "voiceless")) %>%
  filter(!is.na(prosodic_boundary)) %>%
  mutate(
    prosodic_boundary = rescale(prosodic_boundary),
    voicing = rescale(voicing_fact),
    item_pair = as.factor(item_pair),
    subject = as.factor(subject)
  )

contrasts(neutralization$voicing_fact) <- contr.helmert(2)
contrasts(neutralization$place) <- contr.helmert(3)
contrasts(neutralization$vowel) <- contr.helmert(5)

givenness <- read.csv("data/givenness_rmld.csv",
  stringsAsFactors = TRUE
) %>%
  mutate(
    clabel.williams = rescale(conditionLabel),
    npType.pronoun = rescale(npType),
    voice.passive = rescale(voice),
    order.std = rescale(order),
    shifted = (as.numeric(stressshift) - 1),
    item = as.factor(item),
    participant = as.factor(participant)
  )
```

The `turkish_if0` dataset

This discussion also assumes you have loaded the `turkish_if0` dataset:

```
turkish_if0 <- read.csv("data/turkish_if0_rmld.csv",
  stringsAsFactors = TRUE
)
```

This dataset comes from a cross-linguistic corpus study of "intrinsic F0" (IF0; Sonderegger, McAuliffe, and Bang [2017]): small perturbations in a vowel's pitch (F0) due to the identity of the vowel (V) or the preceding consonant (C). The dataset contains 4347 measurements of F0 (`f0`, in semitones) from the first 50 msec of the vowel in CV-initial, utterance-initial words (1265 types) from the GlobalPhone Turkish corpus of read sentences from 98 speakers.

Of primary interest is the effect of `Voicing` of the C—F0 may be higher after VOICELESS versus VOICED consonants—and how robust this effect is across speakers. The C is one of 14 consonants (column `consonant`). Of secondary interest is how V (`base_vowel`: levels A, I, U) affects F0, which may be higher after high (I, U) versus low (A) vowels and may modulate the effect of `Voicing`. Therefore, let's code `base_vowel` using Helmert contrasts `base_vowel_AvIU` (= vowel height) and `base_vowel_IvU`:

```
## Redefine contrast matrix for base_vowel
contrMat <- matrix(c(-2 / 3, 1 / 3, 1 / 3, 0, -1 / 2, 1 / 2), ncol = 2)
rownames(contrMat) <- c("A", "I", "U")
colnames(contrMat) <- c("AvIU", "IvU")
contrasts(turkish_if0$base_vowel) <- contrMat
```

IF0 effects are small perturbations from the F0 that is expected given the speaker (which we will capture by `gender` and a by-speaker random intercept) and the intonational context, which we capture using two predictors: F0 averaged over the next three syllables in the utterance (`local_f0`) and length of the utterance (`utterance_num_syllables`), given that longer utterances may start with higher pitch (e.g., Cooper and Sorensen 1980).

Figure 10.1 (left) shows the empirical effect of `Voicing` and `base_vowel` on `f0`. It looks like `Voicing` may have the expected effect, possibly modulated by vowel height. There may also be substantial interspeaker variability in the `Voicing` effect (figure 10.1, right). But these effects are all very small relative to other factors, such as `local_f0` (figure 10.1, center), so a statistical model is necessary for accurate estimates.

We'll also use numeric and standardized versions of all predictors:

```
turkish_if0 <- turkish_if0 %>%
  mutate(
    utterance_num_syllables = rescale(utterance_num_syllables),
    local_f0 = rescale(local_f0),
    ## voiceless = pos, voiced = neg
    Voicing.vl = rescale(Voicing),
    ## male = pos, female = neg
    gender.male = rescale(gender)
  )
```

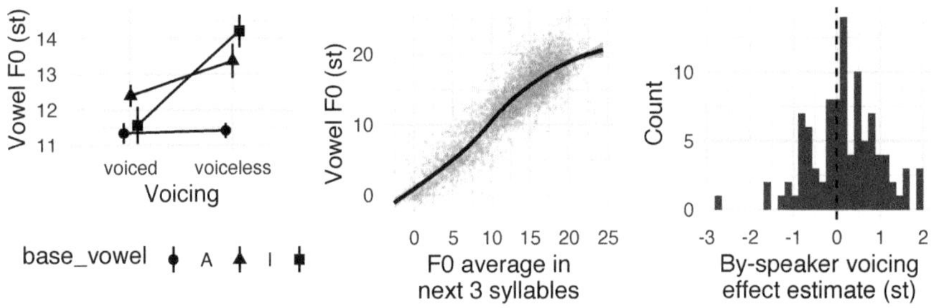

Figure 10.1
Empirical effects on `f0` for the `turkish_if0` dataset. *Left*: `Voicing` and `base_vowel` interaction calculated over by-word means. *Center*: `local_f0` effect calculated over observations. *Right*: By-speaker estimate of `Voicing` effect (VOICELESS-VOICED) calculated over means for each speaker-word pair.

10.2 More on Random Effects

So far we have used random effects to account for the data coming from many "random" levels of one grouping factor or two or more "crossed" grouping factors, such as speakers or words, resulting in nonindependent clusters of observations. We want to generalize to the population, rather than these random samples of speakers or words (e.g., beyond the 98 Turkish speakers in this dataset).

Random effects can also be used to model nonindependence of observations; nonindependence does not (conceptually) come from drawing random levels from a population or additional kinds of grouping structure. Here we'll use the `turkish_if0` dataset to illustrate two cases (see section 10.9 for others).

10.2.1 Controlling for a Nuisance Factor

A minimal model for these data, given the variables just introduced, could be

```
if0_m00 <- lmer(f0 ~ Voicing.vl * base_vowel + gender.male +
  local_f0 + (1 | word) + (1 | speaker), data = turkish_if0)
summary(if0_m00, correlation = FALSE)
...
## Random effects:
##  Groups   Name        Variance Std.Dev.
##  word     (Intercept) 0.534    0.731
##  speaker  (Intercept) 2.585    1.608
##  Residual             2.337    1.529
## Number of obs: 4224, groups:  word, 1238; speaker, 98
##
## Fixed effects:
##                          Estimate Std. Error t value
## (Intercept)               12.5031     0.1692   73.91
## Voicing.vl                 0.7620     0.0931    8.18
## base_vowelAvIU             0.7645     0.0842    9.08
## base_vowelIvU             -0.0308     0.1270   -0.24
## gender.male               -7.1332     0.3952  -18.05
## local_f0                   3.2738     0.1658   19.74
```

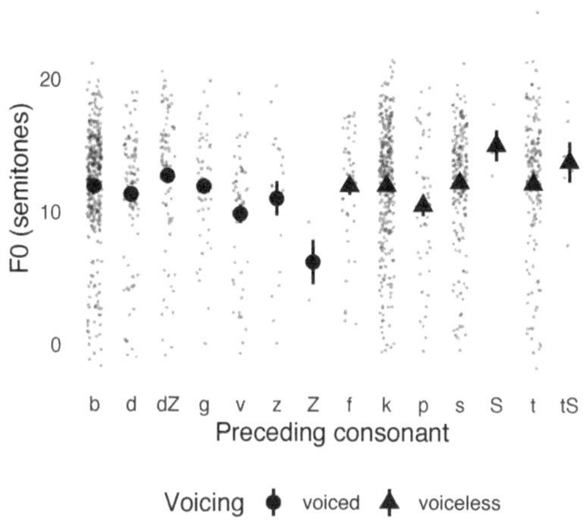

Figure 10.2
Empirical effect of preceding `consonant` on `f0` for the `turkish_if0` data. Means and 95% CIs are over by-word averages.

```
## Voicing.vl:base_vowelAvIU   0.5323    0.1656   3.21
## Voicing.vl:base_vowelIvU    0.2769    0.2592   1.07
```

Of primary interest is the `Voicing.vl` fixed effect and its interaction with `base_vowel`. Note that both are word-level predictors. Including a by-word random intercept means this model asks, "does a word's `Voicing` affect `f0`, after controlling for by-word variation?"

However, this may not be enough to control for variation among words, as figure 10.2 illustrates. It looks like `consonant` affects `f0` beyond just whether the consonant is VOICED or VOICELESS. Not somehow accounting for this in our model could result in the estimate of `Voicing` being overconfident: for example, due to Z having very low F0 or some consonants being much more common than others (as these are corpus data).[1]

We could do this using more fixed-effect terms coding consonant identity, like `Manner` or `Place` of articulation, but these effects are not of direct interest, and not all manner/place/voicing combinations exist in Turkish (or most languages). We could alternatively include `consonant` as a factor with many levels (14) and make one contrast correspond to `Voicing`. Either approach loses power to detect the effect of interest and makes the model hard to interpret, when all we want to do is control for consonant-level variation beyond `Voicing`.

A more elegant solution for a "nuisance" factor with many levels is to code it as a grouping factor; here, by including a by-`consonant` random intercept in the model.[2] This is OK

1. Z is the phone [ʒ].

2. This kind of situation is common for corpus data, e.g., Sonderegger, Bane, and Graff (2017) use random intercepts to control for effects of phonological context on vowel formants. Note that this use of random intercepts is conceptually different from (for example) a by-speaker random intercept—the data contain *all* values of `consonant`, not a random draw of consonants from Turkish—but both are valid, they just differ in conceptual interpretation (Bolker 2015, section 13.3).

as long as levels of the nuisance factor can be reasonably approximated as normally distributed deviations from the grand mean, which is all that the model assumes about random intercepts. The `Voicing.vl` fixed effect now asks, "what is the effect of *consonant*-level predictor `Voicing.vl`, controlling for consonant differences?"

```
if0_m0 <- update(if0_m00, . ~ . + (1 | consonant))
```

We can verify, using a likelihood-ratio test of REML fits (following section 8.5.2), that adding this term improves the model:

```
anova(if0_m00, if0_m0, refit = FALSE)
...
##          npar   AIC   BIC logLik deviance Chisq Df Pr(>Chisq)
## if0_m00    11 16471 16541  -8224    16449
## if0_m0     12 16461 16537  -8219    16437  11.8  1    0.00058
```

The new model's output is

```
summary(if0_m0, correlation = FALSE)
...
## Random effects:
##  Groups    Name        Variance Std.Dev.
##  word      (Intercept) 0.5050   0.711
##  speaker   (Intercept) 2.6041   1.614
##  consonant (Intercept) 0.0682   0.261
##  Residual              2.3332   1.527
## Number of obs: 4224, groups:
## word, 1238; speaker, 98; consonant, 14
##
## Fixed effects:
##                           Estimate Std. Error t value
## (Intercept)                12.4916     0.1889   66.14
## Voicing.vl                  0.7583     0.1863    4.07
## base_vowelAvIU              0.7493     0.0853    8.79
## base_vowelIvU              -0.0237     0.1265   -0.19
## gender.male                -7.1501     0.3964  -18.04
## local_f0                    3.2497     0.1657   19.61
## Voicing.vl:base_vowelAvIU   0.5062     0.1680    3.01
## Voicing.vl:base_vowelIvU    0.2447     0.2586    0.95
```

Note the `Voicing.vl` fixed effect in particular: the SE is higher than in `if0_m00`, as expected, resulting in a more conservative estimate (lower *t*-value).

This example gets at a more general point on avoiding "pseudoreplication" in grouped data: *if there is nonindependence in the data as a function of variable x, this should somehow be accounted for in the model*, whether by coding x as a fixed effect or random intercept (boxes 8.2, 10.1). Otherwise we run the risk of overconfident estimates (including type I errors), especially for x-level predictors. Our first model of the `turkish_if0` data did not account for nonindependence of data from the same `consonant` (= x), so the `Voicing.vl` coefficient estimate was overconfident (t-value too high). As another example, a model containing a fixed effect of speaker `gender` requires a by-speaker random intercept (Judd, Westfall, and Kenny 2017; Baayen, Davidson, and Bates 2008).

10.2.2 Nested Random Effects

This example also shows a *nested* random effect, reflecting hierarchical grouping structure in the data. Each observation comes from one word (and each word can contain several observations), and each word begins with one consonant. The grouping factor `word` is "nested" within grouping factor `consonant`: F0 is modeled as varying by-consonant and then by-word within consonant.

The estimated $\hat{\sigma}$ (variance components) for these two kinds of (normally distributed) variation are 0.26 (by-consonant) and 0.71 (by-word). Note that the by-word random intercept estimate $\hat{\sigma}$ is lower than for the previous model (0.73), because some by-word variation is now estimated as by-consonant variation.

Such hierarchical structure is common in linguistic data, for example, from corpus linguistics (e.g., text nested within register; see Gries [2015]) or dialectology (e.g., speaker nested within dialect; see Tanner et al. [2020]), though explicitly modeling it with nested random effects remains uncommon.[3]

These three ways of writing nested random effects in an lme4 formula are equivalent:

1. `(1|x) + (1|y)` (as in `if0_m0`)
2. `(1|x:y) + (1|y)`
3. `(1|y/x)`

(For our example, `x` is `word` and `y` is `consonant`.) Expression (2) makes the model structure explicit: `word` modulates the effect of `consonant`. Expression (1) is equivalent, as long as `x` and `y` are coded as factors. Expression (3) is the notation for "nesting" and just expands to (2). This chapter uses (1) for simplicity.

Box 10.1
Broader Context: Coding as a Fixed versus Random Effect

The text mentioned that grouping structure indexed by a variable x (e.g., `consonant`) can, in principle, be accounted for using a fixed effect of x or a random intercept. These options are called "no pooling" and "partial pooling" by Gelman and Hill (2007, section 11.4):

- No pooling: `f0 ~ Voicing.vl*base_vowel+ gender.male + local_f0 + consonant + (1 | word) + (1 | speaker)`
- Partial pooling: replace `consonant` by `(1|consonant)`

There are conceptual and practical reasons to choose one option or the other, including whether one wants the model to generalize beyond observed levels of x, whether x has many levels, and whether x is highly imbalanced across levels—all of which favor partial pooling. A natural question is, how many levels does a factor need to have to be included as a random effect? Guidelines vary, but suggest a minimum of 5–10 levels, to avoid inflated type I and type II errors (lower power) and unstable estimates. This issue matters

3. More than two levels of nesting are possible. In Gries (2015), texts are actually nested within subregisters, nested within registers, nested within modes.

more if x is related to the research questions. See Bolker (2015, section 13.3), Gelman and Hill (2007, section 11.4), and Brauer and Curtin (2018).

For example, for the `neutralization` data, three item-level predictors are relevant: the effect of the consonant's `voicing` (two levels) is of primary interest, and we want to control for `vowel` (five levels) and for consonant `place` (three levels). Our models so far do this using fixed effects of `voicing`, `vowel`, and `place`, and a random intercept for `item_pair`, so there are six "nuisance" rows of the fixed-effect table. We could alternatively account for idiosyncrasies of the item's final VC (vowel-consonant) by coding `vowel:place` as a new variable `VC` (15 levels) and coding `VC` as a random intercept, within which `item_pair` is nested (exercise 10.1). This gives a model with cleaner output and with near-identical results for all terms of interest (terms involving `voicing`).

10.3 Model Convergence

Two kinds of fitting problems are common for lme4 models: convergence issues and singular models. We'll first cover sources and fixes for each (sections 10.3 and 10.4), then turn to model selection (section 10.5), which in practice means selecting among convergent and (ideally) nonsingular models. These sections show model fitting messages/warnings, which typically were suppressed in previous chapters. These examples use the "final" models for several datasets from chapters 8 and 9—`neut_mod3`, `vot_mod3`, `diatones_melr`, `givenness_m1`—which you should review and refit if you're reading this chapter in isolation (see sections 8.4.2, 8.7, 9.2.1, 9.3.1).

Throughout this discussion, we will assume a distinction between *critical predictors*—those whose effects are directly related to the research questions or associated with them (i.e., interactions, random slopes)—and *control predictors*. For example, in our model `givenness_m1` of the `givenness` data (section 9.3.1), because the research question involves the condition label, `clabel.williams` and its interaction with `voice.passive` are critical; other predictors are controls. This means that the random slopes for these two predictors, as well as all correlation terms with these random slopes, are "related to critical predictors," whereas other random-effect terms are not.

We'll cover these issues in some depth both because they come up frequently in practice using lme4 and because the underlying issues are common to fitting any complex statistical model: solving a complex optimization problem using off-the-shelf algorithms, defining optimal and valid parameter values, and the consequences of a too-simple or too-complex model given the data.

Author's note Giving worked examples for these topics (fitting issues, model selection) requires refitting many models and examining code output. These steps are not hard, but take a lot of space. Much of this work has been moved to an online appendix to streamline the presentation in the text. You should be able to work through this chapter just executing code in the text, as usual, and refer to the appendix as needed to see full worked examples.

10.3.1 Sources of Nonconvergence and Fixes

To understand fixes for nonconvergent models, it is useful to know a bit about optimization in lme4 (see Bates, Mächler, et al. 2015; Bolker 2015 for more).

Optimizing an lme4 model means estimating the fixed-effect and random-effect parameters of the model $\hat{M}$ that maximizes a likelihood function (ML or REML), by iterating from a start model (M_0) to fit a model progressively closer to the $\hat{M}$, until convergence (defined by some criterion) is reached. Fitting the random effects (= variance components) is particularly tricky because they are restricted to be positive, and a large number of latent variables are involved (the random intercept for each participant, and so on).

This is a complex but standard optimization problem, and lme4 supports several powerful general-purpose nonlinear optimizers, including the Nelder-Mead and the Bound Optimization by Quadratic Approximation (BOBYQA) algorithms, each of which has a range of parameters, including

- The *starting values* of model parameters (M_0)
- *Tolerances* specifying what a small-enough change in the model (likelihood or parameter values) is to stop
- *Maximum iterations* to try before giving up

See `?lmerControl` for more.

Most of these parameters relate to convergence, but others relate to model validity (parameters of the fitted model taking on valid values), which in this case mostly means ensuring the model is not singular (see section 10.4). Both convergence and validity are nontrivial issues at play when one fits any sufficiently complex statistical model, because the likelihood function is complex and different datasets vary hugely. In addition there is a trade-off between accuracy and speed: smaller tolerances or more iterations lead to better estimates of $\hat{M}$, but at the expense of longer running time.

All this means that no one choice of optimizer and parameter values will work for all cases; they will sometimes need to be adjusted. The lme4 maintainers have put great care into choosing default settings that work in common cases and balance accuracy and speed.

Another common reason for nonconvergence or singular fits is that the model is too complex for the data, or *overparametrized*. Intuitively, this makes it hard to find a unique optimal solution. We'll discuss further for singular fits (section 10.4), where overparametrization is a more common cause. It is a common belief among users that nonconvergence is a sign the model is overparametrized, especially in the random effects, but this is incorrect (Seedorff, Oleson, and McMurray 2019, 32): models can fail to converge for other reasons (box 10.2).

The upshot is that you should only report convergent models, but nonconvergence is not a good basis for ruling a model out.

Box 10.2
Broader Context: Convergence Is Not a Metric of Model Quality

It is worth emphasizing that *whether a model converges is not a good measure of model quality*, even though model convergence has come to be seen as meaningful in practice in language sciences. Papers (including my own work) routinely use model convergence to decide which terms to add to a model, which model to report, or to justify using alternative methods. But whether a model converges with the lme4 default settings is not in itself a good tool for these things, because nonconvergence can result from many sources, and false-positive convergence warnings are common. Per `?convergence`, "[lme4] fits may produce convergence warnings; these do not necessarily mean the fit is incorrect"—or that there is any issue with your data or model. They

Table 10.1
Possible fixes for non-convergent (non-intrusive + intrusive) and singular models (intrusive only), ordered by which to try first (adapted from Brauer and Curtin 2018). Fixes 2(a) and 2(b) are tied.

1. Nonintrusive
   a. Check your data and model
   b. Standardize predictors (center, possibly scale)
   c. Increase number of iterations
   d. Change the optimizer
   e. Give the optimizer better start values
2. Intrusive
   a. Remove random effects involving control predictors (must not be in interactions with critical predictors)
   b. Selectively remove random-effect correlations: for control predictors, then correlations that are probably close to 0
   c. Remove random intercept (leaving slope terms in)
   d. Remove random slopes for critical predictors

---

just mean that further steps are needed to figure out if this is the case and to determine how to resolve the warnings.

For example, the default `lmer()` optimizer changed in February 2019 from one version of the BOBYQA algorithm (`bobyqa`, used since 2013; from the minqa package) to `nloptwrap` (which uses a version from the nloptr package). This led to faster model fits, but anecdotal reports (e.g., lme4 GitHub issue #501) suggest more false-positive convergence warnings, which may go away when the previous default optimizer is used—as in our example in section 10.3.2.5. This kind of (very) technical change doesn't affect what the optimal model is, just how it is fitted. As a simpler example, it is often necessary for larger datasets or more complex models to increase the maximum number of iterations (section 10.3.2.3); nonconvergence is meaningless when the optimizer simply needs to run longer.

Note that convergence/singularity issues often reflect more general facts about the data, which are not particular to lme4. For example, Bates, Kliegl, et al. (2015) show a case where the same fundamental issue, of the model being too complex for the data, applies regardless of whether an lme4 or Bayesian model is used.

Two kinds of steps can be taken to resolve nonconvergence or singular fits:

- *Nonintrusive*: data/model checking, transformations, optimizer
- *Intrusive*: changing the model, typically by simplifying random-effect structure, or running several models

The same fixes often address other fitting issues: very slow fits or false-negative models (convergent, but with major problems).

We will discuss some fixes of each type, shown in table 10.1, with examples for those that I have found come up most in practice. This discussion draws a lot on `?convergence`; Seedorff, Oleson, and McMurray (2019); and Brauer and Curtin (2018). I highly recommend reading more if you regularly encounter nonconvergent or singular models (box 10.3).

Box 10.3
Broader Context: Addressing Nonconvergence

Given that dealing with convergent and singular models (along with model selection) is the most common problem users have in fitting mixed-effects models, there is surprisingly little truly practical guidance, showing worked examples and discussing which solutions to try before others. One reason is how difficult it is to give general advice: the steps to take can depend on the particular case and type of data. Nonetheless, guidance is urgently needed (Seedorff, Oleson, and McMurray 2019), and this is an active research area. All work so far in the language sciences/psychology literature (see section 10.9) focuses on data from balanced experiments, with a low number of fixed-effect and random-effect terms. This makes sense (simple before complex cases), but most linguistic data do not fit this description. One goal of this chapter is to give practical advice on convergence/singularity and model selection for corpus data, with which I am most familiar.

Brauer and Curtin (2018) give the most comprehensive discussion to date of practical steps to resolve convergence issues. Their table 17 lists 20 remedies, in rank order of which to try first and which are least likely to affect parameter estimates. All "nonintrusive" remedies we discuss here come before "intrusive" remedies. Their exact ordering is not universally agreed upon, but it forms a great starting point.

The list of convergence fixes we cover is partial: `?convergence` discusses more nonintrusive steps; for intrusive steps, we do not discuss solutions besides simplifying random-effect structure. More drastic analysis changes that can be appropriate, such as fitting subset models or aggregating data, are discussed by Brauer and Curtin (2018); Seedorff, Oleson, and McMurray (2019); and Barr et al. (2013). Most importantly, we do not discuss collecting more or different data, which are often the best options when possible.

10.3.2 Nonintrusive Methods

10.3.2.1 Checking the data

A model may not converge if the *structure of your data* makes the model impossible to fit: the distribution of a predictor or the response may be highly skewed, some combinations of predictor values may have few or no observations, or some participants/items may have few data in a cell.

For example, recall from section 6.8.2.2 that the `diatones` data has second-syllable coda originally coded as a factor (four values: `syll2_coda_orig`), rather than the numeric version (`syll2_coda`) that we have mostly used. Fitting a regular logistic regression predicting `stress_shifted` using `syll2_coda_orig` did not work because the data are separable: there are very few observations for `syll2_coda_orig` = CCC, all with `stress_shifted` = 0. If we try to refit our final mixed-effects regression model for this data (`diatones_melr`; see section 9.2.1), using this predictor, the model doesn't converge:

```
## > diatones_melr_bad <- update(diatones_melr, . ~ . - syll2_coda +
## +   syll2_coda_orig)

## Warning message:
## unable to evaluate scaled gradient
## Warning message:
## Model failed to converge: degenerate  Hessian with 1 negative eigenvalues
```

We can see the problem in lines 1 and 7 of the fixed-effects table:

```
tidy(diatones_melr_bad) %>%
  filter(effect == "fixed") %>%
```

```
  select(-effect, -group)
##                    term estimate std.error statistic p.value
## 1           (Intercept)      1.3   1.2e+00   1.07524  0.2823
## 2              syll2_td      2.7   1.0e+00   2.68394  0.0073
## 3             frequency     -1.3   6.9e-01  -1.90966  0.0562
## 4            syll1_coda      3.7   1.4e+00   2.67503  0.0075
## 5      syll2_coda_origC     -4.3   1.4e+00  -3.07576  0.0021
## 6     syll2_coda_origCC     -4.8   1.6e+00  -3.00755  0.0026
## 7    syll2_coda_origCCC    -25.2   7.5e+04  -0.00034  0.9997
## 8    syll2_td:frequency      4.0   1.5e+00   2.69198  0.0071
## 9 frequency:syll1_coda      6.1   2.4e+00   2.54298  0.0110
```

The issue is the same as for the non-mixed model: very high SEs, because the effect of `syll2_coda_orig` = CCC can't be determined.

More generally, convergence issues can result from extreme cases of most "problems" we have discussed for model validation (e.g., non-normal residuals, multicollinearity), and the solutions to address them will be similar. For example:

- Combining some predictor values or coding the predictor differently (as in this case: code as numeric `syll2_coda`)
- Discarding or imputing values for a subset of the data (e.g., a single participant with unrealistic values of *y* or extreme "outlier" observations)
- Transforming a predictor or the response to meet distributional assumptions

In practice, mixed-effects models are much more robust to such issues than non-mixed models are—for example, outlier participant/items will just be modeled by the random effects—and convergence issues reflect truly dire problems (i.e., complete separation).

10.3.2.2 Checking the model

The structure of the model needs to make sense, given the data. The most common issues are

- Including a random slope for a factor with uncorrelated random effects (shown in section 8.7.4)
- Including a by-unit random slope for a predictor that varies by unit (an "impossible" random effect; see section 8.4.5)
- Including a predictor both as a grouping factor and a fixed effect

For example, consider `vot_mod3`, the final model of a subset of the VOT data (section 8.7), whose formula was

```
formula(vot_mod3)
## log_vot ~ speaking_rate_dev + foll_high_vowel + cons_cluster +
##     log_corpus_freq + place + gender + (1 | word) +
##     (1 + speaking_rate_dev | speaker)
```

Suppose that `place` were a critical predictor. We would then want to add all possible random slopes for `place` to the model. We might think that this means adding by-speaker and by-word random slopes:

```
## > vot_mod3_bad <- lmer(log_vot ~ speaking_rate_dev +
## + foll_high_vowel + cons_cluster + log_corpus_freq + place + gender +
## + (1+speaking_rate_dev+place|speaker) + (1+place|word),
## +   data=vot_voiced_core)
```

```
## Warning message:
## unable to evaluate scaled gradient
## Warning message:
## Model failed to converge: degenerate  Hessian with 1 negative eigenvalues
```

The model does not converge because `place` is a word-level predictor, so a by-word random slope of `place` does not make sense. When refitted without this term (after another fix), the model converges (online appendix, section 2.2).

Interestingly, there is nothing in the actual model output that suggests this is a nonsensical model (output shown in online appendix, section 2.2):

```
summary(vot_mod3_bad)
```

This is because the model does not actually know that `place` is word-level: from the random-effect structure, it assumes that `place` varies within word, but for each word you happen to have only observed one value of `place`. This is technically a valid model, but it does not make sense given the structure of the data.

In practice, this kind of misspecified random-effect structure, or the other two just mentioned, often do *not* lead to convergence problems (or singular models), and it is up to you to notice the nonsensical model structure.

10.3.2.3 Changing optimizer parameters

Recall that `vot_mod3` was fitted for just a subset of voiced stops (`voicing` = VOICED): only observations from certain speakers and for monosyllabic words. Let's refit the model for all voiced stops, without these restrictions:

```
## All observations from voiced stops
vot_voiced <- filter(vot, voicing == "voiced")
```

We have also added a `stress` fixed effect, which is relevant now that words can have more than one syllable.

```
## > vot_mod3_full <- update(vot_mod3, . ~ . + stress, data=vot_voiced)
```

```
## Warning message:
## Model failed to converge with max|grad| = 0.00359033 (tol = 0.002,
component 1)
```

Changing tolerances This model does not converge. ?convergence suggests the first fix to try is decreasing the optimizer's stopping tolerances; this will always lead to more accurate but slower optimization.

```
strict_tol <- lmerControl(optCtrl = list(xtol_abs = 1e-8,
  ftol_abs = 1e-8))
## Refit the model with these tolerances
vot_mod3_full_3 <- update(vot_mod3_full, control = strict_tol)
```

The model now converges.

Changing start values Another option is to simply restart the fit using as start values the estimates from the nonconvergent model or slightly perturbed versions (see ?convergence):[4]

```
## Refit starting from the final values
fittedVals <- getME(vot_mod3_full, "theta")
vot_mod3_full_1 <- update(vot_mod3_full, start = fittedVals)

## Refit starting from the final values + perturbaton
pars <- getME(vot_mod3_full, "theta")
fittedVals_perturbed <- runif(
  length(pars),
  pars / 1.01, pars * 1.01
) * sign(pars)
# sign(pars) makes this work when some fixed effects are negative
vot_mod3_full_2 <- update(vot_mod3_full, start = fittedVals_perturbed)
```

To be safe, we can verify that the models fitted with different start values are identical, for example, in their log-likelihoods and fixed effects (output shown in online appendix, section 3.1):

```
logLik(vot_mod3_full) # start values = defaults
logLik(vot_mod3_full_1) # start values from vot_mod3_full
compareCoefs(vot_mod3_full_1, vot_mod3_full, zvals = TRUE)
```

Increasing number of iterations Another fix, which will always lead to more accurate but slower optimization, is to increase the maximum number of iterations.

For example, consider model givenness_m1 of the givenness data, which included all possible random slopes, uncorrelated. Refitting this model with "maximal" random-effect structure (correlations included):

```
## > giv_maximal <- glmer(stressshift ~ clabel.williams*voice.passive +
## + npType.pronoun + (1+clabel.williams+ npType.pronoun|item) +
## + (1 +clabel.williams*voice.passive + npType.pronoun |participant),
## + data=givenness, family='binomial',
```

4. Note that for a GLMM the syntax is different: getME(mod, c("theta","fixef")).

```
## +   control=glmerControl(optimizer = 'bobyqa'))
## Warning message:
## convergence code 1 from bobyqa: bobyqa -- maximum number of function
evaluations exceeded
## boundary (singular) fit: see ?isSingular
```

This model does not converge with the default number of maximum iterations (10,000), but if we increase this parameter (to 100,000), as suggested by the warning message, the model does converge:

```
giv_maximal <- update(giv_maximal,
  control = glmerControl(
    optimizer = "bobyqa",
    optCtrl = list(maxfun = 100000)))


## Iterations to convergence
giv_maximal@optinfo$feval
## [1] 21536
```

It is common to need to increase the maximum number of observations, especially for larger datasets or models with complex random-effect structure.

10.3.2.4 Standardizing predictors

It is useful to distinguish here between centering and scaling, the two aspects of standardizing predictors.

Centering predictors has the same effect, of reducing unnecessary collinearity and making coefficients more interpretable (for fixed effects), as for non-mixed-effects models—this includes using a centered coding scheme for any factors. Centering is more important in mixed-effects models because it also reduces collinearity in the random effects, which are more often a source of nonconvergence than collinearity in the fixed effects.

It is also helpful to normalize predictors (dividing by one or two SDs). Mixed-effects models fit slower, and in extreme cases will not converge, when predictors are on very different scales.

Given these benefits, it makes sense to *always standardize predictors* in mixed-effect models by default, minimally by centering.

This matters especially for fitting models with complex random-effect structure. For example, consider refitting the new VOT model (`vot_mod3_full`), now with maximal random-effect structure:

```
vot_mod_maximal <- update(vot_mod3_full, . ~ . - (1 | word) -
  (1 + speaking_rate_dev | speaker) +
  (1 + gender + speaking_rate_dev | word) +
  (1 + speaking_rate_dev + foll_high_vowel + cons_cluster +
    log_corpus_freq + place + stress | speaker))
```

This model, where all predictors are standardized, does converge.

```
## speaking_rate_dev row of fixed effect table:
tidy(vot_mod_maximal) %>%
  filter(term == "speaking_rate_dev") %>%
  select(-group)
##    effect              term estimate std.error statistic
## 1  fixed speaking_rate_dev   -0.052     0.018        -3
```

In contrast, the same model fitted using the nonstandardized predictors does not converge (online appendix, section 4).

10.3.2.5 Changing the optimizer

Sometimes a model converges when the optimizer is changed. We saw this for MELR models, such as `givenness_m1`, where using the `bobyqa` optimizer often helps (section 9.3.1).

The assumption is that both models are very similar/identical, and the nonconvergent model was a "false positive." If you want to be sure this is the case, the gold standard (see `?convergence`) is to refit the model with all possible optimizers, using `allFit()` from lme4 (online appendix, section 5):

```
giv_m1.all <- allFit(givenness_m1)
```

If the fits are very similar, it is safe to use the new fit—the nonconvergent fit was a false positive. It is also usually OK to use the new fit if you find that all optimizers except the default give very similar fits, while the default fit looked nonsensical. (In this case, nonconvergence was not a false positive, but we assume that the convergent models are OK.)

We can compare fits by their log-likelihoods and fixed effects (full output shown in online appendix, section 5):[5]

```
summary(giv_m1.all)$llik
##                      bobyqa                      Nelder_Mead
##                        -160                             -160
##                  nlminbwrap                            nmkbw
##                        -160                             -160
##           optimx.L-BFGS-B nloptwrap.NLOPT_LN_NELDERMEAD
##                        -160                             -160
##     nloptwrap.NLOPT_LN_BOBYQA
##                        -160
```

```
summary(giv_m1.all)$fixef
##                             (Intercept) clabel.williams
## bobyqa                             -1.2               4
```

5. We could also check random effects, as shown in `?allFit`, but it is less clear whether these need to be identical across models.

```
## Nelder_Mead                      -1.2          4
## nlminbwrap                       -1.2          4
## nmkbw                            -1.2          4
## optimx.L-BFGS-B                  -1.2          4
## nloptwrap.NLOPT_LN_NELDERMEAD    -1.2          4
## nloptwrap.NLOPT_LN_BOBYQA        -1.2          4
...
```

We see that the values are essentially identical across optimizers, including `nloptwrap` (the default) and `bobyqa` (which we used instead).

10.3.3 Intrusive Methods

Often, nonconvergence is due to the model being too complex for the data. It is reasonable to assume this is the case if nonintrusive methods don't work. Typically this is because the *random-effect structure is too complex*. We will focus on this case, which comes up frequently in practice (see box 10.3 for other cases).

The possible steps to take are 2(a)–(d) in table 10.1. These are in priority order: 2(a) and 2(b) make sense in most cases, and there is not consensus on which one is preferable (removing terms related to control predictors vs. removing correlations; see Brauer and Curtin [2018]; Seedorff, Oleson, and McMurray [2019]; Bates, Kliegl, et al. [2015]).

Step 2(c), removing a random intercept, occasionally makes sense, by the logic that the top priority is having random *slopes* for critical predictors in the model, to avoid type I errors (Brauer and Curtin 2018, 16; Barr et al. 2013). This is also why 2(d), removing random effects for critical predictors, almost never makes sense.

As an example, consider the `diatones_melr` model, but refit with maximal random-effect structure:

```
## > diatones_melr_max <- update(diatones_melr, . ~ . - (1 | prefix) +
## + (1+syll2_coda + syll2_td + frequency + syll2_td:frequency|prefix),
## + control=glmerControl( optCtrl=list(maxfun=100000)))

## Warning message:
## Model failed to converge with max|grad| = 0.0614046 (tol = 0.002,
component 1)
```

```
summary(diatones_melr_max)
...
##
## Random effects:
##  Groups Name               Variance Std.Dev. Corr
##  prefix (Intercept)        16.93    4.11
##         syll2_coda         16.27    4.03      0.89
##         syll2_td           11.53    3.40     -1.00 -0.89
##         frequency           3.69    1.92      0.99  0.82 -0.99
##         syll2_td:frequency 18.06    4.25     -1.00 -0.90  1.00
##
##
```

```
##
##
##
##   -0.99
## Number of obs: 130, groups:  prefix, 13
...
```

(We have also increased the maximum iterations; the model doesn't converge otherwise.)

As we can see just from the random-effect structure, this model is far too complex for the data: 15 variance components, corresponding to 75 random effects (5 terms $\times$ 13 `prefix` values), are being fitted to just 130 observations. Let's just assume this is why the model doesn't converge (technically we should first try the noninvasive methods) and simplify the random-effect structure.

For this example, 2(a) is not an option, because there are no control predictors (all are of interest for the research questions). We instead begin removing correlation terms (2(c)). One can iteratively remove these, in an order respecting the hierarchy principle (box 4.7): correlations with the `syll2_td:frequency` term, then correlations with `frequency` or `syll2_td` (if the model doesn't yet converge), and so on. Online appendix, section 5 shows this process, which results in a model with all correlation terms excluded:

```
## > diatones_melr_max_uncorr <- update(diatones_melr, . ~ .
## + - (1 | prefix) + (1+syll2_coda + syll2_td + frequency +
## +    syll2_td:frequency||prefix))
## boundary (singular) fit: see ?isSingular
```

```
summary(diatones_melr_max_uncorr)
...
## Random effects:
##  Groups   Name               Variance Std.Dev.
##  prefix   (Intercept)        2.52e+00 1.59e+00
##  prefix.1 syll2_coda         2.94e+00 1.71e+00
##  prefix.2 syll2_td           9.62e-05 9.81e-03
##  prefix.3 frequency          2.43e-09 4.93e-05
##  prefix.4 syll2_td:frequency 4.89e-06 2.21e-03
## Number of obs: 130, groups:  prefix, 13
##
...
```

We may or may not choose to simplify the random-effect structure further, because the model is singular, but this model at least converges.

10.4 Singular Models

Recall that a model is *singular* if a "dimension" of the random effects has been estimated as 0 (section 8.7.2). (See `?isSingular`, which we summarize here, for more details.) Typically this means variances that are (near) 0, or correlation parameters near 1 or –1, in which case there is a linear combination of random effects which does not vary (e.g., across participants). Intuitively these values are unlikely to be correct: in reality there is probably some by-group variation in the intercept and every predictor, which will never be perfectly correlated. For example in `neut_mod_4` (section 8.7.2), both correlation term estimates are near 1/–1:

```
## boundary (singular) fit: see ?isSingular
```

```
summary(neut_mod_4)
...
## Random effects:
##  Groups    Name        Variance Std.Dev. Corr
##  item_pair (Intercept)  76.57    8.75
##            voicing       8.35    2.89    -1.00
##  subject   (Intercept) 674.27   25.97
##            voicing      18.86    4.34    1.00
##  Residual              395.35   19.88
## Number of obs: 749, groups:  item_pair, 24; subject, 16
...
```

And in our `diatones_melr_max_uncorr`, several random slope terms have variances near 0 (`syll2_td`, `frequency`, `syll2_td:frequency`).

For correlations sufficiently near 1/–1 (or variances near 0) lme4 detects the model as singular, and you get a warning message. It is also possible for the model to be singular without perfect correlations or zero variances (section 10.4.2 shows an example).

But sometimes a model is near-singular and no warning will be given, even though a random-effect term's estimate is unrealistic. For example, refitting `neut_mod_4` without the by-item random-effect correlation gives a near-perfect correlation, but the fit is not flagged as singular:[6]

```
neut_mod_4_noItemCorr <- update(neut_mod_4, . ~ .
- (1 + voicing | item_pair) + (1 + voicing || item_pair))
neut_mod_4_noItemCorr
...
## Random effects:
##  Groups      Name        Std.Dev. Corr
##  item_pair   voicing      1.57
##  item_pair.1 (Intercept)  8.51
##  subject     (Intercept) 25.98
##              voicing      4.44    0.98
##  Residual                19.92
## Number of obs: 749, groups:  item_pair, 24; subject, 16
...
```

6. Bates, Kliegl, et al. (2015) suggest a more thorough way to see whether the random effects are overparametrized, by examining principal components to gauge the dimensionality of the estimated random effects. The online appendix shows some examples, using the `rePCA()` function from lme4.

Singular or near-singular fits are well-defined models (they maximize [RE]ML), but there are a few reasons why they are suspect (copied from `?isSingular`):

1. Singular fits may correspond to overfitted models that may have poor power.
2. Chances of numerical problems and misconvergence are higher for singular models (e.g., when computing CIs).
3. Standard inferential procedures such as Wald statistics and likelihood-ratio tests may be inappropriate.

Reason (1) is the biggest worry for linguistic datasets in practice: *the random-effect structure is often too complex to be estimated given the data* (it is overparametrized). In other words, we often want to fit more complex models than the size and structure of our data allow.

10.4.1 Fixes

Models that are singular due to (near-)*perfect correlations*, or that are *without a clear source* (no perfect correlations or zero variances), should not be used as a final model. You can take steps 2(a)-(d) (table 10.1) to simplify the random-effect structure until the model is nonsingular.

If a model is singular due to (near-)*zero variances*: technically, these terms should be removed. But in practice, they should not affect model fit—including the values/significances of fixed effects—so it is fine to either remove these terms or leave them in, bearing in mind that there may be a small amount of variability that cannot be detected because the model is overparametrized (Brauer and Curtin 2018). It may especially make sense to leave zero-variance terms in if they are related to critical predictors (e.g., the by-speaker random slope of `clabel.williams` in `givenness_m1`).

10.4.2 Examples

10.4.2.1 Model `neut_mod_4`

Because `voicing` is a critical predictor, the only possibility is to remove correlation terms (table 10.1, 2(b)). Removing the by-item correlation gives the near-singular model `neut_mod_4_noItemCorr`. Removing the by-speaker correlation results in `neut_mod3`, our model with uncorrelated random effects from chapter 8 (section 8.4.2).

10.4.2.2 Model `giv_maximal`

The maximal model of the `givenness` data fitted previously is singular due to perfect correlations among the by-item random effects:

```
summary(giv_maximal)$varcor
## Groups      Name                            Std.Dev. Corr
## participant (Intercept)                      1.50
##             clabel.williams                  2.62    -0.85
##             voice.passive                    5.46     0.06 -0.12
##             npType.pronoun                   2.01    -0.31  0.50
##             clabel.williams:voice.passive   10.66     0.08  0.02
## item        (Intercept)                      1.03
```

```
##               clabel.williams                  2.48    -1.00
##               npType.pronoun                   1.45     1.00 -1.00
##
##
##
##
##   -0.20
##   -0.95 -0.08
...
```

We first remove the correlations with the random slope of `npType.pronoun`, a control predictor:

```
## > giv_ch10_2 <- update(giv_maximal, . ~ . -
## + (1 + clabel.williams + npType.pronoun | item) +
## + (1 + clabel.williams|item) + (0+npType.pronoun|item))
## boundary (singular) fit: see ?isSingular
```

```
summary(giv_ch10_2)$varcor
##  Groups      Name                            Std.Dev. Corr
##  participant (Intercept)                     1.192
##              clabel.williams                 2.096    -0.80
##              voice.passive                   4.390     0.40 -0.56
##              npType.pronoun                  2.019    -0.24  0.49
##              clabel.williams:voice.passive   8.642    -0.24  0.45
##  item        (Intercept)                     0.804
##              clabel.williams                 1.909    -1.00
##  item.1      npType.pronoun                  1.381
##
##
##
##
##   -0.23
##   -0.92 -0.09
...
```

The resulting model still has a perfect correlation among the by-item random effects, which we remove, along with the by-item random intercept with near-zero variance (output shown in online appendix, section 7.1):

```
## > giv_ch10_3 <- update(giv_ch10_2, . ~ . -
## + (1 + clabel.williams| item) + (1 + clabel.williams||item))
## boundary (singular) fit: see ?isSingular
## > giv_ch10_4 <- update(giv_ch10_3, . ~ . -
## + (1 + clabel.williams|| item) + (0 + clabel.williams||item))
## boundary (singular) fit: see ?isSingular
```

The resulting model (giv_ch10_4) is still singular, despite having no remaining zero variances or perfect correlations:

```
summary(giv_ch10_4)$varcor
##  Groups       Name                         Std.Dev. Corr
##  participant (Intercept)                   0.780
##              clabel.williams               1.607    -0.61
##              voice.passive                 2.981     0.68 -0.82
##              npType.pronoun                1.809    -0.25  0.56
##              clabel.williams:voice.passive 5.780    -0.44  0.69
##  item        npType.pronoun                1.060
##  item.1      clabel.williams               0.832
##
##
##
##
##  -0.37
##  -0.85 -0.09
...
```

This is because the by-participant random effects are overparametrized—which makes sense, as this is a small dataset—in some way that is not clear from the model table. It would be possible to proceed more incrementally (discussed in online appendix, section 9.1), but let's just try removing all correlation terms:

```
## > giv_ch10_5 <- update(giv_ch10_4, . ~ . -
## + (1 + clabel.williams * voice.passive + npType.pronoun | participant)  +
## + (1 + clabel.williams*voice.passive + npType.pronoun || participant))

## boundary (singular) fit: see ?isSingular
```

```
summary(giv_ch10_5)$varcor
##  Groups        Name                          Std.Dev.
##  participant   clabel.williams:voice.passive 2.16e+00
##  participant.1 npType.pronoun                9.20e-01
##  participant.2 voice.passive                 1.45e+00
##  participant.3 clabel.williams               4.08e-07
##  participant.4 (Intercept)                   4.16e-01
##  item          clabel.williams               8.10e-01
##  item.1        npType.pronoun                7.39e-01
```

This model is still singular, but only due to the zero `clabel.williams` by-participant slope—which we leave in, as it involves the effect of central interest. (This also explains the underlying reason that giv_ch10_4 was overparametrized.)

10.5 Model Selection

Actually selecting a model—deciding which terms to include—is often the hardest part of mixed-effects modeling. This is backed up by a survey by Meteyard and Davies (2020), of 163 researchers in psycholinguistics and other language sciences, where "selecting and specifying models" was the most common worry about using mixed-effects models, along with a lack of standardized procedures. Both get at the same basic fact: model selection is a hard topic, because there is no one way to choose a model. On why, I paraphrase Snijders and Bosker (2011, section 6.2):

Model selection is difficult because there are two steering wheels: substantive (subject matter–related) and statistical considerations. It is harder still for mixed-effects models, where the second wheel involves specifying both fixed and random effects.

The goal of model selection is a model that describes the data well but without unnecessary complications, that is of scientific interest, and that does not contain spurious effects that are interpreted as meaningful. Put simply, we want to arrive at a model that makes sense, given our data and research questions, via a sensible procedure that we can describe to a reader.

This all means that there are no fixed rules to follow to find the "best" model based on purely statistical considerations. Instead, model selection is guided by *principles* (section 10.5.2), which describe what makes a model better or worse. To motivate these principles, I'll first show examples of models we do not want to end up at, by mechanically applying simple recipes that are not uncommon in published work (shown in section 10.5.1).

We'll then turn to *practical heuristics* that you can use to guide your model building process (section 10.5.3). These motivate three *possible model selection approaches* (section 10.5.4): "maximal," "data-driven," and "uncorrelated first," which were introduced in chapter 8. These are demonstrated in *case studies* (sections 10.5.5–10.5.7). If you would prefer to focus on practical application, you can skip to section 10.5.3.

10.5.1 Bad Models

Model 1: Random intercepts only For a number of years after mixed-effects models were introduced to language sciences (circa 2008), the default recipe was to fit a model with by-participant and by-item random intercepts (or by-speaker/-word for corpus data, etc.), following introductions that emphasized the importance of these terms as improvements on other methods (e.g., repeated measures ANOVA; see box 8.3).

As we have now seen a number of times, mixed-effects models without random slope terms are usually problematic, because they can easily lead to "overconfident" conclusions, given the data (Schielzeth and Forstmeier 2009). For example, in section 8.4.3.1, we discussed two models of a subset of the VOT data, to test the hypothesis that speaking rate affects VOT. The model with just by-speaker and by-word random intercepts (`vot_mod_1`) suggests strong support for the hypothesis ($|t| = 3$, narrow CI), whereas a model with a by-speaker random slope (`vot_mod_3`) does not ($|t| < 2$, wide CI).

The conclusion from the intercepts-only model is incorrect, because it is overconfident. In the data the models were fitted to (`vot_voiced_core`), speakers vary greatly in the `speaking_rate_dev` effect, and the results are very influenced by a couple of speakers. The correct conclusion is that we cannot rule out a `speaking_rate_dev` effect of 0.

Overconfidence is a serious problem, regardless of whether a type I error has in reality been committed. In this case, the true `speaking_rate_dev` effect is probably not 0, as we saw in the model fitted to the full dataset (section 10.3.2.4)—that is, not a type I error. But the degree of evidence the data give against the null—the Fisherian definition of a *p*-value—is still drastically wrong.

Model 2: Too-maximal random effects Barr et al. (2013) pointed out this problem for linguistic data (independently of Schielzeth and Forstmeier [2009], for ecological data), and they recommended defaulting to "maximal" random-effect structure—all possible random slopes, and ideally all possible random-effect correlations—to avoid overconfident estimates. The actual discussion in Barr et al. (2013) is more nuanced and careful, but the lesson the field took away was the title: "keep it maximal."

As we've now seen many times, the main problem with defaulting to maximal random-effect structure is practical: it is often too complex for the data, leading to convergence problems, as in our attempt to fit a maximal model for the `diatones` data (section 10.3.3). But there is a more important conceptual problem, which we can see by returning to this model.

We can in fact get this model to converge by changing the optimizer, as we've often done for MELR models:

```
## > diatones_melr_max_bobyqa <- update(diatones_melr, . ~ . -
## + (1 | prefix) + (1+syll2_coda + syll2_td + frequency +
## + syll2_td:frequency|prefix),
## +    control=glmerControl(optimizer = 'bobyqa'))

## boundary (singular) fit: see ?isSingular
```

The model is singular, but this is a less serious issue than nonconvergence. The fixed effects for this model are

```
summary(diatones_melr_max_bobyqa, correlation = FALSE)
...
## Fixed effects:
##                       Estimate Std. Error z value Pr(>|z|)
## (Intercept)             -4.265      2.161   -1.97    0.048
## syll2_coda              -4.479      2.720   -1.65    0.100
## syll2_td                 4.305      2.572    1.67    0.094
## frequency               -0.168      1.607   -0.10    0.917
## syll1_coda               5.321      2.842    1.87    0.061
## syll2_td:frequency       5.306      3.558    1.49    0.136
## frequency:syll1_coda    12.630      8.725    1.45    0.148
...
```

Recall that for these data, the effect of word `frequency` and its modulation by syllable structure (other predictors) are of interest (section 6.1.2). From this model, we would conclude that there are *no* significant predictors of whether a word shifts stress and no evidence that `frequency` plays a role.

This conclusion is almost certainly wrong (a type II error). Based on prior work in this area, we would expect at least some effect, in line with what we have found in more realistic models of these data in previous chapters. The random-effect structure of the maximal model is far too complex for these data ($n = 130$), leading to greatly reduced power and *underconfident* estimates.

Model 3: Too-data-driven random effects Bates, Kliegl, et al. (2015) and Matuschek et al. (2017) pointed out that mechanical application of "keep it maximal" can easily lead to such overfitted models and associated type II errors and model fitting problems. They recommended balancing type I and type II errors by using hypothesis tests (or another criterion, such as comparing AIC/BIC) to decide which random-effect terms to include. Again, their discussion is more subtle and careful, but this "data-driven" recipe is what the field took away, as an alternative to "keep it maximal."

When applied mechanically, this approach has the same drawbacks as any stepwise approach to model selection (section 5.10.1), including that it ignores the "substantive" wheel. As an example, consider again the `neutralization` data, where the `voicing` effect is of primary interest; this includes by-participant and by-item variability in the effect. As fixed effects, we'll include those from `neut_mod_3`, as well as a `voicing:prosodic_boundary` interaction, which is plausible for this data:[7]

```
neut_ch10_m1 <- lmer(vowel_dur ~ voicing * prosodic_boundary + place +
  vowel + (1 | subject) + (1 | item_pair), data = neutralization)
```

(This is the same as `neut_mod_1` from section 8.3.6.)

One common data-driven method to build up random-effect structure is a forward stepwise algorithm: start with an intercepts-only model, then add in the random-slope term Z that gives the lowest p-value for an LRT (comparing models with and without Z). Repeat, until there is no Z that can be added with $p < 0.05$.

As shown in online appendix, section 8.2, applying this algorithm leads to a model with only a by-participant random slope for `vowel`:

```
neut_ch10_m2 <- lmer(vowel_dur ~ voicing * prosodic_boundary + place +
  vowel + (1 + vowel | subject) + (1 | item_pair),
data = neutralization)
```

In particular, although adding a by-speaker random slope of `voicing` improves `neut_ch10_m1` ($p < 0.05$), the `vowel` random slope is added first because it has a lower p-value, and adding the `voicing` slope to this new model gives $p > 0.05$.

There are no random slopes of `voicing` in this model, so we would conclude that neither participants nor items (significantly) differ in the effect. This is implausible, given both the empirical data (figure 8.3), which show great by-participant variability, and prior work, where such "incomplete neutralization" effects typically vary a lot by participant and item.

7. As in section 8.8.1. This is theoretically plausible because (phonetic) contrasts are often enhanced in prosodically strong positions.

Here, we reached the wrong conclusion by not prioritizing terms related to `voicing`, the predictor of primary interest.

Models 4–5: Misspecified fixed effects Most discussion of model selection has focused on random-effect structure, but choosing appropriate fixed effects is also nontrivial. The issues here are the same as in our discussion for linear regression (sections 5.8.2, 5.10): there are dangers to either underfitting or overfitting the fixed effects. This is a particular risk for corpus (or other observational) data, where the set of fixed effects is not usually clear a priori. Online appendix, section 8.3 shows two examples using the `turkish_if0` data.

For corpus data, it is not uncommon to see "minimal" models fitted, which account for a single fixed effect. For these data, `Voicing` is of primary interest; we have seen that random intercepts for word/consonant/speaker are appropriate; and minimally a by-speaker random slope for `Voicing` is needed. This model gives a `Voicing` fixed effect with $p = 0.03$: this is clearly underconfident, given how strong the effect is in a model where control predictors are included (e.g., `if0_m0`). The minimal model is *underfitted.*

At the other extreme, given our interest in `Voicing`, we might choose a maximal fixed-effect structure (in the sense introduced in section 10.5.3), accounting for all possible interactions with control predictors, up to three-way interactions. The resulting model has 30 fixed-effect terms, of which 15 involve `Voicing` and most are not significant. This model is *overfitted* and too complex to give insight into the research question ("effect of `Voicing`").

10.5.2 General Principles

Laying out principles for model selection lets us articulate what is wrong with such models, and how to arrive at better ones. This discussion follows Snijders and Bosker (2011, section 6.4), with some edits from my own perspective.[8] Principles 1–4 are most important, but all are violable in some situations.

Principle 1: Prioritize subject matter considerations Model selection must be guided by the research questions and study design, theoretical considerations, previous work, and common sense.

For example, in bad model 4, the effect of interest (`Voicing.vl`) is expected to be very small, relative to other factors affecting F0, so our model should control for these. This principle often comes into play in deciding what terms to *not* add to a model: unless there are substantive reasons to consider additional predictors, it often does not make sense to add them, to keep the analysis confirmatory rather than exploratory—even if doing so would improve the model's fit (Winter 2019, section 16.4).

Principle 2: Distinguish between terms on which the research is focused and terms included for fit These are called "critical" and "control" terms. Often, the critical terms are a subset of the fixed effects; all other terms are controls. Our model is designed to estimate the critical terms as accurately as possible.

8. I have omitted their principles 4 and 5, on "doing justice to the multilevel nature of the problem" and random-effect correlations, to simplify the discussion. My principles are numbered differently from theirs.

This principle has a range of consequences. Random-effect selection, whether maximal or data-driven, should prioritize critical terms (as recommended by Barr et al. 2013; Bates, Kliegl, et al. 2015). For fixed-effect selection, we should always be reluctant to drop critical predictors from the model, and if we need to select from many possible control predictors, we should prioritize those that let us better estimate the critical terms (e.g., confounds).

For example, in bad model 3, interspeaker variation in the `voicing` effect is a research question, so we should prioritize by-speaker/-item random slopes for `voicing`, even if this doesn't follow from a pure stepwise procedure. If adding a random slope for (critical predictor) x gives a singular model, with perfect random-effect correlations, we should try a model with uncorrelated random effects.[9]

Principle 3: Fixed effects need appropriate error terms This primarily means appropriate random slopes, especially for critical predictors. For group-level predictors (e.g., gender, lexical class), this also means random intercepts (as discussed in section 10.2.1). Otherwise there is a serious risk of type I (and type II) errors, as in our bad model 1. This principle underlies the "keep it maximal" approach to selecting random effects.

Principle 4: Larger models have lower power to detect individual effects This is the case, all else being equal. In addition, our power is often low in mixed-effects models, especially for group-level predictors or interactions. It is thus easy to build up models with low power. This principle underlies the data-driven approach to random-effect selection.

Principle 5: The model's "hierarchical" structure should make sense The presence of an $x_1 : x_2$ interaction implies main effects of x_1 and x_2 (in fixed effects or random slopes). A by-z random slope of x implies a fixed effect of x and a by-z random intercept.

Principle 6: Be reluctant to overfit/include nonsignificant effects in the model For mixed-effects models, this can apply to either fixed effects (as in bad model 5) or random effects (as in bad model 2).

Other principles can override principle 6, most importantly subject matter/research question considerations (principle 1). For example, in modeling the `givenness` data, where the `clabel.williams` effect is of primary interest, it would not make sense to exclude fixed effects of `voice.passive` or `npType.pronoun`, regardless of significance. These predictors are part of the experimental design. In our final model (`givenness_m1`), there is a nonsignificant main effect of `voice.passive`, which cannot be excluded because it participates in an interaction with `clabel.williams` (principle 5).

Principle 7: Be reluctant to underfit This usually just means, don't drop any significant terms. Exceptions could include truly large datasets (where everything may be significant, regardless of importance) and the desire to keep models confirmatory and substantively motivated, rather than exploratory (see principle 1).

9. It is very common in current practice for by-x random slopes to not be included, because the resulting model is singular, even when x is a critical predictor. This results in overconfident fixed-effect coefficient estimates.

10.5.3 Practical Heuristics

Principles are great, but what should we actually do in practice? This section lays out a general procedure and practical heuristics to inform your model-building process. Each heuristic is violable in some settings: given how much model selection depends on the data and research questions, my goal is to give a set of general guidelines rather than strict rules. We'll focus on points on which there seems to be broad agreement, informed by the sources in section 10.9, paragraph 1.

The heuristics assume that you have hypotheses/research questions about fixed-effect terms (possibly some, not all) and that random-effect terms are to prevent errors on fixed-effect estimates. This is the most common case, and some adjustment would be needed if, for example, individual differences (= by-participant random effects) were of primary interest.

The high-level procedure I assume is[10]

- Step 1. Perform exploratory analyses, visualization, selecting possible predictors, etc.
- Step 2. Build maximal *fixed-effects* structure.
- Step 3. Select *random-effects* structure, using those fixed effects.
- Step 4. Do any pruning of fixed effects.

Three first heuristics inform this whole procedure:

Heuristic 1: Theoretical/subject matter considerations are most important.

Heuristic 2: Nonconvergent models are invalid. Singular models are dispreferred.

Heuristic 3: Practical considerations can play a role For example, it is fine to use different model selection procedures in settings where each model takes hours to fit versus seconds to fit (e.g., large corpus datasets vs. small laboratory datasets).

Let's assume here that step 1 is complete. The next heuristics involve fixed-effect selection, step 2.

Heuristic 4: Choose fixed effects before random effects The motivation for step 2 coming before step 3 is to get accurate residuals, because the random effects are modeling structure among the residuals—the leftover variance after accounting for fixed effects (Fitzmaurice et al. 2008, section 7.5).

In some cases, such as a classic factorial-design laboratory experiment, the fixed-effect structure is known in advance. But usually, there are many possible predictors, and we have to choose what terms to include. It is important that the fixed-effect structure err on the side of including more terms, in case there are (fixed) effects that only become clear once by-group variability is accounted for. This *maximal* fixed-effect structure is "the most elaborate or complex model for the mean response that we would consider from a subject matter point of view" (section 7.5). You can then always choose to prune terms (e.g., using model comparison) once the random effects have been selected (step 3).

10. This roughly follows the "top-down" method of Zuur (2009, section 5.7); see also West, Welch, and Galecki (2014, section 2.7). Snijders and Bosker (2011) describe a "bottom-up" strategy that is more widely used in some fields, where the model is built up alternating between fixed-effect and random-effect terms at different levels. All heuristics except heuristic 4 would hold for the bottom-up strategy.

In practice, choosing a maximal model is a subjective matter, with no automatic recipe. The issues here are similar to model selection for (non-mixed-effects) linear regression, and different approaches are possible. The maximal model could be "all explanatory variables and as many interactions as possible" (Zuur 2009, section 5.7) or something more conservative, like a "maximum reasonable fixed-effect structure" chosen on theoretical grounds—similarly to the Gelman and Hill method discussed for linear regression (section 5.10.3.3), but without dropping any terms.

One common approach, which is strongly dispreferred, is selection of fixed effects by a purely stepwise procedure (e.g., including all terms that lower the model's AIC); this has the same issues as stepwise regression (section 5.10.1). However, it may be reasonable to do some model comparison in reaching your maximal model; this should be done using models with random intercepts only, as these control for basic nonindependence of observations without assuming anything about fixed effects.

Because the issues here are similar to variable selection for non-mixed-effects models from chapter 5, we will just assume that a suitable maximal fixed-effect structure has been chosen in our case studies, where we will also not do the optional term-dropping step 4. In summary:

Heuristic 5: Prefer a larger fixed-effect structure; do not choose by a purely automatic procedure Selecting random-effect structure (step 3) is harder and has been the focus of most work in the linguistics/psychology literature. Different approaches are possible, which I will describe after articulating broad points on which there is consensus. These heuristics follow from principles 1–4.

Heuristic 6: Random intercepts are almost always necessary.

Heuristic 7: Random slopes must be considered at least for critical predictors, whether included a priori or via hypothesis testing.

Heuristic 8: Including random slopes is more important than including correlations between random effects.

Heuristic 9: Random-effect terms (intercepts, slopes, correlations) related to critical predictors should be prioritized over those related to control predictors.

Heuristic 10: Random-effect correlations for intrinsically related predictors (factors, nonlinear effects) are more important than other correlation terms.

10.5.4 Possible Procedures

A high-level procedure is outlined as steps 1–4. In practice, especially when models take a long time to fit, a simpler method is often used: choose the final fixed-effect structure in step 1, erring on the side of including more terms, and skip step 3. This (lazier) procedure simplifies random-effect selection (step 2), avoids having to refit models, and avoids the temptation to boost p-values by dropping terms (step 3). This is the method illustrated in case studies that follow.

Within this (lazier) procedure, different model selection procedures essentially come down to *random-effect selection* (step 3). Heuristics 5–9 suggest fitting models with rich random-effect structure, at minimum including random intercepts and appropriate random slopes, to guard against (primarily) type I errors in critical predictors.

There are two common approaches, within these parameters, which we have seen:[11]

1. *Maximal* approach includes random-effect terms based on the study design. Prioritizes type I error. (Barr et al. 2013)
2. *Data-driven* approach includes random-effect terms if they improve the model, using a liberal α-value, for example, $\alpha = 0.2$. Tries to balance type I error and power. (Bates, Kliegl, et al. 2015; Matuschek et al. 2017)

I often use a third approach, outlined in chapter 8, given the difficulties of fitting either fully maximal models or many models (required for the data-driven approach) for corpus data (many fixed effects and/or large datasets):

3. *Uncorrelated first* approach fits a model with all possible random slopes, without any random-effect correlations, then iteratively adds in correlations, proceeding heuristically.

"Proceeding heuristically" could mean prioritizing correlations involving critical predictors or intrinsically related predictors (heuristics 9 and 10), or it could mean examining pairwise plots of random effects to see which correlations are largest. Adding correlation terms can be justified using model comparison or in another way.

The general ideas of this approach are to heavily prioritize random slopes over correlation terms in building up random-effect structure (heuristic 8) and to acknowledge practical considerations (heuristic 3).

Whichever procedure you use, there are several choice points to arrive at an actual algorithm, including

- "Backwards" (start with maximal, reduce) or "forwards" (start with minimal, build up)?
- How to do each step: assess every possible term that could be added/dropped or consider terms sequentially (e.g., add the random slope with lowest p-value vs. the first one which has $p < 0.2$)?
- How much to prioritize trying uncorrelated random-effect structure, as a practical matter. (The maximal and data-driven papers do discuss dropping correlation terms, they just emphasize it less than the uncorrelated first approach does.)

The literature describes a range of concrete algorithms and assesses them using simulation studies (box 10.4). Now let's review three case studies, each showing one approach.

11. Seedorff, Oleson, and McMurray (2019) argue based on simulations that a "model space" version of the data-driven approach, where all possible random-effect structures are compared using (for example) AIC, performs better than a hypothesis-testing approach. This is feasible for models with a small number of fixed effects.

Box 10.4
Broader Context: Random-Effect Selection

Random-effect selection is an active research area (Barr et al. 2013; Bates, Kliegl, et al. 2015; Matuschek et al. 2017; Seedorff, Oleson, and McMurray 2019), and end users worry a lot about this issue when analyzing data (Brauer and Curtin 2018; Meteyard and Davies 2020).

But as Seedorff, Oleson, and McMurray (2019) emphasize, it is still unclear how much the exact procedure matters—in terms of type I error and power—across different kinds of data (e.g., corpus data) and types of models. For example, the datasets considered in this literature are fairly simple, from (psycholinguistic) laboratory experiments, and most work only considers linear mixed-effects models. Even once there is more work in this area, given the diversity of linguistic data, there will never be a one-size-fits-all recipe. I have in mind here the example of ecology, which has similarly diverse data, where a large literature exists on model selection approaches in different settings. In my experience, (language) researchers are anxious to follow a recipe, and they will extract one from a paper cited or follow a procedure they or a colleague have used before. It is better to *think about what method makes sense* for the kind of data you are dealing with and your research questions . However you do model selection, it is crucial to *report and justify your strategy*.

Although different model selection strategies are possible, it is important to choose your strategy in advance and adjust as little as possible once you are analyzing the data. It is too easy to choose a strategy that happens to support your hypotheses. Reporting can be anything from verbal description of general principles to enumerating the models tried and posting full analysis scripts. Meteyard and Davies (2020) give an excellent and reassuring discussion of these issues.

10.5.5 Case Study 1: Maximal Approach

This example uses the `givenness` data.

Fixed effects For our maximal fixed-effect structure, we'll include the three predictors that are part of the experimental design (`clabel.williams`, `npType.pronoun`, `voice.passive`), as well as the one interaction with the predictor of primary interest (`clabel.williams:voicing.passive`) that looks promising from exploratory plots (exercise 10.2). This is the fixed-effect structure of `givenness_m1` (section 9.3.1).

In this case, another sensible maximal fixed-effect structure could be to include all two-way interactions with `clabel.williams`.

Random effects We'll use a backward-selection version of the maximal approach:

1. Fit the model with maximal random effects.
2. Deal with any convergence or singular model issues.
3. If this requires simplifying random-effect structure, remove as few terms as possible.

We already did (1) in section 10.3.2.3, resulting in the singular model `giv_maximal`. Online appendix, section 9.1 shows the full process of (3), iteratively removing random-effect terms following steps 2(a)–(d) from table 10.1. We'll eventually end up either at the model with uncorrelated maximal random effects shown in section 10.4.2.2 (`giv_ch10_5`) or at that model with a single correlation term (not shown here).

```
summary(giv_ch10_5)
...
## Random effects:
##  Groups       Name                          Variance Std.Dev.
##  participant  clabel.williams:voice.passive 4.67e+00 2.16e+00
```

```
## participant.1 npType.pronoun                   8.47e-01 9.20e-01
## participant.2 voice.passive                    2.10e+00 1.45e+00
## participant.3 clabel.williams                  1.67e-13 4.08e-07
## participant.4 (Intercept)                      1.73e-01 4.16e-01
## item          clabel.williams                  6.56e-01 8.10e-01
## item.1        npType.pronoun                   5.46e-01 7.39e-01
## Number of obs: 382, groups:  participant, 27; item, 16
##
## Fixed effects:
##                               Estimate Std. Error z value Pr(>|z|)
## (Intercept)                     -1.184      0.234   -5.07  4.0e-07
## clabel.williams                  4.026      0.594    6.77  1.3e-11
## voice.passive                    0.407      0.456    0.89    0.373
## npType.pronoun                   0.870      0.426    2.04    0.041
## clabel.williams:voice.passive    2.328      0.941    2.47    0.013
...
```

This model is still singular, but only due to the zero `clabel.williams` by-participant slope, which we leave in, because it involves the effect of central interest, by heuristic 1. If we wanted, we could add back in the by-item intercept, which would give our final model of the `givenness` data from chapter 9 (`givenness_m1`): maximal with uncorrelated random effects. Because this term has zero variance, the fit will be identical.

In sum, for this case, the maximal model has all random slopes, uncorrelated.

10.5.6 Case Study 2: Data-Driven

This example uses the `turkish_if0` data. We'll start with the intercepts-only model `if0_m0` from section 10.2.1:

```
formula(if0_m0)
## f0 ~ Voicing.vl + base_vowel + gender.male + local_f0 + (1 |
##     word) + (1 | speaker) + (1 | consonant) + Voicing.vl:base_vowel
```

Fixed effects The fixed effects here are chosen based on the research question (`Voicing.vl+ Voicing.vl:base_vowel`) with most possible control predictors from the dataframe added. The maximal fixed-effect structure could in principle contain further interactions with `Voicing.vl`. Because we don't have any clear expectations for other interactions, we'll omit these for simplicity.

Random effects Let's use a "forward" version of the data-driven approach, prioritizing critical predictors (heuristic 9).

We'll start with the intercepts-only model `if0_m0` and test whether each possible random slope related to the critical predictors improves the model, defined as a likelihood-ratio test (comparing REML fits) with $p < 0.2$. If a slope doesn't improve the model when added with correlation terms, we'll try adding it without correlation terms. The possible random slopes are

1. By-speaker random slopes of `Voicing.vl`, `base_vowel`

2. By-speaker random slope for their interaction
3. By-consonant random slope of `base_vowel`

We can then consider random slopes involving control predictors. The rationale here, rather than a pure data-driven method (just consider all possible random slopes, add the one with the lowest p), is to avoid the kind of problem we saw in bad model 3, by prioritizing random effects for critical predictors.

Both the `Voicing.vl` and `base_vowel` slopes improve the model with $p < 0.2$:

```
## Test by-speaker random slope of Voicing
newMod1 <- update(if0_m0, . ~ . - (1 | speaker) +
    (1 + Voicing.vl | speaker))
anova(if0_m0, newMod1, refit = FALSE)
...
##          npar   AIC   BIC logLik deviance Chisq Df Pr(>Chisq)
## if0_m0     12 16461 16537  -8219    16437
## newMod1    14 16417 16506  -8194    16389  48.3  2    3.2e-11
```

```
## Test by-speaker random slope of base_vowel
newMod2 <- update(if0_m0, . ~ . - (1 | speaker) +
    (1 + base_vowel | speaker))
anova(if0_m0, newMod2, refit = FALSE)
...
##          npar   AIC   BIC logLik deviance Chisq Df Pr(>Chisq)
## if0_m0     12 16461 16537  -8219    16437
## newMod2    17 16431 16538  -8198    16397  40.5  5    1.2e-07
```

```
## Model with both random slopes added
if0_m1 <- update(if0_m0, . ~ . - (1 | speaker) +
  (1 + Voicing.vl + base_vowel | speaker))
```

Further model comparisons (in online appendix, section 9.2; see box 10.5) show that adding term (2) from the preceding list improves the model ($p < 0.2$), but this is only possible with uncorrelated random effects and adding term (3) does not improve the model ($p > 0.2$). The interim model is thus:

```
if0_m2 <- update(if0_m1, . ~ . +
  (1 + Voicing.vl + base_vowel_AvIU + base_vowel_IvU | speaker) +
  (0 + Voicing.num:base_vowel_AvIU + Voicing.vl:base_vowel_IvU || speaker))
```

```
summary(if0_m2, correlation = FALSE)
...
## Random effects:
##  Groups    Name                      Variance Std.Dev. Corr
##  word      (Intercept)               0.4838   0.696
##  speaker   Voicing.vl:base_vowel_IvU 1.5570   1.248
```

```
## speaker.1 Voicing.vl:base_vowel_AvIU 0.0689  0.262
## speaker.2 (Intercept)                2.6912  1.640
##           Voicing.vl                 0.2625  0.512     0.12
##           base_vowelAvIU             0.1573  0.397     0.17
##           base_vowelIvU              0.1884  0.434     0.02
## consonant (Intercept)                0.0696  0.264
## Residual                             2.1603  1.470
##
##
##
##
##
##
## -0.30
##  0.18 0.75
##
##
## Number of obs: 4224, groups:
## word, 1238; speaker, 98; consonant, 14
##
## Fixed effects:
##                         Estimate Std. Error t value
## (Intercept)               8.6866     0.2741   31.69
## utterance_num_syllables   0.2039     0.0566    3.60
## Voicing.vl                0.7536     0.1946    3.87
## base_vowelAvIU            0.7141     0.0940    7.60
## base_vowelIvU            -0.0288     0.1343   -0.21
## gender.male              -7.1821     0.3978  -18.06
## local_f0                  0.3004     0.0158   18.99
## Voicing.vl:base_vowelAvIU 0.4391     0.1692    2.59
## Voicing.vl:base_vowelIvU  0.2017     0.2906    0.69
```

For the remaining fixed effects (control predictors), there are eight possible random slopes:

- `gender`: by-word, by-consonant
- `utterance_num_sylls`: by-speaker, by-word, by-consonant
- `local_f0`: by-speaker, by-word, by-consonant

Online appendix, section 9.2 shows the full process of considering each one for addition to the model. Only a by-word random slope of `utterance_num_sylls` improves the model with $p < 0.2$, so our final model includes this term:

```
if0_m3 <- update(if0_m2, . ~ . + (0 + utterance_num_sylls | word))
```

```
summary(if0_m3, correlation = FALSE)
...
## Random effects:
##  Groups    Name                      Variance Std.Dev. Corr
```

```
## word      utterance_num_syllables  0.0243   0.156
## word.1    (Intercept)              0.4837   0.695
## speaker   Voicing.vl:base_vowel_IvU 1.5928  1.262
## speaker.1 Voicing.vl:base_vowel_AvIU 0.0708 0.266
## speaker.2 (Intercept)              2.6886   1.640
##           Voicing.vl               0.2606   0.510     0.12
##           base_vowelAvIU           0.1576   0.397     0.17
##           base_vowelIvU            0.1874   0.433     0.03
## consonant (Intercept)              0.0695   0.264
## Residual                           2.1537   1.468
##
##
##
##
##
##
##
## -0.29
##  0.20  0.74
##
##
## Number of obs: 4224, groups:
## word, 1238; speaker, 98; consonant, 14
##
## Fixed effects:
##                           Estimate Std. Error t value
## (Intercept)                 8.6846     0.2740   31.69
## utterance_num_syllables     0.2301     0.0597    3.86
## Voicing.vl                  0.7513     0.1945    3.86
## base_vowelAvIU              0.7154     0.0940    7.61
## base_vowelIvU              -0.0232     0.1343   -0.17
## gender.male                -7.1846     0.3977  -18.07
## local_f0                    0.3006     0.0158   19.01
## Voicing.vl:base_vowelAvIU   0.4360     0.1693    2.57
## Voicing.vl:base_vowelIvU    0.1985     0.2913    0.68
```

Results Figure 10.3 shows the model-predicted `f0` as a function of `Voicing.vl` and `base_vowel`, which address the main research question (effect of voicing on F0). The model suggests that there is a significant consonant voicing effect on F0 in Turkish, of ~0.75 semitones (averaging across vowels); it is larger for high vowels (I/U) than for low vowels (A). We could alternatively report the results in terms of the VOICELESS>VOICED difference for each vowel, using post hoc tests (exercise 10.3).

There is also substantial speaker variability in the `Voicing.vl` effect, as will be discussed in section 10.6.

Box 10.5
Practical Note: Common Sense in Random-Effect Selection

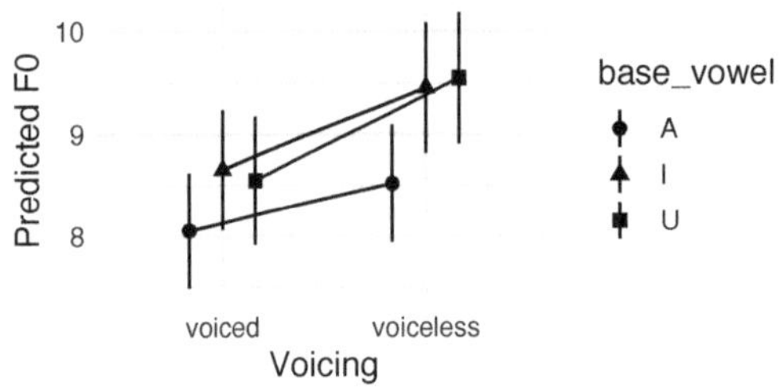

Figure 10.3
Interaction plot of `Voicing.vl` and `base_vowel` for model `if0_m2`, with 95% CIs, marginalizing over other predictors.

> Whatever method you use for random-effect selection must be tempered by common sense: what is possible given the structure of the data and how long models take to fit. This is especially true for models with many predictors (e.g., corpus data) or that are fitted to particularly large (or small) datasets.
>
> For example, for the Turkish IF0 data, truly maximal random-effect structure—with all possible slopes, including correlations—would be impossible (try it). This model contains more by-word random-effect terms than observations, making it literally impossible to fit. Common sense would dictate excluding all by-word random slopes, as there are only approximately 3 observations per word. More generally, when you have many predictors relative to sample size, fitting models with maximal random-effect structure (including correlations) will often result in nonconvergent models or massively overfitting ones. Because most predictors are controls, it isn't worth the effort to spend hours trying to fit overly complex models.
>
> On the other hand, as the Turkish IF0 example in online appendix, section 9.2 illustrates, fully data-driven methods are extremely time-consuming, even for a modest (~5–10) number of predictors. The sheer number of possible model comparisons is huge; there are not currently good tools to automate the process; and pure stepwise methods are dangerous in any case. For models that take an hour to fit—which is not uncommon for corpus data—a pure data-driven method is impossible. More common is the situation in the Turkish IF0 example: fitting and evaluating models to evaluate random effects unlikely to be relevant for the research questions (random slopes for control predictors) took me hours, and the final model (`if0_m3`) is almost identical to the model before these steps (`if0_m2`).
>
> The upshot is that practical considerations (heuristic 3) are an important reason to not just mechanically apply a recipe. It helps to use the *goals of your analysis* and the structure of the data central to choose between different possibilities; typically this means prioritizing terms related to the research questions. There is no shame in this; you just have to *report when writing up your analysis what you did, and why* ("more complex random-effect structure was not considered due to...").
>
> In the Turkish IF0 example, it would have been reasonable to not consider random slopes for control predictors and simply stop after having considered all possible random effects for critical predictors. Our justification could be that we are taking a mild risk of underfitting the data for convenience, but with good confidence about our estimates for critical effects. It would also be reasonable to only try adding by-speaker random slopes for control predictors, with the justification that by-word random slopes are unreasonable (by the preceding logic), and the only purpose of by-consonant random effects is to get a more accurate estimate for `Voicing.vl`, which requires a random intercept, but no random slopes.

10.5.7 Case Study 3: Uncorrelated First

This example returns to the `vot_voiced` data from the `vot_mod3_full` model (section 10.3.2.3). The research question remains "is there an effect of speaking rate?" In models fitted to these data so far, we have operationalized speaking rate as within-speaker deviation (`speaking_rate_dev`). To make things more interesting, let's now consider the speaker's mean rate (`speaking_rate_mean`), which we saw briefly in section 8.10.2.3. So speaking rate could affect VOT via either faster speech (within-speaker) or faster speakers.

We'll first standardize this variable and extract numeric contrasts for `place`, which we will need for uncorrelated random effects:

```
vot_voiced <- mutate(vot_voiced,
  speaking_rate_mean_std = rescale(speaking_rate_mean)
)
mm <- model.matrix(~place, vot_voiced)
vot_voiced$place1 <- mm[, 2] ## contrast 1
vot_voiced$place2 <- mm[, 3] ## contrast 2
```

Fixed effects As fixed effects we will use `speaking_rate_dev`, `speaking_rate_mean`, and the same set of six control predictors we used previously in the `vot_mod3_full` model. The base (intercepts-only) model is

```
vot_voiced_m0 <- lmer(log_vot ~ speaking_rate_dev + foll_high_vowel +
    cons_cluster + stress + log_corpus_freq + place + gender +
    speaking_rate_mean_std + (1 | word) + (1 | speaker),
  data = vot_voiced)
```

At this point we could consider adding interaction terms for a larger maximal model, expanding "the effect of speaking rate" to include modulation by other predictors. Exercise 10.4 asks you to consider this case, which gives interesting results, but for the current example we'll use the fixed effects in `vot_voiced_m0` as "maximal."

Note that this model contains several nonsignificant fixed effects ($|t| < 2$):

```
tidy(vot_voiced_m0) %>%
  select(-effect, -group, t = statistic) %>%
  filter(abs(t) < 2)
##                      term estimate std.error     t
## 1                  stress   -0.030     0.033 -0.88
## 2         log_corpus_freq    0.031     0.024  1.28
## 3 speaking_rate_mean_std    0.107     0.107  1.00
```

If we wanted to consider removing these terms, we would first need to build up the random effects.

Random effects Let's start with the model with all possible random slopes, uncorrelated:

```
vot_voiced_m1 <- update(vot_voiced_m0, . ~ . - (1 | speaker) -
  (1 | word) +
  (1 + gender + speaking_rate_mean_std + speaking_rate_dev || word) +
  (1 + speaking_rate_dev + foll_high_vowel + cons_cluster + stress +
    log_corpus_freq + place1 + place2 || speaker),
data = vot_voiced)
```

The model's output is shown in online appendix, section 9.3:

```
summary(vot_voiced_m1, correlation = FALSE)
```

We'll now try adding in correlation terms holistically. One way to do this, following heuristics 9–10, is to consider terms in the following order:

(a) *Critical predictors*: correlations with each random slope
(b) *Intrinsically related predictors* (here, `place`): correlations between related predictors
(c) *Control predictors*: a subset of correlations, suggested by pairwise plots of estimated random effects

At each step, we'll use model comparison (likelihood-ratio test of REML fits, $p<0.2$) to decide whether to keep the correlation terms in the (convergent/nonsingular) model. We could alternatively not do hypothesis testing and just keep the terms in the model at each step as long as the model converges and is nonsingular.

Steps (a) and (b) are shown in online appendix, section 9.3. Both improve the model, though for (b) the optimizer needs to be changed for convergence. The resulting model is

```
vot_voiced_m3 <- update(vot_voiced_m0, . ~ . - (1 | speaker) -
  (1 | word) +
  (1 + speaking_rate_mean_std + speaking_rate_dev | word) +
  (0 + gender || word) +
  (1 + speaking_rate_dev + place1 + place2 | speaker) +
  (0 + foll_high_vowel + cons_cluster + stress +
    log_corpus_freq || speaker),
control = lmerControl(optimizer = "bobyqa"))
```

The model's output is

```
summary(vot_voiced_m3, correlation = FALSE)
```

We could now do step (c): examining pairwise plots (e.g., `pairscor.fnc(ranef(vot_voiced_m3)$speaker)`) to see which correlations have the largest magnitudes, which we could consider adding. In this case, shown in online appendix, section 9.3, we'd end up trying correlations with the by-speaker random slopes for `cons_cluster` and `log_corpus_freq` (r around 0.5–0.6), but both give singular models. So `vot_voiced_m3` is our final model.

Results It is worth taking a step back to compare the maximal no-correlations model and the final model with some correlations added. Adding the correlation terms significantly improves the model:

```
anova(vot_voiced_m3, vot_voiced_m1, refit = FALSE)
...
##               npar   AIC   BIC logLik deviance Chisq Df Pr(>Chisq)
## vot_voiced_m1   23 21879 22051 -10917    21833
## vot_voiced_m3   32 21875 22114 -10905    21811  22.5  9     0.0075
```

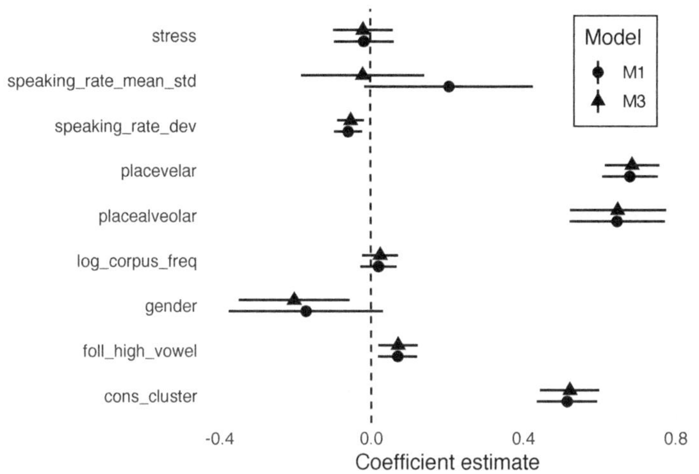

Figure 10.4
Fixed-effect coefficient estimates and 95% Wald CIs (omitting the intercept) from models `vot_voiced_m1` and `vot_voiced_m3` of the `vot_voiced` data with all possible random slopes, differing in which random-effect correlations are included: no correlations in `m1`, all correlations in `m3`.

However, the fixed-effect estimates for the models, shown in figure 10.4, are very similar—all terms are qualitatively identical.

This reiterates the point that we saw for simpler models in section 8.7.3, in a more realistic case: in practice, whether random-effect correlations are added rarely affects the qualitative conclusions with respect to fixed effects (assuming that predictors are all centered, etc.; see box 8.10). This is a reasonable, if purely empirical, argument for using maximal uncorrelated random effects as your default.

In terms of the actual research question, the model suggests that VOT decreases with faster speech, within-speaker (`speaking_rate_dev`), but does not significantly decrease for faster speakers (`speaking_rate_mean_std`).

10.6 Predictions and Uncertainty for Individual Levels

The remaining sections turn from model selection to additional advanced topics for mixed-effects models.

We saw in section 8.8 how to predict different quantities from a fitted mixed-effects model, including predictions for each level (e.g., by-participant/-item), assuming we didn't want to include a measure of uncertainty (CI or PI).

The more general case is complex, for the reason discussed there: how to quantify uncertainty depends on what sources of variability to capture, which depends on exactly what we are trying to predict (section 8.8.2). Quantifying and communicating uncertainty from fitted models is an area where best practices and software are evolving.

This section shows some examples illustrating different ways to address the same research question: robustness of the `Voicing.vl` effect across speakers in the `turkish_if0` data. These are

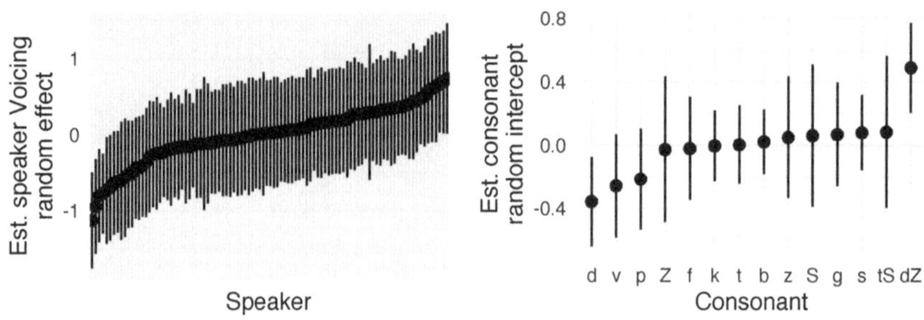

Figure 10.5
BLUPs (with 95% CIs) for random intercept for each speaker (*left*) and consonant (*right*) for model `if0_m2`.

- Uncertainties on *predicted random effects* (each speaker's deviation)
- Uncertainties on *model predictions* (each speaker's VOICELESS, VOICED F0)
- Uncertainties on *effects* (each speaker's `Voicing.vl` effect)

These examples introduce some general methods for quantifying uncertainty, which are progressively more general/accurate/slower:

- Using information from the fitted model (very fast)
- Simulating from the fitted model (fast)
- Performing parametric bootstrapping from the fitted model (very slow)

10.6.1 Example 1: Predicted Random Effects and Uncertainties

Section 8.3.4 introduced predicted random effects, BLUPs —the deviations of each speaker (etc.) from the population effect. These could be of interest for studying "individual differences," for example, where of interest is which speakers have higher and lower values. The fitted model also contains a variance-covariance matrix for BLUPs (see `?ranef`); these can be used to define SEs and hence to define Wald 95% CIs. BLUPs can be extracted using `tidy()` from broom.mixed (or `ranef()` from lme4):

```
##  Each speaker's BLUP for voicing term
tidy(if0_m2, effects = "ran_vals", conf.int = TRUE) %>%
  filter(group == "speaker" & term == "Voicing.vl") %>%
  rename(speaker = level) %>%
  select(-effect)
##       group speaker       term estimate std.error conf.low
## 1 speaker    TU001 Voicing.vl   0.1762      0.35   -0.519
## 2 speaker    TU002 Voicing.vl   0.0131      0.37   -0.715
## 3 speaker    TU003 Voicing.vl  -0.0237      0.36   -0.724
## 4 speaker    TU004 Voicing.vl  -0.1024      0.34   -0.771
...
```

These are shown in figure 10.5 (left). Note how large the CIs are: for most speakers, we cannot even be sure whether they are above or below the mean.

This is not unusual; errors in estimated random effects tend to be large and should be accounted for in any subsequent analyses, such as predicting individual differences using other variables.[12]

In that plot, each speaker shows similar uncertainty (CI width). This is only because speakers have similar sample sizes; in general, uncertainty will be proportional to the amount of data per level. For example, consider the estimated random intercepts for each `consonant` (figure 10.5, right), which is very unbalanced:

```
xtabs(~consonant, data = turkish_if0)
## consonant
##    b    d   dZ    f    g    k    p    s    S    t   tS    v    z
## 1883  235  228  141  100  522  136  479   50  361    7   98   61
##    Z
##   46
```

The CIs are widest for level TS, with the fewest observations, and narrowest for level B, with the most observations.

10.6.2 Example 2: Predicted Values via Simulation from Fitted Model

We saw in section 8.8.1 how to obtain predictions from a fitted model for any new observation, including for each level of a grouping factor. For example, using the `neut_mod_2` model of the `neutralization` data, we predicted `vowel_dur` for each `voicing` value for each participant, averaging across items:

```
neut_mod_2 <- lmer(vowel_dur ~ voicing + (1 + voicing || subject) +
  (1 | item_pair), data = neutralization)
## ggpredict(neut_mod_2, terms = c("voicing", "subject"), type = "re")
```

These correspond to adding together fixed-effect and random-effect estimates. When predicting for a new *observation*, we generally want to incorporate uncertainty beyond just in the fixed effects, to give PIs.

There are different ways to do this, which incorporate different sources of uncertainty, as discussed in section 8.8.2.

The simplest way was to simulate from the fitted model (using `simulate()`): this incorporates uncertainties in the fixed effects and the residual variance. We first set up a prediction dataframe corresponding to both `voicing` values for every `subject`:

```
pred_df <- tidyr::expand(neutralization, voicing, subject)
## Makes predictions using population-level pred for item_pair
pred_df$item_pair <- "XX"
```

The simulation is otherwise similar to the example in section 8.8.2:

12. For example, for the `turkish_if0` data, we could check whether the size of a speaker's `Voicing` effect is predicted by their "macro" prosody: their mean F0, pitch range, etc. We would need to use a form of linear regression that allows for y to have errors (e.g., the `weights` argument of `lm`).

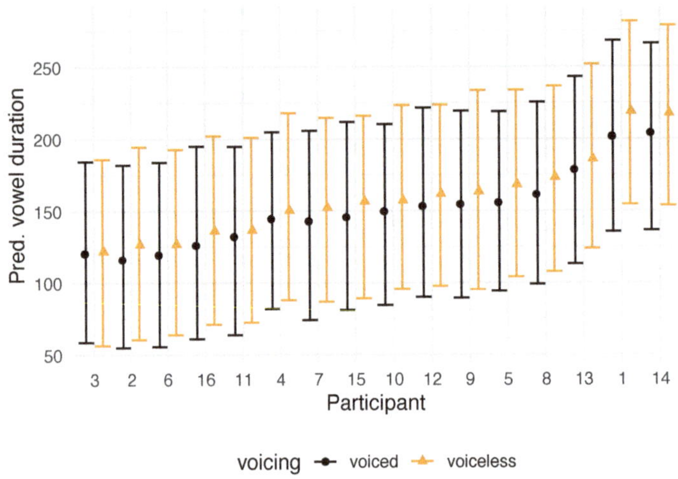

Figure 10.6
By-participant predictions for `vowel_dur` as a function of `voicing`, for `neut_mod2`, with 95% PIs.

```
sims <- simulate(neut_mod_2,
  newdata = pred_df,
  re.form = ~ ((1 | subject) + (0 + voicing | subject)),
  allow.new.levels = TRUE, nsim = 1000
)

quantile_fun <- function(x) {
  quantile(x, c(0.025, 0.5, 0.975))
}
## get prediction (= median) + 95% PIs
simPredsDf <- t(apply(sims, 1, quantile_fun))
colnames(simPredsDf) <- c("lower", "pred", "upper")

pred_df <- cbind(pred_df, simPredsDf) %>%
  mutate(voicing = factor(voicing, labels = c("voiced", "voiceless")))
```

The arguments to `simulate()` specify that $n = 1,000$ new values should be simulated from the model conditional on the current random-effect estimates (BLUPs) for by-speaker terms, but assuming an average item (`item_pair = XX`, `allow.new.levels = TRUE`).

These predictions and 95% PIs are shown in figure 10.6. The uncertainty on each prediction is large, reflecting (primarily) the large residual variance. Comparing to figure 8.5, the version without uncertainties, we get a different picture: by-participant variability in both overall `vowel_dur` and the `voicing` effect are tiny compared to the overall degree of variability in the data. Speakers are internally rather consistent.

10.6.3 Example 3: Predicted Effects via Simulation from the Fitted Model

Often, what we want to extract from the model are not individual predictions, but *effects*, such as "the `Voicing.vl` effect for each speaker." This can be done using `coefficients()`:

```
coefficients(if0_m1)$speaker$Voicing.vl
```

Figure 10.7 (top left) shows a histogram of these values, which is one way to address the research question of by-speaker robustness. Most speakers have a positive `Voicing.effect`, which we can quantify as this percentage:

```
# Percentage of speakers with positive Voicing.vl effects
sum(coefficients(if0_m1)$speaker$Voicing.vl > 0) /
  nrow(coefficients(if0_m1)$speaker)
## [1] 0.96
```

We would also like a way to take uncertainties in the effects into account here. One way to do this, which works very generally, is to state "effects" in terms of model predictions and use these to calculate estimates and uncertainties. For example, the output of `simulate()` could be used to calculate n VOICELESS-VOICED values for each speaker and to define a 95% PI.

It is often simpler to state effects in terms of model parameters, which have already been defined to be quantities of interest. For example, in the Turkish IF0 model, we can address the research question by estimating each speaker's `Voicing.vl` effect—which is already defined to average across different `base_vowel` levels, and so on—with associated uncertainties.

The easiest way to do this is to simulate new parameter values from the fitted model, giving n simulated values of the fixed-effect coefficient value for `Voiced.vl`, n values of speaker 1's `Voiced.vl` random slope (which are added to give n values of speaker 1's `Voiced.vl` effect), and so on.[13] The `sim()` function in the arm package is one implementation of this, which takes into account uncertainty in (a) the fixed-effect coefficients, (b) BLUPs (as in example 1), and (c) the residual variance; the random-effect variance components are held at their estimated values.[14]

The `sim()` function does not work easily for models with uncorrelated random-effect terms. Instead we use model `if0_m1`, where the random effects for each grouping factor are correlated. We first draw $n = 1,000$ values of the `Voicing.vl` BLUP for each speaker and of the `Voicing.vl` fixed-effect coefficient:

13. See Bolker (2021, "Confidence intervals on conditional means/BLUPs/random effects") for an even simpler method, which does not require simulation—just adding together the uncertainties for the fixed effect and BLUPs. This requires assuming their estimates are independent, which is not in general true.

14. Technically, `sim()` does this by treating the fitted model as quasi-Bayesian: it uses the model's likelihood for (a)–(c), assumes a flat prior for each model parameter, and samples from the resulting posterior. This means that the fixed effects, random effects, and residual variance are sampled from distributions centered at their estimates in the fitted model. See Gelman and Hill (2007, sections 7.2, 18.3–18.4) and `simmer.R` in the arm source code.

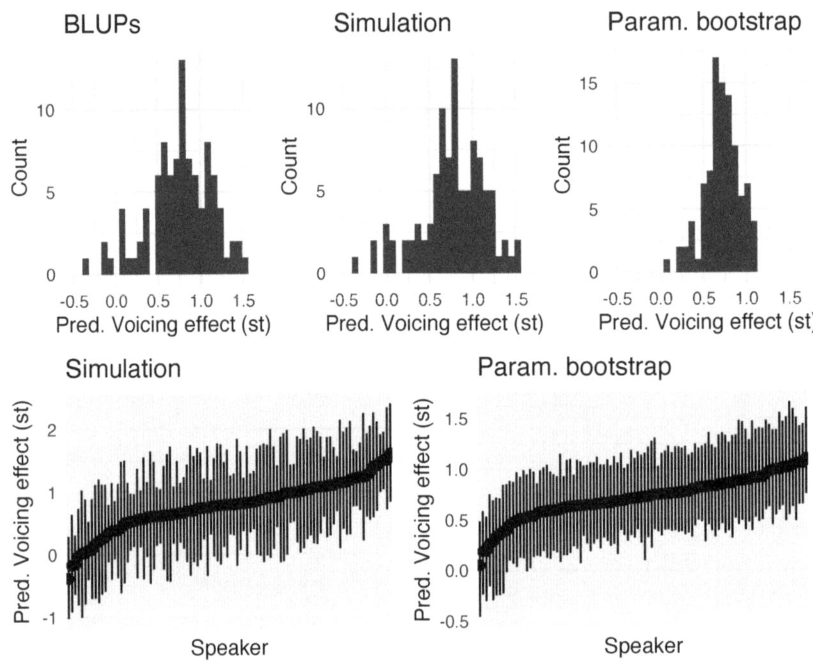

Figure 10.7
Top: Histograms of by-speaker `Voicing.vl` effects estimated by three methods, for model `if0_m1`: fitted model values (BLUPs; *top left*), simulation from the fitted model (*top center*), and parametric bootstrapping (*top right*). *Bottom*: By-speaker `Voicing.vl` effects with 95% PIs, estimated by two methods.

```
# Simulate all model parameters
if0_m1_sim <- sim(if0_m1, n.sims = 1000)
## Extract just those we need
voicingRanefDraws <- ranef(if0_m1_sim)$speaker[, , "Voicing.vl"]
voicingFixefDraws <- fixef(if0_m1_sim)[, "Voicing.vl"]
```

We add these together for each speaker to get n values of the `Voicing.vl` coefficient, which are then used to define by-speaker estimates (medians) and 95% PIs:

```
voicing_coef_draws <- apply(voicingRanefDraws, 2,
  function(x) {
    x + voicingFixefDraws
  })

voicingCoefPreds <- t(apply(voicing_coef_draws, 2, quantile_fun))
colnames(voicingCoefPreds) <- c("lower", "pred", "upper")
```

These are shown in figure 10.7 (bottom left). The PIs are large, reflecting the several sources of uncertainty taken into account; note that the PIs overlap zero for most speakers. We can now calculate the same percentage measure as that of speakers with positive `Voicing.vl` effects, but with a PI:

```
# Calculate the % speakers with pos. Voicing.vl effect for each draw
posVlPercVec <- apply(voicing_coef_draws, 1, function(x) {
  sum(x > 0) / length(x)
})

## Calculate median + 95% PI of percentages
quantile_fun(posVlPercVec)
##   2.5%   50% 97.5%
##   0.78  0.93  0.99
```

A histogram of the by-speaker estimates using the simulation method (figure 10.7, top center) looks very similar to the histogram of by-speaker estimates using BLUPs (top left), but with the simulation method we have a better sense of what the model can and cannot tell us about by-speaker variability in the `Voicing.vl` effect (the research question). For most *individual* speakers, we can't be confident the effect is positive, but we can be confident that a large majority of speakers have a positive effect (95% PI: 0.78–0.99).

10.6.4 Example 4: Predicted Effects via Parametric Bootstrap

The gold standard for estimating any quantity X from a fitted model, with uncertainty, is parametric bootstrapping (PB). The `bootMer()` implementation repeatedly (n times) does the following: simulates a new dataset from the model (using `simulate.merMod()`), refits the model, and calculates X. The resulting distribution of X (n values) can be used to calculate an estimate and PI. This method takes all sources of uncertainty into account, and is more accurate than other methods, but it is very computationally intensive.

As an example, we'll use PB to recalculate an estimate and 95% PI for the `Voicing.vl` effect for each speaker. This is now X, calculated by this function:

```
## bootMer requires X as a function to be computed using fitted model
mySumm <- function(mod) {
  coefficients(mod)$speaker$Voicing.vl
}
```

To perform the parametric bootstrap, with $n = 100$:[15]

```
## This takes ~10 min on my laptop.  Change ncpus for your computer.
## If you run this code, make sure to save the output --
## See chap. 9 ``Saving/Loading models'' box
if0_m1_boot <- bootMer(if0_m1, mySumm,
  nsim = 100, use.u = TRUE,
  parallel = "multicore", ncpus = 8
)
```

The `use.u = TRUE` option makes the simulated datasets conditional on the BLUPs for the fitted model—for example, the `Voicing.vl` effect for speaker 1 in the simulated datasets will be a distribution centered around the value from `coefficients(if0_m1)`,

15. This n is low if you are actually reporting these results; at least 500–1,000 would be better.

rather than simulating a new speaker 1 value centered on the fixed effect of `Voicing.vl` for each dataset. This is necessary for us to obtain uncertainties for the `Voicing.vl` effects for *these speakers*, and parallels what `sim()` does.

The raw values are stored in `t`:

```
## Voicing.vl coeffs for speakers 1-8 for iterations 1-2
if0_m1_boot$t
##        [,1]  [,2] [,3] [,4] [,5]  [,6]   [,7]  [,8] [,9] [,10]
##   [1,] 0.70 0.190 0.25 0.39 0.64 0.430  1.054 0.199 0.92  0.87
##   [2,] 1.07 0.668 1.15 0.72 0.97 0.715  0.724 1.006 1.07  0.96
...
```

These can be used to calculate estimates and 95% PIs for each speaker:

```
bootPredDf <- data.frame(
  speaker = rownames(coefficients(if0_m1)$speaker),
  pred = apply(if0_m1_boot$t, 2, function(x) {
    quantile(x, 0.5)
  }),
  lower = apply(if0_m1_boot$t, 2, function(x) {
    quantile(x, 0.025)
  }),
  upper = apply(if0_m1_boot$t, 2, function(x) {
    quantile(x, 0.975)
  })
)
```

The by-speaker estimates are shown in a histogram in figure 10.7 (top right) and with 95% PIs (bottom right). We can again calculate the percentage of speakers with a positive `Voicing.vl` effect, with 95% PI:

```
posV1PercVec2 <- apply(if0_m1_boot$t, 1, function(x) {
  sum(x > 0) / length(x)
})

quantile(posV1PercVec2, c(0.025, 0.5, 0.975))
##  2.5%   50% 97.5%
##  0.94  0.99  1.00
```

With respect to the research question, we would conclude that almost all speakers have a positive `Voicing.vl` effect (PI: 0.94–1) and most *individual* speakers have a clearly positive effect (figure 10.7, bottom right). As seen in the histogram, the basic picture is that the `Voicing.vl` effect is robust across speakers, but varies greatly in magnitude. This is a stronger conclusion than the one we made from the simulation method.

Thus, in this case, a more accurate method of quantifying uncertainty (PB) gives a stronger conclusion with respect to the research question: robustness of the `Voicing.vl` effect across speakers.[16]

10.7 Nonlinear Effects

Section 9.7.2 introduced nonlinear effects in mixed models, in an example of a model of the `french_cdi_24` data with a nonlinear *fixed* effect. Here, a more realistic case is shown: fitting a mixed-effects model with random slopes for nonlinear terms and making model predictions. Both are complicated by similar issues to random slopes for factors.

First, let's refit the random intercepts-only model for the `vot_voicing` data from section 10.5.7, but now with a nonlinear effect of `speaking_rate_dev`:

```
vot_voiced_m0_ns2 <- lmer(log_vot ~ ns(speaking_rate_dev, 2) +
  foll_high_vowel + cons_cluster + stress + log_corpus_freq +
  place + gender + speaking_rate_mean_std + (1 | word) +
  (1 | speaker), data = vot_voiced)
```

This could be a reasonable maximal fixed-effect structure to consider, because the speaking rate effect is of primary interest, if there is some reason to suspect a nonlinear effect. In this case, there isn't really—the model doesn't improve on `vot_voiced_m0` by AIC, and the empirical effect of speaking rate looks roughly linear—but we'll use this model anyway as an example.

Minimally, this model would need to include random slopes for the speaking rate terms of primary interest:

```
## Doesn't converge unless optimizer changed
vot_voiced_m1_ns2 <- update(vot_voiced_m0_ns2, . ~ . - (1 | speaker) +
  (1 + ns(speaking_rate_dev, 2) | speaker) -
  (1 | word) + (1 + speaking_rate_mean_std | word),
control = lmerControl(optimizer = "bobyqa"))

## boundary (singular) fit: see ?isSingular
```

This model is singular, from a perfect correlation in the random effects:

```
summary(vot_voiced_m1_ns2)$varcor
## Groups   Name                      Std.Dev. Corr
## word     (Intercept)               0.104
##          speaking_rate_mean_std    0.157    0.46
## speaker  (Intercept)               0.106
##          ns(speaking_rate_dev, 2)1 0.232    1.00
##          ns(speaking_rate_dev, 2)2 0.257    0.56 0.56
## Residual                           0.565
```

16. The simulation method seems to be overestimating the amount of uncertainty, and the degree of difference between speakers. This may be because there are correlations between different kinds of uncertainty that `sim` doesn't account for in simulating from the posterior.

This often happens with nonlinear terms in the random slopes, especially because of the ns spline basis functions (= the two speaking_rate_dev predictors are not centered), but there is no simple way to standardize the predictors in this case.

To fit a model with *uncorrelated* random effects, we would need to extract individual components of the ns term as numeric variables, analogously to extracting the contrasts as numeric variables for a factor to use uncorrelated random effects:

```
mm <- model.matrix(~ ns(speaking_rate_dev, 2), vot_voiced)
vot_voiced$speaking_rate_dev1 <- mm[, 2]
vot_voiced$speaking_rate_dev2 <- mm[, 3]
```

We can then fit the model without random-effect correlations:

```
vot_voiced_m1_ns2_uncorr <- update(vot_voiced_m1_ns2, . ~ . -
  (1 + ns(speaking_rate_dev, 2) | speaker) +
  (1 + speaking_rate_dev1 + speaking_rate_dev2 || speaker),
data = vot_voiced
)
```

It is not clear to me how legitimate this is, given that spline components are not in general centered and orthogonal, and it doesn't make conceptual sense to use uncorrelated random effects for predictors that are known to be correlated. But at least in this case, the models give extremely similar results, so we will examine the results of each model:

```
## Correlation between fitted values from both models
cor(predict(vot_voiced_m1_ns2_uncorr), predict(vot_voiced_m1_ns2))
## [1] 1
```

Understanding the nonlinear effect requires model predictions, as the coefficients are uninterpretable (section 7.5.5). We would like predictions because speaker and speaking_rate_dev are varied for an average word. We also vary the factor place, which we'll marginalize over later. First, we'll set up the dataframe:

```
## Varying predictors + speaker
pred_df <- expand.grid(
  speaker = unique(vot_voiced$speaker),
  speaking_rate_dev = seq(-2, 2, 0.1),
  voicing = unique(vot_voiced$voicing),
  place = unique(vot_voiced$place)
)

## Set other predictors to mean values
pred_df <- mutate(pred_df,
  cons_cluster = 0, foll_high_vowel = 0, stress = 0,
  log_corpus_freq = 0, gender = 0, speaking_rate_mean_std = 0
)
```

We obtain predicted values with `predict()`, using the fixed effects and only the by-speaker random effects, and marginalize over `place`:

```
pred_df$pred <- predict(vot_voiced_m1_ns2,
  newdata = pred_df,
  re.form = ~ (1 + ns(speaking_rate_dev, 2) | speaker)
)

## Average across levels of place
pred_df_corr <- pred_df %>%
  group_by(speaker, speaking_rate_dev) %>%
  summarise(pred_marg = mean(pred))
```

These predictions are shown in figure 10.8 (left).

To do the same for the model with uncorrelated random effects, we need to manually fill in the values of the spline components in the prediction dataframe:

```
## Make splines for the *full* dataset:
sr_splines <- ns(vot_voiced$speaking_rate_dev, 2)

## Get values of two components for pred_df values of speaking_rate_dev:
sr_splines_pred <- predict(sr_splines, pred_df$speaking_rate_dev)
pred_df$speaking_rate_dev1 <- sr_splines_pred[, 1]
pred_df$speaking_rate_dev2 <- sr_splines_pred[, 2]
```

We then obtain predictions (and marginalize) as before; these are shown in figure 10.8 (right).

```
pred_df$pred <- predict(vot_voiced_m1_ns2_uncorr,
  newdata = pred_df,
  re.form = ~ (1 + speaking_rate_dev1 + speaking_rate_dev2 || speaker)
)

pred_df_uncorr <- pred_df %>%
  group_by(speaker, speaking_rate_dev) %>%
  summarise(pred_marg = mean(pred))
```

The predictions from the two models are qualitatively similar, but with more curvature in the model without random-effect correlations. Presumably if we added 95% PIs, which could be done using `simulate` rather than `predict` (as in section 10.6.2), we would find that differences between the two models lie within the PI.

10.8 Power for Mixed-Effects Models

How to calculate and think about power is a very important topic for mixed-effects models, but enough of an active research area that it makes more sense to refer you to recent

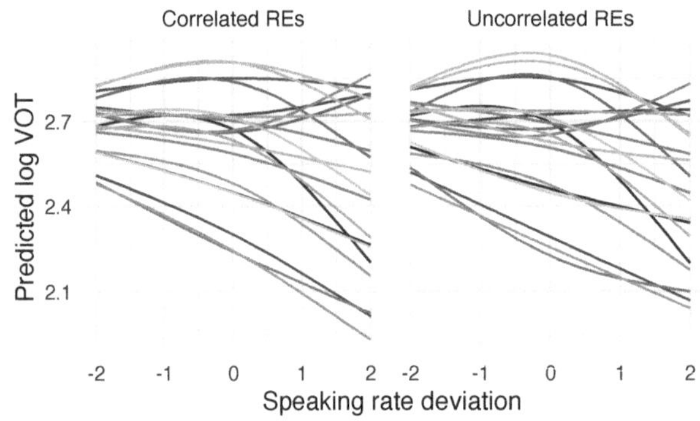

Figure 10.8
Predicted effect of `speaking_rate_dev`, by-speaker, for models `vot_voiced_m1_ns2` and `vot_voiced_m1_ns2_uncorr`, which have correlated and uncorrelated random effects.

articles than to offer a treatment that will be dated by the time you read this.[17] My thoughts as of a couple years ago are in Kirby and Sonderegger (2018a), a case study illustrating design analysis for the `neutralization` data, using power (and effect size) calculations for mixed-effects models of these data.

The high-level considerations are similar to those for non-mixed-effects models (introduced in chapter 3): power is a function of sample size, degree of variability, true effect size, and analysis method (including significance level), and power is relevant for both study design and for interpreting fitted models ("design analysis"). Power calculations are more complex for mixed-effects models, because there are now several sample sizes and degrees of variability (the number of observations, participants, etc.; the residual variance, by-participant random intercept variance, etc.), each of which affects power.

For simple cases it is possible to calculate power using online calculators, such as those by Jake Westfall (`http://jakewestfall.org/`), which include calculators for the crossed random-effects case. In general, power needs to be calculated via simulation; the two best current packages for this are simr and mixedpower (Green and MacLeod 2016; Kumle, Vo, and Draschkow 2020).

Kumle, Vo, and Draschkow (2020) is an excellent recent tutorial, showing practical examples in common situations, using these two packages. Their summary (as of 2020) is that "power analysis for mixed-effects models is still a largely uncharted terrain containing many open and unresolved issues." By the time you read this, more options should be available.

Other recent discussions or tutorials of power for mixed-effects models, in the context of different kinds of linguistic data, include Brysbaert and Stevens (2018), Brysbaert (2020), Sprouse and Almeida (2017).

17. I agree with a reviewer that power "for most of our work in linguistics is rather central (or should be) and ought to be presented as such"—this is just not the place for that presentation.

10.9 Further Reading

There is an influential literature in statistics for language scientists/psychology discussing model selection and fitting (convergence/singularity) for lme4 models, primarily linear mixed-effects models (Barr et al. 2013; Bates, Kliegl, et al. 2015; Matuschek et al. 2017); Seedorff, Oleson, and McMurray (2019) also consider logistic regression. This literature tends to focus on selecting random-effect structure. Several papers discuss these issues as motivation for using Bayesian models (e.g., Kimball et al. 2019; Eager and Roy 2017; Vasishth, Nicenboim, et al. 2018). There is a large literature on these issues in other fields, which tends to focus less narrowly on random-effect structure. I have found Snijders and Bosker (2011), Zuur (2009), Gelman and Hill (2007), and McElreath (2015) helpful, as well as lme4 help pages (`?convergence`, `?isSingular`).

Random effects can be used to capture many additional kinds of nonindependence not discussed here, such as temporal/spatial autocorrelation, and multimembership models (each observation can come from more than one level) (e.g., Zuur 2009; Baayen et al. 2018; Pinheiro and Bates 2000; Bolker 2019b).

In addition to power, another important topic for mixed-effects models we do not have space to cover is dealing with missing/incomplete data (Van Buuren 2018).

Bayesian regression models are a natural next step after you are familiar with mixed-effects models. "Hierarchical Bayesian models," which can be seen as a generalization of frequentist GLMMs, are the most widely used type of Bayesian model. Bayesian models allow for much more general model structure than GLMMs do, greatly expanding the set of research questions you can address. They also have the practical advantage, discussed by the references herein, of avoiding the convergence/singularity issues involving random-effect structure that have been a major theme of this chapter, by using prior distributions over random effects. These advantages come at the cost of a modeling framework and software packages that are decidedly more complex than (G)LMMs using lme4 and require serious study (including a good understanding of GLMMs!) before you can confidently fit and interpret models. This is why this book has not touched on Bayesian models, even briefly, although they are increasingly used to analyze linguistic data and will become more widely used in coming years. Some places to start for linguists and cognitive scientists are Vasishth, Nicenboim, et al. (2018) and Nicenboim, Schad, and Vasishth (2022). McElreath (2020) and Kruschke (2014) are comprehensive but approachable general introductions.

Exercises

10.1. Fit the model to the `neutralization` data described in box 10.1 ("We could alternatively...") as an alternative to model `neut_mod3`. Compare these two models. Do they give the same qualitative conclusions for terms of interest (= involving `voicing`)? Is there any advantage you see to one over the other?

10.2. Make exploratory plots of all possible (two-way) interactions for the `givenness` data. If we were deciding what interactions to include in a maximal fixed-effect structure of a model predicting `stressshift` for these data, based purely on exploratory plots, which interactions look most promising?

10.3. In section 10.5.6, the fitted model `if0_m2` is interpreted with respect to the research question—whether there is a `Voicing.vl` effect, and if it is modulated by `base_vowel`—using contrast coding (model coefficients). Reinterpret the model using post hoc tests that address four questions: whether there is a `Voicing.vl` effect for each vowel (A, I, U), and whether it is larger for high than for low vowels (I/U VS. A). (It may be helpful to first refit the model with `Voicing` coded as a factor.) Can you think of reasons to prefer one way of reporting the results over the other?

10.4. The case studies for model selection in the text use fairly restricted fixed-effect structures for simplicity. This exercise asks you to redo case study 3 (section 10.5.7) using a larger maximal fixed-effect structure and examine how the results change.

(a) Given the research question, it would be reasonable to consider some interactions with `speaking_rate_dev` and `speaking_rate_mean_std` in the maximal fixed effects. We have no a priori hypotheses about these interactions. We could try adding all such interactions, and prune later, or be more selective now. Let's follow Gelman and Hill's strategy (section 5.10.2) and consider just interactions with predictors whose effects are largest. Determine which two or three predictors have the largest effects (e.g., using `Anova()` from the car package). Add interactions of these with `speaking_rate_dev` and `speaking_rate_mean_std` to the maximal fixed effects and fit a random-intercepts-only model (analogous to `vot_voiced_m0`).

(b) Build up the random-effect structure for this model, using the uncorrelated first strategy. You can choose to consider correlation terms or not, for this exercise. The resulting model is analogous to `vot_voiced_m3`.

(c) Which effects involving `speaking_rate_dev` and `speaking_rate_mean_std` are important in this model by some criterion? Make model prediction plots illustrating these effects.

(d) You should find that both `speaking_rate_dev` and `speaking_rate_mean_std` participate in significant interactions, and at least one speaking rate predictor has a significant main effect. Discuss what the model predicts about how speaking rate affects VOT. How does this conclusion differ from `vot_voiced_m3`?

A Appendix: Datasets

Most datasets used in this book are described in the text: `english` and `regularity` from the languageR package associated with Baayen (2008) (sections 4.1.2 and 2.2.3), as well as `neutralization` (section 3.1.2), `diatones` (section 6.1.2), `french_cdi_24` (section 7.1.2), `givenness`, and `turkish_if0` (section 10.1.2). This appendix provides additional detail on two remaining datasets.

The dataset descriptions use `teletype` for names of predictors (including factors) and SMALL CAPS to refer to levels of a factor.

A.1 `transitions`

This dataset is from Roberts, Torreira, and Levinson (2015), a study examining the factors that determine the speed of turn-taking in conversation. The `transitions_rmld.csv` dataset is identical to the version of Roberts, Torreira, and Levinson's data posted in an Open Science Foundation (OSF) project (Roberts, Torreira, and Levinson 2018).

The dataset contains ~20,000 conversational transitions between speakers engaged in spontaneous conversation during telephone calls from the Switchboard Corpus (Godfrey and Holliman 1997). The dataset contains 50 variables (columns), only some of which are used in this book. Here is a brief explanation of some relevant variables:

- `dur`: the duration (milliseconds) of the floor transition between turn A (the turn preceding the floor transition) and turn B (the turn following the transition)
- `spkA`, `spkB`: the id of the speaker of the turn preceding/following the floor transition.
- `sexA`, `sexB`: the sex of the speaker of the turn preceding/following the floor transition.
- `dialActA`, `dialActB`: the dialogue act (e.g., yes-no question, wh-question, statement, answer, backchannel) of the turn preceding/following the floor transition.
- `uttNSylA`, `uttNSylB`: the number of syllables in the turn preceding/following the floor transition.
- `uttDurA`, `uttDurB`: the duration (milliseconds) of the turn preceding/following the floor transition.

A.2 `vot`

This dataset, posted as `vot_rmld.csv` in an OSF project (Sonderegger, Bane, and Graff 2018) and on this book's website, contains VOTs measured for speech from a corpus of speakers of different British English dialects—most of the data analyzed by Sonderegger, Bane, and Graff (2017), with a subset of predictors included.

The dataset contains VOT measurements (in mseconds; column `vot`, log-transformed in `log_vot`) for 25154 word-initial voiceless and voiced stops, corresponding to 1752 word types and 20 word types and 21 speakers. Like most corpus data, this dataset is very unbalanced: there are between 1 and 2814 tokens per word (column `word`) and 118–3318 tokens per speaker (column `speaker`). Note that VOT as defined for this dataset can only be positive, which is important.[1]

The dataframe contains columns corresponding to a number of variables, at the level of the speaker, the word, and individual observations:

- Speaker-level
  - `gender` of the speaker (levels F, M)
  - `speakingRateMean`: mean speaking rate (syllables/second) for data points from this speaker
- Word-level
  - `phone` (levels P, T, K, B, D, G): stop consonant for which VOT was measured
  - `voicing` (levels VOICED, VOICELESS): the phonological voicing of the stop (b/d/g vs. p/t/k)
  - `place` of articulation—levels ALVEOLAR, LABIAL, VELAR
  - `foll_high_vowel` (levels NO, YES): whether the following vowel is high or non-high (e.g., 'tease', 'tack')
  - `stress` (levels NO, YES): whether the syllable ontaining the stop is stressed/unstressed (including function words)
  - `log_corpus_freq`: log-transformed frequency (words/million) of word in the BigBrother Corpus
  - `log_celex_freq`: log-transformed frequency (words/million) of word in the CELEX corpus (Baayen, Piepenbrock, and Gulikers 1996)
  - `cons_cluster` (levels NO, YES): whether the stop is in a consonant cluster (e.g., 'tick', 'trick')
  - `seg_length`: length of the word, in phonemes (averaged across entries in CELEX)
- Observation-level
  - `speaking_rate` (in syllables/second) of the surrounding phrase `speakingRateDev`: speaking rate for this observation, minus `speakingRateMean` for this speaker
  - `pause` (levels NO, YES): does a (hand-annotated) pause precede the stop?

Most predictors are expected to affect VOT from previous work (e.g., Docherty 1992; Lisker and Abramson 1967; Stuart-Smith et al. 2015). The most robust findings across languages are that VOT is larger for VOICELESS stops (Cho and Ladefoged 1999) and that `place` is further from the lips (BILABIAL < ALVEOLAR < VELAR), though the source of the place effect remains unclear. Some predictors' effects can be intuitively understood as consequences of reduction or prosodic position. Faster speaking rate means (by definition) that speech is compressed, including the part of stop consonants corresponding to VOT. This compression might take place across speakers (those who talk faster, on average, have lower VOTs), within speakers (faster speech ⇒ lower VOT, for a single speaker), or both. Phones in unstressed syllables or higher-frequency words are often realized as more reduced phonetically, corresponding to shorter segments (and hence shorter VOT). Stressed syllables and post pause syllables may have longer VOT because they are in prosodically strong positions. Men may have lower VOT than women, or they may just speak faster than women, on average (Morris, Mccrea, and Herring 2008).

1. If you know about VOT, you have learned that it can be positive or negative. VOT in this dataset is always positive, because 'prevoicing' almost never occurs in running speech (e.g. Stuart-Smith et al. 2015; Sonderegger, Bane, and Graff 2017).

B Appendix: R Packages

This book was compiled in R 4.1.2, with the following packages loaded:

```
...
## attached base packages:
## [1] parallel  splines   stats     graphics  grDevices utils
## [7] datasets  methods   base
##
## other attached packages:
##  [1] evaluate_0.18         marginaleffects_0.3.1
##  [3] partR2_0.9.1          DHARMa_0.4.4
##  [5] afex_1.0-1            binom_1.1-1
##  [7] GGally_2.1.2          effsize_0.8.1
##  [9] pwr_1.3-0             lmerTest_3.1-3
## [11] RLRsim_3.1-6          pbkrtest_0.5.1
## [13] broom.mixed_0.2.7     emmeans_1.7.2
## [15] rms_6.2-0             SparseM_1.81
## [17] Hmisc_4.6-0           Formula_1.2-4
## [19] survival_3.2-13       lattice_0.20-45
## [21] margins_0.3.26        ModelMetrics_1.2.2.2
## [23] MuMIn_1.43.17         performance_0.9.2
## [25] arm_1.12-2            lme4_1.1-27.1
## [27] Matrix_1.3-4          MASS_7.3-54
## [29] car_3.0-12            carData_3.0-5
## [31] sjPlot_2.8.10         broom_0.7.11
## [33] patchwork_1.1.1       ggeffects_1.1.1
## [35] ggthemes_4.2.4        ggtext_0.1.1
## [37] feather_0.3.5         forcats_0.5.1
## [39] stringr_1.4.0         dplyr_1.0.7
## [41] purrr_0.3.4           readr_2.1.1
## [43] tidyr_1.1.4           tibble_3.1.6
## [45] ggplot2_3.3.5         tidyverse_1.3.1
## [47] languageR_1.5.0       xtable_1.8-4
## [49] knitr_1.40
...
```

Packages loaded by a namespace (and not attached) are not shown.

Selected Abbreviations

AIC	Akaike information criterion
AICc	Corrected Akaike information criterion
AME	Average marginal effect
ANCOVA	Analysis of covariance
ANOVA	Analysis of variance
APA	American Psychological Association
ASA	American Statistical Association
AUC	Area under the ROC curve
BIC	Bayesian information criterion
BLUP	Best linear unbiased predictor
BOBYQA	Bound Optimization by Quadratic Approximation (algorithm)
CD	Cook's distance
CDA	Confirmatory data analysis
CDI	MacArthur–Bates Communicative Development Inventory
CI	Confidence interval
df	Degrees of freedom
EDA	Exploratory data analysis
ESS	Explained sum of squares
FDR	False discovery rate
FWER	Family-wise error rate
GAM	Generalized additive model
GLM	Generalized linear model
GLMM	Generalized linear mixed model
GVIF	Generalized variance inflation factor
IN	Incomplete neutralization
IF0	Intrinsic pitch

IQR	Interquartile range
LOESS	Locally estimated scatterplot smoothing
LMM	Linear mixed model
log	Natural logarithm (base e)
LR	Likelihood ratio
LRT	Likelihood ratio test
MELR	Mixed-effects logistic regression
ML	Maximum likelihood
NHST	Null hypothesis significance testing
OR	Odds ratio
OSF	Open Science Foundation
PI	Prediction interval
PB	Parametric bootstrapping
Q-Q plot	Quantile-quantile plot
REML	Restricted maximum likelihood
RSE	Residual standard error
RSS	Residual sum of squares
RT	Reaction time
SD	Standard deviation
SE	Standard error
SIF	Standard error inflation factor
TSS	Total sum of squares
VIF	Variance inflation factor
VOT	Voice onset time

Bibliography

Agresti, Alan. 2003. *Categorical data analysis.* Wiley.

Agresti, Alan. 2007. *An introduction to categorical data analysis.* Wiley.

Agresti, Alan. 2015. *Foundations of linear and generalized linear models.* Wiley.

Aickin, Mikel, and Helen Gensler. 1996. "Adjusting for multiple testing when reporting research results: The Bonferroni vs Holm methods." *American Journal of Public Health* 86 (5): 726–728.

American Psychological Association. 2020. *Publication manual of the American Psychological Association.* 7th ed. American Psychological Association.

Anscombe, Francis J. 1973. "Graphs in statistical analysis." *American Statistician* 27 (1): 17–21.

Antonakis, John, Nicolas Bastardoz, and Mikko Rönkkö. 2021. "On ignoring the random effects assumption in multilevel models: Review, critique, and recommendations." *Organizational Research Methods* 24 (2): 443–483.

Appelbaum, Mark, Harris Cooper, Rex B. Kline, Evan Mayo-Wilson, Arthur M. Nezu, and Stephen M Rao. 2018. "Journal article reporting standards for quantitative research in psychology: The APA Publications and Communications Board task force report." *American Psychologist* 73 (1): 3–25.

Arel-Bundock, Vincent. 2021. *marginaleffects: Marginal effects, marginal means, predictions, and contrasts.* R Package Version 0.2.0. https://CRAN.R-project.org/package=marginaleffects.

Baayen, R. Harald. 2008. *Analyzing linguistic data.* Cambridge University Press.

Baayen, R. Harald, D. J. Davidson, and D. M. Bates. 2008. "Mixed-effects modeling with crossed random effects for subjects and items." *Journal of Memory and Language* 59 (4): 390–412.

Baayen, R. Harald, Richard Piepenbrock, and Leon Gulikers. 1996. *CELEX2.* CD-ROM. Linguistic Data Consortium.

Baayen, R. Harald, and Fermin Moscoso del Prado Martin. 2005. "Semantic density and past-tense formation in three Germanic languages." *Language* 81 (3): 666–698.

Baayen, R. Harald, Jacolien van Rij, Cecile de Cat, and Simon Wood. 2018. "Autocorrelated errors in experimental data in the language sciences: Some solutions offered by generalized additive mixed models." In *Mixed-effects regression models in linguistics,* edited by Dirk Speelman, Kris Heylen, and Dirk Geeraerts, 49–69. Springer.

Baayen, R. Harald, and Elnaz Shafaei-Bajestan. 2019. *languageR: Analyzing linguistic data: A practical introduction to statistics.* R package version 1.5.0. https://CRAN.R-project.org/package=languageR.

Baguley, Thomas. 2012. *Serious stats: A guide to advanced statistics for the behavioral sciences.* Macmillan.

Balota, David A., Melvin J. Yap, Keith A. Hutchison, Michael J. Cortese, Brett Kessler, Bjorn Loftis, James H. Neely, Douglas L. Nelson, Greg B. Simpson, and Rebecca Treiman. 2007. "The English lexicon project." *Behavior Research Methods* 39 (3): 445–459.

Barr, Dale J., Roger Levy, Christoph Scheepers, and Harry J. Tily. 2013. "Random effects structure for confirmatory hypothesis testing: Keep it maximal." *Journal of Memory and Language* 68 (3): 255–278. https://doi.org/10.1016/j.jml.2012.11.001.

Barton, Kamil. 2020. *MuMIn: Multi-model inference.* R package version 1.43.17. https://CRAN.R-project.org/package=MuMIn.

Bates, Douglas, Reinhold Kliegl, Shravan Vasishth, and R. Harald Baayen. 2015. "Parsimonious mixed models." ArXiv:1506.04967.

Bates, Douglas, Martin Mächler, Benjamin M. Bolker, and Steve Walker. 2015. "Fitting linear mixed-effects models using lme4." *Journal of Statistical Software* 67 (1): 1–48. https://doi.org/10.18637/jss.v067.i01.

Belsley, D. A., E. Kuh, and R. E. Welsch. 2004. *Regression diagnostics: Identifying influential data and sources of collinearity.* 3rd ed. John Wiley & Sons.

Berry, William D., Jacqueline H. R. DeMeritt, and Justin Esarey. 2010. "Testing for interaction in binary logit and probit models: Is a product term essential?" *American Journal of Political Science* 54 (1): 248–266.

Bolker, Benjamin M. 2015. *Linear and generalized linear mixed models.* 309–333. Oxford University Press.

Bolker, Benjamin M. 2019a. *Contrasts.* Ms. dated November 6, 2019. http://bbolker.github.io/mixedmodels-misc/notes/contrasts.pdf.

Bolker, Benjamin M. 2019b. *Estimating group/multi-membership models with lme4.* Ms. dated March 9, 2019. https://bbolker.github.io/mixedmodels-misc/notes/multimember.html.

Bolker, Benjamin M. 2021. *GLMM FAQ.* June 28, 2021 version. https://bbolker.github.io/mixedmodels-misc/glmmFAQ.html.

Bolker, Benjamin M., Mollie E. Brooks, Connie J. Clark, Shane W. Geange, John R. Poulsen, M. Henry H. Stevens, and Jada-Simone S. White. 2009. "Generalized linear mixed models: A practical guide for ecology and evolution." *Trends in Ecology and Evolution* 24 (3): 127–135.

Bolker, Benjamin M., and David Robinson. 2021. *broom.mixed: Tidying methods for mixed models.* R package version 0.2.7. https://CRAN.R-project.org/package=broom.mixed.

Boudreault, M. C., E. A. Cabirol, D. Poulin-Dubois, A. Sutton, and N. Trudeau. 2007. "MacArthur Communicative Development Inventories: Validity and preliminary normative data." *Revue Canadienne d'Orthophonie et d'Audiologie* 31:27–37.

Braginsky, Mika, Daniel Yurovsky, Virginia A. Marchman, and Michael C. Frank. 2019. "Consistency and variability in children's word learning across languages." *Open Mind* 3:52–67.

Brauer, Markus, and John J. Curtin. 2018. "Linear mixed-effects models and the analysis of nonindependent data: A unified framework to analyze categorical and continuous independent variables that vary within-subjects and/or within-items." *Psychological Methods* 23 (3): 389–411.

Brezina, Vaclav. 2018. *Statistics in corpus linguistics: A practical guide.* Cambridge University Press.

Brysbaert, Marc. 2007. *The language-as-fixed-effect-fallacy: Some simple SPSS solutions to a complex problem.* Royal Holloway, University of London.

Brysbaert, Marc. 2020. "Power considerations in bilingualism research: Time to step up our game." *Bilingualism: Language and Cognition* 24 (5): 813–818.

Brysbaert, Marc, and Michaël Stevens. 2018. "Power analysis and effect size in mixed effects models: A tutorial." *Journal of Cognition* 1 (1): 1–20.

Burnham, Kenneth P., and David R. Anderson. 2002. *Model selection and multimodel inference.* 2nd ed. Springer.

Button, Katherine S., John P. A. Ioannidis, Claire Mokrysz, Brian A. Nosek, Jonathan Flint, Emma S. J. Robinson, and Marcus R. Munafò. 2013. "Power failure: Why small sample size undermines the reliability of neuroscience." *Nature Reviews Neuroscience* 14 (5): 365–376. https://doi.org/10.1038/nrn3475.

Canty, Angelo, and B. D. Ripley. 2017. *boot: Bootstrap R (S-Plus) functions.* R package version 1.3-20.

Carlson, Matthew, Morgan Sonderegger, and Max Bane. 2014. "How children explore the phonological network in child-directed speech: A survival analysis of children's first word productions." *Journal of Memory and Language* 75:159–180.

Champely, Stephane. 2018. *pwr: Basic functions for power analysis.* R package version 1.2-2. https://CRAN.R-project.org/package=pwr.

Chatterjee, Samprit, and Ali Hadi. 2012. *Regression analysis by example.* 5th ed. John Wiley and Sons.

Chen, Henian, Patricia Cohen, and Sophie Chen. 2010. "How big is a big odds ratio? Interpreting the magnitudes of odds ratios in epidemiological studies." *Communications in Statistics—Simulation and Computation* 39 (4): 860–864.

Cho, Taehong, and Peter Ladefoged. 1999. "Variation and universals in VOT: Evidence from 18 languages." *Journal of Phonetics* 27 (2): 207–229. https://doi.org/10.1006/jpho.1999.0094.

Chuang, Yu-Ying, Janice Fon, Ioannis Papakyritsis, and Harald Baayen. 2021. "Analyzing phonetic data with generalized additive mixed models." In *Manual of Clinical Phonetics,* edited by Martin J. Ball, 108–138. Routledge.

Clark, Herbert H. 1973. "The language-as-fixed-effect fallacy: A critique of language statistics in psychological research." *Journal of Verbal Learning and Verbal Behavior* 12 (4): 335–359.

Cohen, Jacob. 1968. "Multiple regression as a general data-analytic system." *Psychological Bulletin* 70 (6 part 1): 426–443.

Cohen, Jacob. 1969. *Statistical power analysis for the behavioral sciences.* 1st ed. Academic Press.

Cohen, Jacob. 1988. *Statistical power analysis for the behavioral sciences.* 2nd ed. Academic Press.

Cohen, Jacob. 1992. "A power primer." *Psychological Bulletin* 112 (1): 155–159.

Cohen, Jacob, Patricia Cohen, Stephen G. West, and Leona S. Aiken. 2002. *Applied multiple regression/correlation analysis for the behavioral sciences.* Routledge.

Cohen Priva, Uriel, and Emily Gleason. 2018. "The role of fast speech in sound change." In *Proceedings of the 40th Annual Conference of the Cognitive Science Society,* 1512–1517. Cognitive Science Society.

Cooper, William E., and John M. Sorensen. 1980. *Fundamental frequency in sentence production.* Springer.

Coupé, Christophe. 2018. "Modeling linguistic variables with regression models: Addressing non-Gaussian distributions, non-independent observations, and non-linear predictors with random effects and generalized additive models for location, scale, and shape." *Frontiers in Psychology* 9:513.

Cox, David R. 1958. *Planning of experiments.* Wiley.

Crawley, Michael J. 2015. *Statistics: An introduction using R.* 2nd ed. Wiley.

Cross Validated. 2016. *Is normality testing "essentially useless"?* Version 2016-01-16, accessed November 21, 2021. https://stats.stackexchange.com/q/2492.

Cumming, Geoff. 2012. *Understanding the new statistics: Effect sizes, confidence intervals, and meta-analysis.* Routledge.

Cumming, Geoff. 2014. "The new statistics: Why and how." *Psychological Science* 25 (1): 7–29.

Cumming, Geoff, and Sue Finch. 2005. "Inference by eye: Confidence intervals and how to read pictures of data." *American Psychologist* 60 (2): 170–180.

Dalal, Dev K., and Michael J. Zickar. 2012. "Some common myths about centering predictor variables in moderated multiple regression and polynomial regression." *Organizational Research Methods* 15 (3): 339–362.

DeBruine, Lisa M., and Dale J. Barr. 2021. "Understanding mixed-effects models through data simulation." *Advances in Methods and Practices in Psychological Science* 4 (1): 1–15.

Dienes, Zoltan. 2008. *Understanding psychology as a science: An introduction to scientific and statistical inference.* Macmillan International Higher Education.

Docherty, Gerard. 1992. *The timing of voicing in British English obstruents.* Foris.

Eager, Christopher, and Joseph Roy. 2017. "Mixed effects models are sometimes terrible." ArXiv:1701.04858.

Eddington, David. 2016. *Statistics for linguists: A step-by-step guide for novices.* Cambridge Scholars Publishing.

Efron, Bradley, and Robert J. Tibshirani. 1994. *An introduction to the bootstrap.* CRC Press.

Faraway, Julian J. 2015. *Linear models with R.* Chapman and Hall/CRC Press.

Faraway, Julian J. 2016. *Extending the linear model with R: Generalized linear, mixed effects and nonparametric regression models.* CRC Press.

Field, Andy, Jeremy Miles, and Zoë Field. 2012. *Discovering statistics using R.* Sage Publications.

Fitzmaurice, Garrett, Marie Davidian, Geert Verbeke, and Geert Molenberghs. 2008. *Longitudinal data analysis.* CRC Press.

Fox, John, and Sanford Weisberg. 2019. *An R companion to applied regression.* 3rd ed. Sage Publications.

Frank, Michael C., M. Braginsky, V. A. Marchman, and D. Yurovsky. 2017. "Wordbank: An open repository for developmental vocabulary data." *Journal of Child Language* 44 (3): 677–694.

Frank, Michael C., M. Braginsky, V. A. Marchman, and D. Yurovsky. 2021. *Variability and consistency in early language learning: The wordbank project.* MIT Press. This reference refers to the online version, accessed September 2020: https://langcog.github.io/wordbank-book/.

Fraundorf, Scott. 2021. *psycholing: R functions for common psycholinguistic and cognitive designs.* R package version 0.5.4. https://github.com/sfraundorf/psycholing.

Frisch, Stefan A., Janet B. Pierrehumbert, and Michael B. Broe. 2004. "Similarity avoidance and the OCP." *Natural Language and Linguistic Theory* 22 (1): 179–228.

Garcia, Guilherme D. 2021. *Data visualization and analysis in second language research.* Routledge.

Gelman, Andrew. 2004. "Exploratory data analysis for complex models." *Journal of Computational and Graphical Statistics* 13 (4): 755–779.

Gelman, Andrew, and John Carlin. 2014. "Beyond power calculation: Assessing type S (sign) and type M (magnitude) errors." *Perspectives on Psychological Science* 9 (6): 641–651.

Gelman, Andrew, and Jennifer Hill. 2007. *Data analysis using regression and multilevel/hierarchical models.* Cambridge University Press.

Gelman, Andrew, and Yu-Sung Su. 2021. *arm: Data analysis using regression and multilevel/hierarchical models.* R package version 1.12-2. https://CRAN.R-project.org/package=arm.

Gelman, Andrew, and Francis Tuerlinckx. 2000. "Type S error rates for classical and Bayesian single and multiple comparison procedures." *Computational Statistics* 15 (3): 373–390. https://doi.org/10.1007/s001800000040.

Gigerenzer, Gerd. 2004. "Mindless statistics." *Journal of Socio-Economics* 33 (5): 587–606.

Godfrey, J. J., and E. Holliman. 1997. *Switchboard-1 release 2.* Linguistic Data Consortium.

Gómez, Pascual Cantos. 2014. *Statistical methods in language and linguistic research.* Equinox.

Good, Phillip. 2006. *Resampling methods: A practical guide to data analysis.* Springer.

Green, Peter, and Catriona J. MacLeod. 2016. "SIMR: An R package for power analysis of generalized linear mixed models by simulation." *Methods in Ecology and Evolution* 7 (4): 493–498. https://doi.org/10.1111/2041-210X.12504.

Greenland, Sander, Stephen J. Senn, Kenneth J. Rothman, John B. Carlin, Charles Poole, Steven N. Goodman, and Douglas G. Altman. 2016. "Statistical tests, P values, confidence intervals, and power: A guide to misinterpretations." *European Journal of Epidemiology* 31 (4): 337–350.

Gries, Stefan Th. 2013. *Statistics for linguistics with R: A practical introduction.* 2nd ed. Walter de Gruyter.

Gries, Stefan Th. 2015. "The most under-used statistical method in corpus linguistics: Multi-level (and mixed-effects) models." *Corpora* 10 (1): 95–125.

Gries, Stefan Th. 2021. *Statistics for linguistics with R.* 3rd ed. De Gruyter Mouton.

Grieve, Jack. 2017. "Spatial statistics for dialectology." In *The handbook of dialectology,* edited by Charles Boberg, John Nerbonne, and Dominic Watt, 415–433. Wiley.

Grolemund, Garrett, and Hadley Wickham. 2016. *R for data science.* 1st ed. http://r4ds.had.co.nz/.

Halekoh, Ulrich, and Søren Højsgaard. 2014. "A Kenward-Roger approximation and parametric bootstrap methods for tests in linear mixed models—the R package pbkrtest." *Journal of Statistical Software* 59 (9): 1–30. http://www.jstatsoft.org/v59/i09/.

Hallahan, Mark, and Robert Rosenthal. 1996. "Statistical power: Concepts, procedures, and applications." *Behaviour Research and Therapy* 34 (5): 489–499.

Haller, Heiko, and Stefan Krauss. 2002. "Misinterpretations of significance: A problem students share with their teachers." *Methods of Psychological Research* 7 (1): 1–20.

Harrell, F. E. 2015. *Regression modeling strategies: With applications to linear models, logistic and ordinal regression, and survival analysis.* 2nd ed. Springer Verlag.

Hartig, Florian. 2021. *DHARMa: Residual diagnostics for hierarchical (multi-level/mixed) regression models.* R package version 0.4.4. https://CRAN.R-project.org/package=DHARMa.

Hastie, Trevor, Robert Tibshirani, and Jerome Friedman. 2009. *The elements of statistical learning: Data mining, inference, and prediction.* Springer Science / Business Media.

Herron, Michael C. 1999. "Postestimation uncertainty in limited dependent variable models." *Political Analysis* 8 (1): 83–98.

Hoekstra, Rink, Richard D. Morey, Jeffrey N. Rouder, and Eric-Jan Wagenmakers. 2014. "Robust misinterpretation of confidence intervals." *Psychonomic Bulletin and Review* 21 (5): 1157–1164.

Hoenig, John M., and Dennis M. Heisey. 2001. "The abuse of power: The pervasive fallacy of power calculations for data analysis." *American Statistician* 55 (1): 1–6.

Hunt, Tyler. 2020. *ModelMetrics: Rapid calculation of model metrics.* R package version 1.2.2.2. https://CRAN.R-project.org/package=ModelMetrics.

Hurlbert, Stuart H. 1984. "Pseudoreplication and the design of ecological field experiments." *Ecological Monographs* 54 (2): 187–211.

Ioannidis, John P. A. 2008. "Why most discovered true associations are inflated." *Epidemiology* 19 (5): 640–648.

Jaeger, T. Florian. 2008. "Categorical data analysis: Away from ANOVAs (transformation or not) and towards logit mixed models." *Journal of Memory and Language* 59 (4): 434–446. https://doi.org/10.1016/j.jml.2007.11.007.

Jaeger, T. Florian, Peter Graff, William Croft, and Daniel Pontillo. 2011. "Mixed effect models for genetic and areal dependencies in linguistic typology." *Linguistic Typology* 15 (2): 281–319.

Jäger, Lena A., Felix Engelmann, and Shravan Vasishth. 2017. "Similarity-based interference in sentence comprehension: Literature review and Bayesian meta-analysis." *Journal of Memory and Language* 94:316–339.

Johnson, Daniel Ezra. 2009. "Getting off the GoldVarb standard: Introducing Rbrul for mixed-effects variable rule analysis." *Language and Linguistics Compass* 3 (1): 359–383.

Johnson, Keith. 2008. *Quantitative methods in linguistics.* Wiley-Blackwell.

Johnson, Paul C. D. 2014. "Extension of Nakagawa & Schielzeth's R2GLMM to random slopes models." *Methods in Ecology and Evolution* 5 (9): 944–946.

Judd, Charles M., Jacob Westfall, and David A. Kenny. 2017. "Experiments with more than one random factor: Designs, analytic models, and statistical power." *Annual Review of Psychology* 68:601–625.

Kelley, Ken, and Kristopher J. Preacher. 2012. "On effect size." *Psychological Methods* 17 (2): 137–152.

Kimball, Amelia E., Kailen Shantz, Christopher Eager, and Joseph Roy. 2019. "Confronting quasi-separation in logistic mixed effects for linguistic data: A Bayesian approach." *Journal of Quantitative Linguistics* 26 (3): 231–255.

Kirby, James, and Morgan Sonderegger. 2018a. "Mixed-effects design analysis for experimental phonetics." *Journal of Phonetics* 70:70–85.

Kirby, James, and Morgan Sonderegger. 2018b. "Model selection and phonological argumentation." In *Shaping phonology,* edited by Diane Brentari and Jackson Lee, 234–252. University of Chicago Press.

Kirby, James, and Morgan Sonderegger. 2020. *Mixed-effects design analysis for experimental phonetics.* Updated January 11, 2020. https://doi.org/10.17605/OSF.IO/E4G5T. osf.io/e4g5t.

Kline, R. 2013. *Beyond significance testing.* 2nd ed. American Psychological Association.

Knowles, Jared E., and Carl Frederick. 2020. *merTools: Tools for analyzing mixed effect regression models.* R package version 0.5.2. https://CRAN.R-project.org/package=merTools.

Kruschke, John. 2014. *Doing Bayesian data analysis: A tutorial with R, JAGS, and Stan.* 2nd ed. Academic Press.

Kumle, Leah, Melissa L.-H. Vo, and Dejan Draschkow. 2020. "Estimating power in (generalized) linear mixed models: An open introduction and tutorial in R." *Behaviour Research Methods* 53 (6): 2528–2543.

Kuznetsova, Alexandra, Per B. Brockhoff, and Rune H. B. Christensen. 2017. "lmerTest package: Tests in linear mixed effects models." *Journal of Statistical Software* 82 (13): 1–26. https://doi.org/10.18637/jss.v082.i13.

Labov, William. 1994. *Principles of linguistic change. Vol. 1: Internal factors.* Blackwell.

Lazic, Stanley E. 2010. "The problem of pseudoreplication in neuroscientific studies: Is it affecting your analysis?" *BMC Neuroscience* 11 (1): 1–17.

Leeper, Thomas J. 2018. *Interpreting regression results using average marginal effects with R's margins.* Vignette: R package version 0.3.23. https://cran.r-project.org/web/packages/margins/vignettes/TechnicalDetails.pdf.

Leeper, Thomas J. 2021. *margins: Marginal effects for model objects.* R package version 0.3.26.

Lenth, Russell. 2021. *emmeans: Estimated marginal means, aka least-squares means.* R package version 1.7.1-1. https://CRAN.R-project.org/package=emmeans.

Levshina, Natalia. 2015. *How to do linguistics with R: Data exploration and statistical analysis.* John Benjamins.

Levy, R. 2012. *Probabilistic methods in the study of language.* Textbook in progress, available at http://www.mit.edu/~rplevy/pmsl_textbook/text.html.

Liddell, Torrin M., and John K. Kruschke. 2018. "Analyzing ordinal data with metric models: What could possibly go wrong?" *Journal of Experimental Social Psychology* 79:328–348.

Lisker, Leigh, and Arthur S. Abramson. 1967. "Some effects of context on voice onset time in English stops." *Language and Speech* 10 (1): 1–28.

Lo, Steson, and Sally Andrews. 2015. "To transform or not to transform: Using generalized linear mixed models to analyse reaction time data." *Frontiers in Psychology* 6:1171.

Long, J. Scott, and Jeremy Freese. 2006. *Regression models for categorical dependent variables using Stata.* Stata Press.

Loy, Adam, and Heike Hofmann. 2014. "HLMdiag: A suite of diagnostics for hierarchical linear models in R." *Journal of Statistical Software* 56 (5): 1–28. https://www.jstatsoft.org/article/view/v056i05.

Loy, Adam, and Heike Hofmann. 2015. "Are you normal? The problem of confounded residual structures in hierarchical linear models." *Journal of Computational and Graphical Statistics* 24 (4): 1191–1209.

Loy, Adam, Heike Hofmann, and Dianne Cook. 2017. "Model choice and diagnostics for linear mixed-effects models using statistics on street corners." *Journal of Computational and Graphical Statistics* 26 (3): 478–492.

Lu, Jiannan, Yixuan Qiu, and Alex Deng. 2019. "A note on Type S/M errors in hypothesis testing." *British Journal of Mathematical and Statistical Psychology* 72 (1): 1–17.

Lüdecke, Daniel. 2018. "ggeffects: Tidy data frames of marginal effects from regression models." *Journal of Open Source Software* 3 (26): 772. https://doi.org/10.21105/joss.00772.

Lüdecke, Daniel. 2021. *sjPlot: Data visualization for statistics in social science.* R package version 2.8.9. https://CRAN.R-project.org/package=sjPlot.

Lüdecke, Daniel, Mattan S. Ben-Shachar, Indrajeet Patil, Philip Waggoner, and Dominique Makowski. 2021. "performance: An R package for assessment, comparison and testing of statistical models." *Journal of Open Source Software* 6 (60): 3139. https://doi.org/10.21105/joss.03139.

Luke, Steven G. 2017. "Evaluating significance in linear mixed-effects models in R." *Behavior Research Methods* 49 (4): 1494–1502.

Maas, Cora J. M., and Joop J. Hox. 2004. "Robustness issues in multilevel regression analysis." *Statistica Neerlandica* 58 (2): 127–137.

Mahowald, Kyle, Ariel James, Richard Futrell, and Edward Gibson. 2016. "A meta-analysis of syntactic priming in language production." *Journal of Memory and Language* 91:5–27.

Matuschek, Hannes, Reinhold Kliegl, Shravan Vasishth, Harald Baayen, and Douglas Bates. 2017. "Balancing type I error and power in linear mixed models." *Journal of Memory and Language* 94:305–315.

Maxwell, Scott E. 2004. "The persistence of underpowered studies in psychological research: Causes, consequences, and remedies." *Psychological Methods* 9 (2): 147–163.

McElreath, Richard. 2015. *Statistical rethinking: A Bayesian course with examples in R and Stan.* Chapman and Hall/CRC.

McElreath, Richard. 2020. *Statistical rethinking: A Bayesian course with examples in R and Stan.* 2nd ed. Chapman and Hall/CRC.

Meteyard, Lotte, and Robert A. I. Davies. 2020. "Best practice guidance for linear mixed-effects models in psychological science." *Journal of Memory and Language* 112:104092.

Mize, Trenton D. 2019. "Best practices for estimating, interpreting, and presenting nonlinear interaction effects." *Sociological Science* 6:81–117.

Morris, Richard J., Christopher R. Mccrea, and Kaileen D. Herring. 2008. "Voice onset time differences between adult males and females: Isolated syllables." *Journal of Phonetics* 36 (2): 308–317. https://doi.org/10.1016/j.wocn.2007.06.003.

Morrissey, Michael Blair, and Graeme Douglas Ruxton. 2018. "Multiple regressions: The meaning of multiple regression and the non-problem of collinearity." *Philosophy, Theory and Practice in Biology* 10:3.

Müller, Samuel, Janice L. Scealy, and Alan H. Welsh. 2013. "Model selection in linear mixed models." *Statistical Science* 28 (2): 135–167.

Nakagawa, Shinichi, Paul C. D. Johnson, and Holger Schielzeth. 2017. "The coefficient of determination R2 and intra-class correlation coefficient from generalized linear mixed-effects models revisited and expanded." *Journal of the Royal Society Interface* 14 (134): 20170213.

Nakagawa, Shinichi, and Holger Schielzeth. 2013. "A general and simple method for obtaining R2 from generalized linear mixed-effects models." *Methods in Ecology and Evolution* 4 (2): 133–142.

Navarro, Daniel. 2016. *Learning statistics with R.* Version 0.6. [Lecture notes]. School of Psychology, University of Adelaide.

Nearey, Terrance M., and Bernard L. Rochet. 1994. "Effects of place of articulation and vowel context on VOT production and perception for French and English stops." *Journal of the International Phonetic Association* 24 (1): 1–19.

Nicenboim, Bruno, Timo B. Roettger, and Shravan Vasishth. 2018. "Using meta-analysis for evidence synthesis: The case of incomplete neutralization in German." *Journal of Phonetics* 70:39–55.

Nicenboim, Bruno, D. J. Schad, and Shravan Vasishth. 2022. *An introduction to Bayesian data analysis for cognitive science.* September 2022 version. https://vasishth.github.io/bayescogsci/book/.

Nicenboim, Bruno, and Shravan Vasishth. 2016. "Statistical methods for linguistic research: Foundational ideas—part II." *Language and Linguistics Compass* 10 (11): 591–613.

Nicenboim, Bruno, Shravan Vasishth, Felix Engelmann, and Katja Suckow. 2018. "Exploratory and confirmatory analyses in sentence processing: A case study of number interference in German." *Cognitive Science* 42:1075–1100.

Nickerson, Raymond S. 2000. "Null hypothesis significance testing: A review of an old and continuing controversy." *Psychological Methods* 5 (2): 241–301.

Nieuwenhuis, Sander, Birte U. Forstmann, and Eric-Jan Wagenmakers. 2011. "Erroneous analyses of interactions in neuroscience: A problem of significance." *Nature Neuroscience* 14 (9): 1105–1107. https://doi.org/10.1038/nn.2886.

O'Brien, Robert M. 2007. "A caution regarding rules of thumb for variance inflation factors." *Quality and Quantity* 41 (5): 673–690.

O'Keefe, Daniel J. 2007. "Post hoc power, observed power, a priori power, retrospective power, prospective power, achieved power: Sorting out appropriate uses of statistical power analyses." *Communication Methods and Measures* 1 (4): 291–299.

Pek, Jolynn, and David B. Flora. 2018. "Reporting effect sizes in original psychological research: A discussion and tutorial." *Psychological Methods* 23 (2): 208.

Perezgonzalez, Jose D. 2015. "Fisher, Neyman-Pearson or NHST? A tutorial for teaching data testing." *Frontiers in Psychology* 6:223.

Pinheiro, José, and Douglas Bates. 2000. *Mixed-effects models in S and S-PLUS.* Springer Science and Business Media.

Preacher, Kristopher J., Patrick J. Curran, and Daniel J. Bauer. 2006. "Computational tools for probing interactions in multiple linear regression, multilevel modeling, and latent curve analysis." *Journal of Educational and Behavioral Statistics* 31 (4): 437–448.

Quené, Hugo, and Huub Van den Bergh. 2004. "On multi-level modeling of data from repeated measures designs: A tutorial." *Speech Communication* 43 (1-2): 103–121.

Rice, John A. 2006. *Mathematical statistics and data analysis.* Cengage Learning.

Rietveld, Toni, and Roeland Van Hout. 1993. *Statistical techniques for the study of language and language behaviour.* De Gruyter Mouton.

Rietveld, Toni, and Roeland Van Hout. 2005. *Statistics in language research: Analysis of variance.* Walter de Gruyter.

Rights, Jason D., and Sonya K. Sterba. 2019. "Quantifying explained variance in multilevel models: An integrative framework for defining R-squared measures." *Psychological Methods* 24 (3): 309–338.

Rights, Jason D., and Sonya K. Sterba. 2020. "New recommendations on the use of R-squared differences in multilevel model comparisons." *Multivariate Behavioral Research* 55 (4): 568–599.

Roberts, Seán G., Francisco Torreira, and Stephen C. Levinson. 2015. "The effects of processing and sequence organization on the timing of turn taking: A corpus study." *Frontiers in Psychology* 6:1–16. https://doi.org/10.3389/fpsyg.2015.00509.

Roberts, Seán G., Francisco Torreira, and Stephen C. Levinson. 2018. *The effects of processing and sequence organization on the timing of turn taking: A corpus study,* September. https://osf.io/dve6h.

Roberts, Seán G., and James Winters. 2013. "Linguistic diversity and traffic accidents: Lessons from statistical studies of cultural traits." *PloS One* 8 (8): e70902.

Robinson, David, Alex Hayes, and Simon Couch. 2021. *broom: Convert statistical objects into tidy tibbles.* R package version 0.7.9. https://CRAN.R-project.org/package=broom.

Rodgers, Joseph Lee. 2010. "The epistemology of mathematical and statistical modeling: A quiet methodological revolution." *American Psychologist* 65 (1): 1–12.

Roettger, Timo B. 2019. "Researcher degrees of freedom in phonetic research." *Laboratory Phonology: Journal of the Association for Laboratory Phonology* 10 (1): 1–27.

Roettger, Timo B., Bodo Winter, Sven Grawunder, James Kirby, and Martine Grice. 2014. "Assessing incomplete neutralization of final devoicing in German." *Journal of Phonetics* 43:11–25.

Saefken, Benjamin, David Ruegamer, Thomas Kneib, and Sonja Greven. 2018. "Conditional model selection in mixed-effects models with cAIC4." ArXiv: 1803.05664.

Schad, Daniel J., Shravan Vasishth, Sven Hohenstein, and Reinhold Kliegl. 2020. "How to capitalize on a priori contrasts in linear (mixed) models: A tutorial." *Journal of Memory and Language* 110:104038.

Scheipl, Fabian, Sonja Greven, and Helmut Kuechenhoff. 2008. "Size and power of tests for a zero random effect variance or polynomial regression in additive and linear mixed models." *Computational Statistics and Data Analysis* 52 (7): 3283–3299.

Schielzeth, Holger. 2010. "Simple means to improve the interpretability of regression coefficients." *Methods in Ecology and Evolution* 1 (2): 103–113.

Schielzeth, Holger, Niels J. Dingemanse, Shinichi Nakagawa, David F. Westneat, Hassen Allegue, Céline Teplitsky, Denis Réale, Ned A. Dochtermann, László Zsolt Garamszegi, and Yimen G. Araya-Ajoy. 2020. "Robustness of linear mixed-effects models to violations of distributional assumptions." *Methods in Ecology and Evolution* 11 (9): 1141–1152.

Schielzeth, Holger, and Wolfgang Forstmeier. 2009. "Conclusions beyond support: Over-confident estimates in mixed models." *Behavioral Ecology* 20 (2): 416–420.

Seedorff, Michael, Jacob Oleson, and Bob McMurray. 2019. "Maybe maximal: Good enough mixed models optimize power while controlling type I error." PsyArXiv preprint: https://psyarxiv.com/xmhfr/.

Shmueli, Galit. 2010. "To explain or to predict?" *Statistical Science* 25 (3): 289–310.

Simmons, Joseph P., Leif D. Nelson, and Uri Simonsohn. 2011. "False-positive psychology: Undisclosed flexibility in data collection and analysis allows presenting anything as significant." *Psychological Science* 22 (11): 1359–1366.

Singmann, Henrik, Benjamin M. Bolker, Jake Westfall, Frederik Aust, and Mattan S. Ben-Shachar. 2021. *afex: Analysis of factorial experiments.* R package version 0.28-1. https://CRAN.R-project.org/package=afex.

Singmann, Henrik, and David Kellen. 2019. "An introduction to mixed models for experimental psychology." *New Methods in Cognitive Psychology* 28:4–31.

Snijders, T. A. B., and R. J. Bosker. 2011. *Multilevel analysis.* 2nd ed. Sage Publications.

Sonderegger, Morgan. 2010. "Testing for frequency and structural effects in an English stress shift." In *Proceedings of the Berkeley Linguistics Society 36,* edited by Jessica Cleary-Kemp, Clara Cohen, Stephanie Farmer, Laura Kassner, John Sylak, and Melinda Woodley, 411–425.

Sonderegger, Morgan, Max Bane, and Peter Graff. 2017. "The medium-term dynamics of accents on reality television." *Language* 93 (3): 598–640.

Sonderegger, Morgan, Max Bane, and Peter Graff. 2018. *Medium-term dynamics of accents on reality TV,* September. https://osf.io/dmxuj.

Sonderegger, Morgan, M. McAuliffe, and H. Bang. 2017. *Segmental influences on f0: A large-scale study of cross-linguistic and interspeaker variability.* Talk delivered at Fourth Workshop on Sound Change. Slides at http://people.linguistics.mcgill.ca/~morgan/wsc2017_slides.pdf.

Sonderegger, Morgan, Michael Wagner, and Francisco Torreira. 2018. *Quantitative methods for linguistic data.* Version 1. http://people.linguistics.mcgill.ca/~morgan/qmld-book/.

Sóskuthy, Márton. 2017. *Generalised additive mixed models for dynamic analysis in linguistics: A practical introduction.* ArXiv:1703.05339.

Speed, Terry. 1986. "Questions, answers and statistics." In *Proceedings of the Second International Congress on Teaching Statistics (Vol 2),* 18–28. Citeseer.

Speelman, Dirk. 2014. "Logistic regression." In *Corpus methods for semantics: Quantitative studies in polysemy and synonymy,* 43:487–533. John Benjamins.

Speelman, Dirk, Kris Heylen, and Dirk Geeraerts, eds. 2018. *Mixed-effects regression models in linguistics.* Springer.

Sprouse, Jon, and Diogo Almeida. 2017. "Design sensitivity and statistical power in acceptability judgment experiments." *Glossa* 2 (1): 1–32.

Stoffel, Martin A., Shinichi Nakagawa, and Holger Schielzeth. 2021. "partR2: Partitioning R2 in generalized linear mixed models." *PeerJ* 9:e11414.

Stuart-Smith, Jane, Morgan Sonderegger, Rachel Macdonald, Jeff Mielke, Michael McAuliffe, and Erik Thomas. 2019. "Large-scale acoustic analysis of dialectal and social factors in English /s/-retraction." In *Proceedings of the 19th International Congress of Phonetic Sciences Melbourne.*

Stuart-Smith, Jane, Morgan Sonderegger, Rachel Macdonald, and Tamara Rathcke. 2015. "The private life of stops: VOT in a real-time corpus of spontaneous Glaswegian." *Laboratory Phonology* 6:505–549.

Tagliamonte, Sali A. 2006. *Analysing sociolinguistic variation.* Cambridge University Press.

Tanner, James, Morgan Sonderegger, Jane Stuart-Smith, and Josef Fruehwald. 2020. "Toward 'English' phonetics: Variability in the pre-consonantal voicing effect across English dialects and speakers." *Frontiers in Artificial Intelligence* 3:38.

Tomaschek, Fabian, Peter Hendrix, and R. Harald Baayen. 2018. "Strategies for addressing collinearity in multivariate linguistic data." *Journal of Phonetics* 71:249–267.

Torchiano, Marco. 2018. *effsize: Efficient effect size computation.* R package version 0.7.4. https://CRAN.R-project.org/package=effsize.

Trudeau, Natacha, and Ann Sutton. 2011. "Expressive vocabulary and early grammar of 16- to 30-month-old children acquiring Quebec French." *First Language* 31 (4): 480–507.

Tukey, John W. 1977. *Exploratory data analysis.* Addison-Wesley.

Tukey, John W. 1980. "We need both exploratory and confirmatory." *American Statistician* 34 (1): 23–25.

UCLA Statistical Consulting Group. 2011. *R library contrast coding systems for categorical variables.* Accessed October 7, 2020. https://stats.idre.ucla.edu/r/library/r-library-contrast-coding-systems-for-categorical-variables/.

Vaida, Florin, and Suzette Blanchard. 2005. "Conditional Akaike information for mixed-effects models." *Biometrika* 92 (2): 351–370.

Van Buuren, Stef. 2018. *Flexible imputation of missing data.* CRC Press.

Vanhove, Jan. 2021. "Collinearity isn't a disease that needs curing." *Meta-Psychology* 5. https://doi.org/10.15626/MP.2021.2548.

Vasishth, Shravan, and Michael B. Broe. 2011. *The foundations of statistics: A simulation-based approach.* Springer.

Vasishth, Shravan, Zhong Chen, Qiang Li, and Gueilan Guo. 2013. "Processing Chinese relative clauses: Evidence for the subject-relative advantage." *PloS One* 8 (10): e77006.

Vasishth, Shravan, Daniela Mertzen, Lena A. Jäger, and Andrew Gelman. 2018. "The statistical significance filter leads to overoptimistic expectations of replicability." *Journal of Memory and Language* 103:151–175.

Vasishth, Shravan, and Bruno Nicenboim. 2016. "Statistical methods for linguistic research: Foundational ideas—part I." *Language and Linguistics Compass* 10 (8): 349–369.

Vasishth, Shravan, Bruno Nicenboim, Mary E. Beckman, Fangfang Li, and Eun Jong Kong. 2018. "Bayesian data analysis in the phonetic sciences: A tutorial introduction." *Journal of Phonetics* 71:147–161.

Vasishth, Shravan, Daniel Schad, Audrey Bürki, and Reinhold Kliegl. 2021. *Linear mixed models in linguistics and psychology: A comprehensive introduction.* December 2021 version. https://vasishth.github.io/Freq_CogSci/.

Venables, W. N., and B. D. Ripley. 2002. *Modern applied statistics with S.* Springer.

Wagner, Michael. 2012. "A givenness illusion." *Language and Cognitive Processes* 27 (10): 1433–1458. https://doi.org/http://dx.doi.org/10.1080/01690965.2011.607713.

Warner, Natasha, Erin Good, Allard Jongman, and Joan Sereno. 2006. "Orthographic vs. morphological incomplete neutralization effects." *Journal of Phonetics* 34:285–293.

Wasserstein, Ronald L., and Nicole A. Lazar. 2016. "The ASA's statement on p-values: Context, process, and purpose." *American Statistician* 70 (2): 129–133.

Wasserstein, Ronald L., Allen L. Schirm, and Nicole A. Lazar. 2019. "Moving to a world beyond 'p < 0.05.'" *American Statistician* 73 (S1): 1–19.

West, Brady T., Kathleen B. Welch, and Andrzej T. Galecki. 2014. *Linear mixed models: A practical guide using statistical software.* CRC Press.

Wickham, Hadley. 2016. *ggplot2: Elegant graphics for data analysis.* Springer-Verlag. https://ggplot2.tidyverse.org/.

Wickham, Hadley. 2021. *tidyr: Tidy messy data.* R package version 1.1.4. https://CRAN.R-project.org/package=tidyr.

Wickham, Hadley, Mara Averick, Jennifer Bryan, Winston Chang, Lucy D'Agostino McGowan, Romain François, Garrett Grolemund, et al. 2019. "Welcome to the tidyverse." *Journal of Open Source Software* 4 (43): 1686. https://doi.org/10.21105/joss.01686.

Wickham, Hadley, Romain François, Lionel Henry, and Kirill Müller. 2021. *dplyr: A grammar of data manipulation.* R package version 1.0.7. https://CRAN.R-project.org/package=dplyr.

Wieling, Martijn. 2018. "Analyzing dynamic phonetic data using generalized additive mixed modeling: A tutorial focusing on articulatory differences between L1 and L2 speakers of English." *Journal of Phonetics* 70:86–116.

Winter, Bodo. 2011. "Pseudoreplication in phonetic research." In *Proceedings of the 17th International Congress of Phonetic Sciences Hong Kong,* 2137–2140.

Winter, Bodo. 2019. *Statistics for linguists: An introduction using R.* Routledge.

Winter, Bodo, and Paul-Christian Bürkner. 2021. "Poisson regression for linguists: A tutorial introduction to modeling count data with brms." *Language and Linguistics Compass* 15 (11): e12439. https://doi.org/10.1111/lnc3.12439.

Wood, Simon N. 2017. *Generalized additive models: An introduction with R.* 2nd ed. Chapman and Hall/CRC Press.

Wurm, Lee H., and Sebastiano A. Fisicaro. 2014. "What residualizing predictors in regression analyses does (and what it does not do)." *Journal of Memory and Language* 72:37–48.

Yanagida, Takuya. 2020. *misty: Miscellaneous functions 'T. Yanagida'.* R package version 0.3.2. https://CRAN.R-project.org/package=misty.

York, Richard. 2012. "Residualization is not the answer: Rethinking how to address multicollinearity." *Social Science Research* 41 (6): 1379–1386.

Zuur, Alain F. 2009. *Mixed effects models and extensions in ecology with R.* Springer.

Zuur, Alain F., Elena N. Ieno, and Chris S. Elphick. 2010. "A protocol for data exploration to avoid common statistical problems." *Methods in Ecology and Evolution* 1 (1): 3–14.

Zuur, Alain F., Elena N. Ieno, and Graham M. Smith. 2007. *Analyzing ecological data.* Springer.

Topic Index

Function Index